Vladimir K. Dobrev
Invariant Differential Operators

De Gruyter Studies in Mathematical Physics

Volume 39

Vladimir K. Dobrev

Invariant Differential Operators

Volume 2: Quantum Groups

DE GRUYTER

Mathematics Subject Classification 2010
16TXX, 17BXX, 17B37, 17B35, 17B62, 17B10, 17B81, 20G42, 33D80, 58B32, 81R50, 82D77, 16S30, 22E47, 22E15, 22E60, 81R05, 47A15, 47A46, 53A55, 70H33

Author
Vladimir K. Dobrev
Bulgarian Academy of Sciences
Institute for Nuclear Research
and Nuclear Energy
Tsarigradsko Chaussee 72
1784 Sofia
Bulgaria
dobrev@inrne.bas.bg
http://theo.inrne.bas.bg/_dobrev/

ISBN 978-3-11-043543-6
e-ISBN (PDF) 978-3-11-042770-7
e-ISBN (EPUB) 978-3-11-042778-3
Set-ISBN 978-3-11-042771-4
ISSN 2194-3532

Library of Congress Cataloging-in-Publication Data
A CIP catalog record for this book has been applied for at the Library of Congress.

Bibliographic information published by the Deutsche Nationalbibliothek
The Deutsche Nationalbibliothek lists this publication in the Deutsche Nationalbibliografie; detailed bibliographic data are available on the Internet at http://dnb.dnb.de.

Typesetting: Integra Software Services Pvt. Ltd.
Printing and binding: CPI books GmbH, Leck
♾ Printed on acid-free paper
Printed in Germany

www.degruyter.com

Preface

This is volume 2 of our trilogy on invariant differential operators. In volume 1 we presented our canonical procedure for the construction of invariant differential operators and showed its application to the objects of the initial domain – noncompact semisimple Lie algebras and groups.

In volume 2 we show the application of our procedure to quantum groups. Similarly to the setting of volume 1 the main actors are in duality. Just as Lie algebras and Lie groups are in duality here the dual objects are the main two manifestations of quantum groups: quantum algebras and matrix quantum groups. Actually, quantum algebras typically are deformations of the universal enveloping algebras of semisimple Lie algebras. Analogously, matrix quantum groups typically are deformations of spaces of functions over semisimple Lie groups.

Chapter 1 presents first the necessary general background material on quantum algebras and some generalizations as Yangians. Then we present the necessary material on q-deformations of noncompact semisimple Lie algebras. Chapter 2 is devoted to highest weight modules over quantum algebras, mostly being considered Verma modules and singular vectors. The latter is given for the quantum algebras related to all semisimple Lie algebras. Chapter 3 considers positive energy representations of noncompact quantum algebras on the example of q-deformed anti de Sitter algebra and q-deformed conformal algebra. In Chapter 4 we consider in detail the matrix quantum groups. Many important examples are considered together with the quantum algebras which are constructed using the duality properties. In many cases we consider the representations of quantum algebras that arise due to the duality. In Chapter 5 we consider systematically and construct induced infinite-dimensional representations of quantum algebras using as carrier spaces the corresponding dual matrix quantum groups. These representations are related to the Verma modules over the complexification of the quantum algebras, while the singular vectors produce invariant q-difference operators between the reducible induced infinite-dimensional representations. This generalizes our considerations of volume 1 to the setting of quantum groups. These considerations are carried out for several interesting examples. In Chapter 6 we continue the same considerations for the invariant q-difference operators related to $GL_q(n)$. Finally, in Chapter 7 we consider representations the q-deformed conformal algebra and the deformations of various representations and hierarchies of q-difference equations related in some sense to the q-Maxwell equations. Each chapter has a summary which explains briefly the contents and the most relevant literature. Besides, there are bibliography, author index, and subject index. Material from volume 1, Chapter N, formula n is cited as (I.N.n).

Note that initially we planned our monograph as a dilogy; however, later it turned out that the material on quantum groups deserves a whole volume, this volume.

Volume 3 will cover applications to supersymmetry, the AdS/CFT correspondence, infinite-dimensional (super-)algebras including (super-)Virasoro algebras, and (q-) Schrödinger algebras.

Sofia, December 2016 *Vladimir Dobrev*

Contents

1 Quantum Groups and Quantum Algebras — 1

1.1 Hopf Algebras and Quantum Groups — **1**

1.2 Quantum Algebras — **4**

1.2.1 Drinfeld's Definition — **4**

1.2.2 Universal R-Matrix and Casimirs — **7**

1.2.3 Jimbo's Definition — **11**

1.3 Drinfeld Second Realization of Quantum Affine Algebras — **14**

1.4 Drinfeld's Realizations of Yangians — **17**

1.4.1 The First Drinfeld Realization of Yangians — **17**

1.4.2 The Second Drinfeld Realization of Yangians — **18**

1.5 q-Deformations of Noncompact Lie Algebras — **19**

1.5.1 Preliminaries — **19**

1.5.2 q-Deformation of the Real Forms — **21**

1.5.3 Example so(p,r) — **25**

1.5.4 Example so(2,1) — **26**

1.5.5 q-Deformed Lorentz Algebra $U_q(\mathrm{so}(3,1))$ — **26**

1.5.6 q-Deformed Real Forms of so(5) — **27**

1.5.7 q-Deformed de Sitter Algebra so(4,1) — **27**

1.5.8 q-Deformed Anti de Sitter Algebra so(3,2) — **28**

1.5.9 q-Deformed Algebras $U_q(\mathrm{sl}(4,\mathbb{C}))$ and $U_q(\mathrm{su}(2,2))$ — **28**

1.5.10 q-deformed Poincaré and Weyl Algebras — **32**

2 Highest-Weight Modules over Quantum Algebras — 34

2.1 Verma Modules, Singular Vectors, and Irreducible Subquotients — **34**

2.2 q-Fock Type Representations — **37**

2.3 Vertex Operators — **39**

2.4 Singular Vectors in Chevalley Basis — **40**

2.4.1 $U_q(A_\ell)$ — **42**

2.4.2 $U_q(D_\ell)$ — **43**

2.4.3 $U_q(E_\ell)$ — **43**

2.4.4 $U_q(B_\ell)$ — **44**

2.4.5 $U_q(C_\ell)$ — **44**

2.4.6 $U_q(F_4)$ — **45**

2.4.7 $U_q(G_2)$ — **46**

2.5 Singular Vectors in Poincaré–Birkhoff–Witt Basis — **46**

2.5.1 PBW Basis — **46**

2.5.2 Singular Vectors for $U_q(A_\ell)$ in PBW Basis — **47**

2.5.3 Singular Vectors for $U_q(D_\ell)$ in PBW Basis — **51**

2.6 Singular Vectors for Nonstraight Roots — **56**

2.6.1 Bernstein–Gel'fand–Gel'fand Resolution — **56**

2.6.2 Case of $U_q(D_\ell)$ in PBW Basis — 61
2.6.3 Case of $U_q(D_\ell)$ in the Simple Roots Basis — 63
2.7 Representations at Roots of Unity — 64
2.7.1 Generalities — 64
2.7.2 The Example of $U_q(sl(2))$ at Roots of Unity — 67
2.7.3 Classification in the $U_q(sl(3,\mathbb{C}))$ Case — 68
2.7.4 Cyclic Representations of $U_q(\mathscr{G})$ — 76
2.8 Characters of Irreducible HWMs — 79
2.8.1 Generalities — 79
2.8.2 $U_q(sl(3,\mathbb{C}))$ — 80
2.8.3 $U_q(sl(3,\mathbb{C}))$ at Roots of Unity — 81
2.8.4 Conjectures — 83

3 Positive-Energy Representations of Noncompact Quantum Algebras — 86
3.1 Preliminaries — 86
3.2 Quantum Anti de Sitter Algebra — 87
3.2.1 Representations — 87
3.2.2 Roots of Unity Case — 91
3.2.3 Character Formulae — 95
3.3 Conformal Quantum Algebra — 97
3.3.1 Generic Case — 97
3.3.2 Roots of 1 Case — 100
3.3.3 Massless Case — 102
3.3.4 Character Formulae — 104

4 Duality for Quantum Groups — 107
4.1 Matrix Quantum Groups — 107
4.1.1 Differential Calculus on Quantum Planes — 111
4.2 Duality between Hopf Algebras — 113
4.3 Matrix Quantum Group $GL_{p,q}(2)$ — 113
4.4 Duality for $GL_{p,q}(2)$ — 115
4.5 Duality for Multiparameter Quantum $GL(n)$ — 119
4.5.1 Multiparameter Deformation of $GL(n)$ — 120
4.5.2 Commutation Relations of the Dual Algebra — 122
4.5.3 Hopf Algebra Structure of the Dual Algebra — 125
4.5.4 Drinfeld–Jimbo Form of the Dual Algebra — 128
4.5.5 Special Cases of Hopf Algebra Splitting — 130
4.6 Duality for a Lorentz Quantum Group — 131
4.6.1 Matrix Lorentz Quantum Group — 131
4.6.2 Dual Algebras to the Algebras $\mathscr{L}_q$ and $\widetilde{\mathscr{L}}_q$ — 133
4.6.3 Coalgebra Structure of the Dual Algebras — 137
4.7 Duality for the Jordanian Matrix Quantum Group $GL_{g,h}(2)$ — 139
4.7.1 Jordanian Matrix Quantum Group $GL_{g,h}(2)$ — 140

4.7.2 The Dual of $GL_{g,h}(2)$ — **141**
4.7.3 Algebra Structure of the Dual — **143**
4.7.4 Coalgebra Structure of the Dual — **146**
4.7.5 One-Parameter Cases — **148**
4.7.6 Application of a Nonlinear Map — **149**
4.8 Duality for Exotic Bialgebras — **151**
4.8.1 Exotic Bialgebras: General Setting — **151**
4.8.2 Exotic Bialgebras: Triangular Case 1 — **152**
4.8.3 Exotic Bialgebras: Triangular Case 2 — **158**
4.8.4 Exotic Bialgebras: Triangular Case 3 — **161**
4.8.5 Higher-Order R-matrix Relations and Quantum Planes — **164**
4.8.6 Exotic Bialgebras: Nontriangular Case $S03$ — **167**
4.8.7 Exotic Bialgebras: Nontriangular Case $S14$ — **176**
4.8.8 Exotic Bialgebras: Nontriangular Case $S14o$ — **183**
4.8.9 Exotic Bialgebras: Higher Dimensions — **187**
Conclusions and Outlook — **198**

5 Invariant q-Difference Operators — 199
5.1 The Case of $GL_{p,q}(2)$ — **199**
5.1.1 Left and Right Action of $U_{p,q}(gl(2))$ on $GL_{p,q}(2)$ — **199**
5.1.2 Induced Representations of $\mathcal{U}_{p,q}$ and Intertwining Operators — **203**
5.1.3 The Case $U_q(sl(2))$ — **208**
5.2 The Case of $GL_{g,h}(2)$ — **209**
5.2.1 Left and Right Action of $U_{g,h}(gl(2))$ on $GL_{g,h}(2)$ — **209**
5.2.2 Induced Representations of $\mathcal{U}_{g,h}$ and Intertwining Operators — **212**
5.2.3 Representations of the Jordanian Algebra $U_h(sl(2))$ — **218**
5.2.4 Highest-Weight Modules over $U_h(sl(2))$ — **219**
5.2.5 Singular Vectors of $U_h(sl(2))$ Verma Modules — **221**
5.3 q-Difference Intertwining Operators for a Lorentz Quantum Algebra — **223**
5.3.1 A Matrix Lorentz Quantum Group — **223**
5.3.2 The Lorentz Quantum Algebra — **224**
5.3.3 Representations of the Lorentz Quantum Algebra — **226**
5.3.4 q-Difference Intertwining Operators — **234**
5.3.5 Classification of Reducible Representations — **235**
5.3.6 The Roots of Unity Case — **236**
5.4 Representations of the Generalized Lie Algebra $sl(2)_q$ — **238**
5.4.1 Preliminaries — **238**
5.4.2 The Quantum Lie Algebra $\mathfrak{sl}(2)_q$ — **239**
5.4.3 Highest-Weight Representations — **240**

5.4.4 Highest-Weight Representations of the Restricted Algebra — 244
5.4.5 Highest-Weight Representations at Roots of Unity — 245
5.4.6 Highest-Weight Representations at Roots of Unity of the Restricted Algebra — 248
5.5 Representations of $U_q(so(3))$ of Integer Spin Only — 250
5.5.1 Preliminaries — 250
5.5.2 Matrix Quantum Group $SO_q(3)$ and the Dual $U_q(\mathscr{G})$ — 250
5.5.3 Representations of $U_q(so(3))$ — 253

6 Invariant q-Difference Operators Related to $GL_q(n)$ — 261
6.1 Representations Related to $GL_q(n)$ — 261
6.1.1 Actions of $U_q(gl(n))$ and $U_q(sl(n))$ — 263
6.1.2 Representation Spaces — 266
6.1.3 Reducibility and Partial Equivalence — 272
6.2 The Case of $U_q(sl(3))$ — 274
6.3 Polynomial Solutions of q-Difference Equations in Commuting Variables — 280
6.3.1 Procedure for the Construction of the Representations — 281
6.3.2 Reducibility of the Representations and Invariant Subspaces — 284
6.3.3 Newton Diagrams — 290
6.4 Application of the Gelfand–(Weyl)–Zetlin Basis — 292
6.4.1 Correspondence with the GWZ Basis — 292
6.4.2 q-Hypergeometric Realization of the GWZ Basis — 296
6.4.3 Explicit Orthogonality of the GWZ Basis — 300
6.4.4 Normalized GWZ basis — 303
6.4.5 Scalar Product and Normalized GWZ States — 307
6.4.6 Summation Formulae — 309
6.4.7 Weight Pyramid of the $SU(3)$ UIRs — 311
6.4.8 The Irregular Irreps in Terms of GWZ States — 316
6.5 The Case of $U_q(sl(4))$ — 321
6.5.1 Elementary Representations — 321
6.5.2 Intertwining Operators — 324

7 q-Maxwell Equations Hierarchies — 327
7.1 Maxwell Equations Hierarchy — 327
7.2 Quantum Minkowski Space–Time — 331
7.2.1 q-Minkowski Space–Time — 331
7.2.2 Multiparameter Quantum Minkowski Space–Time — 333
7.3 q-Maxwell Equations Hierarchy — 335
7.4 q-d'Alembert Equations Hierarchy — 340

7.4.1 Solutions of the q-d'Alembert Equation — 340
7.4.2 q-Plane-Wave Solutions — 341
7.4.3 q-Plane-Wave Solutions for Non-Zero Spin — 344
7.5 q-Plane-Wave Solutions of the Potential q-Maxwell Hierarchy — 347
7.6 q-Plane-Wave Solutions of the Full q-Maxwell Equations — 350
7.7 q-Weyl Gravity Equations Hierarchy — 354
7.7.1 Linear Conformal Gravity — 355
7.7.2 q-Plane-Wave Solutions of q-Weyl Gravity — 358

Bibliography — 361

Author Index — 391

Subject Index — 392

1 Quantum Groups and Quantum Algebras

Summary

We start with the q-deformation $U_q(\mathcal{G})$ of the universal enveloping algebras $U(\mathcal{G})$ of simple Lie algebras $\mathcal{G}$ called also quantum groups [251, 253] or quantum universal enveloping algebras [389, 521]. They arose in the study of quantum integrable systems, especially of the algebraic aspects of quantum inverse scattering method in papers by Faddeev, Kulish, Reshetikhin, Sklyanin and Takhtajan [273, 274, 405, 408]. It was observed by Kulish–Reshetikhin [405] for $\mathcal{G} = sl(2, \mathbb{C})$ and by Drinfeld [251, 253], Jimbo [360, 361] in general that the algebras $U_q(\mathcal{G})$ have the structure of a Hopf algebra, cf. Abe [11]. This new algebraic structure was further studied in [441, 532, 588, 598]. Later, inspired by the Knizhnik-Zamolodchikov equations [395], Drinfeld has developed a theory of formal deformations and introduced a new notion of quasi-Hopf algebras [255]. The representations of $U_q(\mathcal{G})$ were considered first in [389, 405, 523, 532] for generic values of the deformation parameter. Actually all results from the representation theory of $\mathcal{G}$ carry over to the quantum group case. This is not so, however, if the deformation parameter q is a root of unity. Thus this case is very interesting from the mathematical point of view (see, e. g., [170–172, 175, 176, 442, 443]). Lately, quantum groups were intensively applied (with special emphasis on the case when q is a root of unity) in rational conformal field theories [30–32, 304, 309, 319, 320, 482, 483, 524, 596].

We start this chapter with the general notions of Hopf algebras and quantum groups. Then we introduce quantum algebras first in Drinfeld's definition and then in Jimbo's definition. We present also the universal R-matrix and the Casimirs. We also give Drinfeld' second realization of quantum affine algebras and Drinfeld's realizations of Yangians. Then we discuss the q-deformations of noncompact Lie algebras. We propose a procedure for q-deformations of the real forms $\mathcal{G}$ of complex Lie (super) algebras associated with (generalized) Cartan matrices. Our procedure gives different q-deformations for the nonconjugate Cartan subalgebras of $\mathcal{G}$. We give several illustrations, for example, q-deformed Lorentz and conformal (super) algebras. The q-deformed conformal algebra contains as a subalgebra a q-deformed Poincaré algebra and as Hopf subalgebras two conjugate 11-generator q-deformed Weyl algebras. The q-deformed Lorentz algebra is Hopf subalgebra of both Weyl algebras.

1.1 Hopf Algebras and Quantum Groups

Let F be a field of characteristic 0; in fact, most of the time we shall work with $F = \mathbb{C}$ or $F = \mathbb{R}$.

An associative algebra $\mathcal{U}$ over F with unity $1_{\mathcal{U}}$ is called a *bialgebra* [11] if there exist two algebra homomorphisms called *comultiplication* (or *coproduct*) δ:

$$\delta : \mathcal{U} \to \mathcal{U} \otimes \mathcal{U} \, , \delta(1_{\mathcal{U}}) = 1_{\mathcal{U}} \otimes 1_{\mathcal{U}} \, , \tag{1.1}$$

and *counit* ε:

$$\varepsilon : \mathcal{U} \to F \, , \varepsilon(1_{\mathcal{U}}) = 1. \tag{1.2}$$

The comultiplication δ fulfills the axiom of *coassociativity*:

$$(\delta \otimes \mathrm{id}) \circ \delta = (\mathrm{id} \otimes \delta) \circ \delta, \tag{1.3}$$

DOI 10.1515/9783110427707-001

where both sides are maps $\mathcal{U} \to \mathcal{U} \otimes \mathcal{U} \otimes \mathcal{U}$; the two homomorphisms fulfil:

$$(\mathrm{id} \otimes \varepsilon) \circ \delta = i_1\,, \quad (\varepsilon \otimes \mathrm{id}) \circ \delta = i_2\,, \tag{1.4}$$

as maps $\mathcal{U} \to \mathcal{U} \otimes F$, $\mathcal{U} \to F \otimes \mathcal{U}$, where i_1, i_2 are the maps identifying $\mathcal{U}$ with $\mathcal{U} \otimes F$, $F \otimes \mathcal{U}$, respectively.

Next a bialgebra $\mathcal{U}$ is called a *Hopf algebra* [11] if there exists an algebra antihomomorphism γ called *antipode*:

$$\gamma : \mathcal{U} \to \mathcal{U}\,, \gamma(1_{\mathcal{U}}) = 1_{\mathcal{U}}\,, \tag{1.5}$$

such that the following axiom is fulfilled:

$$m \circ (\mathrm{id} \otimes \gamma) \circ \delta = i \circ \varepsilon\,, \tag{1.6}$$

as maps $\mathcal{U} \to \mathcal{U}$, where m is the usual product in the algebra: $m(Y \otimes Z) = YZ$, $Y, Z \in \mathcal{U}$ and i is the natural embedding of F into $\mathcal{U}$: $i(c) = c1_{\mathcal{U}}$, $c \in F$.

The antipode plays the role of an inverse although there is no requirement that $\gamma^2 = \mathrm{id}$.

The operations of comultiplication, counit, and antipode are said to give the *coalgebra* structure of a Hopf algebra.

Sometimes we shall use also the notation of Sweedler [570] for the coproduct of a:

$$\delta_{\mathcal{A}}(a) = a_{(1)} \otimes a_{(2)}\,. \tag{1.7}$$

One needs also the *opposite comultiplication* $\delta' = \pi \circ \delta$, where π is the permutation in $\mathcal{U} \otimes \mathcal{U}$, that is, $\pi(X \otimes Y) = Y \otimes X$, $X, Y \in \mathcal{U}$.

The comultiplication is said to be *cocommutative* if $\delta' = \delta$.

If the antipode has an inverse, then one uses also the notion of *opposite antipode*: $\gamma' = \gamma^{-1}$.

A Hopf algebra $\mathcal{U}$ is called *quasi-triangular Hopf algebra* or *quantum group* [251, 253] if there exists an invertible element $R \in \mathcal{U} \otimes \mathcal{U}$, called *universal R-matrix* [251, 253], which intertwines δ and δ':

$$R\delta(Y) = \delta'(Y)R, \forall Y \in \mathcal{U}\,, \tag{1.8}$$

and obeys also the relations:

$$(\delta \otimes \mathrm{id})R = R_{13}R_{23}\,, R = R_{\cdot 3}\,, \tag{1.9a}$$

$$(\mathrm{id} \otimes \delta)R = R_{13}R_{12}\,, R = R_{1\cdot}\,, \tag{1.9b}$$

where the indices indicate the embeddings of R into $\mathcal{U} \otimes \mathcal{U} \otimes \mathcal{U}$. For future use we write down:

$$R = R'_j \otimes R''_j = \sum_j R'_j \otimes R''_j . \tag{1.10}$$

Then in $\mathcal{U} \otimes \mathcal{U} \otimes \mathcal{U}$:

$$R_{12} = R'_j \otimes R''_j \otimes 1_{\mathcal{U}}$$

and analogously for R_{23}, R_{13}. Further we shall denote $1_{\mathcal{U}} \otimes 1_{\mathcal{U}} \otimes 1_{\mathcal{U}}$ just by $1_{\mathcal{U}}$.

From the above it follows that:

$$(\varepsilon \otimes \mathrm{id})R = (\mathrm{id} \otimes \varepsilon)R = 1_{\mathcal{U}} . \tag{1.11}$$

[*Proof:* Apply $\varepsilon \otimes \mathrm{id} \otimes \mathrm{id}$ to both sides of (1.9b). On the LHS we have (using (1.4)):

$$\begin{aligned}(\varepsilon \otimes \mathrm{id} \otimes \mathrm{id}) \circ (\delta \otimes \mathrm{id})\, R &= ((\varepsilon \otimes \mathrm{id}) \circ \delta) \otimes \mathrm{id})\, R = \\ &= (i_2 \otimes \mathrm{id})\, (R'_j \otimes R''_j) = \\ &= 1_{\mathcal{U}} \otimes R'_j \otimes R''_j = R_{23}\end{aligned} \tag{1.12}$$

On the RHS we have:

$$(\varepsilon \otimes \mathrm{id} \otimes \mathrm{id})\, R_{13}\, R_{23} = (\varepsilon(R'_i) \otimes 1_{\mathcal{U}} \otimes R''_i)\, R_{23}. \tag{1.13}$$

Comparing the first and third components of (1.12) and (1.13) we get $(\varepsilon \otimes \mathrm{id})R = 1_{\mathcal{U}}$ from (1.11). Analogously it is proved $(\mathrm{id} \otimes \varepsilon)R = 1_{\mathcal{U}}$ from (1.11).]

Using also (1.11) one has:

$$(\gamma \otimes \mathrm{id})R = R^{-1}\, , (\mathrm{id} \otimes \gamma)R^{-1} = R. \tag{1.14}$$

Proof. For the first equality in (1.14) we consider:

$$\begin{aligned}R\,(\gamma \otimes \mathrm{id})R &= R'_j\, \gamma(R'_k) \otimes R''_j\, R''_k = \\ &= (m \otimes \mathrm{id}) \circ (R'_j \otimes \gamma(R'_k) \otimes R''_j\, R''_k) = \\ &= (m \otimes \mathrm{id}) \circ (\mathrm{id} \otimes \gamma \otimes \mathrm{id})\, (R'_j \otimes R'_k \otimes R''_j\, R''_k) = \\ &= (m \otimes \mathrm{id}) \circ (\mathrm{id} \otimes \gamma \otimes \mathrm{id})\, R_{13}\, R_{23} = \\ &= (m \otimes \mathrm{id}) \circ (\mathrm{id} \otimes \gamma \otimes \mathrm{id}) \circ (\delta \otimes \mathrm{id})\, R = \\ &= ((m \circ (\mathrm{id} \otimes \gamma) \circ \delta) \otimes \mathrm{id})\, R = \\ &= (\varepsilon \otimes \mathrm{id})\, R = 1_{\mathcal{U}}\ . \blacksquare\end{aligned} \tag{1.15}$$

The term *quantum group* is used [253] also if R is not in $\mathcal{U} \otimes \mathcal{U}$ but in some completion of it (cf. next subsection).

From (1.8) and one of (1.9) follows the *Yang–Baxter equation (YBE)* for R:

$$R_{12}R_{13}R_{23} = R_{23}R_{13}R_{12}. \tag{1.16}$$

[*Proof:* Using (1.9b) we have:

$$\begin{aligned} R_{12}R_{13}R_{23} &= R_{12}\,(\delta\otimes \mathrm{id})\,R = \\ &= (R'_j \otimes R''_j \otimes 1_{\mathcal{U}})\,(\delta(R'_k)\otimes R''_k) = \\ &= (\delta'(R'_k)\otimes R''_k)\,(R'_j\otimes R''_j\otimes 1_{\mathcal{U}}) = \\ &= R_{23}\,R_{13}\,R_{12} \end{aligned}$$

where for the last equality one applies π to both sides of (1.9b).]

A quasi-triangular Hopf algebra is called *triangular Hopf algebra* if also the following holds:

$$\pi R^{-1} = R. \tag{1.17}$$

The axiom of coassociativity (1.3) may be relaxed being replaced by:

$$(\delta\otimes \mathrm{id})\circ\delta = \Phi\,\{(\mathrm{id}\otimes\delta)\circ\delta\}\,\Phi^{-1}, \tag{1.18}$$

where $\Phi \in \mathcal{U}\otimes\mathcal{U}\otimes\mathcal{U}$ is invertible. The corresponding objects in which (1.18) holds are called *quasi-bialgebras* and *quasi-Hopf algebras*, respectively (cf. [255]).

1.2 Quantum Algebras

1.2.1 Drinfeld's Definition

From now on (unless specified otherwise) we set $F = \mathbb{C}$. Let $\mathcal{G}$ be a complex simple Lie algebra; then the q-deformation $U_q(\mathcal{G})$ of the universal enveloping algebras $U(\mathcal{G})$ is defined [251, 253] as the associative algebra over $\mathbb{C}$ with generators $X_i^\pm, H_i, i = 1,\dots,\ell = \mathrm{rank}\,\mathcal{G}$ and with commutation relations:

$$\begin{gathered}[H_i\,,H_j] = 0, \quad [H_i\,,X_j^\pm] = \pm a_{ij}X_j^\pm\,, \\ [X_i^+,X_j^-] = \delta_{ij}\frac{q_i^{H_i/2}-q_i^{-H_i/2}}{q_i^{1/2}-q_i^{-1/2}} = \delta_{ij}[H_i]_{q_i}, \quad q_i = q^{(\alpha_i\,,\alpha_i)/2}\,,\end{gathered} \tag{1.19}$$

and *q-Serre relations*:

$$\sum_{k=0}^{n}(-1)^k\binom{n}{k}_{q_i}\left(X_i^\pm\right)^k X_j^\pm \left(X_i^\pm\right)^{n-k} = 0\,, \quad i\neq j\,, \tag{1.20}$$

where $(a_{ij}) = (2(\alpha_i\,,\alpha_j)/(\alpha_i\,,\alpha_i))$ is the *Cartan matrix* of $\mathcal{G}$ and $(\cdot\,,\cdot)$ is the scalar product of the roots normalized so that for the short roots α we have $(\alpha,\alpha) = 2\,, n = 1 - a_{ij}\,,$

$$\binom{n}{k}_q = \frac{[n]_q!}{[k]_q![n-k]_q!}\,, [m]_q! = [m]_q[m-1]_q \dots [1]_q\,, \tag{1.21}$$
$$[m]_q = \frac{q^{m/2}-q^{-m/2}}{q^{1/2}-q^{-1/2}} = \frac{sh(mh/2)}{sh(h/2)} = \frac{sin(\pi m\tau)}{sin(\pi\tau)}\,, \quad q = e^h = e^{2\pi i\tau}, \quad h,\tau \in \mathbb{C},$$
$$q_i^{a_{ij}} = q^{(\alpha_i,\alpha_j)} = q_j^{a_{ji}}\,.$$

Remark 1.1. Expressions like $q^{H/2} = e^{hH/2}$ are made mathematically rigorous in the so-called h-adic topology used in [251, 253] ($q = e^h$). [By standard notation $F[[h]]$ is the ring of formal power series in the indeterminate h over the field F. Every $F[[h]]$ module V (e. g., $U_q(\mathcal{G})$) has the h-adic topology, which is characterized by requiring that $\{h^n V | n \geq 0\}$ is a base of neighbourhoods of 0 in V and that translations in V are continuous.] Physicists work formally with such exponents which is also justified as explained below. ◊

Further we shall omit the subscript q in the *q-number* $[m]_q$ if no confusion can arise. Note also that sometimes instead of q one uses $q' = q^2$, so that $[m]_{q'} = \frac{q^m-q^{-m}}{q-q^{-1}} \equiv [m]'_q$.

In [558] for $\mathcal{G} = sl(2)$ and in [251, 253, 360, 361] in general it was observed that the algebra $U_q(\mathcal{G})$ is a Hopf algebra, the comultiplication, counit, and antipode being defined on the generators of $U_q(\mathcal{G})$ as follows:

$$\begin{aligned} \delta(H_i) &= H_i \otimes 1 + 1 \otimes H_i\,, \\ \delta(X_i^\pm) &= X_i^\pm \otimes q_i^{H_i/4} + q_i^{-H_i/4} \otimes X_i^\pm\,, \\ \varepsilon(H_i) &= \varepsilon(X_i^\pm) = 0\,, \\ \gamma(H_i) &= -H_i\,, \quad \gamma(X_i^\pm) = -q_i^{\hat\rho/2} X_i^\pm q_i^{-\hat\rho/2} = -q_i^{\pm 1/2} X_i^\pm\,, \end{aligned} \tag{1.22}$$

where $\hat\rho \in \mathcal{H}$ corresponds to $\rho = \frac{1}{2}\sum_{\alpha\in\Delta^+}\alpha$, Δ^+ is the set of positive roots and $\hat\rho = \frac{1}{2}\sum_{\alpha\in\Delta^+} H_\alpha$.

The above definition is valid also when $\mathcal{G}$ is an affine Kac-Moody algebra [251]; however, another realization, called Drinfeld's second realization, was given in [254] and will be presented below. It was also generalized to the complex Lie superalgebras with a symmetrizable Cartan matrix (cf., e. g., [385]).

The algebras $U_q(\mathcal{G})$ were called *quantum groups* [251, 253] or *quantum universal enveloping algebras* [389, 521]. For shortness we shall call them *quantum algebras* as it is now commonly accepted in the literature.

For $q \to 1$, $(h \to 0)$, we recover the standard commutation relations from (1.19) and q-Serre relations from (1.20) in terms of the *Chevalley generators* $H_i, X_i^\pm$.

The elements H_i span the Cartan subalgebra $\mathcal{H}$ of $\mathcal{G}$, while the elements $X_i^\pm$ generate the subalgebras $\mathcal{G}^\pm$. We shall use the standard *triangular decomposition* into direct sums of vector subspaces $\mathcal{G} = \mathcal{H} \oplus \underset{\beta\in\Delta}{\oplus} \mathcal{G}_\beta = \mathcal{G}^+ \oplus \mathcal{H} \oplus \mathcal{G}^-$, $\mathcal{G}^\pm = \underset{\beta\in\Delta^\pm}{\oplus} \mathcal{G}_\beta$, where $\Delta = \Delta^+ \cup \Delta^-$ is the root system of $\mathcal{G}$, Δ^+, Δ^-, the sets of positive, negative, roots, respectively; Δ_S will denote the set of simple roots of Δ. We recall that H_j corresponds to the simple roots α_j of $\mathcal{G}$, and if $\beta^\vee = \sum_j n_j \alpha_j^\vee$ and $\beta^\vee \equiv 2\beta/(\beta,\beta)$, then to β corresponds $H_\beta = \sum_j n_j H_j$. The elements of $\mathcal{G}$ which span $\mathcal{G}_\beta$ (dim $\mathcal{G}_\beta = 1$) are denoted by E_β. These *Cartan–Weyl* generators H_β, E_β [198, 360, 361, 575], may be normalized so that:

$$[E_\beta\,, E_{-\beta}] = [H_\beta]_{q_\beta}\,, \qquad q_\beta \equiv q^{(\beta,\beta)/2} \tag{1.23a}$$

$$[H_\beta\,, E_{\pm\beta'}] = \pm(\beta^\vee, \beta')E_{\pm\beta'}\,, \quad \beta, \beta' \in \Delta^+. \tag{1.23b}$$

To display more explicitly the Cartan–Weyl generators we need the notion of normal ordering [385]:

Definition 1.1. We say that the root system Δ is in the *normal ordering* if in the situation $\gamma = \alpha + \beta \in \Delta_+$, where $\alpha \neq \lambda\beta$, $\alpha, \beta \in \Delta_+$, the roots are ordered as $\alpha < \gamma < \beta$. ◊

Then the Cartan–Weyl generators are constructed as follows: Let $\gamma = \alpha + \beta$, $\alpha < \gamma < \beta$, and $[\alpha; \beta]$ is a minimal segment including γ; that is, there do not exist roots α', β', such that $\alpha' > \alpha$, $\beta' < \beta$ and $\alpha' + \beta' = \gamma$. Then the root vectors $E_{\pm\gamma}$ are given as follows:

$$E_\gamma = (\mathrm{ad}_q\, E_\alpha)\, E_\beta \equiv E_\alpha\, E_\beta - q^{(\alpha,\beta)/2} E_\beta\, E_\alpha\,, \tag{1.24}$$

$$E_{-\gamma} = (\mathrm{Ad}_{q^{-1}}\, E_\beta)\, E_\alpha = E_\beta\, E_\alpha - q^{-(\alpha,\beta)/2} E_\alpha\, E_\beta. \tag{1.25}$$

As an example we give the Cartan–Weyl generators for $\mathcal{G} = sl(n)$. Let X^+_{jk}, X^-_{kj} be the Cartan–Weyl generators corresponding to the roots $\alpha_{j,k+1}$, $-\alpha_{j,k+1}$, with $j \le k$; in particular, $X^+_{jj} = X^+_j$ $X^-_{jj} = X^-_j$, correspond to the simple roots α_j.

Here the normal ordering coincides with the lexicographic ordering. In the case of the root $\alpha_{j,k+1}$ we have two minimal segments since:

$$\begin{aligned} &\alpha_{j,k+1} = \alpha_j + \alpha_{j+1,k+1} = \alpha_{jk} + \alpha_k, \quad j < k, \\ &\alpha_{j,j+1} = \alpha_j \end{aligned} \tag{1.26}$$

the orderings being:

$$\begin{aligned} \alpha_j < \alpha_{j,k+1} < \alpha_{j+1,k+1}\,, \qquad & \alpha_{jk} < \alpha_{j,k+1} < \alpha_k\,, \\ \alpha_j < \alpha_{jk}\,, \qquad & \alpha_{j+1,k+1} < \alpha_k \end{aligned} \tag{1.27}$$

Then instead of (I.2.46a,b) we have:

$$\begin{aligned} X^+_{jk} &= (\mathrm{ad}_q\, X^+_j)\, X^+_{j+1,k} \equiv \\ &\equiv X^+_j X^+_{j+1,k} - q^{(\alpha_j,\alpha_{j+1,k+1})/2}\, X^+_{j+1,k} X^+_j = \\ &= (\mathrm{ad}_q\, X^+_{j,k-1})\, X^+_k \equiv \\ &\equiv X^+_{j,k-1} X^+_k - q^{(\alpha_k,\alpha_{jk})/2}\, X^+_k X^+_{j,k-1}\,, \qquad j < k\,, \end{aligned} \tag{1.28a}$$

$$\begin{aligned}X^-_{kj} &= (\mathrm{Ad}_{q^{-1}}\, X^-_k)\, X^-_{k-1,j} \equiv \\ &\equiv X^-_k X^-_{k-1,j} - q^{-(\alpha_k,\alpha_{jk})/2}\, X^-_{k-1,j} X^-_k = \\ &= (\mathrm{Ad}_{q^{-1}}\, X^-_{k,j+1})\, X^-_j \equiv \\ &\equiv X^-_{k,j+1} X^-_j - q^{-(\alpha_j,\alpha_{j+1,k+1})/2}\, X^-_j X^-_{k,j+1}\,, \qquad j < k.\end{aligned} \tag{1.28b}$$

In the affine case, the Cartan–Weyl formulae are as above, though it is useful to write them down the analogues of (I.2.158a) and (I.2.161):

$$[E_{k\bar{d}+\alpha}\,, E_{-(k\bar{d}+\alpha)}] = [H_{k\bar{d}+\alpha}]_{q_\alpha} = [H_\alpha + k\hat{c}]_{q_\alpha}\,, \tag{1.29a}$$

$$[E^i_{k\bar{d}}, E^i_{\ell\bar{d}}] = \delta_{k,-\ell}\,[H^i_{k\bar{d}}]_q = \delta_{k,-\ell}\,\frac{q^{k\hat{c}/2} - q^{-k\hat{c}/2}}{q^{1/2} - q^{-1/2}}\,. \tag{1.29b}$$

The action of δ, ε, γ on the Cartan–Weyl generators is obtained easily from (1.22) since H_β and E_β are given algebraically in terms of the Chevalley generators. (Of course, if $\alpha \notin \Delta_S$ the coalgebra operations δ, γ look more complicated than (1.22).) The axioms in (1.1)–(1.6) are fulfilled by the explicit definition (1.22).

The opposite comultiplication and antipode [253, 361] introduced above define a Hopf algebra $U_q(\mathcal{G})'$, which is related to $U_q(\mathcal{G})$ by:

$$U_q(\mathcal{G})' = U_{q^{-1}}(\mathcal{G}). \tag{1.30}$$

1.2.2 Universal R-Matrix and Casimirs

For $\mathcal{G} = sl(2)$ the universal R-matrix is given explicitly by [253]:

$$R = q^{H\otimes H/4} \sum_{n\geq 0} \frac{(1-q^{-1})^n q^{\frac{n(n-1)}{4}}}{[n]!} (q^{\frac{H}{4}} X^+)^n \otimes (q^{-\frac{H}{4}} X^-)^n \tag{1.31}$$

where $H = H_1\,, X^\pm = X_1^\pm\,, r = 1$. Note that this R-matrix is not in $U_q(sl(2)) \otimes U_q(sl(2))$, since it contains power series involving the generators $X^\pm$, but in some completion of it (in the h–adic topology used in [251, 253]). This is valid for the R-matrices of all $U_q(\mathcal{G})$. Hopf algebras with such an R-matrix are called *pseudo quasi-triangular Hopf algebras* [253] or *essentially quasi-triangular Hopf algebras* [454].

Here we can point out the only serious inequivalence between the Drinfeld and Jimbo definitions. Namely, there is no element in $\tilde{U}_q(\mathcal{G}) \otimes \tilde{U}_q(\mathcal{G})$ corresponding to the factor $q^{H\otimes H/4}$. Nevertheless, the universal R-matrix can act on any tensor product of finite-dimensional $\tilde{U}_q(\mathcal{G})$-modules.

For $\mathcal{G} = sl(n)$ an explicit formula for R was given in [533]. Explicit multiplicative formulas for R were given in [389, 424] for all complex simple Lie algebras $\mathcal{G}$ and in [385] for all finite-dimensional superalgebras with symmetrizable Cartan matrices. Then the universal R-matrix for the untwisted affine Lie algebras was given in [578]. Then this was obtained using the quantum Weyl group for $A_1^{(1)}$ in [426] and for general untwisted case in [167].

We recall results of [389, 447] where were given explicit multiplicative formulas for R for any $U_q(\mathscr{G})$. For this they introduced q-version of the Weyl group for $U_q(\mathscr{G})$. Let us recall that for $\alpha \in \Delta$,

$$s_\alpha(\Lambda) = \Lambda - \frac{2(\lambda, \alpha)}{(\alpha, \alpha)}\alpha, \quad \Lambda \in \mathscr{H}^* \tag{1.32}$$

are the standard reflections in $\mathscr{H}^*$. The *Weyl group* W is generated by the reflections $s_i \equiv s_{\alpha_i}$, where α_i is the simple root. Thus every element $w \in W$ can be written as the product of simple reflections. It is said that w is written in a reduced form if it is written with the minimal possible number of simple reflections; the number of reflections of a reduced form of w is called the length of w, denoted by $\ell(w)$.

The elements of the *q-Weyl group* belong to the completion $\bar{U}_q(\mathscr{G})$ of $U_q(\mathscr{G})$ [389]. They are defined by the action of the generating elements in the irreducible representations of $U_q(\mathscr{G})$.

In the case of $sl(2, \mathbb{C})$ the nontrivial element w of W is defined to act in the representation defined (e. g., [389]):

$$w|j, n >_q = (-1)^{j-n} q^{(n-j(j+1))/2} |j, -n >_q . \tag{1.33}$$

It satisfies the relations [389]:

$$wX^{\pm}w^{-1} = -q^{\pm 1/2}X^{\mp}, \quad wHw^{-1} = -H. \tag{1.34}$$

Since $\bar{U}_q(\mathscr{G})$ is also a Hopf algebra we have [389]:

$$\delta(w) = R^{-1} w \otimes w, \quad \varepsilon(w) = 1, \quad \gamma(w) = wq^{H/2}, \tag{1.35}$$

where R is given by (1.31). Further let us introduce the element

$$u = \sum_i \gamma(a_i) b_i, \tag{1.36}$$

where a_i, b_i are the coordinates of the element R:

$$R = \sum_i a_i \otimes b_i. \tag{1.37}$$

One may show that:

$$\gamma^2(Y) = uYu^{-1}, \tag{1.38}$$

and

$$v = uq^{-\hat{r}/2} \in \text{ centre of } U_q(\mathscr{G}), \tag{1.39}$$

$\hat{r}$ is used in (1.10). Let ϵ be the unipotent central element, that is, $\epsilon|j, n>_q = (-1)^{2j}|j, n>_q$, $\epsilon^2 = \text{id}$. Then [389]

$$w^2 = v\epsilon = uq^{-\hat{r}/2}\epsilon. \tag{1.40}$$

For arbitrary $U_q(\mathscr{G})$ let L_Λ be an irreducible representation of $U_q(\mathscr{G})$. Let $L_\Lambda = \oplus_j(W_\Lambda^j \otimes L_j)$ be the decomposition of L_Λ into irreducible $(U_q(sl(2,\mathbb{C})))_j$ submodules. Define the action of w_i in L_Λ as $w_i = \oplus_j(\text{Id}_{W_\Lambda^j} \otimes (w_i)_j)$, where $(w_i)_j$ is the action of w in L_j as in (1.33). Further one has [389]:

$$w_i H_j w_i^{-1} = H_j - a_{ij}H_i, \quad w_i X_i^{\pm} w_i^{-1} = -q^{\pm 1/2} X_i^{\mp}. \tag{1.41}$$

$$\delta w_i = R(i)^{-1} w_i \otimes w_i, \tag{1.42}$$

where $R(i) = R(H_i, X_i^{\pm}|q_i)$,

$$(w_i w_j)^{2-a_{ij}} = 1, \quad \text{for} \quad i \neq j, \quad (w_i)^2 = 1, \tag{1.43a}$$

$$(\tilde{w}_i \tilde{w}_j)^{2-a_{ij}} = 1, \quad \text{for} \quad i \neq j, \quad (\tilde{w}_i)^2 = 1, \tag{1.43b}$$

$$\tilde{w}_i = w_i q_i^{H_i^2/8}. \tag{1.43c}$$

Further let $s_0 = s_{i_1} \ldots s_{i_k}$ be the reduced form of the element of W with maximal length $\ell(s_0)$. It can be shown that the element

$$\tilde{w}_0 = \tilde{w}_{i_1} \ldots \tilde{w}_{i_k} \tag{1.44}$$

is well defined and does not depend on the choice of decomposition of s_0. Finally the result of [389] for the universal R-matrix is:

$$\begin{aligned} R &= q^{\sum_{i,j=1}^{n}(B^{-1})_{ij}H_i \otimes H_j/4}(\tilde{w}_0 \otimes \tilde{w}_0)\delta(\tilde{w}_0)^{-1}, \\ &\qquad (B_{ij}) = ((\alpha_i, \alpha_j)), \end{aligned} \tag{1.45a}$$

or

$$\begin{aligned} R &= q^{\sum_{i,j=1}^{n}(B^{-1})_{ij}H_i \otimes H_j/4}\tilde{R}(i_k|s_{i_1} \ldots s_{i_{k-1}}) \ldots \\ &\quad \ldots \tilde{R}(i_2|s_{i_1})\tilde{R}(i_1), \end{aligned} \tag{1.45b}$$

where

$$\tilde{R}(i_\ell|s_{i_1} \ldots s_{i_{\ell-1}}) = (T_{i_1}^{-1} \otimes T_{i_1}^{-1}) \ldots (T_{i_{\ell-1}}^{-1} \otimes T_{i_{\ell-1}}^{-1})\tilde{R}(i_\ell), \tag{1.45c}$$

$$T_i(Y) = \tilde{w}_i^{-1} Y \tilde{w}_i. \tag{1.45d}$$

The same construction works for affine Lie algebras [389]. Earlier work in this case includes the explicit construction for $A_1^{(1)}$ in any representation [360, 405, 406, 558]; for $A_n^{(1)}$, $B_n^{(1)}$, $C_n^{(1)}$, $D_n^{(1)}$ in the vector representation [82, 362]; for $B_n^{(1)}$, $D_n^{(1)}$ in the spinor representation [500]; and for $G_2^{(1)}$ [415].

The centre of $U_q(\mathscr{G})$, and for generic q the centre of $\tilde{U}_q(\mathscr{G})$, is generated by q-analogues of the *Casimir operators* [360, 361, 557]. For $\mathscr{G} = sl(2)$ one has:

$$C_2 = [(H+1)/2]^2 + X^- X^+. \tag{1.46}$$

For $\mathscr{G} = sl(n+1,\mathbb{C})$ we shall need more explicit expressions for the Cartan–Weyl generators as in (1.28). Let $\alpha_{j,k+1} \in \Delta^+$, $1 \le j \le k \le n$ be a positive root given explicitly in terms of the simple roots α_j, $j = 1, \dots, n$ (as in (1.26)) by:

$$\alpha_{j,k+1} = \alpha_j + \alpha_{j+1} + \dots + \alpha_k, \quad j < k. \tag{1.47}$$

Then the corresponding root vectors elements $X^\pm_{jk}$, $j < k$ are defined inductively:

$$X^\pm_{jk} = \pm(q^{1/4} X^\mp_j X^\pm_{j+1k} - q^{-1/4} X^\pm_{j+1k} X^\mp_j), \quad j < k. \tag{1.48}$$

Note that there is some inessential ambiguity in the definition (1.48), namely, $X'^\pm_{jk} = q^{\pm n} X^\pm_{jk}$ for generic q is also a good choice. Particularly often are used the choices $n = 1/4$ or $n = -1/4$. Thus, (1.48) differs by such normalization from (1.28). One can check (1.23) with

$$H_{\alpha_{j,k+1}} = H_j + H_{j+1} + \dots + H_k, \quad j < k. \tag{1.49}$$

Now the Casimir operator is given by [478]:

$$\begin{aligned} C_2 = K^0 \Bigg(&\sum_{1\le i\le j\le n} K_1^{-1} \dots K_{i-1}^{-1} K_{j+1} \dots K_n X^-_{ij} X^+_{ij} q^{(i+j-n-1)/2} + \\ &+ \sum_{j=0}^{n} K_1^{-1} \dots K_j^{-1} K_{j+1} \dots K_n q^{-j+n/2} (q^{1/2} - q^{-1/2})^{-2} \Bigg), \end{aligned} \tag{1.50}$$

where

$$\begin{aligned} K^0 &= K_1^{a_1} \dots K_n^{a_n}, \quad a_i = (n+1-2i)/(n+1), \\ K_i^{\pm 1} &= q_i^{\pm H_i/2}. \end{aligned}$$

For $n = 1$ this expression differs from (1.46) by an additive constant.

1.2.3 Jimbo's Definition

In some considerations it is useful to use a subalgebra $\tilde{U}_q(\mathcal{G})$ of $U_q(\mathcal{G})$ generated by $X_i^\pm$ and

$$K_i^{\pm 1} = q_i^{\pm H_i/4}, \tag{1.51}$$

and then (1.19) is replaced by:

$$\begin{aligned} K_i K_i^{-1} = K_i^{-1} K_i = 1\,, \quad [K_i\,, K_j] = 0, \\ K_i X_j^\pm K_i^{-1} = q_i^{\pm a_{ij}/4} X_j^\pm\,, \end{aligned} \tag{1.52a}$$

$$[X_i^+\,, X_j^-] = \delta_{ij} \frac{K_i^2 - K_i^{-2}}{q_i^{1/2} - q_i^{-1/2}}. \tag{1.52b}$$

On the other hand one may forget (1.51) and define $\tilde{U}_q(\mathcal{G})$ with the generators $X_i^\pm$ and $K_i^{\pm 1}$ and relations (1.20) and (1.52). In terms of these generators the coalgebra relations are:

$$\delta(K_i) = K_i \otimes K_i\,, \quad \delta(X_i^\pm) = X_i^\pm \otimes K_i + K_i^{-1} \otimes X_i^\pm \tag{1.53a}$$

$$\varepsilon(K_i) = 1\,, \quad \varepsilon(X_i^\pm) = 0\,, \tag{1.53b}$$

$$\gamma(K_i) = K_i^{-1}\,, \quad \gamma(X_i^\pm) = -q_i^{\pm 1/2} X_i^\pm. \tag{1.53c}$$

This is actually how quantum groups are defined in [360, 361]. This definition has the advantage that $\tilde{U}_q(\mathcal{G})$ is an algebra in the strict sense of the notion. The algebra $\tilde{U}_q(\mathcal{G})$ is also called *rational form* of $U_q(\mathcal{G})$, or *Jimbo quantum algebra.*

Nevertheless, even if not used, relation (1.51) is present in a 'hidden way'. That is why quantum algebras are called quantum groups a la Drinfeld-Jimbo in spite of the fact that the two definitions are not strictly equivalent. (In the mathematical literature (cf., e. g., Chari–Pressley [147]) one starts also by treating $q^{\pm 1/2}$ as formal variables.)

We shall point out now one of the inequivalences, the so-called twisting. Let $(\sigma_1, \ldots, \sigma_n) \in \{\pm 1\}^n$. Then there exists an algebra homomorphism of $\tilde{U}_q(\mathcal{G})$ given by:

$$K_i \mapsto \sigma_i K_i\,, \quad X_i^+ \mapsto \sigma_i X_i^+\,, \quad X_i^- \mapsto X_i^-\,. \tag{1.54}$$

On the other hand, except from the identity automorphism $\sigma_i = 1, \forall i$, there are no analogous automorphisms for $U_q(\mathcal{G})$. We note that this inequivalence is not very important since these automorphisms (except for the identity one) do not respect the coalgebra structure of $\tilde{U}_q(\mathcal{G})$.

For the Cartan–Weyl generators we need also the notation:

$$K_\beta \doteq q_\beta^{H_\beta/4} = \prod_{i=1}^{\ell} K_i^{n_i}\,, \qquad H_\beta = \sum_{i=1}^{\ell} n_i H_i \tag{1.55}$$

Then we have for the analogue of (1.23a):

$$[E_\beta\,, E_{-\beta}] = \frac{K_\beta^2 - K_\beta^{-2}}{q_\beta^{1/2} - q_\beta^{-1/2}} = \frac{q_\beta^{H_\beta/2} - q_\beta^{-H_\beta/2}}{q_\beta^{1/2} - q_\beta^{-1/2}}\,. \tag{1.56}$$

In the affine case, we use (1.55) for $\beta = k\bar{d} + \alpha$, $H_\beta = H_\alpha + k\hat{c}$, and $K_\delta \doteq q^{k\hat{c}/4}$, and then instead of (1.56) we have for the analogues of (1.29)

$$\begin{aligned}
[E_{k\bar{d}+\alpha}\,, E_{-(k\bar{d}+\alpha)}] &= \frac{K^2_{k\bar{d}+\alpha} - K^{-2}_{k\bar{d}+\alpha}}{q_\alpha^{1/2} - q_\alpha^{-1/2}} = \frac{q_\alpha^{(H_\alpha+k\hat{c})/2} - q_\alpha^{-(H_\alpha+k\hat{c})/2}}{q_\alpha^{1/2} - q_\alpha^{-1/2}} \\
[E^i_{k\bar{d}}, E^i_{\ell\bar{d}}] &= \delta_{k,-\ell}\,\frac{K_\delta^{2k} - K_\delta^{-2k}}{q^{1/2} - q^{-1/2}} = \delta_{k,-\ell}\,\frac{q^{k\hat{c}/2} - q^{-k\hat{c}/2}}{q^{1/2} - q^{-1/2}}\,.
\end{aligned} \tag{1.57}$$

One may also use instead of $X_i^\pm$ the generators:

$$E_i = X_i^- q_i^{H_i/4} = X_i^- K_i\,, \quad F_i = X_i^+ q_i^{-H_i/4} = X_i^+ K_i^{-1}\,. \tag{1.58}$$

(A similar change was used in [533].) In terms of the generators $K_i^{\pm 1}, E_i\,, F_i$ the coalgebra relations are rewritten as follows:

$$\begin{aligned}
\delta(K_i) &= K_i \otimes K_i\,, \quad \delta(E_i) = E_i \otimes K_i^2 + 1 \otimes E_i\,, \\
&\quad \delta(F_i) = F_i \otimes 1 + K_i^{-2} \otimes F_i
\end{aligned} \tag{1.59a}$$

$$\varepsilon(K_i) = 1\,, \quad \varepsilon(E_i) = \varepsilon(F_i) = 0\,, \tag{1.59b}$$

$$\begin{aligned}
\gamma(K_i) &= K_i^{-1}\,, \quad \gamma(E_i) = -E_i K_i^{-2}\,, \\
&\quad \gamma(F_i) = -K_i^2 F_i.
\end{aligned} \tag{1.59c}$$

We note for further use:

$$\begin{aligned}
\delta'(K_i) &= K_i \otimes K_i\,, \quad \delta'(E_i) = E_i \otimes 1 + K_i^2 \otimes E_i\,, \\
&\quad \delta'(F_i) = F_i \otimes K_i^{-2} + 1 \otimes F_i
\end{aligned} \tag{1.60a}$$

$$\begin{aligned}
\gamma'(K_i) &= K_i^{-1}\,, \quad \gamma'(E_i) = -K_i^{-2} E_i\,, \\
&\quad \gamma'(F_i) = -F_i K_i^2.
\end{aligned} \tag{1.60b}$$

One may also rewrite the q-Serre relation (1.20) as [533]:

$$(\mathrm{ad}_q E_i)^n(E_j) = 0 = (\mathrm{ad}'_q F_i)^n(F_j), \quad i \neq j, \quad \text{where} \tag{1.61a}$$

$$\mathrm{ad}_q : U_q(\mathscr{G}^+) \to \mathrm{End}(U_q(\mathscr{G}^+))\,,$$

$$\mathrm{ad}_q = m \circ (L \otimes R)(\mathrm{id} \otimes \gamma)\delta\,, \tag{1.61b}$$

$$\mathrm{ad}'_q : U_q(\mathscr{G}^-) \to \mathrm{End}(U_q(\mathscr{G}^-))\,,$$

$$\mathrm{ad}'_q = m \circ (L \otimes R)(\mathrm{id} \otimes \gamma')\delta'\,, \tag{1.61c}$$

and L (respectively, R) is the left (respectively, right) representation. In particular,

$$\begin{aligned}
\mathrm{ad}_q\, E_i &= (L \otimes R)(\mathrm{id} \otimes \gamma')\delta'(E_i) = \\
&= (L \otimes R)(\mathrm{id} \otimes \gamma')(E_i \otimes 1_{\mathscr{U}} + K_i^2 \otimes E_i) = \\
&= (L \otimes R)(E_i \otimes 1_{\mathscr{U}} + K_i^2 \otimes \gamma'(E_i)) = \\
&= (L \otimes R)(E_i \otimes 1_{\mathscr{U}} - K_i^2 \otimes K_i^{-2}E_i) = \\
&= L(E_i) \otimes 1_{\mathscr{U}} - L(K_i^2) \otimes R(K_i^{-2}E_i)
\end{aligned}$$

$$\begin{aligned}
\mathrm{ad}_q\, E_i\,(E_j) &= \left(L(E_i) \otimes 1_{\mathscr{U}} - L(K_i^2) \otimes R(K_i^{-2}E_i)\right)\,(E_j) = \\
&= E_i\, E_j - K_i^2\, E_j\, K_i^{-2}\, E_i = \\
&= E_i\, E_j - q^{a_{ij}/2}\, E_j\, E_i = \\
&= E_i\, E_j - q^{(1-n)/2}\, E_j\, E_i,
\end{aligned}$$

where the action of $m\circ$ on the RHS is understood where appropriately.

Furthermore $\mathrm{ad}_q(E_i)$ acts as a twisted derivation; that is, for $X, Y \in U_q(\mathscr{G}^+)$ homogeneous of degree β, $\gamma \in \mathscr{H}^*$ we have:

$$\mathrm{ad}_q(E_i)(XY) = \mathrm{ad}_q(E_i)(X)Y + q^{(\alpha_i^\vee,\beta)/2}X\,\mathrm{ad}_q(E_i)(Y). \tag{1.62}$$

Proof: the LHS and the RHS of the above equality are:

$$\begin{aligned}
\text{LHS:}\quad \mathrm{ad}_q(E_i)(XY) &= E_i\, XY - K_i^2\, XY\, K_i^{-2}\, E_i = \\
&= E_i\, XY - q^{(\alpha_i^\vee,\beta+\gamma)/2}\, XY\, E_i\,, \\
\text{RHS:}\quad \mathrm{ad}_q(E_i)(X)\, Y + q^{(\alpha_i^\vee,\beta)/2}\, X\, \mathrm{ad}_q(E_i)(Y) &= \\
&= E_i\, X\, Y - K_i^2\, X\, K_i^{-2}\, E_i\, Y + \\
&+ q^{(\alpha_i^\vee,\beta)/2}\, X\, \left(E_i\, Y - K_i^2\, Y\, K_i^{-2}\, E_i\right) = \\
&= E_i\, X\, Y - q^{(\alpha_i^\vee,\beta)/2}\, X\;E_i\, Y + \\
&+ q^{(\alpha_i^\vee,\beta)/2}\, X\, \left(E_i\, Y - q^{(\alpha_i^\vee,\gamma)/2}\, Y\, E_i\right) = \\
&= E_i\, X\, Y - q^{(\alpha_i^\vee,\beta+\gamma)/2}\, X\, Y\, E_i.
\end{aligned} \tag{1.63}$$

The action of $\mathrm{ad}'_q(F_i)$ on $X, Y \in U_q(\mathscr{G}^-)$ is defined analogously.

1.3 Drinfeld Second Realization of Quantum Affine Algebras

In [254] Drinfeld introduced the so-called new realization of quantum affine algebras. Our exposition will follow mostly [386].

Let $\hat{\mathcal{G}}$ be a untwisted affine Lie algebra, and $\Pi = \{\alpha_0, \alpha_1, \ldots, \alpha_\ell\}$ be a system of simple roots for $\hat{\mathcal{G}}$. We assume as in Volume 1 that the roots $\Pi_0 = \{\alpha_1, \alpha_2, \ldots, \alpha_\ell\}$ generate the root system of the corresponding finite-dimensional Lie algebra $\mathcal{G}$.

In this second realization, the algebra $U_q(\hat{\mathcal{G}})$ is generated by an infinite set of generators:

$$K_c,\ \chi_{i,m},\ \xi^{\pm}_{i,m}, \qquad (\text{for } i = 1, 2, \ldots, r; \quad m \in \mathbf{Z}), \tag{1.64}$$

with the defining relations:

$$[K_c, \text{everything}] = 0, \qquad \chi_{i,0}\xi^{\pm}_{j,m} = q^{\pm(\alpha_i,\alpha_j)/2}\xi^{\pm}_{j,m}\chi_{i,0}, \tag{1.65}$$

$$[\chi_{i,m}, \chi_{j,n}] = \delta_{m,-n} a_{ij}(m) \frac{K_c^{2m} - K_c^{-2m}}{q^{1/2} - q^{-1/2}}, \tag{1.66}$$

$$[\chi_{i,m}, \xi^{\pm}_{j,n}] = \pm a_{ij}(m)\, \xi^{\pm}_{j,m+n} K_c^{(-m\pm|m|)}, \tag{1.67}$$

$$\xi^{\pm}_{i,m+1}\xi^{\pm}_{j,n} - q^{\mp(\alpha_i,\alpha_j)/2}\xi^{\pm}_{j,n}\xi^{\pm}_{i,m+1} = q^{\mp(\alpha_i,\alpha_j)/2}\xi^{\pm}_{i,m}\xi^{\pm}_{j,n+1} - \xi^{\pm}_{j,n+1}\xi^{\pm}_{i,m}, \tag{1.68}$$

$$[\xi^{+}_{i,m}, \xi^{-}_{j,n}] = \delta_{i,j} \frac{\phi_{i,m+n} K_c^{2m} - \psi_{i,m+n} K_c^{2m}}{q^{1/2} - q^{-1/2}}, \tag{1.69}$$

$$\mathrm{Sym}\left(\sum_{s=0}^{n'_{ij}} (-1)^s C^s_{n'_{ij}} (q^{(\alpha_i,\alpha_j)/2}) \xi^{\pm}_{i,l_1} \cdots \xi^{\pm}_{i,l_s} \xi^{\pm}_{j,m} \xi^{\pm}_{i,l_{s+1}} \cdots \xi^{\pm}_{i,l_{n'_{ij}}} \right) = 0 \quad \text{for } i \neq j, \tag{1.70}$$

where

$$a_{ij}(m) = \frac{q^{m(\alpha_i,\alpha_j)/2} - q^{-m(\alpha_i,\alpha_j)/2}}{m(q^{1/2} - q^{-1/2})}, \tag{1.71}$$

the elements $\phi_{i,p}$, $\psi_{i,p}$ are defined from the relations:

$$\sum_p \phi_{i,p} u^{-p} = \chi_{i,0} \exp((q^{-1/2} - q^{1/2}) \sum_{p<0} \chi_{i,p} u^{-p}), \tag{1.72}$$

$$\sum_{p} \psi_{i,p} u^{-p} = \chi_{i,0}^{-1} \exp((q^{1/2} - q^{-1/2}) \sum_{p>0} \chi_{i,p} u^{-p}) , \tag{1.73}$$

the *q-binomial coefficients* $C_n^s(q)$ are determined by the formula

$$C_n^s(q) = \frac{[n]_q!}{[s]_q![n-s]_q!} , \tag{1.74}$$

the symbol "Sym" in (1.70) denotes a symmetrization on $l_1, l_2, \ldots, l_{n_{ij}}$, and $n'_{ij} := n_{ij} + 1$.

It should be noted that the matrix $(a_{ij}(m))$ with the elements (1.71) may be considered as a q-analog of the "level m" for the matrix Cartan (a_{ij}^{sym}).

Drinfeld has shown how to express the Chevalley generators e_{α_i}, h_{α_i} in terms of $\chi_{i,0}^{\pm}$ and $\xi_{i,k}^{\pm}$, $k = 0, \pm 1$ (see [254]). He suggested also other formulas of the comultiplication for $U_q(\mathscr{G})$, which originates in a quantization of the corresponding bialgebra structure [254] (different from the usual one):

$$\delta^{(D)}(K_c) = K_c \otimes K_c , \qquad \delta^{(D)}(\chi_{i,0}) = \chi_{i,0} \otimes \chi_{i,0}, \tag{1.75}$$

$$\delta^{(D)}(\chi_{i,m}) = \chi_{i,m} \otimes 1 + K_c^{-1} \otimes \chi_{i,m} , \quad \delta^{(D)}(\chi_{i,-m}) = \chi_{i,-m} \otimes K_c + 1 \otimes \chi_{i,-m}, \tag{1.76}$$

for $m > 0$, and

$$\delta^{(D)}(\xi_{i,m}^{+}) = \xi_{i,m}^{+} \otimes 1 + \sum_{n \geq 0} K_c^n \phi_{i,n} \otimes \xi_{i,m+n}^{+} , \tag{1.77}$$

$$\delta^{(D)}(\xi_{i,m}^{-}) = 1 \otimes \xi_{i,m}^{-} + \sum_{n \geq 0} \xi_{i,m-n}^{-} \otimes \psi_{i,n} K_c^n , \tag{1.78}$$

for any $m \in \mathbf{Z}$.

Next we show how the generators K_c, $\chi_{i,m}$, $\xi_{i,m}^{\pm}$ can be expressed via the Cartan–Weyl generators in Jimbo's realization.

We fix some special normal ordering in $\Delta_+(\hat{\mathscr{G}}) := \Delta_+$, which satisfies the following additional constraint:

$$\ell\bar{d} + \alpha_i < (m+1)\bar{d} < (n+1)\bar{d} - \alpha_j \tag{1.79}$$

for any simple roots $\alpha_i, \alpha_j \in \Pi_0$, and ℓ, m, $n \geq 0$. Here $\bar{d}$ is the minimal positive *imaginary root* (cf. Section I.2.6). Furthermore we put

$$E_{\bar{d}}^{(i)} = [X_i^+, E_{\bar{d}-\alpha_i}]_q , \tag{1.80}$$

$$E_{n\bar{d}+\alpha_i} = (-1)^n ([(\alpha_i, \alpha_i)]_q)^{-n} (\tilde{\mathrm{ad}}\, E_{\bar{d}}^{(i)})^n X_i^+ , \tag{1.81}$$

$$E_{(n+1)\bar{d}-\alpha_i} = ([(\alpha_i, \alpha_i)]_q)^{-n} (\tilde{\mathrm{ad}}\, E_{\bar{d}}^{(i)})^n E_{\bar{d}-\alpha_i} , \tag{1.82}$$

$$E'^{(i)}_{(n+1)\bar{d}} = [E_{n\bar{d}+\alpha_i}, E_{\bar{d}-\alpha_i}]_q , \tag{1.83}$$

(for $n > 0$), where $(\tilde{\mathrm{ad}}\, x)y = [x, y]$ is the usual commutator. The imaginary root vectors $E'^{(i)}_{\pm n\tilde{d}}$ do not satisfy the relation (1.57). We introduce new vectors $E^{(i)}_{\pm n\tilde{d}}$ by the following (Schur) relations:

$$E'^{(i)}_{n\tilde{d}} = \sum_{p_1+2p_2+\ldots+kp_k=n} \frac{(q^{1/2}-q^{-1/2})^{\sum p_i - 1}}{p_1!\cdots p_k!} (E^{(i)}_{\tilde{d}})^{p_1} \cdots (E^{(i)}_{k\tilde{d}})^{p_k} . \tag{1.84}$$

In terms of the generating functions

$$E'_i(z) = (q^{1/2}-q^{-1/2}) \sum_{m\geq 1} E'^{(i)}_{m\tilde{d}} z^m \tag{1.85}$$

and

$$E_i(z) = (q^{1/2}-q^{-1/2}) \sum_{m\geq 1} E^{(i)}_{m\tilde{d}} z^m , \tag{1.86}$$

the relation (1.84) may be rewritten in the form

$$E'_i(z) = -1 + \exp E_i(z) \tag{1.87}$$

or

$$E_i(z) = \ln(1 + E'_i(z)) . \tag{1.88}$$

From this we have the inverse formula to (1.84)

$$E^{(i)}_{n\tilde{d}} = \sum_{p_1+2p_2+\ldots+kp_k=n} \frac{(q^{-1/2}-q^{1/2})^{\sum p_i - 1}(\sum_{i=1}^{k} p_i - 1)!}{p_1!\cdots p_k!} (E'^{(i)}_{\tilde{d}})^{p_1} \cdots (E'^{(i)}_{k\tilde{d}})^{p_k} . \tag{1.89}$$

We construct the rest of the real root vectors using the root vectors $E_{n\tilde{d}+\alpha_i}$, $E_{(n+1)\tilde{d}-\alpha_i}$, $E^{(i)}_{(n+1)\tilde{d}}$ ($i = 1, 2, \ldots, \ell$; $n \in \mathbf{Z}_+$). The root vectors of negative roots are obtained by the *Cartan involution* (*):

$$E_{-\gamma} = (E_\gamma)^* \tag{1.90}$$

for $\gamma \in \Delta(\hat{\mathscr{G}})$.

To proceed further, we introduce two types of root vectors $\hat{E}_\gamma$ and $\check{E}_\gamma$ by the following formulas [386]:

$$\hat{E}_\gamma := E_\gamma , \qquad \hat{E}_{-\gamma} := -K_\gamma^{-1} E_{-\gamma} , \qquad \forall \gamma \in \Delta , \tag{1.91}$$

and

$$\check{E}_{-\gamma} := E_{-\gamma} , \qquad \check{E}_\gamma := -E_\gamma K_\gamma , \qquad \forall \gamma \in \Delta . \tag{1.92}$$

Using the explicit relations (1.80–1.83), (1.89), and (1.90), we can prove the following theorem which states the connection between the Cartan–Weyl and Drinfeld's generators for the quantum untwisted affine algebra $U_q(\hat{\mathscr{G}})$:

Theorem 1.1 ([386]). *Let some function π: $\{\alpha_1, \alpha_2 \ldots, \alpha_\ell\} \mapsto \{0,1\}$ be chosen such that $\pi(\alpha_i) \neq \pi(\alpha_j)$ if $(\alpha_i, \alpha_j) \neq 0$ and let the root vectors $\hat{E}_{\pm\gamma}$ and $\check{E}_{\pm\gamma}$ of the real roots $\gamma \in \Delta_+(\hat{\mathscr{G}})$ be the Cartan–Weyl generators (1.91), (1.92) and $E^{(i)}_{n\bar{d}}$ be imaginary root vectors of $U_q(\hat{\mathscr{G}})$. Then the elements*

$$K_c := K_{\bar{d}}\,, \qquad \chi_{i,0} := K_{\alpha_i}\,, \qquad \chi_{i,n} := (-1)^{n\pi(\alpha_i)} E^{(i)}_{n\bar{d}}, \tag{1.93}$$

$$\xi^{+}_{i,n} = (-1)^{n\pi(\alpha_i)} \hat{E}_{n\bar{d}+\alpha_i}\,, \qquad \xi^{-}_{i,n} = (-1)^{n\pi(\alpha_i)} \check{E}_{n\bar{d}-\alpha_i}\,, \tag{1.94}$$

for $n \in \mathbf{Z}$, and

$$\begin{aligned} \phi_{i,0} &= K_{\alpha_i}\,, & \phi_{i,-n} &= (q^{-1/2} - q^{1/2}) K_{\alpha_i} E'^{(i)}_{-n\bar{d}}\,, \\ \psi_{i,0} &= K^{-1}_{\alpha_i}\,, & \psi_{i,n} &= (q^{1/2} - q^{-1/2}) K^{-1}_{\alpha_i} E'^{(i)}_{n\bar{d}}\,, \end{aligned} \tag{1.95}$$

for $n > 0$ satisfies the relations (1.65–1.70); that is, the elements (1.93–1.95) are the generators of the Drinfeld's second realization of $U_q(\hat{\mathscr{G}})$. ◊

1.4 Drinfeld's Realizations of Yangians

This section follows mostly [577]. Let $\mathfrak{g}$ be a finite-dimensional complex simple Lie algebra. Fix a nonzero invariant bilinear form (,) on $\mathfrak{g}$, and let $\{I_\alpha\}$ be an orthonormal basis of $\mathfrak{g}$ with respect to (,).

1.4.1 The First Drinfeld Realization of Yangians

Definition 1.2 ([251]). The Yangian $Y_\eta(\mathfrak{g})$ is generated as an associative algebra over $\mathbb{C}[[\eta]]$ by the Lie algebra $\mathfrak{g}$ and elements $J(x)$, $x \in \mathfrak{g}$, with the defining relations:

$$\begin{aligned} &J(\lambda x + \mu y) = \lambda J(x) + \mu J(y), \\ &J([x,y]) = [x, J(y)] \qquad \text{for } x, y \in \mathfrak{g},\ \lambda, \mu \in \mathbb{C}, \end{aligned} \tag{1.96}$$

$$\begin{aligned} &\text{if } \sum_i [x_i, y_i] = 0 \quad \text{for } x_i, y_i \in \mathfrak{g} \quad \Rightarrow \\ &\sum_i [J(x_i), J(y_i)] = \frac{\eta^2}{12} \sum_i \sum_{\alpha,\beta,\gamma} ([[x_i, I_\alpha], [y_i, I_\beta]], I_\gamma)\{I_\alpha, I_\beta, I_\gamma\}\,, \end{aligned} \tag{1.97}$$

$$\text{if } \sum_i [[x_i, y_i], z_i] = 0 \quad \text{for } x_i, y_i, z_i \in \mathfrak{g} \quad \Rightarrow$$

$$\sum_i [[J(x_i), J(y_i)], J(z_i)] = \frac{\eta^2}{4} \sum_i \sum_{\alpha,\beta,\gamma} f(x_i, y_i, z_i, I_\alpha, I_\beta, I_\gamma) \times \{I_\alpha, I_\beta, J(I_\gamma)\}, \tag{1.98}$$

where the notations are used $\{a_1, a_2, a_3\} := (1/6)\sum_{i\neq j\neq k} a_i a_j a_k$ and $f(x, y, z, a, b, c) := \underset{x,y}{\text{Alt}}\,\underset{x,z}{\text{Sym}}\,([x, [y, a]], [[z, b], c])$. A comultiplication map ($\delta_\eta : Y_\eta(\mathfrak{g}) \to Y_\eta(\mathfrak{g}) \otimes Y_\eta(\mathfrak{g})$), an antipode ($S_\eta : Y_\eta(\mathfrak{g}) \to Y_\eta(\mathfrak{g})$) and a counit ($\varepsilon_\eta : Y_\eta(\mathfrak{g}) \to \mathbb{C}$) are given by the formulas ($x \in \mathfrak{g}$)

$$\begin{aligned} \delta_\eta(x) &= x \otimes 1 + 1 \otimes x, \\ \delta_\eta(J(x)) &= J(x) \otimes 1 + 1 \otimes J(x) + \frac{\eta}{2}[x \otimes 1, \Omega_2], \end{aligned} \tag{1.99}$$

$$S_\eta(x) = -x, \quad S_\eta(J(x)) = -J(x) + \frac{\eta}{4}\lambda x, \tag{1.100}$$

$$\varepsilon_\eta(x) = \varepsilon_\eta(J(x)) = 0, \quad \varepsilon_\eta(1) = 1, \tag{1.101}$$

where Ω_2 is the Casimir two-tensor ($\Omega_2 = \sum_\alpha I_\alpha \otimes I_\alpha$) and λ is the eigenvalue of the Casimir operator $C_2 = \sum_\alpha I_\alpha I_\alpha$ in the adjoint representation of $\mathfrak{g}$ in $\mathfrak{g}$.

We may specialize the formal parameter η to any complex number $\nu \in \mathbb{C}$; however, the resulting Hopf algebra $Y_\nu(\mathfrak{g})$ (over $\mathbb{C}$) is essentially independent of ν, provided that $\nu \neq 0$. It means that any two Hopf algebras $Y_\nu(\mathfrak{g})$ and $Y_{\nu'}(\mathfrak{g})$ with $\nu \neq \nu'$; $\nu, \nu' \neq 0$ are isomorphic. Thus, we can as well take $\nu = 1$ and drop the parameter η. However, for the convenience of passage to the limit $\eta \to 0$, we shall keep the formal parameter η.

Remark 1.2. In the case $\mathfrak{g} = \mathfrak{sl}_2(\mathbb{C})$ there is a more complicated relation instead (1.97) (see [251]). ◊

1.4.2 The Second Drinfeld Realization of Yangians

Let $A = (a_{ij})_{i,j=1}^l$ be a standard Cartan matrix of $\mathfrak{g}$, $\Pi := \{\alpha_1, \dots, \alpha_l\}$ be a system of simple roots (l is rank of $\mathfrak{g}$), and $B_{ij} := \frac{1}{2}(\alpha_i, \alpha_j)$.

Theorem 1.2 ([254]). *The Yangian $Y_\eta(\mathfrak{g})$ is isomorphic to the associative algebra over $\mathbb{C}[[\eta]]$ with the generators:*

$$\xi_{in}^+, \quad \xi_{in}^-, \quad \varphi_{in} \qquad \text{for} \quad i = 1, 2, \dots, l; \; n = 0, 1, 2, \dots, \tag{1.102}$$

and the following defining relations:

$$\begin{aligned} [\varphi_{in}, \varphi_{jm}] &= 0\,, \\ [\varphi_{i0}, \xi^{\pm}_{jn}] &= \pm 2B_{ij}\, \xi^{\pm}_{jn}\,, \\ [\xi^{+}_{in}, \xi^{-}_{jm}] &= \delta_{ij}\, \varphi_{j\, n+m}\,, \end{aligned} \tag{1.103}$$

$$[\varphi_{i\,n+1}, \xi^{\pm}_{jm}] - [\varphi_{i\,n}, \xi^{\pm}_{j\,m+1}] = \pm\eta\, B_{ij} \left(\varphi_{i\,n}\xi^{\pm}_{jm} + \xi^{\pm}_{jm}\varphi_{in}\right),$$

$$[\xi^{\pm}_{i\,n+1}\, \xi^{\pm}_{jm}] - [\xi^{\pm}_{i\,n}, \xi^{\pm}_{j\,m+1}] = \pm\eta\, B_{ij} \left(\xi^{\pm}_{i\,n}\xi^{\pm}_{jm} + \xi^{\pm}_{jm}\xi^{\pm}_{in}\right),$$

$$\underset{n_1,n_2,\ldots,n_k}{Sym}\, [\xi^{\pm}_{i\,n_1}, [\xi^{\pm}_{in_2}[\ldots[\xi^{\pm}_{i\,n_k}, \xi^{\pm}_{jm}]\ldots]]] = 0 \quad for\ i \neq j,\quad k = 1 - A_{ij}.\ \Diamond$$

Explicit formulas for the action of the comultiplication δ_η on the generates $\xi^{\pm}_{in}$, φ_{in} are rather cumbersome (see [386]), and they are not given here.

For the Yangian $Y_\eta(\mathfrak{sl}(n, \mathbb{C}))$) Drinfeld has also given a third realization. It is presented in terms of *RLL*-relations (see details in [254, 272]).

All these realizations of Yangians are not minimal; that is, they are not given in terms of a Chevalley basis. However, the minimal realization may be given using the connection of Yangians with quantum untwisted affine algebras (cf. [577]).

More information on Yangians may be found in the book [479].

1.5 q-Deformations of Noncompact Lie Algebras

1.5.1 Preliminaries

Noncompact Lie groups and algebras play a very important role in physics – recall. Thus ever since the introduction of quantum groups as deformations $U_q(\mathcal{G})$ of the universal enveloping algebras of complex simple Lie algebras or as matrix quantum groups, one was always asking what would be the deformation of the real forms. Actually, the deformation of *compact* simple Lie algebras is used in the physics literature without much explanation assuming the implementation of the Weyl unitary trick. In [272] Faddeev–Reshetikhin–Takhtajan introduced the compact matrix quantum groups $SU_q(n)$ (for $n = 2$ first in [599]), $SO_q(n)$, $Sp_q(n)$, and the maximally split real noncompact forms $SL_q(n, \mathbb{R})$, $SO_q(n, n)$, $SO_q(n, n+1)$, $Sp_q(n, \mathbb{R})$. From our point of view it is not accidental that these cases were obtained first since the root systems of these real forms coincide (up to multiple of i in the compact case) with the root systems of their complexifications (cf. the description of our approach below). Besides the above among the first noncompact cases were considered: $U_q(su(1, 1))$ in [140], $U_q(su(n, 1))$ in [143], quantum Lorentz groups in [123, 250, 320, 511], quantum deformation of Poincare algebra in [435, 437].

Here we present a universal approach to the q-deformation of real simple algebras. Let $\mathcal{G}$ be a real simple Lie algebra (below we shall need to extend the construction to real reductive Lie algebras). We shall use the standard q-deformation for the simple components of the complexification $\mathcal{G}^{\mathbb{C}}$ of $\mathcal{G}$ to obtain deformation $U_q(\mathcal{G})$ as a real form of $U_q(\mathcal{G}^{\mathbb{C}})$. Though the procedure is described mostly in terms which are known from the undeformed case, we stress which steps are necessitated by the q-deformation. The first basic ingredient of our approach relies on the fact that the real forms $\mathcal{G}$ of a complex simple Lie algebra $\mathcal{G}^{\mathbb{C}}$ are in one-to-one correspondence with the Cartan automorphisms θ of $\mathcal{G}^{\mathbb{C}}$. This allows to study the structure of the real forms and to find their explicit embeddings as real subalgebras of $\mathcal{G}^{\mathbb{C}}$ invariant under θ and consequently, using the same generators, to find $U_q(\mathcal{G})$. This ingredient is enough for the compact case (up to the choice of the range of q). The second basic ingredient is related to the fact that a real noncompact simple Lie algebra has in general (a finite number of) nonconjugate Cartan subalgebras [10]. This is very important since we have to choose which conjugacy class of Cartan subalgebras will correspond to the unique conjugacy class of Cartan subalgebras of $\mathcal{G}^{\mathbb{C}}$ and will be "freezed" under a q-deformation (cf. (4a) below). For each such choice we shall get a different q-deformation. The third basic ingredient are the *Bruhat decompositions* $\mathcal{G} = \mathcal{A} \oplus \mathcal{M} \oplus \tilde{\mathcal{N}} \oplus \mathcal{N}$, (direct sum of vector subspaces), where $\mathcal{A}$ is a noncompact abelian subalgebra, $\mathcal{M}$ (a reductive Lie algebra) is the centralizer of $\mathcal{A}$ in $\mathcal{G}$ (mod $\mathcal{A}$), and $\tilde{\mathcal{N}}$, $\mathcal{N}$ are nilpotent subalgebras forming the positive, negative, respectively, root spaces of the root system $(\mathcal{G}, \mathcal{A})$. Consistently, the Cartan subalgebras of $\mathcal{G}$ have the decomposition $\mathcal{H} = \mathcal{A} \oplus \mathcal{H}^m$, where $\mathcal{H}^m$ is a Cartan subalgebra of $\mathcal{M}$. A general property of the deformations $U_q(\mathcal{G})$ obtained by our procedure is that $U_q(\mathcal{M})$, $U_q(\tilde{\mathcal{P}})$, $U_q(\mathcal{P})$ are Hopf subalgebras of $U_q(\mathcal{G})$, where $\mathcal{P} = \mathcal{A} \oplus \mathcal{M} \oplus \mathcal{N}$, $\tilde{\mathcal{P}} = \mathcal{A} \oplus \mathcal{M} \oplus \tilde{\mathcal{N}}$ are parabolic subalgebras of $\mathcal{G}$. Our approach is easily generalized for the real forms of the basic classical Lie superalgebras and of the corresponding affine Kac-Moody (super) algebras.

These q-deformations are called canonical because they are obtained by a well-defined procedure presented below. This does not exclude other deformations, for example, multiparameter deformations, or deformation by contaction (cf. also comments in the text). Also as in the undeformed case for each real form there exists an antilinear (anti)involution σ of $U_q(\mathcal{G}^{\mathbb{C}})$ which preserves $U_q(\mathcal{G})$. Unlike the undeformed case it is necessary to consider both involutions and antiinvolutions, since there are two possibilities for the deformation parameter q, that is, either $|q| = 1$ or $q \in \mathbb{R}$. For instance, $U_q(su(2))$ has $|q| = 1$ when σ is an involution and $q \in \mathbb{R}$ when σ is an antiinvolution. Further, σ is a coalgebra (anti)homomorphism; that is, $\delta \circ \sigma = (\sigma \times \sigma) \circ \delta$, or $\delta \circ \sigma = (\sigma \times \sigma) \circ \delta'$; $\varepsilon(\sigma(X)) = \bar{\varepsilon}(X)\ \forall X \in U_q(\mathcal{G}^{\mathbb{C}})$. Then the relations for the antipode are $\sigma \circ \gamma = \gamma \circ \sigma$ if σ is an algebra involution and a coalgebra homomorphism or if it is an algebra antiinvolution and a coalgebra antihomomorphism and $(\sigma \circ \gamma)^2 = id$ otherwise. One approach to the real forms would be to try to classify directly the possible conjugation σ. Our approach is more constructive, and the conjugation σ is obtained as a by-product of the procedure proposed below (this is pointed out in some examples).

1.5.2 q-Deformation of the Real Forms

Let $\mathscr{G}$ be a real noncompact semisimple Lie algebra, θ be the Cartan involution in $\mathscr{G}$, and $\mathscr{G} = \mathscr{K} \oplus \mathscr{Q}$ be the *Cartan decomposition* of $\mathscr{G}$, so that $\theta X = X, X \in \mathscr{K}$, $\theta X = -X, X \in \mathscr{Q}$; $\mathscr{K}$ is the maximal compact subalgebra of $\mathscr{G}$. Let $\mathscr{A}_0$ be the maximal subspace of $\mathscr{Q}$, which is an abelian subalgebra of $\mathscr{G}$; r_0 =dim $\mathscr{A}_0$ is the *real rank* (or *split rank*) of $\mathscr{G}$, $1 \leq r_0 \leq \ell = \text{rank}\, \mathscr{G}$.

Let Δ_R^0 be the root system of the pair $(\mathscr{G}, \mathscr{A}_0)$, also called ($\mathscr{A}_0$ -) *restricted root system*:

$$\Delta_R^0 = \{\lambda \in \mathscr{A}_0^* | \lambda \neq 0, \mathscr{G}_\lambda^0 \neq 0\}, \tag{1.104}$$

$$\mathscr{G}_\lambda^0 = \{X \in \mathscr{G} | [Y, X] = \lambda(Y)X, \forall Y \in \mathscr{A}_0\}. \tag{1.105}$$

The elements of $\Delta_R^0 = \Delta_R^{0+} \cup \Delta_R^{0-}$ is called ($\mathscr{A}_0$ -) *restricted roots*; if $\lambda \in \Delta_R^0$, $\mathscr{G}_\lambda^0$ is called ($\mathscr{A}_0$-) *restricted root space*, $\dim_R \mathscr{G}_\lambda^0 \geq 1$. Now we can introduce the subalgebras corresponding to the positive (Δ_R^{0+}) and negative (Δ_R^{0-}) restricted roots:

$$\tilde{\mathscr{N}}_0 = \bigoplus_{\lambda \in \Delta_R^{0+}} \mathscr{G}_\lambda^0 = \tilde{\mathscr{N}}_0^1 \oplus \tilde{\mathscr{N}}_0^2, \tag{1.106}$$

$$\mathscr{N}_0 = \bigoplus_{\lambda \in \Delta_R^{0-}} \mathscr{G}_\lambda^0 = \mathscr{N}_0^1 \oplus \mathscr{N}_0^2 = \theta \tilde{\mathscr{N}}_0, \tag{1.107}$$

where $\tilde{\mathscr{N}}_0^1$, $\tilde{\mathscr{N}}_0^2$ is the direct sum of $\mathscr{G}_\lambda^0$ with $\dim_R \mathscr{G}_\lambda^0 = 1$, $\dim_R \mathscr{G}_\lambda^0 > 1$, respectively, and analogously for $\mathscr{N}_0^a = \theta \tilde{\mathscr{N}}_0^a$. Then we have the Bruhat decompositions which we shall use for our q-deformations:

$$\mathscr{G} = \tilde{\mathscr{N}}_0 \oplus \mathscr{A}_0 \oplus \mathscr{M}_0 \oplus \mathscr{N}_0 = \tilde{\mathscr{N}}_0^1 \oplus \tilde{\mathscr{N}}_0^2 \oplus \mathscr{A}_0 \oplus \mathscr{M}_0 \oplus \mathscr{N}_0^1 \oplus \mathscr{N}_0^2, \tag{1.108}$$

where $\mathscr{M}_0$ is the centralizer of $\mathscr{A}_0$ in $\mathscr{K}$; that is, $\mathscr{M}_0 = \{X \in \mathscr{K} | [X, Y] = 0, \forall Y \in \mathscr{A}_0\}$. In general $\mathscr{M}_0$ is a compact reductive Lie algebra, and we shall write $\mathscr{M}_0 = \mathscr{M}_0^s \oplus \mathscr{Z}_0^m$, where $\mathscr{M}_0^s = [\mathscr{M}_0, \mathscr{M}_0]$ is the semisimple part of $\mathscr{M}_0$, and $\mathscr{Z}_0^m$ is the centre of $\mathscr{M}_0$. Note that $\tilde{\mathscr{P}}_0^0 \equiv \tilde{\mathscr{N}}_0 \oplus \mathscr{A}_0 \oplus \mathscr{M}_0$, $\mathscr{P}_0^0 \equiv \mathscr{A}_0 \oplus \mathscr{M}_0 \oplus \mathscr{N}_0$ are subalgebras of $\mathscr{G}$, the so-called minimal parabolic subalgebras of $\mathscr{G}$. Identifying $\tilde{\mathscr{P}}_0^0$, $\mathscr{P}_0^0$ is the first step of our procedure.

Further, let $\mathscr{H}_0^m$ be the Cartan subalgebra of $\mathscr{M}_0$; that is, $\mathscr{H}_0^m = \mathscr{H}_0^{ms} \oplus \mathscr{Z}_0^m$, where $\mathscr{H}_0^{ms}$ is the Cartan subalgebra of $\mathscr{M}_0^s$. Then $\mathscr{H}_0 \equiv \mathscr{H}_0^m \oplus \mathscr{A}_0$ is a Cartan subalgebra of $\mathscr{G}$, the most noncompact one; $\dim_R \mathscr{H}_0 = \dim_R \mathscr{H}_0^{ms} + \dim_R \mathscr{Z}_0^m + r_0$. We choose $\mathscr{H}_0$ to be also the Cartan subalgebra of $U_q(\mathscr{G})$. Let $\mathscr{H}^{\mathbb{C}}$ be the complexification of $\mathscr{H}_0$ ($\ell = \text{rank}\, \mathscr{G}^{\mathbb{C}} = \dim_C \mathscr{H}^{\mathbb{C}}$); then it is a Cartan subalgebra of the complexification $\mathscr{G}^{\mathbb{C}}$ of $\mathscr{G}$.

The second step in our procedure is to choose consistently the basis of the rest of $\mathscr{G}$ and $\mathscr{G}^{\mathbb{C}}$, and thus of $U_q(\mathscr{G})$. For this we use the classification of the roots from Δ with respect to $\mathscr{H}_0$. The set $\Delta_r^0 \equiv \{\alpha \in \Delta | \alpha|_{\mathscr{H}_0^m} = 0\}$ is called the set of *real roots*,

$\Delta_i^0 \equiv \{\alpha \in \Delta | \alpha|_{\mathscr{A}_0} = 0\}$ - the set of *compact roots*, $\Delta_c^0 \equiv \Delta\backslash(\Delta_r^0 \cup \Delta_i^0)$ - the set of *complex roots* (cf. Bourbaki [109]). Thus $\Delta = \Delta_r^0 \cup \Delta_i^0 \cup \Delta_c^0$. Further, let $\alpha \in \Delta^+$; let $\mathscr{L}_\alpha^c$ be the complex linear span of $H_\alpha, X_\alpha, X_{-\alpha}$; and let $\mathscr{L}_\alpha = \mathscr{L}_\alpha^c \cap \mathscr{G}$. Then $\dim_R \mathscr{L}_\alpha = 3$ if the $\alpha \in \Delta_r^0 \cup \Delta_i^0$ [10]. If $\alpha \in \Delta_r^0$ then $X_\alpha \in \mathscr{P}^{\mathbb{C}}$ and $\mathscr{L}_\alpha$ is noncompact. Since the Cartan subalgebra is $\mathscr{H}_0$, then $X_\alpha \in \mathscr{K}^{\mathbb{C}}$ and $\mathscr{L}_\alpha$ is compact if $\alpha \in \Delta_i^0$. The algebras $\mathscr{L}_\alpha$ are given by:

$$\mathscr{L}_\alpha = r.l.s.\{H_\alpha, X_\alpha, X_{-\alpha}\}, \quad \alpha \in \Delta_r^{0+}, \tag{1.109a}$$

$$\mathscr{L}_\alpha = r.l.s.\{iH_\alpha, X_\alpha - X_{-\alpha}, i(X_\alpha + X_{-\alpha})\}, \quad \alpha \in \Delta_i^{0+}, \tag{1.109b}$$

where r.l.s. stands for real linear span.

Note that there is a one-to-one correspondence between the real roots $\alpha \in \Delta_r^0$ and the restricted roots $\lambda \in \Delta_R^0$ with $\dim_R \mathscr{G}_\lambda^0 = 1$ and naturally this correspondence is realized by the restriction: $\lambda = \alpha|_{\mathscr{A}_0}$. Thus the elements in (8a) $X_\alpha^\pm$ for $\alpha \in \Delta_r^0$ we take also as elements of $U_q(\mathscr{G})$. Thus, following (1.19),(1.23) these generators obey:

$$[X_\alpha, X_{-\alpha}] = [H_\alpha]_{q_\alpha}, \quad [H_\alpha, X_{\pm\alpha}] = \pm\alpha(H_\alpha)X_{\pm\alpha}, \quad \text{for} \quad \alpha \in \Delta_r^{0+}, \tag{1.110}$$

and the Hopf algebra structure is given exactly as for $\alpha \in \Delta$ (cf. (1.22) and the text after that).

Remark 1.3. Formulae (1.109a) and (1.110) determine completely a q-deformation of any maximally split real form (or normal real form), when all roots are real, $\mathscr{M}_0 = 0$, and $\mathscr{H}_0 = \mathscr{A}_0$. In this case the Bruhat decomposition is just

$$\mathscr{G} = \tilde{\mathscr{N}}_0 \oplus \mathscr{A}_0 \oplus \mathscr{N}_0, \tag{1.111}$$

that is, this is the restriction to $\mathbb{R}$ of the standard triangular decomposition $\mathscr{G}^{\mathbb{C}} = \mathscr{G}_+^{\mathbb{C}} \oplus \mathscr{H}^{\mathbb{C}} \oplus \mathscr{G}_-^{\mathbb{C}}$, and hence $U_q(\mathscr{G})$ is just the restriction of $U_q(\mathscr{G}^{\mathbb{C}})$ to $\mathbb{R}$ with $q \in \mathbb{R}$. Thus we also inherit the property that $U_q(\tilde{\mathscr{N}}_0 \oplus \mathscr{A}_0)$, $U_q(\mathscr{N}_0 \oplus \mathscr{A}_0)$ are Hopf subalgebra of $U_q(\mathscr{G})$, since $U_q(\mathscr{G}_\pm^{\mathbb{C}} \oplus \mathscr{H}^{\mathbb{C}})$ is Hopf subalgebra of $U_q(\mathscr{G}^{\mathbb{C}})$. Note that σ here is an antilinear involution and co-algebra homomorphism such that $\sigma(Y) = Y$ $\forall Y \in U_q(\mathscr{G}^{\mathbb{C}})$. For the classical complex Lie algebras these forms are $U_q(sl(n, \mathbb{R}))$, $U_q(so(n, n))$, $U_q(so(n+1, n))$, $U_q(sp(n, \mathbb{R}))$, which are dual to the matrix quantum groups $SL_q(n, \mathbb{R})$, $SO_q(n, n)$, $SO_q(n, n+1)$, $Sp_q(n, \mathbb{R})$, introduced in [272] from another point of view than ours. ◊

Further note that the set of the compact roots Δ_i^0 may be identified with the root system of $\mathscr{M}_0^{s\mathbb{C}}$. Thus the elements in (1.109b) give the Hopf algebra $U_q(\mathscr{M}_0^s)$ by the formulae:

$$
\begin{aligned}
&[C^+_\alpha, C^-_\alpha] = \frac{sinh(\tilde{H}_\alpha h_\alpha/2)}{sin(h_\alpha/2)}, \\
&[\tilde{H}_\alpha, \quad C^\pm_\alpha] = \pm C^\mp_\alpha, \quad q_\alpha = q^{(\alpha,\alpha)/2} = e^{-ih_\alpha}, \\
&C^+_\alpha = (i/\sqrt{2})(X_\alpha + X_{-\alpha}), \quad C^-_\alpha = (1/\sqrt{2})(X_\alpha - X_{-\alpha}) \\
&\tilde{H}_\alpha = -iH_\alpha, \\
&\delta(C^\pm_\alpha) = C^\pm_\alpha \otimes e^{\tilde{H}_\alpha h_\alpha/4} + e^{-\tilde{H}_\alpha h_\alpha/4} \otimes C^\pm_\alpha, \quad \alpha \in \Delta^+_i \cap \Delta_S.
\end{aligned}
\tag{1.112}
$$

Since $\mathscr{M}_0 = \mathscr{M}^s_0 \oplus \mathscr{Z}^m_0$ is a compact reductive Lie algebra we have to choose how to do the deformation in such cases. Our choice is to preserve the reductive structure, that is, writing in more detail $\mathscr{M}_0 = \oplus_j \mathscr{M}^{sj}_0 \oplus \oplus_k \mathscr{Z}^{mk}_0$, where $\mathscr{M}^{sj}_0$ is simple and $\mathscr{Z}^{mk}_0$ is one-dimensional; then we shall have the Hopf algebra $U_q(\mathscr{M}_0) = \otimes_j U_q(\mathscr{M}^{sj}_0) \otimes \otimes_k U_q(\mathscr{Z}^{mk}_0)$, where we also have to specify that if $\mathscr{Z}^{mk}_0$ is spanned by K, then $U_q(\mathscr{Z}^{mk}_0)$ is spanned by $K, q^{\pm K/4}$.

Remark 1.4. Formulae (1.109b) and (1.112) (with $h_\alpha \in \mathbb{R}$) determine completely a Drinfeld-Jimbo q-deformation of any compact semisimple Lie algebra [251] (when all roots of Δ are compact). Here one may take σ as an antilinear involution and coalgebra homomorphism such that $\sigma(X^\pm_\alpha) = -X^\mp_\alpha$, $\forall \alpha \in \Delta$, $\sigma(H) = -H$, $\forall H \in \mathscr{H}$. Note that in this case the q-deformation inherited from $U_q(\mathscr{G}^\mathbb{C})$ is often used in the physics literature without the basis change (1.112). ◊

Returning to the general situation, so far we have chosen consistently the generators of $\tilde{\mathscr{N}}^1_0 \oplus \mathscr{A}_0 \oplus \mathscr{M}_0 \oplus \mathscr{N}^1_0$ (cf. (1.106)) as linear combinations of the generators of $\mathscr{H}_0 \oplus \oplus_{\alpha \in \Delta^0_r \cup \Delta^0_i} \mathscr{G}_\alpha$. Now it remains to choose consistently the generators of $\tilde{\mathscr{N}}^2_0$, $\mathscr{N}^2_0$ as linear combinations of the generators of the rest of $\mathscr{G}^\mathbb{C}$, that is, of $\oplus_{\alpha \in \Delta^{0+}_c} \mathscr{G}_\alpha$, $\oplus_{\alpha \in \Delta^{0-}_c} \mathscr{G}_\alpha$, respectively. If $\alpha \in \Delta^0_c$, $\lambda = \alpha|_{\mathscr{A}_0}$, then $\dim_R \mathscr{G}^0_\lambda > 1$. Let $\Delta_\lambda = \{\alpha \in \Delta | \alpha|_{\mathscr{A}_0} = \lambda\}$. If $\alpha \in \Delta^0_c$, then we have $X_\alpha = Y_\alpha + Z_\alpha$, where $Y_\alpha \in \mathscr{Q}^\mathbb{C}$, $Z_\alpha \in \mathscr{K}^\mathbb{C}$. Now we can see that $\mathscr{G}^0_\lambda$ = r.l.s. $\{\tilde{X}_\alpha = Y_\alpha + iZ_\alpha, \forall \alpha \in \Delta_\lambda\}$. The actual choice of basis in $\mathscr{G}^0_\lambda$ is a matter of convenience (cf. the examples below) and is related to the choice of σ and q, and to the general property that $U_q(\tilde{\mathscr{P}}^0_0)$, $U_q(\mathscr{P}^0_0)$ are Hopf subalgebras of $U_q(\mathscr{G})$.

1.5.2.1 *q*-Deformations with Other Cartan Subalgebras

For the purposes of q-deformations we need also to consider Cartan subalgebras $\mathscr{H}$ which are not conjugate to $\mathscr{H}_0$. Cartan subalgebras which represent different conjugacy classes may be chosen as $\mathscr{H} = \mathscr{H}_k \oplus \mathscr{A}$, where $\mathscr{H}_k$ is compact, $\mathscr{A}$ is noncompact, $\dim\mathscr{A} < \dim\mathscr{A}_0$ if $\mathscr{H}$ is nonconjugate to $\mathscr{H}_0$. The Cartan subalgebras with maximal dimension of $\mathscr{A}$ are conjugate to $\mathscr{H}_0$; also those with minimal dimension of $\mathscr{A}$ are conjugate to each other.

All notions introduced until now are easily generalized for $\mathscr{H} = \mathscr{H}_k \oplus \mathscr{A}$ nonconjugate to $\mathscr{H}_0$. We note the differences, and notationwise we drop all 0 subscripts and superscripts. One difference is that the algebra $\mathcal{M}$ is the centralizer of $\mathcal{A}$ in $\mathcal{G}$ (mod $\mathcal{A}$) and thus is in general a noncompact reductive Lie algebra which has the compact $\mathcal{H}_k$ as Cartan subalgebra (besides, in general, other noncompact Cartan subalgebras); in particular, if $\mathcal{G}$ has a compact Cartan subalgebra then for the choice $\mathcal{A} = 0$ one has $\mathcal{M} = \mathcal{G}$. For the purposes of the q-deformation we shall use this compact Cartan subalgebra, that is, we set $\mathcal{H}^m = \mathcal{H}_k$. Further, the classification of the roots of Δ with respect to $\mathcal{H}$ goes as before. The difference is that if $\alpha \in \Delta_i$ then $\mathcal{L}_\alpha$ may also be noncompact. Thus for $\alpha \in \Delta_i$ the root α is called *singular root*, $\alpha \in \Delta_s$, if $\mathcal{L}_\alpha$ is noncompact, and α is called as before compact root, $\alpha \in \Delta_k$, if $\mathcal{L}_\alpha$ is compact. Thus $\Delta_i = \Delta_s \cup \Delta_k$. Formulae (1.109b) hold for Δ_k, while for $\alpha \in \Delta_s$ we have:

$$\begin{aligned}
&\mathcal{L}_\alpha = r.l.s.\{iH_\alpha, i(X_\alpha - X_{-\alpha}), X_\alpha + X_{-\alpha}\}, \quad \alpha \in \Delta_s^+,\\
&[S_\alpha^+, S_\alpha^-] = \frac{sinh(\tilde{H}_\alpha h_\alpha/2)}{sin(h_\alpha/2)},\\
&[\tilde{H}_\alpha, S_\alpha^\pm] = \mp S_\alpha^\mp, \quad q_\alpha = q^{(\alpha,\alpha)/2} = e^{-ih_\alpha},\\
&S_\alpha^+ = (1/\sqrt{2})(X_\alpha + X_{-\alpha}), \quad S_\alpha^- = (i/\sqrt{2})(X_\alpha - X_{-\alpha}),\\
&\tilde{H}_\alpha = -iH_\alpha,\\
&\delta(S_\alpha^\pm) = S_\alpha^\pm \otimes e^{\tilde{H}_\alpha h_\alpha/4} + e^{-\tilde{H}_\alpha h_\alpha/4} \otimes S_\alpha^\pm, \quad \alpha \in \Delta_s^+ \cap \Delta_S.
\end{aligned} \tag{1.113}$$

Further as before the set of the compact roots in Δ may be identified with the root system of $\mathcal{M}^{s\mathbb{C}}$. Thus formulae (1.109b),(1.112), and (1.113) give also the deformation $U_q(\mathcal{M}^s)$. Since the centre of $\mathcal{M}$ is compact (it is in the Cartan subalgebra $\mathcal{H}^m$ which is compact), then the deformation $U_q(\mathcal{Z}^m)$ is given as after (1.112). Thus the Hopf algebra $U_q(\mathcal{M})$ is given. Otherwise, the considerations for the factors $\mathcal{N}$, $\tilde{\mathcal{N}}$ go as for $\mathcal{N}_0$, $\tilde{\mathcal{N}}_0$.

Thus our scheme provides a different q*-deformation for each conjugacy class of Cartan subalgebras.*

1.5.2.2 *q*-Deformations for Arbitrary Parabolic Subalgebras and Reductive Lie Algebras

Until now our data are the nonconjugate Cartan subalgebras $\mathcal{H} = \mathcal{H}_k \oplus \mathcal{A}$ and the related with Bruhat decompositions (1.106). In these decompositions special role for the q-deformations is played by the minimal parabolic subalgebras $\mathcal{P}_0$ $\tilde{\mathcal{P}}_0$. A standard parabolic subalgebra is any subalgebra $\mathcal{P}'$ of $\mathcal{G}$ such that $\mathcal{P}_0 \subseteq \mathcal{P}'$. The number of standard parabolic subalgebras, including $\mathcal{P}_0$ and $\mathcal{G}$, is 2^r, $r = \dim\mathcal{A}$. They are all of the form $\mathcal{P}' = \mathcal{M}' \oplus \mathcal{A}' \oplus \mathcal{N}'$, $\mathcal{M}' \supseteq \mathcal{M}$, $\mathcal{A}' \subseteq \mathcal{A}$, $\mathcal{N}' \subseteq \mathcal{N}$; $\mathcal{M}'$ is the centralizer of $\mathcal{A}'$ in $\mathcal{G}$ (mod $\mathcal{A}'$); $\mathcal{N}'$ (resp. $\tilde{\mathcal{N}}' = \theta\mathcal{N}'$) is comprised from the negative (resp. positive) root spaces of the restricted root system Δ_R' of $(\mathcal{G}, \mathcal{A}')$. One also has the corresponding Bruhat decompositions:

$$\mathcal{G} = \tilde{\mathcal{N}}' \oplus \mathcal{A}' \oplus \mathcal{M}' \oplus \mathcal{N}'. \tag{1.114}$$

Note that $\mathcal{M}'$ is a noncompact reductive Lie algebra which has a noncompact Cartan subalgebra $\mathcal{H}'^m \cong \mathcal{H}_k \oplus \mathcal{H}_n$, where $\mathcal{H}_n$ is noncompact and $\mathcal{A} \cong \mathcal{H}_n \oplus \mathcal{A}'$. This Cartan subalgebra $\mathcal{H}'^m$ of $\mathcal{M}'$ will be chosen for the purposes of the q-deformation.

Thus we need to extend our scheme to noncompact reductive Lie algebras. Let $\hat{\mathcal{G}} = \mathcal{G} \oplus \mathcal{Z} = \hat{\mathcal{K}} \oplus \hat{\mathcal{Q}}$ be a real reductive Lie algebra, where $\mathcal{G}$ is the semisimple part of $\hat{\mathcal{G}}$; $\mathcal{Z}$ is the centre of $\hat{\mathcal{G}}$; $\hat{\mathcal{K}}, \hat{\mathcal{Q}}$ are the +1, −1 eigenspaces of the Cartan involution $\hat{\theta}$; $\hat{\mathcal{A}}' = \mathcal{A}' \oplus \mathcal{Z}_p$ is the analogue of $\mathcal{A}'$; $\mathcal{Z}_p = \mathcal{Z} \cap \hat{\mathcal{Q}}$. The root system of the pair $(\hat{\mathcal{G}}; \hat{\mathcal{A}}')$ coincides with Δ'_R, and the subalgebras $\tilde{\mathcal{N}}'$ and $\mathcal{N}'$ are inherited from $\mathcal{G}$. The decomposition (1.114) then is:

$$\hat{\mathcal{G}} = \tilde{\mathcal{N}}' \oplus \hat{\mathcal{A}}' \oplus \hat{\mathcal{M}}' \oplus \mathcal{N}', \tag{1.115}$$

where $\hat{\mathcal{M}}' = \mathcal{M}'^s \oplus \hat{\mathcal{Z}}'^m$, $\hat{\mathcal{Z}}'^m = \mathcal{Z}'^m \oplus \mathcal{Z} \cap \hat{\mathcal{K}}$. As in the compact reductive case we choose a deformation which preserves the splitting of $\hat{\mathcal{G}}$, that is, $U_q(\hat{\mathcal{G}}) = U_q(\mathcal{G}) \otimes U_q(\mathcal{Z})$, and even further into simple Lie subalgebras and one-dimensional central subalgebras.

Remark 1.5. A general property of the deformations $U_q(\mathcal{G})$ obtained by the above procedure is that $U_q(\mathcal{M}_0)$, $U_q(\tilde{\mathcal{P}}_0)$, $U_q(\mathcal{P}_0)$ are Hopf subalgebras of $U_q(\mathcal{G})$. ◊

1.5.3 Example so(p,r)

Let $\mathcal{G} = so(p, r)$, with $p \geq r \geq 2$ or $p > r = 1$ with generators: $M_{AB} = -M_{BA}$, $A, B = 1, \ldots, p + r$, $\eta_{AB} = \text{diag}\,(- \cdots - + \cdots +)$, ($p$ times minus, r times plus) which obey:

$$[M_{AB}, M_{CD}] = i(\eta_{BC}M_{AD} - \eta_{AC}M_{BD} - \eta_{BD}M_{AC} + \eta_{AD}M_{BC})$$

Besides the "physical" generator M_{AB} we shall also use the "mathematical" generator $Y_{AB} = -iM_{AB}$. One has: $\mathcal{K} \cong so(p) \oplus so(r)$ if $r \geq 2$ and $\mathcal{K} \cong so(p)$ if $r = 1$. The generators of $\mathcal{K}$ are M_{AB} with $1 \leq A < B \leq p$ and $p + 1 \leq A < B \leq p + r$. The split rank is equal to r; $\mathcal{M}_0 \cong so(p - r)$, if $p - r \geq 2$, and $\mathcal{M}_0 = 0$ if $p - r = 0, 1$, $\dim \tilde{\mathcal{N}} = \dim \mathcal{N} = r(p - 1)$. Furthermore the dimensions of the roots in the root system Δ of $so(p + r, \mathbb{C})$ and in Δ_R depending on the parity of $p + r$ are given by:

roots	$p + r$ even	$p + r$ odd
$\lvert\Delta_r^\pm\rvert$	$r(r-1)$	r^2
$\lvert\Delta_i^\pm\rvert$	$(p-r)(p-r-2)/4$	$(p-r-1)^2/4$
$\lvert\Delta_c^\pm\rvert$	$r(p-r)$	$r(p-r-1)$
$\lvert\Delta_R^\pm\rvert$	r^2	$r(r+1)$

Note that the algebra $so(2n+1,1)$ has only one conjugacy class of Cartan subalgebras. Thus in these cases our q-deformation is unique. The algebra $so(2n,1)$ has two conjugacy classes of Cartan subalgebras, and in these cases there are two q-deformations which we ilustrate below for $n=1$.

1.5.4 Example so(2,1)

Using notation from above $A,B = 1,2,0$, $(--+)$; Y_{12} is the generator of $\mathscr{K}$, and we may choose Y_{20} for the generator of $\mathscr{A}$; $\mathscr{M}_0 = 0$. Thus we can choose either Y_{20} or Y_{12} as a generator of $\mathscr{H}$ and $\mathscr{H}^{\mathbb{C}}$. Let $\Delta^{\pm} = \{\pm\alpha\}$ be the root system of $\mathscr{G}^{\mathbb{C}} = sl(2,\mathbb{C})$. If $\mathscr{H}^{\mathbb{C}}$ is generated by Y_{20} (and $\mathscr{H} = \mathscr{H}_0 = \mathscr{A}$), then α is a real root, and this deformation, denoted by $U_q^0(so(2,1))$, is given by formulae (1.110) and (1.22) over $\mathbb{R}$. If $\mathscr{H}^{\mathbb{C}}$ is generated by Y_{12}, then α is a singular compact root, and the deformation, denoted, $U_q^1(so(2,1))$, is given by formula (1.113) with $h_\alpha \in \mathbb{R}$.

1.5.5 q-Deformed Lorentz Algebra U_q(so(3,1))

With $A,B = 1,2,3,0$, $(---+)$, choose $\tilde{D} = M_{30}$ for the generator of $\mathscr{A}$ and $H = M_{12}$ for the generator of $\mathscr{M}$. From the above table we see that all roots are complex (as is also verified by a simple calculation). It is convenient to use the generators $M^{\pm} = -M_{23} \pm iM_{13} \in \mathscr{K}^{\mathbb{C}}$, $N^{\pm} = -M_{10} \mp iM_{20} \in \mathscr{Q}^{\mathbb{C}}$. We recall that $\mathscr{G}^{\mathbb{C}} = so(4,\mathbb{C}) \cong so(3,\mathbb{C}) \oplus so(3,\mathbb{C})$. The generators of the two commuting $so(3,\mathbb{C})$ algebras are $X_1^{\pm}, H_1$ and $X_2^{\pm}, H_2$, where

$$\begin{aligned} X_1^{\pm} &= (1/2)(M^{\pm} - iN^{\pm}), \quad H_1 = H - i\tilde{D}, \\ X_2^{\pm} &= (1/2)(M^{\pm} + iN^{\pm}), \quad H_2 = H + i\tilde{D}. \end{aligned} \tag{1.116}$$

We use $U_q(so(4,\mathbb{C})) = U_q(so(3,\mathbb{C})) \otimes U_q(so(3,\mathbb{C}))$ given by:

$$[X_a^+, X_a^-] = [H_a], \quad [H_a, X_a^{\pm}] = \pm 2X_a^{\pm}, \quad a = 1,2, \tag{1.117}$$

and the Hopf algebra structure is given just by (1.22). Using this we obtain the following $U_q(so(3,1))$ relations with $q = e^h \in \mathbb{R}$:

$$\begin{aligned} &[H, M^{\pm}] = \pm M^{\pm}, \quad [H, N^{\pm}] = \pm N^{\pm}, \\ &[(\tilde{D}, M^{\pm}] = \pm N^{\pm}, \quad [(\tilde{D}, N^{\pm}] = \mp M^{\pm}, \\ &[M^+, M^-] = [N^-, N^+] = 2[H]cos(\tilde{D}h/2), \\ &[M^{\pm}, N^{\mp}] = \pm 2\frac{cosh(Hh/2)sin(\tilde{D}h/2)}{sinh(h/2)} \end{aligned} \tag{1.118}$$

$$\begin{aligned}\delta(M^{\pm}) &= M^{\pm}\otimes e^{Hh/4}cos(\tilde{D}h/4) - N^{\pm}\otimes e^{Hh/4}sin(\tilde{D}h/4)\ + \\ &+ e^{-Hh/4}cos(\tilde{D}h/4)\otimes M^{\pm} + e^{-Hh/4}sin(\tilde{D}h/4)\otimes N^{\pm} \\ \delta(N^{\pm}) &= N^{\pm}\otimes e^{Hh/4}cos(\tilde{D}h/4) + M^{\pm}\otimes e^{Hh/4}sin(\tilde{D}h/4)\ + \\ &+ e^{-Hh/4}cos(\tilde{D}h/4)\otimes N^{\pm} - e^{-Hh/4}sin(\tilde{D}h/4)\otimes M^{\pm}\end{aligned} \tag{1.119}$$

$$\begin{aligned}\gamma(H) &= -H, \quad \gamma(M^{\pm}) = -q^{\pm 1/2}M^{\pm}, \\ \gamma(\tilde{d}) &= -\tilde{d}, \quad \gamma(N^{\pm}) = -q^{\pm 1/2}N^{\pm}.\end{aligned} \tag{1.120}$$

1.5.6 q-Deformed Real Forms of so(5)

The algebras $so(4,1)$ and $so(3,2)$ have the same complexification $\mathscr{G}^{\mathbb{C}} = so(5,\mathbb{C})$. The root system of $so(5,\mathbb{C})$ is given by $\Delta^{\pm} = \{\pm\alpha_1, \pm\alpha_2, \pm\alpha_3, \pm\alpha_4\}$; the simple roots are α_1, α_2, while $\alpha_3 = \alpha_1 + \alpha_2, \quad \alpha_4 = 2\alpha_1 + \alpha_2$; the products between the simple roots are $(\alpha_1,\alpha_1) = 2 = -(\alpha_1,\alpha_2)$, $(\alpha_2,\alpha_2) = 4$. The Cartan–Weyl basis for the nonsimple roots is given by:

$$\begin{aligned}X_3^+ &= X_1^+X_2^+ - q^{(\alpha_1,\alpha_2)/2}X_2^+X_1^+ = X_1^+X_2^+ - q^{-1}X_2^+X_1^+ \equiv \\ &\equiv [X_1^+, X_2^+]_{q^{-1}}, \\ X_3^- &= X_2^-X_1^- - q^{-(\alpha_1,\alpha_2)/2}X_1^-X_2^- = X_2^-X_1^- - qX_1^-X_2^- = \\ &= [X_2^-, X_1^-]_q, \\ X_4^+ &= X_1^+X_3^+ - q^{(\alpha_1,\alpha_3)/2}X_3^+X_1^+ = X_1^+X_3^+ - X_3^+X_1^+, \\ X_4^- &= X_3^-X_1^- - X_1^-X_3^-\end{aligned} \tag{1.121}$$

All other commutation relations follow from these definitions. We shall mention only:

$$[X_4^{\pm}, X_2^{\pm}] = \pm(q^{\pm 1} - 1)(X_3^{\pm})^2, \quad [X_4^{\pm}, X_2^{\mp}] = \pm(1 - q^{-2})(1 - q^{\pm 1})(X_1^{\pm})^2 q^{\pm H_2}. \tag{1.122}$$

1.5.7 q-Deformed de Sitter Algebra so(4,1)

Let $\mathscr{G} = so(4,1)$. With $A, B = 1,2,3,4,0$, $(- - - - +)$, choose Y_{30} for the generator of $\mathscr{A}$; $\mathscr{M} \cong so(3)$ with generators Y_{ab}, $a, b = 1,2,4$, and we choose Y_{12} for the generator of its Cartan subalgebra. The algebra $\mathscr{G} = so(4,1)$ has two nonconjugate Cartan subalgebras; besides $\mathscr{H}_0$ generated by Y_{30}, Y_{12}, we have a compact Cartan subalgebra $\mathscr{H}_1$ generated, say, by Y_{12}, Y_{34}.

In the case of $\mathscr{H} = \mathscr{H}_0$ the generators of $\mathscr{G}$ are expressed in terms of those of $so(5,\mathbb{C})$ by:

$$\begin{aligned}
&Y_{30} = -H_1, \quad Y_{12} = i(H_1 + H_2), \quad Y_{14} = (1/\sqrt{2})(X_3^+ + X_3^-),\\
&Y_{24} = (i/\sqrt{2})(X_3^- - X_3^+), \quad Y_{34} = (1/\sqrt{2})(X_1^+ + X_1^-),\\
&Y_{40} = (1/\sqrt{2})(X_1^+ - X_1^-)\\
&Y_{13} = (1/2)(X_2^- + X_2^+ + X_4^+ + X_4^-),\\
&Y_{23} = (i/2)(X_2^- - X_2^+ - X_4^+ + X_4^-)\\
&Y_{10} = (1/2)(X_2^- - X_2^+ + X_4^+ - X_4^-),\\
&Y_{20} = (i/2)(X_2^- + X_2^+ - X_4^+ + X_4^-)
\end{aligned} \tag{1.123}$$

Now we can give all commutation relations and Hopf algebra operations for Y_{AB} as generators of q-deformed $so(4,1)$ as inherited from $U_q(so(5,\mathbb{C}))$. The deformation obtained in this way is denoted by U_{41}^0.

In the case of the Cartan subalgebra $\mathscr{H}_1$ we have $Y_{34} = iH_1$, $Y_{12} = -i(H_1 + H_2)$. To save space we omit the other generators. We denote the deformation obtained in this way by U_{41}^1.

1.5.8 q-Deformed Anti de Sitter Algebra so(3,2)

Let $\mathscr{G} = so(3,2)$. With $A, B = 1,2,3,4,0$, $(- - - + +)$, choose Y_{20} and Y_{34} as generators of $\mathscr{H}_0 = \mathscr{A}$. The algebra $\mathscr{G} = so(3,2)$ has three nonconjugate Cartan subalgebras; besides $\mathscr{H}_0$ we have $\mathscr{H}_1$ generated, say, by Y_{12}, Y_{30} and $\mathscr{H}_2$ generated, say, by Y_{12}, Y_{40}. Thus $\mathscr{H}_a$, $a = 0,1,2$, is a Cartan subalgebra with a compact generators.

For the Cartan subalgebra $\mathscr{H}_0$ we identify: $Y_{34} = H_1$, $Y_{20} = H_1 + H_2$; for $\mathscr{H}_1$ we have: $Y_{12} = -iH_1$, $Y_{34} = H_1 + H_2$; for $\mathscr{H}_2$ one uses $M_{12} = H_1$, $M_{40} = H_1 + H_2$. We shall denote the deformation using the Cartan subalgebra $\mathscr{H}_a$ by U_{32}^a.

1.5.9 q-Deformed Algebras U_q(sl(4,ℂ)) and U_q(su(2,2))

The root system of the complexification $sl(4,\mathbb{C})$ of $su(2,2)$ is given by $\Delta^\pm = \{\pm\alpha_1, \pm\alpha_2, \pm\alpha_3, \pm\alpha_{12}, \pm\alpha_{23}, \pm\alpha_{13}\}$; the simple roots are $\alpha_1, \alpha_2, \alpha_3$, while $\alpha_{12} = \alpha_1 + \alpha_2$, $\alpha_{23} = \alpha_2 + \alpha_3$, $\alpha_{13} = \alpha_1 + \alpha_2 + \alpha_3$; all roots are of length 2 and the nonzero products between the simple roots are $(\alpha_1, \alpha_2) = (\alpha_2, \alpha_3) = -1$. The Cartan–Weyl basis for the nonsimple roots is given by:

$$\begin{aligned}
&X_{jk}^\pm = \pm q^{\mp 1/4}(q^{1/4} X_j^\pm X_k^\pm - q^{-1/4} X_k^\pm X_j^\pm), \quad (jk) = (12), (23),\\
&X_{13}^\pm = \pm q^{\mp 1/4}(q^{1/4} X_1^\pm X_{23}^\pm - q^{-1/4} X_{23}^\pm X_1^\pm) =\\
&- \pm q^{\mp 1/4}(q^{1/4} X_{12}^\pm X_3^\pm - q^{-1/4} X_3^\pm X_{12}^\pm).
\end{aligned} \tag{1.124}$$

All other commutation relations follow from these definitions. Besides those in (1.23) we have ($X_{aa}^\pm \equiv X_a^\pm$):

$$
\begin{aligned}
[X_a^+, X_{ab}^-] &= -q^{H_a/2} X_{a+1b}^-, \\
[X_b^+, X_{ab}^-] &= X_{ab-1}^- q^{-H_b/2}, \quad 1 \le a < b \le 3, \\
[X_a^-, X_{ab}^+] &= X_{a+1b}^+ q^{-H_a/2}, \\
[X_b^-, X_{ab}^+] &= -q^{H_b/2} X_{ab-1}^+, \quad 1 \le a < b \le 3, \\
X_a^\pm X_{ab}^\pm &= q^{1/2} X_{ab}^\pm X_a^\pm, \quad 1 \le a < b \le 3, \\
[X_2^\pm, X_{13}^\pm] &= 0, \quad [X_2^\pm, X_{13}^\mp] = 0, \quad [X_{12}^+, X_{13}^-] = -q^{H_1+H_2} X_3^-, \\
[X_{12}^-, X_{13}^+] &= X_3^+ q^{-H_1-H_2}, \\
[X_{23}^+, X_{13}^-] &= X_1^- q^{-H_2-H_3}, \quad [X_{23}^-, X_{13}^+] = -q^{H_2+H_3} X_1^+, \\
[X_{12}^\pm, X_{23}^\pm] &= \tilde{\lambda} X_2^\pm X_{13}^\pm, \quad [X_{12}^\pm, X_{23}^\mp] = -\tilde{\lambda} q^{\pm H_2/2} X_1^\pm X_3^\mp, \\
\tilde{\lambda} &\equiv q^{1/2} - q^{-1/2}.
\end{aligned}
\tag{1.125}
$$

Further we consider the conformal algebra $\mathcal{G} = su(2,2) \cong so(4,2)$. It has three nonconjugate classes of Cartan subalgebras represented, say, by $\mathcal{H}^a$, $a = 0,1,2$ with a noncompact generators. Thus according to our procedure it has five different deformations – three in the case of $\mathcal{H}^2$ (since there are three nontrivial parabolic subalgebras) and one each for the other two choices of Cartan subalgebras. We shall work with the most noncompact Cartan subalgebra $\mathcal{H} = \mathcal{H}_0 = \mathcal{H}^2$ and with the maximal parabolic subalgebra. Using the notation from Section 1.5.3 with $A, B = 1,2,3,5,6,0$, $(- - - - ++)$, choose Y_{30} and Y_{56} as generators of $\mathcal{A}$ and Y_{12} for the generator of $\mathcal{M}$. Since $su(2,2)$ is the conformal algebra of four-dimensional Minkowski space – time we would like to deform it consistently with the subalgebra structure relevant for the physical applications. These subalgebras are the *Lorentz subalgebra* $\mathcal{M}' \cong so(3,1)$ generated by $Y_{\mu\nu}$, $\mu,\nu = 1,2,3,0$; the subalgebra $\tilde{\mathcal{N}}'$ of *translations* generated by $P_\mu = Y_{\mu 5} + Y_{\mu 6}$; the subalgebra $\mathcal{N}'$ of *special conformal transformations* generated by $K_\mu = Y_{\mu 5} - Y_{\mu 6}$; the *dilatations* subalgebra $\mathcal{A}'$ generated by $D = Y_{56}$. The commutation relations besides those for the Lorentz subalgebra are:

$$
\begin{aligned}
&[D, Y_{\mu\nu}] = 0, \quad [D, P_\mu] = P_\mu, \quad [D, K_\mu] = -K_\mu, \\
&[Y_{\mu\nu}, P_\lambda] = \eta_{\nu\lambda} P_\mu - \eta_{\mu\lambda} P_\nu, \quad [Y_{\mu\nu}, K_\lambda] = \eta_{\nu\lambda} K_\mu - \eta_{\mu\lambda} K_\nu, \\
&[P_\mu, K_\nu] = 2Y_{\mu\nu} + 2\eta_{\mu\nu} D.
\end{aligned}
\tag{1.126}
$$

The algebra $\mathcal{P}_{max} = \mathcal{M}' \oplus \mathcal{A}' \oplus \mathcal{N}'$ (or equivalently $\tilde{\mathcal{P}}_{max} = \mathcal{M}' \oplus \mathcal{A}' \oplus \tilde{\mathcal{N}}'$) is the so-called *maximal parabolic subalgebra* of $\mathcal{G}$, where $\tilde{\mathcal{N}}'$, $\mathcal{N}'$, is the root vector space of the restricted root system $\Delta_R' = \{\pm\lambda; \lambda(D) = 1\}$ of $(\mathcal{G}, \mathcal{A}')$, corresponding to λ, $-\lambda$, respectively.

For the Lorentz algebra generators we have the following expressions (which are inverse to (1.116)):

$$
\begin{aligned}
H &= -Y_{30} = (1/2)(H_1 + H_3),\\
&\quad M^{\pm} = -iY_{13} \pm iY_{10} = X_1^{\pm} + X_3^{\pm},\\
\tilde{D} &= -Y_{12} = (i/2)(H_1 - H_3),\\
&\quad N^{\pm} = -iY_{20} \pm iY_{23} = i(X_1^{\pm} - X_3^{\pm}).
\end{aligned}
\tag{1.127}
$$

For the dilatations, translations, and special conformal transformations we have:

$$D = (1/2)(H_1 + H_3) + H_2, \tag{1.128}$$

$$
\begin{aligned}
P_0 &= i(X_{13}^{+} + X_2^{+}), \quad P_1 = i(X_{12}^{+} + X_{23}^{+}),\\
P_2 &= X_{12}^{+} - X_{23}^{+}, \quad P_3 = i(X_2^{+} - X_{13}^{+}),
\end{aligned}
\tag{1.129}
$$

$$
\begin{aligned}
K_0 &= -i(X_{13}^{-} + X_2^{-}), \quad K_1 = i(X_{12}^{-} + X_{23}^{-}),\\
K_2 &= X_{23}^{-} - X_{12}^{-}, \quad K_3 = i(X_2^{-} - X_{13}^{-}).
\end{aligned}
\tag{1.130}
$$

Now we can derive the relations in $U_q(su(2,2))$:

1) According to our general scheme the deformed Lorentz subalgebra is a Hopf subalgebra; its deformation is described by formulae (1.118) and (1.119).
2) The commutation relations of the generators $H, \tilde{D}, D$ of the Cartan subalgebra $\mathscr{H} = \mathscr{H}_0$ are not deformed.
3) The deformation of the translations and special conformal transformations subalgebras is given by:

$$
\begin{aligned}
&P_a(P_1 \pm iP_2) = q^{\mp 1/2}(P_1 \pm iP_2)P_a, \quad a = 0, 3; \quad [P_0, P_3] = 0,\\
&[P_1 + iP_2, P_1 - iP_2] = \tilde{\lambda}(P_0^2 - P_3^2).
\end{aligned}
\tag{1.131}
$$

$$
\begin{aligned}
&K_a(K_1 \pm iK_2) = q^{\pm 1/2}(K_1 \pm iK_2)K_a, \quad a = 0, 3; \quad [K_0, K_3] = 0,\\
&[K_1 + iK_2, K_1 - iK_2] = \tilde{\lambda}(K_0^2 - K_3^2).
\end{aligned}
\tag{1.132}
$$

4) The commutation relations of $M^{\pm}$ with P_μ are given by:

$$
\begin{aligned}
&M^{+}(P_1 - iP_2) - q^{-1/2}(P_1 - iP_2)M^{+} = P_0 - P_3,\\
&M^{+}(P_1 + iP_2) - q^{1/2}(P_1 + iP_2)M^{+} = q^{1/2}(P_0 - P_3)\\
&M^{+}(P_0 - P_3) - \frac{[2]}{2}(P_0 - P_3)M^{+} = \frac{i\tilde{\lambda}}{2}(P_0 - P_3)N^{+},\\
&M^{+}(P_0 + P_3) - \frac{[2]}{2}(P_0 + P_3)M^{+} = \frac{i\tilde{\lambda}}{2}(P_0 + P_3)N^{+} +\\
&+(P_1 + iP_2) - q^{1/2}(P_1 - iP_2),
\end{aligned}
$$

$$[M^-, P_1 - iP_2] = -q^{(i(\tilde{D}+H)/2}(P_0 + P_3),$$
$$[M^-, P_1 + iP_2] = (P_0 + P_3)q^{(i(\tilde{D}-H)/2}$$
$$[M^-, P_0 - P_3] = (P_1 - iP_2)q^{(i(\tilde{D}-H)/2} - q^{(i(\tilde{D}+H)/2}(P_1 + iP_2),$$
$$[M^-, P_0 + P_3] = 0. \tag{1.133}$$

The commutation relations between $M^\pm$ and K_μ are obtained from the above by the following changes: $M^\pm \mapsto M^\mp$, $N^+ \mapsto -N^-$, $H \mapsto -H$, $\tilde{D} \mapsto \tilde{D}$, $P_\mu \mapsto \eta_{„}K_\mu$, $q^{1/2} \mapsto q^{-1/2}$. These follow from the automorphism of $U_q(\mathcal{G}^{\mathbb{C}})$: $X_1^\pm \longrightarrow X_3^\mp$, $H_1 \longrightarrow -H_3$, $X_2^\pm \mapsto -X_2^\mp$, $H_2 \mapsto -H_2$, $q^{1/2} \mapsto q^{-1/2}$ (then $X_{12}^\pm \longrightarrow -X_{23}^\mp$, $X_{13}^\pm \mapsto -X_{13}^\mp$). The commutation relations between $N^\pm$ and P_μ, K_μ are obtained from those between $M^\pm$ and P_μ by the changes $M^\pm \longrightarrow iN^\pm$, $P_0 \longrightarrow -P_3$, $P_1 \longrightarrow -iP_2$ and from those between $M^\pm$ and K_μ by the changes $M^\pm \longrightarrow -iN^\pm$, $K_0 \longrightarrow K_3$, $K_1 \longrightarrow -iK_2$.

5) For $[P_\mu, K_\nu]$ we have:

$$P_1 \pm iP_2, K_1 \pm iK_2] = \pm\tilde{\lambda}q^{\mp(H-D)/2}(M^+ \pm iN^+)(M^- \mp iN^-),$$
$$[P_1 \pm iP_2, K_1 \mp iK_2] = 4[\pm i(\tilde{D} - D]$$
$$[P_0 \pm P_3, K_3 \mp K_0] = \pm 4[\pm H - D], \quad [P_0 \pm P_3, K_3 \pm K_0] = 0,$$
$$[P_1 - iP_2, K_3 - K_0] = 2(M^+ - iN^+)q^{(H-D)/2},$$
$$[P_1 - iP_2, K_0 + K_3] = 2(M^- + iN^-)q^{-(D+i(\tilde{D})/2}$$
$$[P_1 + iP_2, K_3 - K_0] = -2q^{(D-H)/2}(M^+ + iN^+),$$
$$[P_1 + iP_2, K_0 + K_3] = -2q^{(D-i(\tilde{D})/2}(M^- - iN^-). \tag{1.134}$$

and four more relations which are obtained from (36c,d) by the first set of changes described after formula (1.133) and by $D \mapsto -D$.

The comultiplication for the Lorentz subalgebra is given by (1.119); for the dilatation generator $D \in \mathcal{H} \subset \mathcal{H}^{\mathbb{C}}$ it is trivial and for the translations and special conformal transformations we have:

$$\delta(T^\pm) = \begin{cases} T^\pm \otimes q^{(D\pm i(\tilde{D})/4} + q^{-(D\pm i(\tilde{D})/4} \otimes T^\pm + \delta_1(T^\pm), \\ \qquad T^\pm = P_1 \mp iP_2, \quad K_1 \pm iK_2 \\ T^\pm \otimes q^{(D\pm H)/4} + q^{-(D\pm H)/4} \otimes T^\pm + \delta_1(T^\pm), \\ \qquad T^\pm = P_0 \mp P_3, \quad K_3 \pm K_0 \end{cases}$$

$$\delta_1(T^\pm) = \begin{cases} \pm(\tilde{\lambda}/2)(M^\pm \mp iN^\pm)q^{(H-D)/4} \otimes q^{(H\pm i(\tilde{D})/4}\tilde{T}^\pm, \\ \qquad T^+ = P_1 - iP_2, \quad T^- = K_1 - iK_2 \\ \pm(\tilde{\lambda}/2)\tilde{T}^\pm q^{(-H\pm i(\tilde{D})/4} \otimes q^{(D-H)/4}(M^\pm \mp iN^\pm), \\ \qquad T^+ = P_1 + iP_2, \quad T^- = K_1 + iK_2, \end{cases}$$

$$\delta_1(T^\pm) = (\tilde{\lambda}/2)(\tilde{T}'^{\pm} q^{(-H\pm i(\tilde{D})/4} \otimes q^{(D\pm i(\tilde{D})/4}(M^\pm \pm iN^\pm) + \\ +(M^\pm \mp iN^\pm) q^{(-D\pm i(\tilde{D})/4} \otimes q^{(H\pm i(\tilde{D})/4} \tilde{T}''^{\pm}). \tag{1.135}$$

$$T^+ = P_0 - P_3, T^- = K_0 + K_3, \tilde{T}'^+ = P_1 - iP_2, \tilde{T}'^- = K_1 - iK_2, \\ \tilde{T}''^+ = -(P_1 + iP_2), \tilde{T}''^- = K_1 + iK_2$$

$\delta_1(\tilde{T}^\pm) = 0$, $\tilde{T}^+ = P_0 + P_3$, $\tilde{T}^- = K_3 - K_0$.

The antipode for the Lorentz subalgebra is given by (1.120); for the translations, special conformal transformations and dilataions we have:

$$\gamma(P_0 \pm P_3) = -q^{1\pm 1/2}(P_0 \pm P_3) + \frac{q^{1/4\pm 1/4}(q-1)}{2}(P_1 - iP_2)(M^+ \mp iN^+)$$
$$\gamma(P_1 + iP_2) = -q^{3/2}(P_1 + iP_2) + \frac{q(q-1)}{2}(P_0 + P_3)(M^+ + iN^+) + \\ + \frac{q^{1/2}(q-1)}{2}(P_0 - P_3)(M^+ - iN^+) - \\ - \frac{(q-1)^2}{4}(P_1 - iP_2)((M^+)^2 + (N^+)^2),$$
$$\gamma(P_1 - iP_2) = -q^{1/2}(P_1 - iP_2); \tag{1.136}$$
$$\gamma(K_0 \pm K_3) = -q^{-1\mp 1/2}(K_0 \pm K_3) - \frac{q^{-1/4\mp 1/4}(q^{-1}-1)}{2}(K_1 + iK_2)(M^- \pm iN^-)$$
$$\gamma(K_1 - iK_2) = -q^{-3/2}(K_1 - iK_2) - \frac{q^{-1/2}(q^{-1}-1)}{2}(K_0 - K_3)(M^- + iN^-) - \\ - \frac{q^{-1}(q^{-1}-1)}{2}(K_0 + K_3)(M^- - iN^-) - \\ - \frac{(q^{-1}-1)^2}{4}(K_1 + iK_2)((M^-)^2 + (N^-)^2),$$
$$\gamma(K_1 + iK_2) = -q^{-1/2}(K_1 + iK_2)$$
$$\gamma(D) = -D. \tag{1.137}$$

Consistently with the general scheme (cf. Remark 1.3.), formulae (1.135) and (1.136) tell us that the deformed subalgebras of translations and special conformal transformations are not Hopf subalgebras of $\mathcal{G}$.

1.5.10 *q*-deformed Poincaré and Weyl Algebras

The Poincaré algebra is not a semisimple (or reductive) Lie algebra, and our procedure is not directly applicable. One may try to use the fact that it is a subalgebra of the conformal algebra. Indeed, there is a q-deformed Poincaré algebra with generators $M^\pm$, $N^\pm$, H, $\hat{D} = i\tilde{D}$, P_μ, and with commutations relations given by (1.118), (1.131), and

(1.133) and those obtained from the latter two by the changes $M^{\pm} \longrightarrow iN^{\pm}$, $P_0 \longrightarrow -P_3$, $P_1 \longrightarrow -iP_2$. However, from formula (1.135) follows that the deformation of the Poincaré subalgebra of $su(2,2)$ is not a Hopf subalgebra; rather the deformation $U_q(\tilde{\mathscr{P}}_{max})$ of the 11 – generator *Weyl subalgebra* = Poincaré & dilatations = $\tilde{\mathscr{P}}_{max}$ – is a Hopf subalgebra of $U_q(\mathscr{G})$. Another Weyl algebra conjugate to this is $U_q(\mathscr{P}_{max})$ with generators $M^{\pm}$, $N^{\pm}$, H, $\hat{D} = i\tilde{D}$, K_μ, D, and with commutations relations given by (1.118), (1.132), and those obtained from (1.133) as explained in the text thereafter.

Other deformed Poincaré algebras may be obtained from the contraction of U^a_{41} and U^a_{32} discussed above. Only for U^0_{41} and U^1_{32} one may expect to obtain a deformed Lorentz subalgebra as a Hopf subalgebra after contracting $Y_{4\mu} \rightarrow RP_\mu$, $R \rightarrow \infty$, since $Y_{4\mu}$ are not Cartan generators. However, if $q \neq 1$, this limit is not consistent with the commutation relations which are inherited from relation (1.122). The other possibility is to make contractions which involve Cartan generators. This may be a noncompact generator which is possible for U^0_{41} and U^a_{32}, $a = 0, 1$, or a compact generator which is possible for U^a_{41}, $a = 0, 1$, and U^2_{32}. (The last case was studied in [437].) The resulting deformed Poincaré algebras will have a noncompact Hopf subalgebra in the case U^0_{32} and in one of the U^0_{41} cases and a compact Hopf subalgebra in the other four cases.

2 Highest-Weight Modules over Quantum Algebras

Summary

In [198] we began the study of the representation theory of $U_q(\mathscr{G})$ when the deformation parameter q is a root of unity. We consider the induced highest-weight modules (HWMs) over $U_q(\mathscr{G})$, especially Verma modules. In [198] we adapted to $U_q(\mathscr{G})$ the previously developed approach of multiplet classification of Verma modules over (infinite-dimensional) (super-) Lie algebras [193, 194, 196, 197]. In [199–201] we gave the character formulae for the irreducible HWM over $U_q(\mathscr{G})$ when $\mathscr{G} = sl(3, \mathbb{C})$.

The above developments use results on the embeddings of the reducible Verma modules. These embeddings are realized by the so-called *singular* vectors (or *null* or *extremal* vectors). In the classical case, that is, $q = 1$, the singular vectors were discussed in detail in Volume 1. In [198] we gave the general formula for the singular vectors which however was not so explicit. Some explicit formulae for singular vectors for $\mathscr{G} = A_\ell$ and for some rank two subalgebras of $\mathscr{G} \neq A_\ell$ were presented in [205]. In the present chapter following [206] we give explicit formulae for the singular vectors of Verma modules over $U_q(\mathscr{G})$ for arbitrary $\mathscr{G}$ corresponding to a class of positive roots of $\mathscr{G}$, which we shall call straight roots. In some special cases we give singular vectors corresponding to arbitrary positive roots. We use a special basis of $U_q(\mathscr{G}^-)$, where $\mathscr{G}^-$ is the negative roots subalgebra of $\mathscr{G}$, whose basis was introduced in our earlier work in the case $q = 1$ [193, 194, 197]. This basis seems more economical than the Poincaré–Birkhoff–Witt–type of basis used by Malikov, Feigin, and Fuchs [460] for the construction of singular vectors of Verma modules in the case $q = 1$. Furthermore our basis turns out to be part of a general basis introduced recently for other reasons by Lusztig [445] for $U_q(\mathscr{B}^-)$, where $\mathscr{B}^-$ is a Borel subalgebra of $\mathscr{G}$. On the other hand, there are examples [225, 244, 245], where it is convenient to use singular vectors in the Poincaré–Birkhoff–Witt (PBW) basis. In principle, the paper [459] generalizes the results of [460], to the quantum group case, and from there PBW singular vectors may be extracted.

2.1 Verma Modules, Singular Vectors, and Irreducible Subquotients

A *HWM* V over $U_q(\mathscr{G})$ [360] is given (as for $q = 1$) by its highest weight $\Lambda \in \mathscr{H}^*$ and highest-weight vector $v_0 \in V$ so that:

$$X_i^+ v_0 = 0, i = 1, \ldots, \ell, \quad H v_0 = \Lambda(H) v_0, \quad H \in \mathscr{H} \tag{2.1}$$

We define a *Verma module* V^Λ as the HWM over $U_q(\mathscr{G})$ with *highest weight* $\Lambda \in \mathscr{H}^*$ and *highest-weight vector* $v_0 \in V^\Lambda$, induced from the one-dimensional representation $V_0 \cong \mathbb{C}v_0$ of $U_q(\mathscr{B})$, where $\mathscr{B} = \mathscr{B}^+$, $\mathscr{B}^\pm = \mathscr{H} \oplus \mathscr{G}^\pm$ are Borel subalgebras of $\mathscr{G}$, such that $U_q(\mathscr{G}^+)v_0 = 0$, $Hv_0 = \Lambda(H)v_0$, $H \in \mathscr{H}$. (Note that the algebras $U_q(\mathscr{B}^\pm)$ with generators $H_i, X_i^\pm$ are Hopf subalgebras of $U_q(\mathscr{G})$ [521].) Thus one has $V^\Lambda \cong U_q(\mathscr{G}) \otimes_{U_q(\mathscr{B})} v_0 \cong U_q(\mathscr{G}^-) \otimes v_0$.

DOI 10.1515/9783110427707-002

The representation theory of $U_q(\mathcal{G})$ parallels the theory over $\mathcal{G}$ when q is not a root of unity.

We recall several facts from [198]. The Verma module V^Λ is reducible if there exists a root $\beta \in \Delta^+$ and $m \in \mathbb{N}$ such that

$$[(\Lambda + \rho, \beta^\vee) - m]_{q_\beta} = [(\Lambda + \rho)(H_\beta) - m]_{q_\beta} = 0\,, \quad \beta^\vee \equiv 2\beta/(\beta, \beta) \tag{2.2}$$

holds.

If q is not a root of unity then (2.2) is also a necessary condition for reducibility, and then it may be rewritten as $2(\Lambda + \rho, \beta) = m(\beta, \beta)$. (In that case it is the generalization of the (necessary and sufficient) reducibility conditions for Verma modules over finite-dimensional semisimple Lie algebras $\mathcal{G}$ [96] and affine Lie algebras [373], cf. also (I.2.205).) For uniformity we shall write the reducibility condition in the general form (2.2).

Now follow several properties which are as in the case $q = 1$.

If (2.2) holds then there exists a vector $v_s \in V^\Lambda$, called a *singular vector*, such that $v_s \neq v_0$, $X_i^+ v_s = 0, i = 1, \ldots, \ell$, $Hv_s = (\Lambda(H) - m\beta(H))v_s$, $\forall H \in \mathcal{H}$. (In the case of affine Lie algebras when $(\beta, \beta) = 0$, there are $p(n)$ independent singular vectors for each $n \in \mathbb{N}$, $p(\cdot)$ being the partition function [460].) The space $U_q(\mathcal{G}^-)v_s$ is a proper submodule of V^Λ isomorphic to the Verma module $V^{\Lambda - m\beta} = U_q(\mathcal{G}^-) \otimes v_0'$ where v_0' is the highest-weight vector of $V^{\Lambda - m\beta}$; the isomorphism being realized by $v_s \mapsto 1 \otimes v_0'$. This situation is again denoted by $V^\Lambda \to V^{\Lambda - m\beta}$.

The singular vector is given by [198]:

$$v_s = v^{m\,b} = \mathcal{P}_{m,\beta}(X_1^-, \ldots, X_\ell^-) \otimes v_0\,, \tag{2.3}$$

where $\mathcal{P}_{m\beta}$ is a homogeneous polynomial in its variables of degrees mn_i, where $n_i \in \mathbb{Z}_+$ comes from $\beta = \sum n_i \alpha_i$, and α_i – the system of simple roots. The polynomial $\mathcal{P}_{m\beta}$ is unique up to a nonzero multiplicative constant.

If (2.2) holds for several pairs $(m, \beta) = (m_i, \beta_i)$, $i = 1, \ldots, k$, there are other Verma modules $V^{\Lambda - m_i\beta_i}$, all of which are isomorphic to submodules of V^Λ.

The Verma module V^Λ contains a unique proper maximal submodule I^Λ.

Among the HWM with highest weight Λ there is a unique irreducible one, denoted by L_Λ, that is,

$$L_\Lambda = V^\Lambda / I^\Lambda. \tag{2.4}$$

If V^Λ is irreducible then $L_\Lambda = V^\Lambda$.

Suppose that q is not a root of 1. Then the representations of $U_q(\mathcal{G})$ are deformations of the representations of $U(\mathcal{G})$, and the latter are obtained from the former for $q \to 1$ [441, 532].

Consider V^Λ reducible w.r.t. every simple root (and thus w.r.t. all positive roots):

$$[(\Lambda + \rho, \alpha_i^\vee) - m_i]_{q_i} = [\Lambda(H_i) + 1 - m_i]_{q_i} = 0\,, \quad m_i \in \mathbb{N}\,, i = 1, \ldots, \ell\,, \tag{2.5}$$

where we used $\rho(\alpha_i^\vee) = 1$. Then L_Λ is a finite-dimensional HWM over $U_q(\mathscr{G})$, and all such modules may be obtained in this way [389, 405, 441, 532]. If we restrict $U_q(\mathscr{G})$ to its compact real form $U_q(\mathscr{G}_k)$ then the set of all L_Λ coincides with the set of all finite-dimensional unitary irreducible representations of $U_q(\mathscr{G}_k)$.

Example 2.1 Let us consider the example of $\mathscr{G} = sl(2,\mathbb{C})$; $r = 1$, $X_1^\pm = X^\pm$, $H_1 = H$, $\alpha_1 = \alpha = \alpha^\vee = 2\rho$:

$$[H, X^\pm] = \pm 2X^\pm\,, \tag{2.6}$$

$$[X^+, X^-] = \frac{q^{H/2} - q^{-H/2}}{q^{1/2} - q^{-1/2}} = [H]_q\,. \tag{2.7}$$

In this case the Verma module is given explicitly by $V^\Lambda \cong U_q(\mathscr{G}^-) \otimes v_0$, with basis

$$(X^-)^k \otimes v_0\,, \quad k = 0, 1, \ldots. \tag{2.8}$$

Let us consider again the reducible case when (2.2) is holding. Note that the highest weight is $\Lambda = ((m-1)/2)\alpha$, $(\alpha, \alpha) = 2$. The singular vector is given precisely by (2.3):

$$v_s = (X^-)^m \otimes v_0\,. \tag{2.9}$$

The submodule $U_q(\mathscr{G}^-)v_s \cong V^{\Lambda - m\alpha}$ has the basis

$$(X^-)^{m+k} \otimes v_0\,, \quad k = 0, 1, \ldots. \tag{2.10}$$

The irreducible HWM $L_m \equiv L_\Lambda$ is obtained by factorizing the submodule $U_q(\mathscr{G}^-)v_s$, that is, by the condition

$$(X^-)^m|0> = 0\,, \tag{2.11}$$

where $|0>$ is the highest-weight vector of L_m. Explicitly, all vectors of L_m are given by:

$$v_{m,k} \equiv (X^-)^k|0>\,, \quad k = 0, 1, \ldots, m-1\,, \tag{2.12}$$

which transform as follows:

$$Hv_{m,k} = (m-k-1)v_{m,k}\,, \quad k = 0, 1, \ldots, m-1\,, \tag{2.13a}$$

$$X^+v_{m,k} = [k][m-k]v_{m,k-1}\,, \quad k = 0, 1, \ldots, m-1\,, \tag{2.13b}$$

$$X^-v_{m,k} = v_{m,k+1}\,, \quad k = 0, 1, \ldots, m-2\,, \quad X^-v_{m,m-1} = 0\,. \tag{2.13c}$$

We note that as usual theirs is also a lowest-weight state which is annihilated by X^-, namely, $v_{m,m-1}$; thus, the lowest weight is $\Lambda = 0$. (One can also introduce normalized vectors.) Thus we obtain the usual result for the finite-dimensional irreducible

HWMs over $sl(2, \mathbb{C})$, or equivalently for the finite-dimensional unitary irreducible representations of $su(2)$, namely, that they are parametrized by the positive integers $m \in \mathbb{N}$, or equivalently, by the nonnegative half integers $j = (m-1)/2$, and their dimensions are:

$$\dim L_m = m = 2j + 1 , \quad m = 1, 2, \dots , \quad j = 0, 1/2, 1, \dots \ \lozenge \tag{2.14}$$

De Concini and Kac [175] have given a formula for the determinant of the contravariant form on the Verma modules V^λ. For $Y = Y(q) \in \mathbb{C}(q)$, let $\bar{Y} = Y(q^{-1})$. A $\mathbb{C}$-bilinear form $\mathscr{F}$ on a vector spave V over $\mathbb{C}(q)$ with values in $\mathbb{C}(q)$ is called Hermitian if:

$$\begin{aligned} \mathscr{F}(Yu, v) &= \bar{Y}\mathscr{F}(u, v) , \quad \mathscr{F}(u, Yv) = Y\mathscr{F}(u, v) , \\ \mathscr{F}(u, v) &= \bar{\mathscr{F}}(v, u) , \quad Y \in \mathbb{C}(q) , \quad u, v \in V . \end{aligned} \tag{2.15}$$

The Verma module V^λ carries a unique contravariant Hermitian form $\mathscr{F}$ such that:

$$\mathscr{F}(v_0, v_0) = 1 , \quad \mathscr{F}(Yu, v) = \mathscr{F}(u, \omega Yv) , \quad Y \in U_q(\mathscr{G}) , \quad u, v \in V^\lambda , \tag{2.16}$$

where ω is the involutive antiautomorphism such that $\omega X_i^\pm = X_i^\mp$, $\omega H_i = H_i$.

Let Γ (respectively, Γ_+) be the set of all integral elements (respectively, *integral dominant elements*), of $\mathscr{H}^*$, that is, $\lambda \in \mathscr{H}^*$ such that $(\lambda, \alpha_i^\vee) \in \mathbb{Z}$ (respectively, $\mathbb{Z}_+$), for all simple roots α_i. For each invariant subspace $V \subset U_q(\mathscr{G}^-) \otimes v_0 \cong V^\lambda$, we have the following decomposition

$$V = \underset{\mu \in \Gamma_+}{\oplus} V_\mu , \quad V_\mu = \{u \in V | H_k u = (\lambda - \mu)(H_k) u, \ \forall H_k\} . \tag{2.17}$$

(Note that $V_0 = \mathbb{C}v_0$.) We have $\mathscr{F}(V_\mu, V_\nu) = 0$ if $\mu \neq \nu$. Let $\mathscr{F}_\mu$ be the restriction of $\mathscr{F}$ to V_μ, $\mu \in \Gamma_+$, and let $\det_\mu^\lambda$ denote the determinant of the matrix $\mathscr{F}_\mu$. Then we have [175]:

$$\det{}_\mu^\lambda = \prod_{\beta \in \Delta^+} \prod_{k=1}^{\infty} ([k]_{q_\beta} [H_\beta - \rho(H_\beta) - k(\beta, \beta)/2]_{q_\beta})^{P(\mu - k\beta)} , \tag{2.18}$$

where q_β is defined as above in (1.23), $P(\mu)$ is a *generalized partition function*, $P(\mu)$ = # of ways μ can be presented as a sum of positive roots β_j, each root taken with its multiplicity $m_j = \dim \mathscr{G}_{\beta_j}$ (here $m_j = 1$), $P(0) \equiv 1$.

This result implies in the usual way the description of irreducible subquotients of V^Λ. In particular, this confirms results on the embeddings of the reducible modules V^Λ [198] summarized partially here.

2.2 *q*-Fock Type Representations

In the previous subsection we considered the irreducible HWM L_Λ over $U_q(\mathscr{G})$ as factor modules V^Λ / I^Λ, where I^Λ is the maximal submodule of V^Λ. As in the undeformed case there is a dual way of directly describing at least the finite-dimensional irreducible representations by the so-called Fock-type representations. One particular example is the so-called bosonic realization in the Jordan–Schwinger approach [102].

Let us recall this approach on the example of $sl(2,\mathbb{C})$, or, equivalently, $su(2)$. One takes a Heisenberg algebra of a pair of independent boson operators $a_i, \bar{a}_i, i = 1, 2$ with commutation relations

$$[\bar{a}_i, a_j] = \delta_{ij} \tag{2.19}$$

and all other commutators vanishing. Then the approach is to map as follows:

$$X^+ \mapsto a_1\bar{a}_2, \quad X^- \mapsto a_2\bar{a}_1, \quad H \mapsto a_1\bar{a}_1 - a_2\bar{a}_2. \tag{2.20}$$

The analogue of this construction in the deformed case was given by [99] (see also [139, 404, 448]). Relations (2.19) are replaced by

$$\bar{a}_i^q a_j^q - q^{1/2} a_j^q \bar{a}_i^q = \delta_{ij} q^{-\mathcal{N}_i/2}, \tag{2.21}$$

where $\mathcal{N}_i$ are *number operators* such that

$$[\mathcal{N}_i, a_j^q] = \delta_{ij} a_i^q, \quad [\mathcal{N}_i, \bar{a}_j^q] = -\delta_{ij} \bar{a}_i^q. \tag{2.22}$$

This algebra is the deformation of the Heisenberg algebra (2.19) which is obtained for $q = 1$. The mapping (2.20) is replaced by [99]:

$$X^+ \mapsto a_1^q \bar{a}_2^q, \quad X^- \mapsto a_2^q \bar{a}_1^q, \quad H \mapsto \mathcal{N}_1 - \mathcal{N}_2. \tag{2.23}$$

One uses the vacuum vector $|0>_q$ such that

$$\bar{a}_i^q |0>_q = 0, \quad \mathcal{N}_i |0>_q = 0. \tag{2.24}$$

Now one can introduce the eigenstates which are analogues of the undeformed angular momentum states [99]:

$$|j, n>_q \equiv ([j+n]_q! [j-n]_q!)^{-1/2} (a_1^q)^{j+n} (a_2^q)^{j-n} |0>_q, \tag{2.25}$$

$$j = 0, 1/2, 1, \dots, \quad n = -j, -j+1, \dots, j. \tag{2.26}$$

One easily verifies that:

$$\begin{aligned} H|j, n>_q &= 2n|j, n>_q, \\ X^\pm |j, n>_q &= ([j \mp n][j \pm n + 1])^{1/2} |j, n \pm 1>_q . \end{aligned} \tag{2.27}$$

Comparing with the states in (2.12) we see that $|j, n>_q, n = -j, -j+1, \dots, j$ corresponds to $v_{m,k}$, $k = 0, 1, \dots, m-1 = 2j$. One should note that the mapping (2.23) satisfies the commutation relation (2.6) only on the vectors $|j, n>_q$. For this one uses also the formula [99]:

$$[H]|j, n>_q = ([j+n][j-n+1] - [j-n][j+n+1])|j, n>_q \,. \tag{2.28}$$

The matrix elements of the these q-Fock-type representations can be expressed in terms of little q-Jacobi polynomials [446, 584].

Other q-Fock-type representations were constructed in [569] for $U_q(su(n))$, in [H] for $U_q(sl(n,\mathbb{C})^{(1)})$, and in [169, 379, 477]. In [169] the Gel'fand–Tsetlin bases become monomes in the tensor algebra of the fundamental representation of $U_q(sl(n,\mathbb{C}))$ at $q = 0$; that is, this strange choice of q provides the most simple basis. This was called crystal base [379] and was generalized for the integrable representations of $U_q(\mathcal{G})$ for $\mathcal{G} = A_n, B_n, C_n, D_n$ in [379] and for the basic representation of $U_q(sl(n,\mathbb{C})^{(1)})$ in [477].

2.3 Vertex Operators

Let us consider the affine quantum group $U_q(\mathcal{G}^{(1)})$ where $\mathcal{G}^{(1)}$ is the untwisted affinization of $\mathcal{G}$, rank $\mathcal{G} = r$. Let $a_i^\vee$ be the dual Kac labels; that is, $\sum_{j=0}^r a_j^\vee a_{jk} = 0$, normalized so that min $a_j^\vee = 1$. The element

$$\hat{K} = \prod_{j=0}^{r} K_j^{a_j^\vee} \tag{2.29}$$

belongs to the centre of $U_q(\mathcal{G}^{(1)})$.

Let us introduce bosonic variables $y^j, x^j(n), j = 1, \ldots r = \text{rank}\,\mathcal{G}, n \in \mathbb{Z}$ satisfying the Heisenberg relations:

$$[x^j(m), x^k(n)] = m\delta_{jk}\delta_{m+n,0}, \tag{2.30a}$$

$$[x^j(0), y^k] = i\delta_{jk}. \tag{2.30b}$$

Further let for $q = e^h, h \in \mathbb{C}$ [299]

$$\Delta_{jk}(n) = (q^{n(\alpha_j,\alpha_k)/2} - q^{-n(\alpha_j,\alpha_k)/2})(\mathcal{K}^n - \mathcal{K}^{-n})/n^2h^2. \tag{2.31}$$

Then the deformed Heisenberg algebra generators are defined by:

$$a^j(n) = \sum_{j=1}^{r} (\Delta_{jk}(n))^{1/2} x^k(n). \tag{2.32}$$

Now for each simple root $\alpha_j, j = 1, \ldots, r$ the q-deformed vertex operators are defined by [299]:

$$V^j_\pm(z) = :\exp\left(\pm ihQ^j_\pm(z)\right):= \\ = \exp\left(\pm ihQ^{\pm;j}_<(z)\right)\exp\left(\pm ihQ^{\pm;j}_>(z)\right)e^{\pm i(\alpha_j,y)}z^{\pm(\alpha_j,x(0))}, \tag{2.33}$$

where

$$Q^j_\pm = (\alpha_j, y - ix(0)logz) + Q^{\pm;j}_>(z) + Q^{\pm;j}_<(z), \tag{2.34a}$$

$$Q^{\pm;j}_>(z) = i\sum_{n>0}\frac{q^{\mp|n|/4}}{q^{n/2}-q^{-n/2}}\alpha^j(n)z^{-n}, \tag{2.34b}$$

$$Q^{\pm;j}_<(z) = i\sum_{n<0}\frac{q^{\mp|n|/4}}{q^{n/2}-q^{-n/2}}\alpha^j(n)z^{-n}. \tag{2.34c}$$

This construction is valid for the simply laced algebras $\mathcal{G}$, for which all roots have equal length; that is, for $\mathcal{G} = A_n, D_n, E_6, E_7, E_8$. Later this construction was generalized for $\mathcal{G} = B_n$ [92]. Another construction in terms of screened vertex operators was introduced in [319].

2.4 Singular Vectors in Chevalley Basis

Here we give explicit formulae for singular vectors of Verma modules over $U_q(\mathcal{G})$, where $\mathcal{G}$ is any complex simple Lie algebra. The vectors we present correspond exhaustively to a class of positive roots of $\mathcal{G}$ which we call *straight roots*. In some special cases we give singular vectors corresponding to arbitrary positive roots. For our vectors we use a special basis of $U_q(\mathcal{G}^-)$, where $\mathcal{G}^-$ is the negative roots subalgebra of $\mathcal{G}$, whose basis was introduced in our earlier work in the case $q = 1$. This basis seems more economical than the Poincaré–Birkhoff–Witt–type of basis used by Malikov, Feigin, and Fuchs for the construction of singular vectors of Verma modules in the case $q = 1$. Furthermore this basis turns out to be part of a general basis introduced recently for other reasons by Lusztig for $U_q(\mathcal{B}^-)$, where $\mathcal{B}^-$ is a Borel subalgebra of $\mathcal{G}$.

It is well known [109] that every root may be expressed as the result of the action of an element of the Weyl group W on some simple root. More explicitly, for any $\beta \in \Delta^+$ we have:

$$\beta = w(\alpha_u) = s_{i_1}s_{i_2}\cdots s_{i_u}(\alpha_v), \tag{2.35}$$

and consequently

$$s_\beta = ws_vw^{-1} = s_{i_1}\ldots s_{i_u}s_vs_{i_u}\ldots s_{i_1}, \tag{2.36}$$

where α_v is a simple root; the element $w \in W$ is written in a reduced form, that is, in terms of the minimal possible number of the (generating W) simple reflections $s_i \equiv s_{\alpha_i}$; and the action of $s_\alpha, \alpha \in \Delta$ on $\mathcal{H}^*$ is given by $s_\alpha(\lambda) = \lambda - (\lambda, \alpha^\vee)\alpha$. The positive root β is called a *straight root* if all numbers $i_1, \ldots i_u, v$ in (2.35) are different. Note that there may exist different forms of (2.35) involving other elements w' and $\alpha_{v'}$; however, this definition does not depend on the choice of these elements. Obviously, any simple root is a straight root. Other easy examples of straight roots are those which are sums of simple roots with coefficients not exceeding 1; that is, $\beta = \sum_k n_k\alpha_k$, with $n_k = 1$ or 0. All straight roots of the simply laced algebras A_ℓ, D_ℓ, E_ℓ are of this form.

Note that for any $\mathcal{G}$ it is enough to consider roots for which $n_k \neq 0$ for $1 \leq k \leq \ell$. Any other root β' may be considered as a root of a complex simple Lie algebra $\mathcal{G}'$ isomorphic to a subalgebra of $\mathcal{G}$ of rank $\ell' < \ell$, so that $\beta' = \sum_k n'_k\alpha'_k$ and $n'_k \neq 0$ for $1 \leq k \leq \ell'$ (α'_k being the simple roots of $\mathcal{G}'$). Thus in the case of the straight roots we shall consider always the case when $u = \ell - 1$, and $\{i_1, \dots i_u, v\}$ will be a permutation of $\{1, \dots \ell\}$.

In what follows we shall use also the following notion. A root $\gamma' \in \Delta^+$ is called a *subroot* of $\gamma'' \in \Delta^+$ if $\gamma'' - \gamma' \neq 0$ may be expressed as a linear combination of simple roots with nonnegative coefficients.

In this section we consider $U_q(\mathcal{G})$ when the deformation parameter q is not a nontrivial root of unity. This generic case is very important for two reasons. First, for $q = 1$ all formulae are valid also for the undeformed case, and most formulae first given in [206] were new at the time also for $q = 1$ (especially in our basis). Second, the formulae for the case when q is a root of unity use the formulae for generic q as important input as will be explained in Section 2.7.

We prove a statement which presents results from [206] in one uniform formula.

Proposition 1. *Let $\mathcal{G}$ be a complex simple Lie algebra and let α_k, $1 \leq k \leq \ell$, be the simple roots of the root system Δ of $\mathcal{G}$. Let $\beta = n_1\alpha_1 + n_2\alpha_2 + \cdots + n_\ell\alpha_\ell$, where $n_k \in \mathbb{Z}_+$ be a straight root (cf. (2.35)) of the positive root system Δ^+ of $\mathcal{G}$, and m a positive integer. Let $\lambda \in \mathcal{H}^*$ be such that (2.2) is fulfilled with this choice of β and m, but is not fulfilled for any subroot of β. Then the singular vector of the Verma module V^λ corresponding to β and m is given by:*

$$
\begin{aligned}
v_\lambda^{\beta,m} &= \mathcal{P}_\lambda^{\beta,m} \otimes v_0 = \sum_{k_1=0}^{mn_{i_1}} \cdots \sum_{k_u=0}^{mn_{i_u}} c_{k_1\dots k_u} (X_{i_1}^-)^{mn_{i_1}-k_1} \dots (X_{i_u}^-)^{mn_{i_u}-k_u} \times \\
&\quad \times (X_v^-)^{mn_v} (X_{i_u}^-)^{k_u} \dots (X_{i_1}^-)^{k_1} \otimes v_0, \qquad (2.37)\\
c_{k_1\dots k_u} &= (-1)^{k_1+\dots+k_u} c_u \begin{pmatrix} mn_{i_1} \\ k_1 \end{pmatrix}_{q_{i_1}} \cdots \begin{pmatrix} mn_{i_u} \\ k_u \end{pmatrix}_{q_{i_u}} \times \\
&\quad \times \frac{[(\lambda+\rho)(\tilde{H}_{i_1})]_{q_{i_1}}}{[(\lambda+\rho)(\tilde{H}_{i_1}) - k_1]_{q_{i_1}}} \cdots \frac{[(\lambda+\rho)(\tilde{H}_{i_u})]_{q_{i_u}}}{[(\lambda+\rho)(\tilde{H}_{i_u}) - k_u]_{q_{i_u}}}, \qquad (2.38)
\end{aligned}
$$

where the indices $i_1, \dots i_u, v$ come from the presentation (2.35), and $\tilde{H}_{i_1} \dots \tilde{H}_{i_u}$ are linear combinations of the basis elements H_i of the Cartan subalgebra $\mathcal{H}$ of $\mathcal{G}$, which can be computed explicitly in all cases. ◊

The *Proof* of this statement takes the rest of this section. We first treat the case of the simple roots. Then in the following subsections for all complex simple Lie algebras we give their straight roots with explicit presentations of type (2.35), and then we give explicitly the elements $\tilde{H}_{i_1} \dots \tilde{H}_{i_u}$.

We start with the case of the *simple roots*. Let $\beta = \alpha_j$; then from the expression (2.3) we have:

$$v^{j,m} = (X_j^-)^m \otimes v_0. \qquad (2.39)$$

Using (1.19) we obtain:

$$\begin{aligned}[X_j^+, (X_j^-)^m] &= (X_j^-)^{m-1} \sum_{k=0}^{m-1} [H_j - 2k]_{q_j} = \\ &= (X_j^-)^{m-1} [m]_{q_j} [H_j - m + 1]_{q_j}.\end{aligned}$$

If $v^{j,m}$ is a singular vector we should have $X_j^+ v^{j,m} = [X_j^+, (X_j^-)^m] \otimes v_0 = (X_j^-)^{m-1}[m]_{q_j}[\lambda(H_j) - m + 1]_{q_j} \otimes v_0$ 0. If $q_j = q^{(\alpha_j,\alpha_j)/2}$ is not a root of unity then the last equality gives just condition (2.2). (Note that $X_k^+ v^{j,m} = 0$, for $k \neq j$.)

To check (2.37) we use also formulae involving the q-hypergeometric function ${}_2F_1^q$:

$$\begin{aligned}{}_2F_1^q(-k, s; s+1-p; q^{(p-k)/2}) &= \delta_{p0} \frac{[k]![s]!}{[k+s]!} q^{ks/2}, \quad k > 0, p \leq k, s \\ {}_2F_1^q(a, b; c; z) &\equiv \sum_{n \in \mathbb{Z}_+} \frac{\Gamma_q(a+n)\Gamma_q(b+n)\Gamma_q(c)}{\Gamma_q(a)\Gamma_q(b)\Gamma_q(c+n)[n]!} z^n, \end{aligned} \qquad (2.40)$$

where for integer arguments the *q-Gamma function* Γ_q is defined as:

$$\begin{aligned}\Gamma_q(m) &\doteq [m-1]_q! \,, \quad m \in \mathbb{N} \\ 1/\Gamma_q(m) &\doteq 0 \,, \quad m \in \mathbb{Z}_- \end{aligned} \qquad (2.41)$$

Such *q-special functions* are in use from XIX century – for a review see [37].

We turn now to the nonsimple straight roots for the different simple Lie algebras.

2.4.1 $U_q(A_\ell)$

Let $\mathcal{G} = A_\ell$, $(\alpha_i, \alpha_j) = -1$ for $|i - j| = 1$, $(\alpha_i, \alpha_j) = 2\delta_{ij}$ otherwise. Then every root $\beta \in \Delta^+$ is given by $\beta = \beta_{in} = \alpha_i + \alpha_{i+1} + \dots + \alpha_{i+n-1}$, where $1 \leq i \leq \ell$, $1 \leq n \leq \ell - i + 1$. Note that *every root is straight* since $\beta_{i,n} = s_i(\beta_{i+1,n}) = s_i \cdots s_{i+n-2}(\alpha_{i+n-1}) = s_{i+n-1} \cdots s_{i+1}(\alpha_i) = s_i \cdots s_{i+t-1} s_{i+n-1} \cdots \cdots s_{i+t+1}(\alpha_{i+t}) = s_{i+n-1} \cdots s_{i+t+1} s_i \cdots s_{i+t-1}(\alpha_{i+t})$, $0 \leq t \leq n-1$, where we have demonstrated different forms of (2.35) in this case. For A_ℓ the highest root is given by $\tilde{\alpha} = \alpha_1 + \alpha_2 + \cdots + \alpha_\ell$. Thus every root $\beta \in \Delta^+$ is the highest root of a subalgebra of A_ℓ; explicitly β_{in} is the highest root of the subalgebra A_n with simple roots $\alpha_i, \alpha_{i+1}, \ldots, \alpha_{i+n-1}$. This means that it is enough to give the formula for the singular vector corresponding to the highest root. Thus in formula (2.37) with $\beta = \tilde{\alpha}$ we have $n_k = 1$, $1 \leq k \leq \ell$, and for the sets $i_1, \ldots i_u, v$ we obtain from $\tilde{\alpha} = s_1 s_2 \cdots s_t s_\ell s_{\ell-1} \cdots s_{t+2}(\alpha_{t+1})$ the following:

$$\{i_1, \ldots i_{\ell-1}; v\} = \{1, 2, \ldots, t, \ell, \ell - 1, \ldots, t + 2; t + 1\},$$
$$\tilde{H}_{i_s} = \begin{cases} H^s, & 1 \leq s \leq t \\ H'^{\ell+t+1-s}, & t + 1 \leq s \leq j = \ell - 1 \end{cases} \tag{2.42}$$
$$H^k \equiv H_1 + H_2 + \cdots + H_k, H'^k \equiv H_\ell + H_{\ell-1} + \cdots + H_k.$$

Formula (2.37) for A_2 was given in [198] and for arbitrary A_ℓ in [201].

2.4.2 $U_q(D_\ell)$

Let $\mathscr{G} = D_\ell$, $\ell \geq 4$, $(\alpha_i, \alpha_j) = -1$ for $|i - j| = 1, i, j \neq \ell$ and for $ij = \ell(\ell - 2)$, $(\alpha_i, \alpha_j) = 2\delta_{ij}$ otherwise. First we note that if $n_{\ell-2} + n_{\ell-1} + n_\ell \leq 2$, then the root β is a positive root of a subalgebra of D_ℓ of type A_n, $n < \ell$. Thus it remains to consider straight roots $\beta_i \in \Delta^+$ given by $\beta_i = \alpha_i + \alpha_{i+1} + \cdots + \alpha_\ell$. Note that β_i is a root of the subalgebra $D_{\ell-i+1}$ with simple roots $\alpha_i, \alpha_{i+1}, \ldots, \alpha_\ell$. This means that in order to account for all roots β_i it is enough to consider the root $\tilde{\beta} = \beta_1 = \alpha_1 + \alpha_2 + \cdots + \alpha_\ell = s_1 s_2 \ldots s_{\ell-3} s_{\ell-1} s_\ell(\alpha_{\ell-2}) = s_\ell \ldots s_2(\alpha_1) = s_1 s_2 \ldots s_{\ell-3} s_{\ell-1} s_{\ell-2}(\alpha_\ell) = s_1 s_2 \ldots s_{\ell-3} s_\ell s_{\ell-2}(\alpha_{\ell-1})$. Thus in formula (11) with $\beta = \tilde{\beta}$ we have $n_k = 1$, $1 \leq k \leq \ell$, and for the set $i_1, \ldots i_u, v$ we give only the values corresponding to the first presentation of $\tilde{\beta}$ above, namely, we have:

$$\{i_1, \ldots i_{\ell-1}; v\} = \{1, 2, \ldots, \ell - 3, \ell - 1, \ell; \ell - 2\}, \tag{2.43}$$
$$\tilde{H}_{i_s} = \begin{cases} H^s, & 1 \leq s \leq \ell - 3 \\ H_{s+1}, & s = \ell - 2, \ell - 1 \end{cases}$$

2.4.3 $U_q(E_\ell)$

Let $\mathscr{G} = E_\ell$, $\ell = 6, 7, 8$, $(\alpha_i, \alpha_{i+1}) = -1$, $i = 1, \ldots, \ell - 2$ $(\alpha_3, \alpha_\ell) = -1$, $(\alpha_i, \alpha_j) = 2\delta_{ij}$ otherwise. First we note that if $n_2 + n_4 + n_\ell \leq 2$ then the root β is a positive root of a subalgebra of E_ℓ of type A_n, $n < \ell$. Analogously, if $n_2 + n_4 + n_\ell = 3$ and $n_1 + n_5 \leq 1$, the root β is a positive root of a subalgebra of E_ℓ of type D_n, $n < \ell$. Thus it remains to consider the straight root $\tilde{\beta} = \alpha_1 + \cdots + \alpha_\ell = s_1 s_2 s_\ell s_{\ell-1} \ldots s_4(\alpha_3) = s_\ell \ldots s_2(\alpha_1) = s_1 s_2 s_{\ell-1} \ldots s_4 s_3(\alpha_\ell) = s_1 s_2 s_\ell s_3 \ldots s_{\ell-2}(\alpha_{\ell-1})$. Thus in formula (2.37) with $\beta = \tilde{\beta}$, we have $n_k = 1$, $1 \leq k \leq \ell$, and for the set $i_1, \ldots i_u, v$ we give only the values corresponding to the first presentation of $\tilde{\beta}$ above, namely, we have:

$$\{i_1,\dots i_{\ell-1};\nu\} = \{1,2,\ell,\ell-1,\dots,4;3\}, \tag{2.44}$$

$$\tilde{H}_{i_s} = \begin{cases} H_s, & s=1,2 \\ H_\ell, & s=3 \\ H''^{\ell+3-s}, & s=4,\dots,\ell-1 \end{cases}$$

$$H''^k \equiv H_{\ell-1}+\dots+H_k.$$

2.4.4 $U_q(B_\ell)$

$\mathcal{G} = B_\ell$, $\ell \geq 2$, $(\alpha_i,\alpha_j) = -2$ if $|i-j| = 1$, $(\alpha_i,\alpha_j) = 2\delta_{ij}(2-\delta_{i\ell})$ otherwise. The straight roots are of two types: $\beta_{in} = \alpha_i + \alpha_{i+1} + \cdots + \alpha_{i+n-1}$, $1 \leq i \leq \ell$, $1 \leq n \leq \ell - i + 1$, and $\beta'_i = \alpha_i + \cdots + \alpha_{\ell-1} + 2\alpha_\ell$, $1 \leq i < \ell$. If $i+n-1 < \ell$ then β_{in} is a positive root of a subalgebra of B_ℓ of type A_n, $n < \ell$ (with the scalar products scaled by 2 and q replaced by q^2). Thus we are left with two types of straight roots $\beta_i = \beta_{i,\ell+1-i} = \alpha_i + \alpha_{i+1} + \cdots + \alpha_\ell$, $1 \leq i < \ell$, and β'_i. As above it is enough to account for the roots with $i = 1$. Thus we consider $\tilde{\beta} = \beta_1 = \alpha_1 + \cdots + \alpha_\ell = s_1 \dots s_{\ell-1}(\alpha_\ell)$, and $\tilde{\beta}' = \beta'_1 = \alpha_1 + \cdots + \alpha_{\ell-1} + 2\alpha_\ell = s_1 \dots s_{\ell-2}s_\ell(\alpha_{\ell-1})$ $(= s_\ell \dots s_2(\alpha_1))$. We note that $(\tilde{\beta},\tilde{\beta}) = 2$, $\tilde{\beta}^\vee = \tilde{\beta} = 2\alpha_1^\vee + \cdots + 2\alpha_{\ell-1}^\vee + \alpha_\ell^\vee$, $(\tilde{\beta}',\tilde{\beta}') = 4$, $\tilde{\beta}'^\vee = (1/2)\tilde{\beta}' = \alpha_1^\vee + \cdots + \alpha_\ell^\vee$.

Thus in formula (2.37) with $\beta = \tilde{\beta}$ we have $n_k = 1$, $1 \leq k \leq \ell$, and

$$\{i_1,\dots i_{\ell-1};\nu\} = \{1,\dots,\ell-1;\ell\}, \quad \tilde{H}_{i_s} = H^s, q_{i_s} = q^2, \quad s = 1,\dots,\ell-1; \tag{2.45}$$

while for $\beta = \tilde{\beta}'$ we have $n_k = 1 + \delta_{k\ell}$, $1 \leq k \leq \ell$, and

$$\{i_1,\dots i_{\ell-1};\nu\} = \{1,\dots,\ell-2,\ell;\ell-1\}, \tag{2.46}$$

$$\tilde{H}_{i_s} = \begin{cases} H^s, & s = 1,\dots,\ell-2 \\ H_\ell, & s = \ell-1 \end{cases}$$

$$q_{i_s} = q^{2-\delta_{s\ell-1}}.$$

The case $\ell = 2$ was given first in [202].

2.4.5 $U_q(C_\ell)$

Let $\mathcal{G} = C_\ell$, $\ell \geq 3$, $(C_2 \cong B_2)$, $(\alpha_i,\alpha_j) = -1$ if $|i-j| = 1$ and $i,j < \ell$, $(\alpha_i,\alpha_j) = -2$ if $ij = \ell(\ell-1)$, $(\alpha_i,\alpha_j) = 2\delta_{ij}(1+\delta_{i\ell})$ otherwise. The straight roots are of two types:

$\beta_{in} = \alpha_i + \alpha_{i+1} + \cdots + \alpha_{i+n-1}$, $1 \le i \le \ell$, $1 \le n \le \ell - i + 1$, and $\beta_i'' = 2\alpha_i + \cdots + 2\alpha_{\ell-1} + \alpha_\ell$, $1 \le i < \ell$. If $i + n - 1 < \ell$ then β_{in} is a positive root of a subalgebra of C_ℓ of type A_n, $n < \ell$. Thus we are left with two types of straight roots $\beta_i = \beta_{i,\ell+1-i} = \alpha_i + \alpha_{i+1} + \cdots + \alpha_\ell$, $1 \le i < \ell$, and β_i''. As above it is enough to account for the roots with $i = 1$. Thus we consider $\tilde{\beta} = \beta_1 = \alpha_1 + \cdots + \alpha_\ell = s_\ell \ldots s_2(\alpha_1)$ $(= s_1 \ldots s_{\ell-2}s_\ell(\alpha_{\ell-1}))$ and $\tilde{\beta}'' = \beta_1'' = 2\alpha_1 + \cdots + 2\alpha_{\ell-1} + \alpha_\ell = s_1 \ldots s_{\ell-1}(\alpha_\ell)$. We note that $(\tilde{\beta}, \tilde{\beta}) = 2$, $\tilde{\beta}^\vee = \tilde{\beta} = \alpha_1^\vee + \cdots + \alpha_{\ell-1}^\vee + 2\alpha_\ell^\vee$, $(\tilde{\beta}'', \tilde{\beta}'') = 4$, $\tilde{\beta}''^\vee = (1/2)\tilde{\beta}'' = \alpha_1^\vee + \cdots + \alpha_\ell^\vee$.

Thus in formula (2.37) with $\beta = \tilde{\beta}$ we have $n_k = 1$, $1 \le k \le \ell$, and

$$\{i_1, \ldots i_{\ell-1}; \nu\} = \{\ell, \ldots, 2; 1\}, \quad \tilde{H}_{i_s} = H'^{\ell+1-s}, \quad q_{i_s} = q^{1+\delta_{s1}}, \quad s = 1, \ldots, \ell - 1, \tag{2.47}$$

while for $\beta = \tilde{\beta}''$ we have $n_k = 2 - \delta_{k\ell}$, $1 \le k \le \ell$, and

$$\{i_1, \ldots i_{\ell-1}; \nu\} = \{1, \ldots, \ell - 1; \ell\}, \quad \tilde{H}_{i_s} = H^s, \quad q_{i_s} = q, \quad s = 1, \ldots, \ell - 1. \tag{2.48}$$

2.4.6 $U_q(F_4)$

Let $\mathcal{G} = F_4$, $(\alpha_1, \alpha_1) = (\alpha_2, \alpha_2) = 2(\alpha_3, \alpha_3) = 2(\alpha_4, \alpha_4) = 4$, and $(\alpha_1, \alpha_2) = (\alpha_2, \alpha_3) = 2(\alpha_3, \alpha_4) = -2$ are the nonzero products between the simple roots. We have straight roots of type A_2: $\alpha_1 + \alpha_2$, $\alpha_3 + \alpha_4$; B_2: $\alpha_2 + \alpha_3$, $\alpha_2 + 2\alpha_3$; B_3: $\alpha_1 + \alpha_2 + \alpha_3$, $\alpha_1 + \alpha_2 + 2\alpha_3$; C_3: $\alpha_2 + \alpha_3 + \alpha_4$, $\alpha_2 + 2\alpha_3 + 2\alpha_4$. Thus we are left with the two roots $\tilde{\beta} = \alpha_1 + \alpha_2 + \alpha_3 + \alpha_4 = s_1s_2s_4(\alpha_3)$ and $\tilde{\beta}' = \alpha_1 + \alpha_2 + 2\alpha_3 + 2\alpha_4 = s_1s_4s_3(\alpha_2)$. We note that $(\tilde{\beta}, \tilde{\beta}) = 2$, $\tilde{\beta}^\vee = \tilde{\beta} = 2\alpha_1^\vee + 2\alpha_2^\vee + \alpha_3^\vee + \alpha_4^\vee$, $(\tilde{\beta}'', \tilde{\beta}'') = 4$, $\tilde{\beta}''^\vee = (1/2)\tilde{\beta}'' = \alpha_1^\vee + \alpha_2^\vee + \alpha_3^\vee + \alpha_4^\vee$.

Thus in formula (2.37) with $\beta = \tilde{\beta}$, we have $n_k = 1$, $1 \le k \le 4$, and

$$\{i_1, \ldots i_3; \nu\} = \{1, 2, 4; 3\}, \tag{2.49}$$

$$\tilde{H}_{i_s} = \begin{cases} H^s, & s = 1, 2 \\ H_4, & s = 3 \end{cases}$$

$$q_{i_s} = q^{2-\delta_{s3}}, \tag{2.50}$$

while for $\beta = \tilde{\beta}'$ we have $n_k = 1$, $k = 1, 2$, $n_k = 2$, $k = 3, 4$, and

$$\{i_1, \ldots i_3; \nu\} = \{1, 4, 3; 2\}, \quad q_{i_s} = q^{1+\delta_{s1}}, \quad \tilde{H}_{i_s} = \begin{cases} H_1, & s = 1 \\ H'^{6-s}, & s = 2, 3. \end{cases} \tag{2.51}$$

2.4.7 $U_q(G_2)$

Let $\mathcal{G} = G_2$, $(\alpha_1, \alpha_1) = 3(\alpha_2, \alpha_2) = -2(\alpha_1, \alpha_2) = 6$. The nonsimple straight roots are the two roots $\tilde{\beta} = \alpha_1 + \alpha_2 = s_1(\alpha_2)$ and $\tilde{\beta}''' = \alpha_1 + 3\alpha_2 = s_2(\alpha_1)$. We note that $(\tilde{\beta}, \tilde{\beta}) = 2$, $\tilde{\beta}^\vee = \tilde{\beta} = 3\alpha_1^\vee + \alpha_2^\vee$, $(\tilde{\beta}''', \tilde{\beta}''') = 6$, $\tilde{\beta}'''^\vee = (1/3)\tilde{\beta}''' = \alpha_1^\vee + \alpha_2^\vee$.

Thus in formula (2.37) with $\beta = \tilde{\beta}$ we have $n_k = 1$, $k = 1, 2$, and

$$\{i_1; \nu\} = \{1; 2\}, \quad \tilde{H}_{i_1} = H_1, \quad q_{i_1} = q^3. \tag{2.52}$$

while for $\beta = \tilde{\beta}'''$ we have $n_1 = 1$, $n_2 = 3$, and

$$\{i_1; \nu\} = \{2; 1\}, \quad \tilde{H}_{i_1} = H_2, \quad q_{i_1} = q. \tag{2.53}$$

Note that for the nonstraight root $\tilde{\beta}'' = \alpha_1 + 2\alpha_2 = s_2 s_1(\alpha_2)$, $(\tilde{\beta}'', \tilde{\beta}'') = 2$, $\tilde{\beta}''^\vee = \tilde{\beta}'' = 3\alpha_1^\vee + 2\alpha_2^\vee$ and with condition (2.2) fulfilled for $m = 1$:

$$[(\Lambda + \rho, \tilde{\beta}'^\vee) - 1]_{q''_{\tilde{\beta}}} = [3\Lambda(H_1) + 2\Lambda(H_2) + 4]_q = 0 \tag{2.54}$$

the formula for the singular vector is given as for B_2 and $m = 1$.

2.5 Singular Vectors in Poincaré–Birkhoff–Witt Basis

In the present section we give explicit expressions for the singular vectors of $U_q(\mathcal{G})$ in terms of the PBW basis. We also relate these expressions to those in terms of the simple root vectors. The first result may be compared for $q = 1$ with the formulae of [460] (not without problems, cf. below), and here we should stress that our derivation is independent from that of [459], [460]. The second result is not known also for $q = 1$, except for $\ell = 2$.

2.5.1 PBW Basis

Let $\mathcal{G}$ be a complex simple Lie algebra with Chevalley generators $X_i^\pm$, H_i, $i = 1, \ldots, \ell = \operatorname{rank} \mathcal{G}$. Here we take the Jimbo version of the quantum algebra $U_q(\mathcal{G})$, though with slightly different normalization than in Section 1.2.3, with generators $X_i^\pm, K_i \equiv q_i^{H_i}$, $K_i^{-1} \equiv q_i^{-H_i}$, and with relations [360, 361]:

$$[K_i, K_j] = 0, \quad K_i K_i^{-1} = K_i^{-1} K_i = 1, \quad K_i X_j^\pm K_i^{-1} = q_i^{\pm a_{ij}} X_j^\pm,$$
$$[X_i^+, X_j^-] = \delta_{ij} \frac{K_i - K_i^{-1}}{q_i - q_i^{-1}} \tag{2.55}$$

We use also the q-Serre relations (1.20); however, the q-numbers are taken as: $[m]_q = \frac{q^m - q^{-m}}{q - q^{-1}}$.

For the PBW basis of $U_q(\mathcal{G})$ besides $X_i^\pm$, $K_i^{\pm 1}$, we need also the Cartan–Weyl (CW) generators $X_\beta^\pm$ corresponding to the nonsimple roots $\beta \in \Delta^+$. Naturally, we shall use uniform notation, so that $X_{\alpha_i}^\pm \equiv X_i^\pm$. The CW generators $X_\beta^\pm$ are normalized so that [202, 360, 361]:

$$\begin{aligned} [X_\beta^+, X_\beta^-] &= \frac{K_\beta - K_\beta^{-1}}{q_\beta - q_\beta^{-1}}, \qquad q_\beta \equiv q^{(\beta,\beta)/2} \\ K_\beta &\equiv \textstyle\prod_j K_j^{n_j(\beta,\beta)/(\alpha_j,\alpha_j)} (= q_\beta^{H_\beta}) \end{aligned} \tag{2.56}$$

The HWMs V over $U_q(\mathcal{G})$ are given by their highest weight $\Lambda \in \mathcal{H}^*$ and highest-weight vector $v_0 \in V$ such that

$$K_i\, v_0 = q_i^{\Lambda_i}\, v_0, \quad X_i^+ v_0 = 0, \quad i = 1, \dots, \ell, \quad \Lambda_i \equiv (\Lambda, \alpha_i^\vee) \tag{2.57}$$

We know that the Verma module V^Λ is reducible if there exists a root $\beta \in \Delta^+$ and $m \in \mathbb{N}$ such that (2.2) holds. Then there exists a vector $v_s \in V^\Lambda$, called a *singular vector*, such that $v_s \notin \mathbb{C}\, v_0$, and which obeys in the present setting:

$$K_i\, v_s = q_i^{\Lambda_i - m(\beta, \alpha_i^\vee)}\, v_s, \quad i = 1, \dots, \ell, \tag{2.58a}$$

$$X_i^+ v_s = 0, \quad i = 1, \dots, \ell. \tag{2.58b}$$

We know that the singular vector is given by

$$v_s = v^{m,\beta} = \mathcal{P}^{m,b} \otimes v_0$$

and in the previous subsections we have given explicit expressions of the homogeneous polynomials $\mathcal{P}_m^\beta$ in the simple root basis.

The aim of the present section is to give expressions for the singular vectors in terms of the PBW basis and to relate these expressions with those of [206].

2.5.2 Singular Vectors for $U_q(A_\ell)$ in PBW Basis

Let $\mathcal{G} = A_\ell$. Let $\alpha_i, i = 1, \dots, \ell$ be the simple roots, so that $(\alpha_i, \alpha_j) = -1$ for $|i - j| = 1$ and $(\alpha_i, \alpha_j) = 2\delta_{ij}$ for $|i - j| \neq 1$. Then every root $\alpha \in \Delta^+$ is given by $\alpha = \alpha_{ij} = \alpha_i + \alpha_{i+1} + \cdots + \alpha_j$, where $1 \le i \le j \le \ell$; in particular, the simple roots in this notation are $\alpha_i = \alpha_{ii}$. We recall that for A_ℓ the highest root is given by $\tilde{\alpha} = \alpha_1 + \alpha_2 + \cdots + \alpha_\ell$ and that every root $\alpha \in \Delta^+$ is the highest root of a subalgebra of A_ℓ; explicitly α_{ij} is the highest root of the subalgebra A_{j-i+1} with simple roots $\alpha_i, \alpha_{i+1}, \dots, \alpha_j$. This means that it is enough to give the formula for the singular vector corresponding to the highest root.

Further we shall need the explicit expressions for the nonsimple root Cartan–Weyl generators of $U_q(\mathscr{G})$. Let $X^\pm_{ij}$ be the Cartan–Weyl generators corresponding to the roots $\pm\alpha_{ij}$ with $i \le j$; in particular, $X^\pm_{ii} = X^\pm_i$, correspond to the simple roots α_i. The CW generators corresponding to the nonsimple roots with $i < j$ are given as follows:

$$\begin{aligned} X^\pm_{ij} &= \pm q^{\mp 1/2}\left(q^{1/2}\, X^\pm_i X^\pm_{i+1,j} - q^{-1/2}\, X^\pm_{i+1,j} X^\pm_i\right) = \\ &= \pm q^{\mp 1/2}\left(q^{1/2}\, X^\pm_{i,j-1} X^\pm_j - q^{-1/2}\, X^\pm_j X^\pm_{i,j-1}\right). \end{aligned} \tag{2.59}$$

Now the PBW basis of $U_q(\mathscr{G}^-)$ is given by monomials of the following kind:

$$\begin{aligned} &(X^-_\ell)^{k_{\ell\ell}}\,(X^-_{\ell-1,\ell})^{k_{\ell-1,\ell}} \dots (X^-_{1\ell})^{k_{1\ell}}\,(X^-_{\ell-1})^{k_{\ell-1}} \dots \times \\ &\times \dots (X^-_p)^{k_{pp}} \dots (X^-_{rp})^{k_{rp}} \dots (X^-_{12})^{k_{12}}\,(X^-_1)^{k_{11}} \end{aligned} \tag{2.60}$$

This monomials are in the so-called normal order, cf. Definition 1.1 of the previous chapter. Explicitly, here we put the simple root vectors X^-_j in the order X^-_ℓ , $X^-_{\ell-1}$, ... X^-_1 . Then we put a root vector X^-_α corresponding to the nonsimple root α between the root vectors X^-_β and X^-_γ if $\alpha = \beta + \gamma$, $\alpha, \beta, \gamma \in \Delta^+$. This order is not complete, but this is not a problem, since when two roots are not ordered this means that the corresponding root vectors commute, for example, X^-_i and $X^-_{i-k,i+k}$.

Let us have condition (2.2) fulfilled for $\tilde\alpha$, but not for any other positive root. Standardly, we denote the corresponding singular vector by $v^{m,\tilde\alpha}_s$. We start with an arbitrary linear combination of the PBW basis (2.60):

$$\begin{aligned} v^{m,\tilde\alpha}_s = &\sum_{\substack{k_{ij}\in\mathbb{Z}_+\\ 1\le i\le j\le\ell}} \tilde C_K\,(X^-_\ell)^{k_{\ell\ell}}\,(X^-_{\ell-1,\ell})^{k_{\ell-1,\ell}} \dots (X^-_{1\ell})^{k_{1\ell}}\,(X^-_{\ell-1})^{k_{\ell-1}} \dots \times \\ &\times \dots (X^-_p)^{k_{pp}} \dots (X^-_{rp})^{k_{rp}} \dots (X^-_{12})^{k_{12}}\,(X^-_1)^{k_{11}} \otimes v_0 \end{aligned} \tag{2.61}$$

and impose the conditions (2.58) with $\beta \to \tilde\alpha$. Condition (2.58a) restricts the linear combination to terms of weight $m\tilde\alpha$. In our parametrization these are the following ℓ conditions between the powers k_{ij} :

$$\begin{aligned} k_{1j} &= \sum_{p=j+1}^{\ell} k_{j+1,p} - \sum_{p=2}^{j} k_{pj}, \qquad 1 < j < \ell, \\ k_{1\ell} &= m - k_{11} + \sum_{p=2}^{\ell-1} k_{2p} - \sum_{p=3}^{\ell} k_{p\ell}, \\ k_{2\ell} &= k_{11} - \sum_{p=2}^{\ell-1} k_{2p}. \end{aligned} \tag{2.62}$$

Imposing conditions (2.58b) result in ℓ recurrence relations between the different $\tilde C_K$, namely,

$$\begin{aligned}
&\tilde{C}_{\{k_j+1\}}\,[k_j+1]_q\,[\lambda_j - \sum_{p=1}^{j} k_{pj} + \sum_{p=1}^{j-1} k_{p,j-1}]_q + \\
&+\sum_{p=1}^{j-1} \tilde{C}_{\{k_{pj}+1,k_{p,j-1}-1\}}\,[k_{pj}+1]_q q^{-\lambda_j+1+\sum_{i=1}^{p}(k_{ij}-k_{ij-1})} - \\
&-\sum_{p=j+1}^{\ell} \tilde{C}_{\{k_{jp}+1,k_{j+1,p}-1\}}\,[k_{jp}+1]_q q^{\lambda_j+1-\sum_{i\le r<p}(\alpha_j,\alpha_{ir})k_{ir}} = 0
\end{aligned} \tag{2.63}$$

where by $\tilde{C}_{\{k_j+1\}}$ we denote a $\tilde{C}_K$ in which the parameter k_j is replaced by k_j+1, and so on.

Solving the recurrence relation (2.63) fixes the coefficient $\tilde{C}_K$ up to an overal multiplicative constant, for example, C. In order to streamline the final expression instead of the $\ell(\ell+1)/2$ parameters k_{ij} we introduce $\ell(\ell-1)/2$ independent parameters t_{ij}, $1 \le i < j \le \ell$, as follows:

$$\begin{aligned}
k_{pp} &= m - t_{p-1,p} - \sum_{r=p+1}^{\ell} t_{pr}, \qquad 1 \le p \le \ell \\
k_{pr} &= t_{pr} - t_{p-1,r}, \qquad 1 \le p < r \le \ell,\ (t_{0r} \equiv 0).
\end{aligned} \tag{2.64}$$

We denote by C_T the coefficients $\tilde{C}_K$ with the substitution (2.64). Finally, the expression for the singular vector is:

$$\begin{aligned}
v_s^{m,\bar{\alpha}} = &\sum_{\substack{t_{ij}\in\mathbb{Z}_+\\ 1\le i<j\le\ell}} C_T\,(X_\ell^-)^{m-t_{\ell-1,\ell}}(X_{\ell-1,\ell}^-)^{t_{\ell-1,\ell}-t_{\ell-2,\ell}} \ldots (X_{1\ell}^-)^{t_{1\ell}} \times \\
&\times (X_{\ell-1}^-)^{m-t_{\ell-1,\ell}-t_{\ell-2,\ell-1}} \ldots (X_p^-)^{m-t_{p\ell}-t_{p,\ell-1}-\cdots-t_{p,p+1}-t_{p-1,p}} \times \\
&\times \ldots (X_{rp}^-)^{t_{rp}-t_{r-1,p}} \ldots (X_{12}^-)^{t_{12}}(X_1^-)^{m-t_{1\ell}-t_{1,\ell-1}-\cdots-t_{12}} \otimes v_0
\end{aligned} \tag{2.65}$$

where the summations in the variables t_{ij} are such that all powers are nonnegative, and the coefficients C_T are given by:

$$\begin{aligned}
C_T &= C\, q^{\sum_{i<j}(2m-1)t_{ij}-t_{ij}^2 - \sum_{1\le i<j\le\ell-1}\sum_{p=j+1}^{\ell} t_{ij}(t_{ip}+t_{i+1,p})} \times \\
&\times \frac{q^{-\sum_{r=1}^{\ell-1} t^r(\Lambda+\rho,\beta^r)} \prod_{r=1}^{\ell-1}[m-t^r]!}{\left(\prod_{i=1}^{\ell}[m-t^i-t_{i-1,i}]!\right)\prod_{1\le i<j\le\ell}[t_{ij}-t_{i-1,j}]!} \times \\
&\times \prod_{j=1}^{\ell-1} \frac{\Gamma_q((\Lambda,\beta^j)+j-m+t^j)}{\Gamma_q((\Lambda,\beta^j)+j-m)}, \\
\beta^r &\equiv \alpha_1 + \ldots + \alpha_r, \qquad t^i = \sum_{i+1\le s\le\ell} t_{is}.
\end{aligned} \tag{2.66}$$

For $\ell = 2$ and $t = t_{12}$ this formula is given in [245], and for $q = 1$ and $\ell = 2, 3$ in [194]. For $q = 1$ our formula would coincide with the result of [460], if we correct one of the definitions there, namely, if we use instead of $B_k = \sum_{i\le k\le j} \alpha_{ij}$ in f-la (20) of [460] the quantity $B_k^{corr} = \sum_{i\le k<j} \alpha_{ij}$, where the reader should not confuse the parameters α_{ij} from [460] with our notation of the roots – in fact, their α_{ij} correspond exactly to our k_{ij}; note also that $B_k^{corr} = t^k$.

Next we would like to present the relation between the expressions for the singular vectors in the PBW basis given in (2.65) and the expression using simple root vectors basis in (2.37) (the case $t = \ell - 1$). We just introduce an alternative expression for the coefficient $a_{k_1 \ldots k_{\ell-1}}$:

$$\begin{aligned} a_{k_1\ldots k_{\ell-1}} &= a^{\ell}(-1)^{k_1+\cdots+k_{\ell-1}} \binom{m}{k_1}_q \cdots \binom{m}{k_{\ell-1}}_q \times \\ &\times \frac{[\Lambda^1+1]_q}{[\Lambda^1+1-k_1]_q} \cdots \frac{[\Lambda^{\ell-1}+\ell-1]_q}{[\Lambda^{\ell-1}+\ell-1-k_{\ell-1}]_q}, \qquad a^{\ell} \neq 0 \end{aligned} \tag{2.67}$$

where $\Lambda^s = (\Lambda, \beta^s)$.

To make the connection explicit we give the C-coefficients in terms of the a-coefficients by the following formula [223]:

$$\begin{aligned} C_T &= (-1)^{\sum\limits_{i<j} t_{ij}} q^{\sum\limits_{i<j}((2m-1)t_{ij}-t_{ij}^2) - \sum_{1\le i<j\le(\ell-1)} \sum_{(j+1)\le p\le \ell} t_{ij}(t_{i+1,p}+t_{i,p})} \times \\ &\times \frac{q^{(1-\ell)m^2} \prod\limits_{r=1}^{\ell-1} [m-t^r]!}{\left(\prod\limits_{s=1}^{\ell} [m-t^s-t_{s-1,s}]! \right) \prod\limits_{1\le i<j\le \ell} [t_{ij}-t_{i-1,j}]!} \times \\ &\times \sum_{k_1,k_2,\ldots,k_{\ell-1}} a_{k_1,k_2,\ldots,k_{\ell-1}} q^{\frac{1}{2}\sum\limits_{i=1}^{\ell-1} k_i(m-t^i)} \prod_{r=1}^{\ell-1} \frac{[m-k_r]!}{[m-t^r-k_r]!} \end{aligned} \tag{2.68}$$

where $0 \le k_r \le m - t^r$.

To prove the above formula one can use the following lemmas:

Lemma 1.

$$\sum_{j=0}^{m} \frac{(-1)^j q^{mj}}{[a-j][m-j]![j]!} = \frac{(-1)^m q^{ma}}{\prod\limits_{k=0}^{m} [a-k]} \tag{2.69}$$

Lemma 2.

$$\frac{(X_r^-)^m (X_s^-)^n}{[m]![n]!} = \sum_{0\le p\le \min(m,n)} (-1)^p \, q^{(p-m)(n-p)-p} \frac{(X_s^-)^{n-p} (X_{rs}^-)^p (X_r^-)^{m-p}}{[n-p]![p]![m-p]!} \tag{2.70}$$

Lemma 1 follows from formula (60) of [244], which is:

$$
\begin{aligned}
{}_2F_1{}^q(-\nu,\, b;\, c;\, q^{\pm(b-c+1-\nu)}) &= \sum_{s=0}^{\nu} (-1)^s \binom{\nu}{s}_q \frac{(b)_s^q}{(c)_s^q}\, q^{\pm s(b-c+1-\nu)} = \\
&= \frac{(c-b)_\nu^q}{(c)_\nu^q}\, q^{\pm b\nu}
\end{aligned}
\tag{2.71}
$$

in which here we set $b = -a$, $c = 1-a$, $\nu = m$, and use sign minus. Lemma 2 follows just from (1.20) and (2.59).

Next the a-coefficients are given in terms of the C-coefficients by the following formula:

$$
\begin{aligned}
a_{j_1,\ldots,j_{\ell-1}} &= (-1)^{j_1+\cdots+j_{\ell-1}+l(m-1)} \sum_{\substack{m-t^1\le j_1\\ \vdots \\ m-t^{\ell-1}\le j_{\ell-1}}} C_T \times \\
&\times \prod_{i=1}^{\ell-1} \frac{q^{(m-t^i)(1-j_i)-j_i}}{[m-t^i]![m-j_i]![t^i+j_i-m]!} \times \\
&\times \left(\prod_{i=1}^{\ell}[m-t^i-t_{i-1,i}]!\right) \prod_{1\le i<j\le \ell} [t_{ij}-t_{i-1,j}]! \times \\
&\times q^{\sum_{i<j}((1-2m)t_{ij}+t_{ij}^2) + \sum_{1\le i<j\le \ell-1}\sum_{p=j+1}^{\ell} t_{ij}(t_{i+1,p}+t_{ip}) + (\ell-1)m^2} .
\end{aligned}
\tag{2.72}
$$

To prove (2.72) one can use the relation:

$$
\begin{aligned}
&\sum_{t=0}^{p} (-1)^{p+t} q^{t(k-p+1)-p} \binom{m-k}{m-t}_q \binom{m-t}{m-p}_q = \\
&= (-1)^{p-k} q^{(k+1)(k-p)} \frac{[m-k]!}{[m-p]!} \sum_{s=0}^{p-k} \frac{(-1)^s q^{s(1-p+k)}}{[s]![p-k-s]!} = \delta_{p,k}
\end{aligned}
\tag{2.73}
$$

which also follows from (2.71) setting $b = c$, $\nu = p-k$, and using the fact that $(0)_\nu^q = \Gamma_q(\nu)/\Gamma_q(0) = \delta_{\nu,0}$.

2.5.3 Singular Vectors for $U_q(D_\ell)$ in PBW Basis

Let $\mathcal{G} = D_\ell$, $\ell \ge 4$. Let $\alpha_i, i = 1, \ldots, \ell$ be the simple roots, so that $(\alpha_i, \alpha_j) = -1$ if either $|i-j| = 1, i, j \ne \ell$ or $ij = \ell(\ell-2)$ and $(\alpha_i, \alpha_j) = 2\delta_{ij}$ in other cases.

Then the positive roots are given as follows:

$$
\begin{aligned}
\alpha_{ij} &= \alpha_i + \alpha_{i+1} + \cdots + \alpha_j\,, \quad 1 \le i < j \le \ell-2\,,\\
\beta_j &= \alpha_j + \alpha_{j+1} + \cdots + \alpha_{\ell-2} + \alpha_\ell\,, \quad 1 \le j \le \ell-2\,,
\end{aligned}
$$

$$\begin{aligned}
&\tilde{b}_j = \alpha_j + \alpha_{j+1} + \cdots + \alpha_{\ell-2} + \alpha_{\ell-1}\,, \quad 1 \le j \le \ell-2\,,\\
&\beta_0 = \alpha_{\ell-2} + \alpha_{\ell-1} + \alpha_\ell\,,\\
&\gamma_j = \alpha_j + \alpha_{j+1} + \cdots + \alpha_{\ell-2} + \alpha_{\ell-1} + \alpha_\ell\,, \quad 1 \le j \le \ell-3\,,\\
&\gamma_{ij} = \alpha_i + \alpha_{i+1} + \cdots + 2(\alpha_j + \cdots + \alpha_{\ell-2}) + \alpha_{\ell-1} + \alpha_\ell\,,\\
&\qquad\qquad 1 \le i < j \le \ell-2\,.
\end{aligned} \tag{2.74}$$

We recall that the roots α_{ij}, β_j, $\tilde{b}_j$, β_0 are positive roots of various A_n subalgebras. Thus, we have to consider only the roots γ_j and γ_{ij}. We recall from [206] that γ_j is straight, while γ_{ij} is not straight.

Further we shall need the explicit expressions for the nonsimple root Cartan–Weyl (CW) generators of $U_q(\mathscr{G})$. Let $X^\pm_{i,j}$, $Y^\pm_j$, $\tilde{Y}^\pm_j$, $Y^\pm_0$, $Z^\pm_j$ and $Z^\pm_{i,j}$ be the Cartan–Weyl generators corresponding, respectively, to the roots $\pm\alpha_{ij}$, $\pm\beta_j$, $\pm\tilde{b}_j$, $\pm\beta_0$, $\pm\gamma_j$, and $\pm\gamma_{ij}$. These generators are given recursively as follows (with $X^\pm_{jj} \equiv X^\pm_j$):

$$\begin{aligned}
X^\pm_{ij} &= \pm q^{\mp 1/2}\left(q^{1/2}\, X^\pm_i X^\pm_{i+1,j} - q^{-1/2}\, X^\pm_{i+1,j} X^\pm_i\right) =\\
&= \pm q^{\mp 1/2}\left(q^{1/2}\, X^\pm_{i,j-1} X^\pm_j - q^{-1/2}\, X^\pm_j X^\pm_{i,j-1}\right),\\
&\qquad 1 \le i < j \le \ell-2,\\
Y^\pm_j &= \pm q^{\mp 1/2}\left(q^{1/2}\, X^\pm_\ell X^\pm_{j,\ell-2} - q^{-1/2}\, X^\pm_{j,\ell-2} X^\pm_\ell\right) =\\
&= \pm q^{\mp 1/2}\left(q^{1/2}\, X^\pm_j Y^\pm_{j+1} - q^{-1/2}\, Y^\pm_{j+1} X^\pm_j\right), \qquad 1 \le j \le \ell-2,\\
\tilde{Y}^\pm_j &= \pm q^{\mp 1/2}\left(q^{1/2}\, X^\pm_{\ell-1} X^\pm_{j,\ell-2} - q^{-1/2}\, X^\pm_{j,\ell-2} X^\pm_{\ell-1}\right) =\\
&= \pm q^{\mp 1/2}\left(q^{1/2}\, X^\pm_j \tilde{Y}^\pm_{j+1} - q^{-1/2} \tilde{Y}^\pm_{j+1} X^\pm_j\right), \qquad 1 \le j \le \ell-2,\\
Y^\pm_0 &= \pm q^{\mp 1/2}\left(q^{1/2}\, X^\pm_{\ell-1} Y^\pm_{\ell-2} - q^{-1/2}\, Y^\pm_{\ell-2} X^\pm_{\ell-1}\right) =\\
&= \pm q^{\mp 1/2}\left(q^{1/2}\, X^\pm_\ell \tilde{Y}^\pm_{\ell-2} - q^{-1/2} \tilde{Y}^\pm_{\ell-2} X^\pm_\ell\right),\\
Z^\pm_j &= \pm q^{\mp 1/2}\left(q^{1/2}\, X^\pm_{j,\ell-3} Y^\pm_0 - q^{-1/2}\, Y^\pm_0 X^\pm_{j,\ell-3}\right) =\\
&= \pm q^{\mp 1/2}\left(q^{1/2}\, X^\pm_\ell \tilde{Y}^\pm_j - q^{-1/2} \tilde{Y}^\pm_j X^\pm_\ell\right),\\
&= \pm q^{\mp 1/2}\left(q^{1/2}\, X^\pm_{\ell-1} Y^\pm_j - q^{-1/2} \tilde{Y}^\pm_j X^\pm_{\ell-1}\right), \qquad 1 \le j \le \ell-3\\
Z^\pm_{ij} &= \pm q^{\mp 1/2}\left(q^{1/2}\, Z^\pm_i X^\pm_{j,\ell-2} - q^{-1/2}\, X^\pm_{j,\ell-2} Z^\pm_i\right),\\
&\qquad 1 \le i < j \le \ell-2.
\end{aligned} \tag{2.75}$$

Now the PBW basis of $U_q(\mathscr{G}^-)$ is given by the following monomials:

$$\begin{aligned}
&(X^-_{\ell-2})^{a_{\ell-2}} (X^-_{\ell-3,\ell-2})^{t_{\ell-3,\ell-2}} \ldots (X^-_{1,\ell-2})^{t_{1,\ell-2}} (\tilde{Y}^-_{\ell-2})^{\tilde{t}_{\ell-2}} (Y^-_{\ell-2})^{t_{\ell-2}} \times\\
&\times (Z^-_{\ell-3,\ell-2})^{s_{\ell-3,\ell-2}} (Z^-_{\ell-4,\ell-2})^{s_{\ell-4,\ell-2}} \ldots (Z^-_{1,\ell-2})^{s_{1,\ell-2}} (\tilde{Y}^-_{\ell-3})^{\tilde{t}_{\ell-3}} (Y^-_{\ell-3})^{t_{\ell-3}} \times\\
&\times (Z^-_{\ell-4,\ell-3})^{s_{\ell-4,\ell-3}} \ldots (Z^-_{1,\ell-3})^{s_{1,\ell-3}} \ldots (\tilde{Y}^-_1)^{\tilde{t}_1} (Y^-_1)^{t_1} (Y^-_0)^{t} \times\\
&\times (Z^-_{\ell-3})^{s_{\ell-3}} \ldots (Z^-_1)^{s_1} (X^-_\ell)^{a_\ell} (X^-_{\ell-1})^{a_{\ell-1}} (X^-_{\ell-3})^{a_{\ell-3}} \times
\end{aligned}$$

$$\times(X^-_{\ell-4,\ell-3})^{t_{\ell-4,\ell-3}}\dots(X^-_{1,\ell-3})^{t_{1,\ell-3}}(X^-_{\ell-4})^{a_{\ell-4}}\times$$
$$\times\dots(X^-_2)^{a_2}(X^-_{12})^{t_{12}}(X^-_1)^{a_1}. \tag{2.76}$$

These monomials are in the so-called normal order, cf. Definition 1.1 of the previous chapter. Explicitly, here we put the simple root vectors X^-_j in the order $X^-_{\ell-2}, X^-_\ell, X^-_{\ell-1}, X^-_{\ell-3}, \dots, X^-_2, X^-_1$. Then we put a root vector E^-_α corresponding to the nonsimple root α between the root vectors E^-_β and E^-_γ if $\alpha=\beta+\gamma$, $\alpha,\beta,\gamma\in\Delta^+$. This order is not complete but this is not a problem, since when two roots are not ordered this means that the corresponding root vectors commute, for example, $[X^-_i, X^-_{i-k,i+k}]=0$, and $[Y^-_i,\tilde{Y}^-_i]=0$, $1\le i\le\ell-2$.

2.5.3.1 Singular Vectors in PBW Basis for Straight Roots

In this subsection we deal with the straight roots γ_j. Now we recall that every root γ_j is the highest straight root of a $D_{\ell-j+1}$ subalgebra of D_ℓ. This means that it is enough to give the formula for the singular vector corresponding to the highest straight root γ_1.

Let us have condition (2.2) fulfilled for γ_1, but not for any of its subroots γ_i, $i>1$:

$$[(\Lambda+\rho,\gamma_1{}^\vee)-m]_q=0,\quad m\in\mathbb{N}, \tag{2.77a}$$
$$[(\Lambda+\rho,\gamma_i^\vee)-m']_q\neq 0,\quad \forall m'\in\mathbb{N}. \tag{2.77b}$$

(The necessity of condition (2.77b) was explained in [206].) Let us denote the singular vector corresponding to (2.77a) by:

$$\begin{aligned}v_s^{\gamma_1,m} &= \sum_T D_T^{\gamma_1,m}\,(X^-_{\ell-2})^{a_{\ell-2}}(X^-_{\ell-3,\ell-2})^{t_{\ell-3,\ell-2}}\dots(X^-_{1,\ell-2})^{t_{1,\ell-2}}(\tilde{Y}^-_{\ell-2})^{\tilde{t}_{\ell-2}}\times\\
&\times(Y^-_{\ell-2})^{t_{\ell-2}}(Z^-_{\ell-3,\ell-2})^{s_{\ell-3,\ell-2}}(Z^-_{\ell-4,\ell-2})^{s_{\ell-4,\ell-2}}\dots(Z^-_{1,\ell-2})^{s_{1,\ell-2}}\times\\
&\times(\tilde{Y}^-_{\ell-3})^{\tilde{t}_{\ell-3}}(Y^-_{\ell-3})^{t_{\ell-3}}(Z^-_{\ell-4,\ell-3})^{s_{\ell-4,\ell-3}}\dots(Z^-_{1,\ell-3})^{s_{1,\ell-3}}\dots(\tilde{Y}^-_1)^{\tilde{t}_1}\times\\
&\times(Y^-_1)^{t_1}(Y^-_0)^{t}(Z^-_{\ell-3})^{s_{\ell-3}}\dots(Z^-_1)^{s_1}(X^-_\ell)^{a_\ell}(X^-_{\ell-1})^{a_{\ell-1}}(X^-_{\ell-3})^{a_{\ell-3}}\times\\
&\times(X^-_{\ell-4,\ell-3})^{t_{\ell-4,\ell-3}}\dots(X^-_{1,\ell-3})^{t_{1,\ell-3}}(X^-_{\ell-4})^{a_{\ell-4}}\times\\
&\times\dots(X^-_2)^{a_2}(X^-_{12})^{t_{12}}(X^-_1)^{a_1}\otimes v_0\,,\end{aligned} \tag{2.78}$$

where T denotes the set of summation variables $a_i, t_{ij}, s_{ij}, \tilde{t}_i, t_i, s_i, t$, all of which are nonnegative integers.

The derivation now proceeds as follows. We have to impose condition (2.58) with $\beta\to\gamma_1$, $v_s\to v_s^{\gamma_1,m}$. (Inequalities (2.77b) mean that no other conditions need to be imposed.) First we impose condition (2.58a). This restricts the linear combination to terms of weight $m\gamma_1$. In our parametrization these are the following ℓ conditions:

$$a_p = m-\sum_{i=1}^{p}\left((t_i+\tilde{t}_i+s_i)+\sum_{j=p}^{\ell-2}t_{ij}+\sum_{j=p+1}^{\ell-2}s_{ij}+2\sum_{1\le i<j\le p}s_{ij}\right),$$
$$1\le p\le\ell-3\,;$$

$$\begin{aligned}
a_{\ell-2} &= m - \left(t + \sum_{i=1}^{\ell-3}(t_i + \tilde{t}_i) + \sum_{i=1}^{\ell-3}(s_i + t_{i,\ell-2}) + 2 \sum_{1\le i<j\le \ell-2} s_{ij} \right) \\
a_{\ell-1} &= m - \left(t + \sum_{i=1}^{\ell-2} \tilde{t}_i + \sum_{i=1}^{\ell-3} s_i + \sum_{1\le i<j\le \ell-2} s_{ij} \right) \\
a_{\ell} &= m - \left(t + \sum_{i=1}^{\ell-2} t_i + \sum_{i=1}^{\ell-3} s_i + \sum_{1\le i<j\le \ell-2} s_{ij} \right).
\end{aligned} \tag{2.79}$$

This eliminates the summation in a_i in (2.78) and also restricts further the summation $t_{ij}, s_{ij}, \tilde{t}_i, t_i, s_i, t$ so that the a_i in (2.79) would be all nonnegative.

Next we impose condition (2.58b). These ℓ conditions produce ℓ recursive relations, which are too cumbersome and we omit them. Solving these relations fixes the coefficient $D_T^{\gamma_1,m}$ completely and we obtain:

$$\begin{aligned}
D_T^{\gamma_1,m} = \quad & D^\ell \, (-1)^{\sum\limits_{i\le j} s_{ij}} \frac{\prod\limits_{p=2}^{\ell-3} \frac{[\tilde{a}_p]!}{[a_p]!}}{[t]! \prod\limits_{j=2}^{\ell-2} [s_{1j}]![s_{j-1}]! \prod\limits_{j=1}^{\ell-2} [t_j]![\tilde{t}_j]! \prod\limits_{1\le i<j\le \ell-2} [t_{ij}]!} \times \\
& \times \frac{q^{\mathbf{A}} q^{(\Lambda+\rho, a_\ell \alpha_\ell + a_{\ell-1}\alpha_{\ell-1})}}{[m-2t-\sum\limits_{i=1}^{\ell-2}(t_i+\tilde{t}_i) - 2\sum\limits_{1\le i<j\le \ell-2} s_{ij} - 2\sum\limits_{i=1}^{\ell-3} s_i - \sum\limits_{i=1}^{\ell-3} t_{i,\ell-2}]!} \times \\
& \times \prod_{j=1}^{\ell-3} q^{a_j(\Lambda+\rho)(H^j)} \frac{\Gamma_q(\Lambda^j + j - a_j + t_{j-1,j})}{\Gamma_q(\Lambda^j + j + 1)} \times \\
& \times \frac{\Gamma_q(\Lambda_{\ell-1} + 1 - a_{\ell-1}) \Gamma_q(\Lambda_\ell + 1 - a_\ell)}{\Gamma_q(\Lambda_{\ell-1} + 2) \Gamma_q(\Lambda_\ell + 2)},
\end{aligned} \tag{2.80}$$

$$D^\ell \neq 0, \qquad \Lambda^r := (\Lambda, \beta^r), \quad \text{with} \quad \beta^r := \alpha_1 + \cdots + \alpha_r$$

where

$$\tilde{a}_p = m - \sum_{i=1}^{p} \big((t_i + \tilde{t}_i + s_i) + \sum_{j=p+1}^{\ell-2} (t_{ij} + s_{ij}) + 2 \sum_{1\le i<j\le p} s_{ij}\big), \quad 1 \le p \le \ell - 3$$

and the factor $\mathbf{A}$ is given by:

$$\begin{aligned}
\mathbf{A} = & \sum_{1\le i<j\le \ell-2} \left\{ t_{ij} \sum_{p=0}^{\ell-4} t^{p+j-1} + s_{ij} \sum_{p=0}^{\ell-4} s^{p+j-1} \right\} + \sum_{1\le i<j\le \ell-2} s_{ij}^2 + \sum_{i=1}^{\ell-3} s_i^2 + \\
& + \sum_{1\le i<j\le \ell-2} t_{ij}^2 - \\
& - \left((\ell-2) \sum_{1\le i<j\le \ell-2} t_{ij} + (\ell+1) \sum_{1\le i<j\le \ell-2} s_{ij} + \ell \sum_{i=1}^{\ell-3} s_i \right) m \\
& + \sum_{1\le i<j\le \ell-2} t_{ij} \sum_{i=1}^{\ell-3} (t_i + d_i) + \sum_{i=1}^{\ell-4} (t_i + d_i) \sum_{j=1}^{\ell-3} (t_j + d_j)
\end{aligned}$$

$$
\begin{aligned}
&+ \sum_{p=1}^{\ell-2} \left\{ t_p \left(\sum_{j=1}^{p} t_j + \sum_{1\le i<j\le \ell-2} s_{ij} + \sum_{i=1}^{\ell-3} s_i - (\ell-p)m \right) \right.\\
&\left. + \tilde{t}_p \left(\sum_{j=1}^{p} \tilde{t}_j + \sum_{1\le i<j\le \ell-2} s_{ij} + \sum_{i=1}^{\ell-3} s_i - (\ell-p)m \right) \right\}\\
&+ t\left(t + t_{\ell-2} + \tilde{t}_{\ell-2} + \sum_{i=1}^{\ell-3} s_i + \sum_{1\le i<j\le \ell-3} s_{ij} - 3m\right)\\
&+ \sum_{1\le i<j\le \ell-3} s_i s_j + (\ell-2) \sum_{1\le i<j\le \ell-2} s_{ij} \sum_{k=1}^{\ell-3} s_k\\
&+ \sum_{1\le i<j\le \ell-2} t_{ij} \left(\sum_{1\le i<j\le \ell-2} s_{ij} + \sum_{i=1}^{\ell-3} s_i \right)\\
&+ \sum_{1\le i<j\le \ell-2} (j-i)t_{ij} + \sum_{\substack{1\le i\le \ell-4\\ i<j}} (\ell-j+3)s_{ij} + 4s_{\ell-3,\ell-2}\\
&+ \sum_{i=1}^{\ell-3} (\ell-i)s_i + \sum_{i=1}^{\ell-2} (\ell-i-1)(t_i+\tilde{t}_i)
\end{aligned}
\tag{2.81}
$$

where $t^b := \sum_{k=j+1}^{\ell-3} t_{bk}$.

Finally, we explain how to obtain the singular vectors for the roots $\gamma_i, i > 1$ from the above formulae. For this one has to replace $\ell \to \ell - i + 1$, and then to shift the enumeration of the roots, namely, to replace $1, \ldots, \ell - i + 1$ by $i, \ldots, \ell$.

2.5.3.2 Relation between the Two Expressions for the Singular Vectors

Here we would like to present the relation between the expressions for the singular vectors in the PBW basis given in (2.80) and in the simple root vectors basis given in Section 2.4.2. Thus, we compare with formula (2.37).

The D-coefficients are given in term of the d-coefficients by the following formula:

$$
\begin{aligned}
D_T^{\gamma_1,m} &= \frac{\prod_{p=2}^{\ell-3} \frac{[\tilde{a}_p]!}{[a_p]!}}{[t]! \prod_{j=2}^{\ell-2} [s_{1j}]![s_{j-1}]! \prod_{j=1}^{\ell-2} [t_j]![\tilde{t}_j]! \prod_{1\le i<j\le \ell-2} [t_{ij}]!} \times\\
&\times \frac{(-1)^{\sum_{i=1}^{\ell} a_i} q^{\mathbf{A}}}{[m - 2t - \sum_{i=1}^{\ell-2}(t_i+\tilde{t}_i) - 2\sum_{1\le i<j\le \ell-2} s_{ij} - 2\sum_{i=1}^{\ell-3} s_i - \sum_{i=1}^{\ell-3} t_{i,\ell-2}]!} \times\\
&\times \sum_{k_1,k_2,\ldots,k_{\ell-1}} d_{k_1,k_2,\ldots,k_{\ell-1}} \prod_{p=1}^{\ell-3} \frac{[m-k_p]! q^{k_p(a_p - t_{p-1,p})}}{[a_p - t_{p-1,r} - k_p]!} \times\\
&\times \frac{[m-k_{\ell-1}]!}{[a_{\ell-1}-k_{\ell-1}]!} \frac{[m-k_{\ell-2}]!}{[a_\ell - k_{\ell-2}]!} q^{(k_{\ell-1}a_{\ell-1} + k_{\ell-2}a_\ell)}
\end{aligned}
\tag{2.82}
$$

where $0 \le k_p \le a_p$, $0 \le p \le \ell - 3$, $k_{\ell-1} \le a_{\ell-1}$ and $k_{\ell-2} \le a_\ell$.

To prove the above one can use the formula (following from (1.20) and (2.59)):

$$\frac{V^m\,U^n}{[m]![n]!} = \sum_{0\le p\le min(m,n)} (-1)^p\,q^{(m-p)(n-p)+p}\,\frac{U^{n-p}\,W^p\,V^{m-p}}{[n-p]![p]![m-p]!} \tag{2.83}$$

where the triples U, V, W are given as follows: as W runs over the vectors defined in (2.59), U, V run over the pairs which appear on the corresponding RHS; for example, if $W = X^-_{ij}$ then either $(U,V) = (X^-_i, X^-_{i+1,j})$ or $(U,V) = (X^-_{i,j-1}, X^-_j)$.

2.6 Singular Vectors for Nonstraight Roots

2.6.1 Bernstein–Gel'fand–Gel'fand Resolution

Let us say that condition (2.2) is *almost fulfilled* if it is satisfied for $m = 0$, that is, when $\lambda+\rho$ is on the walls of the dominant Weyl chamber [109]. First we discuss the situation when condition (2.2) is fulfilled for β and is fulfilled or almost fulfilled for any subroot of β. Consider the explicit expansion of $\beta \in \Delta^+$ into simple roots $\beta = \sum_{k=1}^{\ell} n_k\alpha_k$, with $n_k \in \mathbb{Z}_+$, and define $J_\beta \equiv \{k|n_k \neq 0\}$.

In this situation we can give the formula for the singular vector for *arbitrary positive roots*, that is, not only for straight roots. We have:

Proposition 2 ([208]). *Let $\mathcal{G}$ be a complex simple Lie algebra. Let $\lambda \in \mathcal{H}^*$, $\beta \in \Delta^+$ and $m \in \mathbb{N}$ be such that* (2.2) *is fulfilled. Let also:*

$$[(\lambda+\rho,\alpha_k^\vee) - m_k]_{q_k} = 0, \quad k \in J_\beta, \quad m_k \in \mathbb{Z}_+. \tag{2.84}$$

Assume also the presentation (2.35) *of β. Then the singular vector of the Verma module V^λ over $U_q(\mathcal{G})$ corresponding to β, m is given by:*

$$\begin{aligned} v_\lambda^{\beta,m} &= c_{\beta,m}(X^-_{i_1})^{mn_{i_1}-\tilde m_1}\cdots(X^-_{i_u})^{mn_{i_u}-\tilde m_u}(X^-_v)^{mn_v}(X^-_{i_u})^{\tilde m_u}\cdots \\ &\quad\cdots(X^-_{i_1})^{\tilde m_1}\otimes v_0, \end{aligned} \tag{2.85a}$$

$$\tilde m_k \equiv (s_{i_{k-1}}\cdots s_{i_1}(\lambda+\rho),\alpha^\vee_{i_k}) \in \mathbb{Z}_+. \tag{2.85b}$$

$$\begin{aligned} \tilde m_k &= (\lambda+\rho-\sum_{t=1}^{k-1}\tilde m_t\alpha_{i_t},\alpha^\vee_{i_k}),\\ \tilde m_{u+1} &\equiv (s_v s_{i_u}\cdots s_{i_1}(\lambda+\rho),\alpha^\vee_{i_{u+1}}) = mn_v,\\ \tilde m'_k &\equiv (s_{i_{k+1}}\cdots s_{i_u}s_v s_{i_u}\cdots s_{i_1}(\lambda+\rho),\alpha^\vee_{i_k}) =\\ &= (\lambda+\rho-\sum_{t=1}^{u}\tilde m_t\alpha_{i_t}-\tilde m_{u+1}\alpha_v-\sum_{t=k+1}^{u}\tilde m'_t\alpha_{i_t},\alpha^\vee_{i_k}) =\\ &= mn_{i_k}-\tilde m_k \in \mathbb{Z}_+. \end{aligned}$$

◊

Note that $\sum_{k\in J_\beta} m_k \in \mathbb{N}$, and, for example, $\tilde{m}_1 = m_{i_1}$, $\tilde{m}_2 = m_{i_2} - a_{i_2 i_1} m_{i_1}$. Note also that (2.85) follows from the explicit formulae in the previous subsection after suitably renormalizing the coefficients in formula (2.37). Then because of (2.84) *all terms but one* in (2.37) will vanish and we obtain the monomial expression in (2.85a).

We now state the main result of this subsection.

Proposition 3 ([208]). *Let $\mathcal{G}$ be a complex simple Lie algebra. Let $\lambda \in \Gamma_+$. Then all singular vectors of the Verma module V^λ over $U_q(\mathcal{G})$ when q is not a root of 1 are parametrized by the Weyl group W; that is, their weights are given by $w \cdot \lambda$, $e \neq w \in W$. Furthermore, for a fixed $w = s_{i_r} \dots s_{i_1}$ written in reduced form, the corresponding singular vector $v_s(w)$ is given explicitly by:*

$$v_s(w) = (X^-_{i_r})^{\tilde{m}_r} \dots (X^-_{i_1})^{\tilde{m}_1} \otimes v_0, \tag{2.86}$$

where $\tilde{m}_k$ is defined for $k = 1, \dots r$ as in (2.85b). ◊

Proof. It is almost obvious that (2.86) is a singular vector. For fixed $k = 1, \dots r$ formula (2.86) means that condition (2.2) is fulfilled with respect to the root α_{i_k} in a Verma module V^{λ_k} with highest weight shifted by the Weyl dot reflection $\lambda_k = s_{i_{k-1}} \cdots s_{i_1} \cdot \lambda$. For this we have to prove that $\tilde{m}_k \in \mathbb{N}$. Actually from the previous proposition we know that $\tilde{m}_k \in \mathbb{Z}_+$. Suppose now that for some k we have $\tilde{m}_k = 0$. This means that $s_{i_k} s_{i_{k-1}} \cdots s_{i_1} \cdot \lambda = s_{i_{k-1}} \cdots s_{i_1} \cdot \lambda$, which would contradict the fact [109] that the Weyl group acts transitively on the Weyl chambers. Finally we have to prove that (2.86) provides all singular vectors of V^λ. For this we use the fact that when q is not a root of 1 the structure of the Verma module V^λ is the same as for $q = 1$ [441, 531]. In the case $q = 1$ and $\lambda \in \Gamma_+$ the submodule structure of V^λ is completely described by the Weyl group; namely, there is a one-to-one correspondence between the submodules of V^λ and the elements $w \in W$, $w \neq e$. ∎

Corollary: Let $\mathcal{G}$ be a complex simple Lie algebra. Let $\lambda + \rho \in \Gamma_+$. Then formula (2.86) describes a singular vector of the Verma module V^λ over $U_q(\mathcal{G})$ when q is not a root of 1. We also have $\tilde{m}_k \in \mathbb{Z}_+$. ◊

Remark 2.1. The above corollary follows from either of the propositions in this section. Note that if $\lambda+\rho \in \Gamma_+$ and $\lambda \notin \Gamma_+$, that is, when $\lambda+\rho$ is on the walls of the dominant Weyl chamber, the submodule structure of V^λ is not completely described by the singular vectors in (2.86), and furthermore singular vectors corresponding to different elements of W may coincide (the action of W being not transitive). ◊

The results presented so far provide an explicit realization of the **Bernstein–Gel'fand–Gel'fand resolution** [96]. In the multiplet classification approach [193, 194, 196–198], the submodule structure of V^λ for λ integral dominant was described by the maximal multiplet $\mathcal{M}^\lambda$, the elements of which are Verma modules $V^{\lambda'}$ which are in one-to-one correspondence with the elements $w \in W$, namely, $\lambda' = w \cdot \lambda$. Let us define the following submodules of V^λ

$$\mathcal{C}_n^\lambda \equiv \underset{w \in W, \ell(w) = n}{\oplus} V^{w\cdot\lambda}. \tag{2.87}$$

Note that $\mathcal{C}_0^\lambda = V^\lambda$. Recall [109] that the maximal length of an element of W is equal to the number of positive roots, that is, $\ell(w_0) = |\Delta^+|$, where w_0 is the longest element of W. By [28] there exists a resolution of L_λ for $\lambda \in \Gamma_+$ in terms of the above submodules, that is, an exact sequence:

$$0 \leftarrow L_\lambda \leftarrow V^\lambda \leftarrow \mathcal{C}_1^\lambda \leftarrow \cdots \leftarrow \mathcal{C}_{\ell(w_0)}^\lambda \leftarrow 0 \tag{2.88}$$

The map $V^\lambda \to L_\lambda$ is the natural surjection, while for fixed $n = 1, \dots \ell(w_0)$ the map $d^n : \mathcal{C}_n^\lambda \to \mathcal{C}_{n-1}^\lambda$ is a collection of the maps embedding the components $V^{w\cdot\lambda}$, $\ell(w) = n$, of $\mathcal{C}_n^\lambda$ into the components $V^{w\cdot\lambda}$, $\ell(w) = n-1$, of $\mathcal{C}_{n-1}^\lambda$. One has to check $d^{n-1} \circ d^n = 0$. In [28] this was proved by using general properties of the Weyl group and the uniqueness of the embedding between two Verma modules. (The BGG resolution in a similar context was considered in [110] with explicit expressions in the A_2 case using singular vectors in the Poincaré—Birkhoff—Witt basis.)

Here we would like to present an explicit realization of the above uniqueness using our results on the singular vectors. The main ingredient is the commutativity of certain embedding diagrams which involve only subalgebras of rank 2. The reason is that any multiplet of Verma modules, in particular, the maximal one, may be viewed as consisting of submultiplets containing four and six members (for the simply laced algebras), also with eight members (for the nonsimply laced algebras) and with 12 members (for G_2). More explicitly, let $V \equiv V^{\lambda'}$, $\lambda' = w \cdot \lambda$; for some $w \in W$ be such that condition (2.2) is fulfilled for λ' for at least two simple roots; say α_p and α_r, $p \neq r$. Then V is contained in a submultiplet with four members if $a_{rp}a_{pr} = 0$ with weights $V^{w\cdot\lambda'}$, $w = \{e, s_p, s_r, s_p s_r = s_r s_p\} \cong W(A_1 \oplus A_1)$; with six members if $a_{rp}a_{pr} = 1$ with weights $V^{w\cdot\lambda'}$, $w \in \{e, s_p, s_r, s_p s_r, s_r s_p, s_p s_r s_p = s_r s_p s_r\} \cong W(A_2)$; with eight members if $a_{rp}a_{pr} = 2$ with weights $V^{w\cdot\lambda'}$, $w \in \{e, s_p, s_r, s_p s_r, s_r s_p, s_p s_r s_p, s_r s_p s_r, s_p s_r s_p s_r = s_r s_p s_r s_p\} \cong W(B_2)$; with twelve members if $a_{rp}a_{pr} = 3$ with weights $V^{w\cdot\lambda'}$, $W(G_2) \cong \{e, s_p, s_r, s_p s_r, s_r s_p, s_p s_r s_p, s_r s_p s_r, s_p s_r s_p s_r, s_r s_p s_r s_p, s_r s_p s_r s_p s_r, s_p s_r s_p s_r s_p, s_p s_r s_p s_r s_p s_r = s_r s_p s_r s_p s_r s_p\}$.

Remark 2.2. Naturally, if (2.2) is fulfilled for $n > 2$ simple roots, then V will play the same role in $\binom{n}{2}$ submultiplets of the type just described. If (2.2) is fulfilled only with respect to 0, 1, simple roots than V is a member of such a multiplet with weight $\lambda' = w_0 \cdot \lambda$. respectively, $\lambda' = w \cdot \lambda, w \neq e, w_0$, where w_0 is the longest element of the rank two subalgebras used above. Thus no new submultiplets of the type described above arise. ◊

We have to establish commutativity of the embedding diagrams describing the above submultiplets. In the four-member submultiplet this is trivial since $[X_p^-, X_r^-] = 0$. For the six-member submultiplet we use the case A_2 for $\beta = \alpha_1+\alpha_2$, $m_j = (\lambda+\rho, \alpha_j^\vee) \in \mathbb{Z}_+$, $m = m_1 + m_2 \in \mathbb{N}$. The singular vector is given by:

$$\begin{aligned} v^{m,\beta} &= c_1(X_1^-)^{m_2}(X_2^-)^{m}(X_1^-)^{m_1} \otimes v_0 = \\ &= c_2(X_2^-)^{m_1}(X_1^-)^{m}(X_2^-)^{m_2} \otimes v_0 = \qquad (2.89a)\\ &= (X_2^-)^{m_1} \sum_{k=0}^{m_2} a_k^1(X_1^-)^{m_2-k}(X_2^-)^{m_2}(X_1^-)^{k+m_1} \otimes v_0 = \\ &= (X_1^-)^{m_2} \sum_{k=0}^{m_1} a_k^0(X_2^-)^{m_1-k}(X_1^-)^{m_1}(X_2^-)^{k+m_2} \otimes v_0, \qquad (2.89b) \end{aligned}$$

where (2.89a) gives the two forms of (2.85a) in this case, and a_k^1, a_k^0, respectively, is given by (2.37) for $u = 1$, $(i_1, v) = (1,2), (2,1)$, also (2.42), with $\binom{m}{k_1}_q$ replaced by $\binom{m_i}{k}_q$, $i = 1,2$, respectively, and by λ replaced by $\lambda - m\alpha_1 = s_1 \cdot \alpha_1$, $\lambda - m\alpha_2 = s_2 \cdot \alpha_2$, respectively, that is, with $\lambda(H_i) + 1$ replaced by $-m_i$, $i = 1,2$, respectively. The four expressions in (2.89) are used to prove commutativity of certain embedding diagrams, in particular, the hexagon diagram of $U_q(sl(3,\mathbb{C}))$ [198] (or, for $q = 1$, the hexagon diagram of $sl(3,\mathbb{C})$ [195]).

For the eight- and twelve-member submultiplets we need the following:

Lemma: Assume the above setting and also that α_p is shorter than α_r, thus $a_{rp} = -1$, and set $a_{pr} = -1-\varepsilon$, $\varepsilon = 1, 2$, let $m_k \equiv (\lambda' + \rho, \alpha_k^\vee) \in \mathbb{N}$. Then we have:

$$\tilde{c}_1\,(X_r^-)^{m_p+m_r}\,(X_p^-)^{m_p} \otimes v_0 = \mathscr{P}^{\beta,m_p}_{s_r\cdot\lambda'}\,(X_r^-)^{m_r} \otimes v_0\,, \qquad (2.90a)$$

$$\tilde{c}_2\,(X_p^-)^{m_p+(1+\varepsilon)m_r}\,(X_r^-)^{m_r} \otimes v_0 = \mathscr{P}^{\beta',m_r}_{s_p\cdot\lambda'}\,(X_p^-)^{m_p} \otimes v_0\,, \qquad (2.90b)$$

$$(X_p^-)^{m_p+2m_r}\,\mathscr{P}^{\beta,m_p}_{s_r\cdot\lambda'} \otimes v_0 = \mathscr{P}^{\beta,m_p}_{s_ps_r\cdot\lambda'}\,(X_p^-)^{m_p+2m_r} \otimes v_0\,,\ \varepsilon = 1 \qquad (2.90c)$$

$$(X_r^-)^{m_p+m_r}\,\mathscr{P}^{\beta',m_r}_{s_p\cdot\lambda'} \otimes v_0 = \mathscr{P}^{\beta',m_r}_{s_rs_p\cdot\lambda'}\,(X_r^-)^{m_p+m_r} \otimes v_0\,,\ \varepsilon = 1 \qquad (2.90d)$$

$$\tilde{c}_3\,(X_p^-)^{m_p}\,(X_r^-)^{m_p+m_r} \otimes v_0 = (X_r^-)^{m_r}\,\mathscr{P}^{\beta,m_p}_{(s_ps_r)^\varepsilon\cdot\lambda'} \otimes v_0\,, \qquad (2.90e)$$

$$\tilde{c}_4\,(X_r^-)^{m_r}\,(X_p^-)^{m_p+(1+\varepsilon)m_r} \otimes v_0 = (X_p^-)^{m_p}\,\mathscr{P}^{\beta',m_r}_{(s_rs_p)^\varepsilon\cdot\lambda'} \otimes v_0 \qquad (2.90f)$$

where $\tilde{c}_k$ are nonzero constants, $\beta = \alpha_p + \alpha_r = s_r(\alpha_p)$, $\beta' = (1+\varepsilon)\alpha_p + \alpha_r = s_p(\alpha_r)$, $\mathscr{P}^{\beta,m_p}_{\lambda''}$, $\mathscr{P}^{\beta',m_r}_{\lambda''}$ is given by (11) for $j = 1$, and $\{i_1; i\} = \{r; p\}, \{p; r\}$, respectively, $n_{i_1} = 1, 1+\varepsilon$, $q_{i_1} = q^{1+\varepsilon}, q$, respectively; the weight of the highest-weight vector v_0 is λ' in (2.90a,b), $s_r \cdot \lambda'$ in (2.90c), $s_p \cdot \lambda'$ in (2.90d), $(s_rs_p)^\varepsilon \cdot \lambda'$ in (2.90e), $(s_ps_r)^\varepsilon \cdot \lambda'$ in (2.90f).

Proof. Direct calculation using Serre relations and formulae of the type of (2.40). ∎

The above lemma ensures the desired property $d^{n-1} \circ d^n = 0$ by just choosing properly the constants in (2.90). Only the case G_2 requires further work since we have to establish the following relations:

$$(X_p^-)^{2m_p+3m_r} \mathscr{P}^{\beta,m_p}_{s_r\cdot\Lambda'} \otimes v_0 = \mathscr{P}^{\gamma,m_p}_{s_ps_r\cdot\Lambda'} (X_p^-)^{m_p+3m_r} \otimes v_0, \tag{2.91a}$$

$$(X_r^-)^{m_p+2m_r} \mathscr{P}^{\beta',m_r}_{s_p\cdot\Lambda'} \otimes v_0 = \mathscr{P}^{\gamma',m_r}_{s_rs_p\cdot\Lambda'} (X_r^-)^{m_p+m_r} \otimes v_0, \tag{2.91b}$$

$$(X_p^-)^{2m_p+3m_r} \mathscr{P}^{\gamma',m_r}_{s_rs_p\cdot\Lambda'} \otimes v_0 = \mathscr{P}^{\gamma',m_r}_{s_ps_rs_p\cdot\Lambda'} (X_p^-)^{2m_p+3m_r} \otimes v_0 \tag{2.91c}$$

$$(X_r^-)^{m_p+2m_r} \mathscr{P}^{\gamma,m_p}_{s_ps_r\cdot\Lambda'} \otimes v_0 = \mathscr{P}^{\gamma,m_p}_{s_rs_ps_r\cdot\Lambda'} (X_r^-)^{m_p+2m_r} \otimes v_0 \tag{2.91d}$$

$$(X_p^-)^{m_p+3m_r} \mathscr{P}^{\gamma,m_p}_{s_rs_ps_r\cdot\Lambda'} \otimes v_0 = \mathscr{P}^{\beta,m_p}_{(s_ps_r)^2\cdot\Lambda'} (X_p^-)^{2m_p+3m_r} \otimes v_0 \tag{2.91e}$$

$$(X_r^-)^{m_p+m_r} \mathscr{P}^{\gamma',m_r}_{s_ps_rs_p\cdot\Lambda'} \otimes v_0 = \mathscr{P}^{\beta',m_r}_{(s_rs_p)^2\cdot\Lambda'} (X_r^-)^{m_p+2m_r} \otimes v_0 \tag{2.91f}$$

where $\gamma = 2\alpha_p + \alpha_r = s_ps_r(\alpha_p)$, $\gamma' = 3\alpha_p + 2\alpha_r = s_rs_p(\alpha_r)$, the weight of the highest-weight vector v_0 is $s_r\cdot\Lambda'$ in (2.91a), $s_p\cdot\Lambda'$ in (2.91b), $s_rs_p\cdot\Lambda'$ in (2.91c), $s_ps_r\cdot\Lambda'$ in (2.91d), $s_rs_ps_r\cdot\Lambda'$ in (2.91e), $s_ps_rs_p\cdot\Lambda'$ in (2.91f), and we have the following *conjecture* for the singular vectors corresponding to the nonstraight roots γ, γ':

$$v_\Lambda^{\gamma,m} = \mathscr{P}_\Lambda^{\gamma,m} \otimes v_0 = \sum_{k_1=0}^{m}\sum_{k_2=0}^{m} g_{k_1k_2} (X_p^-)^{m-k_1}(X_r^-)^{m-k_2}(X_p^-)^m(X_r^-)^{k_2} \times$$
$$\times (X_p^-)^{k_1} \otimes v_0, \tag{2.92a}$$

$$g_{k_1k_2} = (-1)^{k_1+k_2} g \binom{m}{k_1}_q \binom{m}{k_2}_{q^3} \frac{[(\Lambda+\rho)(\tilde H_{i_1})]_q}{[(\Lambda+\rho)(\tilde H_{i_1}) - k_1]_q} \times$$
$$\times \frac{[(\Lambda+\rho)(H_r)]_{q^3}}{[(\Lambda+\rho)(H_r) - k_2]_{q^3}}, \tag{2.92b}$$

$$v_\Lambda^{\gamma',m} = \mathscr{P}_\Lambda^{\gamma',m} \otimes v_0 = \sum_{k_1=0}^{m}\sum_{k_2=0}^{3m} g_{k_1k_2} (X_r^-)^{m-k_1}(X_p^-)^{3m-k_2}(X_r^-)^m \times$$
$$\times (X_p^-)^{k_2}(X_r^-)^{k_1} \otimes v_0, \tag{2.93a}$$

$$g'_{k_1k_2} = (-1)^{k_1+k_2} g' \binom{m}{k_1}_{q^3} \binom{3m}{k_2}_q \frac{[(\Lambda+\rho)(\tilde H_{i_1})]_{q^3}}{[(\Lambda+\rho)(\tilde H_{i_1}) - k_1]_{q^3}} \times$$
$$\times \frac{[(\Lambda+\rho)(H_p)]_q}{[(\Lambda+\rho)(H_p) - k_2]_q}. \tag{2.93b}$$

For λ obeying the assumptions of Proposition 2, the above polynomials should reduce to monomials as in (2.85); this is one justification for the above conjecture.
Example A_3

In the situation when (2.2) is almost fulfilled there are also available mixed forms of the singular vectors. We considered the example A_2 above. Analogously let us have for A_3

$$[(\lambda+\rho,\alpha_j^\vee)-m_j]_q=0,\quad j=1,2,3,\quad m_j\in\mathbb{Z}_+,\quad m=m_1+m_2+m_3\in\mathbb{N},\tag{2.94}$$

Denoting $m_{ij}=m_i+m_j$ we write down the reduction of formula (2.37):

$$\begin{aligned} v_s^{m,\beta} &= c_1'(X_1^-)^{m_{23}}(X_2^-)^{m_3}(X_3^-)^{m}(X_2^-)^{m_{12}}(X_1^-)^{m_1}\otimes v_0 = \\ &= c_2'(X_1^-)^{m_{23}}(X_3^-)^{m_{12}}(X_2^-)^{m}(X_3^-)^{m_3}(X_1^-)^{m_1}\otimes v_0 = \\ &= c_3'(X_3^-)^{m_{12}}(X_2^-)^{m_1}(X_1^-)^{m}(X_2^-)^{m_{23}}(X_3^-)^{m_3}\otimes v_0, \end{aligned}\tag{2.95}$$

and several other expressions which analogously to (2.89b) use the polynomials corresponding to roots which are the sum of two simple roots (and some expressions which use the trivial commutativity $[X_1^-,X_3^-]=0$).

Remark 2.3. Most results above may be extended to affine Lie algebras. Consider, for example, $U_q(A_1^{(1)})$ and let α_1,α_2 be the simple roots of $A_1^{(1)}$, so that $(\alpha_1,\alpha_1)=(\alpha_2,\alpha_2)=2=-(\alpha_1,\alpha_2)$. There are two nonsimple straight roots: $\beta_{ij}=\alpha_i+2\alpha_j=s_j(\alpha_i)$ for $(i,j)=(1,2),(2,1)$. The singular vector for β_{12} is given by formula for B_2 and for β_{21} by the interchange of indices 1 and 2. ◊

2.6.2 Case of $U_q(D_\ell)$ in PBW Basis

The nonstraight roots of D_ℓ are given in (2.74). We shall also write them as:

$$\begin{aligned} \gamma_{rp} &= \sum_{j=r}^{\ell} n_j\alpha_j, \qquad 1\le r<p\le \ell-2 \\ n_j &= \begin{cases} 1 & for \quad r\le j<p \\ 2 & for \quad p\le j\le \ell-2 \\ 1 & for \quad j=\ell-1,\ell \end{cases} \end{aligned}\tag{2.96}$$

Like in the case of straight toots we could use the fact that every root γ_{rp} can be treated as the root γ_{1p} of a $D_{\ell-r+1}$ subalgebra of D_ℓ. This means that it would be enough to give the formula for the singular vector corresponding to the roots γ_{1p}. However, we shall not do this for these roots, since anyway it is not reduced to one root.

Let us have condition (2.2) fulfilled for γ_{rp}, but not for any of its subroots. The singular vectors corresponding to these roots are given by:

$$\begin{aligned}
v_s^{\gamma_{rp},m} &= \sum_T D_T^{\gamma_{rp},m} (X_{\ell-2}^-)^{2m-b_{\ell-2}} (X_{\ell-3,\ell-2}^-)^{t_{\ell-3,\ell-2}} \dots (X_{r,\ell-2}^-)^{t_{r,\ell-2}} \times \\
&\times (\tilde{Y}_{\ell-2}^-)^{\tilde{t}_{\ell-2}} (Y_{\ell-2}^-)^{t_{\ell-2}} (Z_{\ell-3,\ell-2}^-)^{s_{\ell-3,\ell-2}} (Z_{\ell-4,\ell-2}^-)^{s_{\ell-4,\ell-2}} \dots \times \\
&\times (Z_{r,\ell-2}^-)^{s_{r,\ell-2}} (\tilde{Y}_{\ell-3}^-)^{\tilde{t}_{\ell-3}} (Y_{\ell-3}^-)^{t_{\ell-3}} \times \\
&\times (Z_{\ell-4,\ell-3}^-)^{s_{\ell-4,\ell-3}} \dots (Z_{r,\ell-3}^-)^{s_{r,\ell-3}} \dots (\tilde{Y}_r^-)^{\tilde{t}_r} (Y_r^-)^{t_r} (Y_0^-)^t \times \\
&\times (Z_{\ell-3}^-)^{s_{\ell-3}} \dots (Z_r^-)^{s_r} (X_\ell^-)^{m-b_\ell} (X_{\ell-1}^-)^{m-b_{\ell-1}} (X_{\ell-3}^-)^{mn_{\ell-3}-b_{\ell-3}} \times \\
&\times (X_{\ell-4,\ell-3}^-)^{t_{\ell-4,\ell-3}} \dots (X_{r,\ell-3}^-)^{t_{r,\ell-3}} (X_{\ell-4}^-)^{mn_{\ell-4}-b_{\ell-4}} \times \\
&\times \dots (X_{r+1}^-)^{mn_{r+1}-b_{r+1}} (X_{r,r+1}^-)^{t_{r,r+1}} (X_r^-)^{m-b_r} \otimes v_0
\end{aligned} \tag{2.97}$$

In (2.97) we have already imposed conditions (2.58a) and the summation is only over those elements of the PBW basis which have the weight $m\gamma_{rp}$. Further we impose (2.58b) the procedure being as in the case of the straight roots. Thus, the coefficients $D_T^{\gamma_{rp},m}$ are found to be:

$$\begin{aligned}
D_T^{\gamma_{rp},m} &= D^{ns} (-1)^{\sum_{r\le j} s_{ij}} \frac{\prod_{s=r+1}^{\ell-3} \frac{[mn_s-\tilde{b}_s]!}{[mn_s-b_s]!}}{[t]! \prod_{j=r+1}^{\ell-2} [s_{rj}]![s_{j-1}]! \prod_{j=r}^{\ell-2} [t_j]![\tilde{t}_j]! \prod_{r\le i<j\le \ell-2} [t_{ij}]!} \times \\
&\times \frac{q^{\mathbf{A}^{ns}} q^{(\Lambda+\rho,b_\ell\alpha_\ell+b_{\ell-1}\alpha_{\ell-1})}}{[2m-2t-\sum_{i=r}^{\ell-2}(t_i+\tilde{t}_i) - 2\sum_{r\le i<j\le \ell-2} s_{ij} - 2\sum_{i=r}^{\ell-3} s_i - \sum_{i=r}^{\ell-3} t_{i,\ell-2}]!} \times \\
&\times \prod_{j=r}^{\ell-3} q^{(mn_j-b_j\Lambda'^j)} \frac{\Gamma_q(\Lambda'^j - mn_j + b_j + t_{j-1,j})}{\Gamma_q(\Lambda^j+1)} \times \\
&\times \frac{\Gamma_q(\Lambda_{\ell-1}+1-m+b_{\ell-1})\Gamma_q(\Lambda_\ell+1-m+b_\ell)}{\Gamma_q(\Lambda_{\ell-1}+2)\Gamma_q(\Lambda_\ell+2)}, \\
\Lambda'^j &:= \sum_{i=r}^{j} n_i(\Lambda_i+1), \quad D^{ns} \neq 0
\end{aligned} \tag{2.98}$$

where we have set for $r \le p \le \ell-3$:

$$\begin{aligned}
\tilde{b}_p &= \sum_{i=r}^{p} \left((t_i+\tilde{t}_i+s_i) + \sum_{j=p+1}^{\ell-2}(t_{ij}+s_{ij}) + 2\sum_{r\le i<j\le p} s_{ij}\right), \\
b_p &= \sum_{i=r}^{p} \left((t_i+\tilde{t}_i+s_i) + \sum_{j=p}^{\ell-2} t_{ij} + \sum_{j=p+1}^{\ell-2} s_{ij} + 2\sum_{1\le i<j\le p} s_{ij}\right),
\end{aligned}$$

$$
\begin{aligned}
b_{\ell-2} &= t + \sum_{i=r}^{\ell-3}(t_i + \tilde{t}_i) + \sum_{i=r}^{\ell-3}(s_i + t_{i,\ell-2}) + 2 \sum_{r\le i<j\le \ell-2} s_{ij}\,,\\
b_{\ell-1} &= t + \sum_{i=r}^{\ell-2} \tilde{t}_i + \sum_{i=r}^{\ell-3} s_i + \sum_{r\le i<j\le \ell-2} s_{ij}\,,\\
b_{\ell} &= t + \sum_{i=r}^{\ell-2} t_i + \sum_{i=r}^{\ell-3} s_i + \sum_{r\le i\le<j\le \ell-2} s_{ij}
\end{aligned}
\tag{2.99}
$$

2.6.3 Case of $U_q(D_\ell)$ in the Simple Roots Basis

The singular vectors corresponding to the nonstraight roots, γ_{rp}, $1 \le r < p \le \ell - 2$, in the simple root basis are given by:

$$
\begin{aligned}
v^{\gamma_{rp},m} &= \sum_{k_r=0}^{m} \sum_{k_{r+1}=0}^{mn_{r+1}} \cdots \sum_{k_{\ell-1}=0}^{m} d_{k_1\ldots k_{\ell-1}} (X_r^-)^{m-k_r} (X_{r+1}^-)^{mn_{r+1}-k_{r+1}} \ldots \times\\
&\times (X_{\ell-3}^-)^{2m-k_{\ell-3}} (X_{\ell-1}^-)^{m-k_{\ell-1}} (X_\ell^-)^{m-k_{\ell-2}} (X_{\ell-2}^-)^{2m} (X_\ell^-)^{k_{\ell-2}} \times\\
&\times (X_{\ell-1}^-)^{k_{\ell-1}} (X_{\ell-3}^-)^{k_{\ell-3}} \ldots (X_r^-)^{k_r} \otimes v_0 .
\end{aligned}
\tag{2.100}
$$

The coefficients d were not given in [206], but now using the PBW expression (2.97) for $v^{\gamma_{rp},m}$ we find that they are given by the following formula:

$$
\begin{aligned}
d_{k_r,\ldots,k_{\ell-1}} &= (-1)^{k_r+\cdots+k_{\ell-1}} \times \sum_{\substack{mn_r - b_r \le k_r\\ \vdots \\ mn_{\ell-3}-b_{\ell-3}\le k_{\ell-3}}} \; \sum_{\substack{m-b_{\ell-1}\le k_{\ell-1}\\ m-b_\ell \le k_{\ell-2}}} D^{ns} \times\\
&\times \prod_{j=r}^{\ell-3} \frac{q^{(mn_j-b_j)(1-k_j)-k_j}[mn_j - b_j]!}{[mn_j - k_j]![mn_j - \tau b_j]![k_j - mn_j + b_j]!} \times\\
&\times \frac{q^{(m-b_\ell)(1-k_{\ell-2})-k_{\ell-2}}}{[m-b_\ell]![k_{\ell-2} - m - b_\ell]![m-k_{\ell-2}]!} \times\\
&\times \frac{q^{(m-b_{\ell-1})(1-k_{\ell-1})-k_{\ell-1}}}{[m-b_{\ell-1}]![k_{\ell-1} - m - b_{\ell-1}]![m-k_{\ell-1}]!} \times\\
&\times [2m - 2t - \sum_{i=r}^{\ell-2}(t_i + \tilde{t}_i) - 2 \sum_{r\le i<j\le \ell-2} s_{ij} -\\
&- 2\sum_{i=r}^{\ell-3} s_i - \sum_{i=r}^{\ell-3} t_{i,\ell-2}]! \times\\
&\times [t]! \prod_{j=r+1}^{\ell-2} [s_{rj}]![s_{j-1}]! \prod_{j=r}^{\ell-2} [t_j]![\tilde{t}_j]! \prod_{r\le i<j\le \ell-2} [t_{ij}]! q^{-A^{ns}} .
\end{aligned}
\tag{2.101}
$$

or more explicitly:

$$\begin{aligned}
d_{k_1\ldots k_{\ell-1}} &= d^{ns}(-1)^{k_r+\cdots+k_{\ell-1}} \binom{mn_r}{k_r}_q \cdots \binom{mn_{\ell-1}}{k_{\ell-1}}_q \times \\
&\times \frac{[(\Lambda+\rho,\beta^{r,r})]_q}{[(\Lambda+\rho,\beta^{r,r})-k_r]_q} \cdots \frac{[(\Lambda+\rho,\beta^{r,\ell-3})]_q}{[(\Lambda+\rho,\beta^{r,\ell-3})-k_{\ell-3}]_q} \times \\
&\times \frac{[(\Lambda+\rho,\alpha_\ell)]_q}{[(\Lambda+\rho,\alpha_\ell)-k_{\ell-2}]_q} \frac{[(\Lambda+\rho,\alpha_{\ell-1})]_q}{[(\Lambda+\rho,\alpha_{\ell-1})-k_{\ell-1}]_q} \\
&= d^{ns}(-1)^{k_r+\cdots+k_{\ell-1}} \binom{mn_r}{k_r}_q \cdots \binom{mn_{\ell-1}}{k_{\ell-1}}_q \times \\
&\times \frac{[\Lambda'^r+n^r]_q}{[\Lambda'^r+n^r-k_1]_q} \cdots \frac{[\Lambda'^{\ell-3}+n^{\ell-3}]_q}{[\Lambda'^{\ell-3}+n^{\ell-3}-k_{\ell-3}]_q} \times \\
&\times \frac{[\Lambda_\ell+1]_q}{[\Lambda_\ell+1-k_{\ell-2}]_q} \frac{[\Lambda_{\ell-1}+1]_q}{[\Lambda_{\ell-1}+1-k_{\ell-1}]_q}, \qquad d^{ns} \neq 0 \\
\beta^{r,j} &:= \sum_{i=r}^{j} n_i\alpha_i, \quad \Lambda'^j = (\Lambda,\beta^{r,j}), \quad n^j := \sum_{i=r}^{j} n_i.
\end{aligned} \tag{2.102}$$

In the derivation of these formulae one can use (2.70).

2.7 Representations at Roots of Unity

2.7.1 Generalities

If the deformation parameter q *is a root of unity*, the representation theory of $U_q(\mathscr{G})$ differs very much from the generic case (cf. e. g., [175, 198, 442]).

We start with the case of the simple roots. Let $\beta = \alpha_j$ and we try the same expression (2.9) for the singular vector as in the case $q = 1$ or generic q:

$$v_s = (X_j^-)^m \otimes v_0 . \tag{2.103}$$

We obtain using (1.19):

$$\begin{aligned}
[X_j^+,(X_j^-)^m] &= \sum_{k=0}^{m-1} (X_j^-)^{m-1-k}[H_j](X_j^-)^k = (X_j^-)^{m-1} \sum_{k=0}^{m-1} [H_j - 2k] = \\
&= (X_j^-)^{m-1}[m][H_j - m + 1] .
\end{aligned} \tag{2.104}$$

If v_s is a singular vector we should have

$$0 = X_j^+ v_s = [X_j^+,(X_j^-)^m] \otimes v_0 = (X_j^-)^{m-1}[m][\Lambda(H_j) - m + 1] \otimes v_0. \tag{2.105}$$

(Note that $X_k^+ v_s = 0$, for $k \neq j$.) If $q_j = q^{(\alpha_j,\alpha_j)/2}$ is not a root of unity then (2.105) gives just condition (2.2) rewritten as $\Lambda(H_j) = (\Lambda, \alpha_j^\vee) = m - (\rho, \alpha_j^\vee) = m - 1$, where $\beta^\vee = 2\beta/(\beta,\beta)$ ($(\rho,\alpha_j^\vee) = 1$ for all α_j).

If q_j is a root of unity, $q_j^{N_j} = 1$, $N_j \in \mathbb{N}+1$, then for $k \in \mathbb{Z}$:

$$[kN_j]_{q_j} = \frac{q_j^{kN_j/2} - q_j^{-kN_j/2}}{q_j^{1/2} - q_j^{-1/2}} = \frac{\sin(\pi k)}{\sin(\pi/N_j)} = 0\,, \quad q_j = e^{2\pi i/N_j}. \tag{2.106}$$

In this case (2.105) gives that v_s from (2.103) is a singular vector iff

$$\text{either} \quad [m]_{q_j} = 0\,, \quad \forall \Lambda \in \mathcal{H}^*\,, \tag{2.107a}$$
$$\text{or} \ [\Lambda(H_j) + 1 - m]_{q_j} = 0\,, \quad m \neq \ell N_j. \tag{2.107b}$$

Thus, we see that for $q_j^{N_j} = 1$ the Verma module V^Λ is always reducible.

Thus, if q is a root of unity, then all Verma modules V^Λ are reducible for any $U_q(\mathcal{G})$ and for any $\Lambda \in \mathcal{H}^*$. These generic singular vectors are given explicitly by [198]:

$$v^{k_1,\ldots,k_\ell} = \prod_{j=1}^{\ell} (X_j^-)^{k_j N_j} \otimes v_0, \quad k_j \in \mathbb{Z}_+, \quad \sum_{j=1}^{\ell} k_j > 0, \tag{2.108}$$

where $N_j \in \mathbb{N}+1$ are the smallest integers such that $q_j^{N_j} = 1, j = 1, \ldots \ell$.

There exist also other singular vectors if the highest-weight Λ obeys condition (2.2). We have:

Proposition 4. *(i) Let us have the assumptions of Proposition 1, and in addition let q be a root of 1. Let $N_\beta \in \mathbb{N}+1$ be the smallest integer such that $q_\beta^{N_\beta} = 1$, with q_β as in (1.23a) and let $k, n \in \mathbb{Z}_+$, $n < N_\beta$ be such that $m = kN_\beta + n$. Then the following expression is a singular vector:*

$$v^{\beta,n,k} = \left(\mathcal{P}^{\beta,n} \mathcal{P}^{\beta,N_\beta - n} \right)^k \mathcal{P}^{\beta,n} \otimes v_0, \tag{2.109}$$

where $\mathcal{P}^{\beta,t}(X_1^-,\ldots,X_\ell^-)$ is given by (2.37).
(ii) Let us define the positive integer n_β by $\mathrm{n}_\beta = 1$ if $N_\beta \geq N_j$ $\forall j \in J_\beta$ and $\mathrm{n}_\beta = N_{j_0}/N_\beta$ for $j_0 \in J_\beta$ such that $N_{j_0} > N_\beta$. If we replace in formula (2.109) any of the factors $\left(\mathcal{P}^{\beta,n}\mathcal{P}^{\beta,N_\beta-n}\right)^{\mathrm{n}_\beta}$ by $\prod_{j=1}^{\ell}(X_j^-)^{\mathrm{n}_\beta N_\beta n_j}$ we obtain different in general singular vectors although with the same weight.
(iii) Further let us suppose that condition (2.2) is not fulfilled for any positive root except β. Then the general formula for the singular vectors of V^λ is:

$$v^{\beta,n,k'}_{k_1\dots k_\ell} = \prod_{j=1}^{\ell}(X_j^-)^{k_jN_j}\left(\mathscr{P}^{\beta,n}\mathscr{P}^{\beta,N_\beta-n}\right)^{k'}\mathscr{P}^{\beta,n}\otimes v_0, k', \quad k_j \in \mathbb{Z}_+, \tag{2.110}$$

plus many expressions in which a factor $\prod_{j=1}^{\ell}(X_j^-)^{k'_jN_j}$ *may be introduced before each factor* $\mathscr{P}^{\beta,t}$. ◊

The submodule structure of V^Λ is much more complicated if condition (2.2) is fulfilled for some other positive roots. In the other extreme situation when (2.2) is fulfilled for all simple roots (which means that Λ is either dominant integral weight or may be represented by $\Lambda = \Lambda' + \sum_{j=1}^{\ell} k_jN_j\alpha_j$ with Λ' dominant integral) then *the set of submodules of* V^λ *is in one-to-one correspondence with the elements of the Weyl group* $\hat{W}$ *of the affine Lie algebra* $\hat{\mathscr{G}}$ [195, 198]. As above the singular vectors are obtained by the combination of the factors $\prod_{j=1}^{\ell}(X_j^-)^{k_jN_j}$ with the monomials from (2.85) with the degree restricted by N_β:

$$v_s^{\beta,n,k} = \prod_{j=1}^{r}(X_j^-)^{kn_jN_\beta}\mathscr{P}_n^\beta(X_1^-,\dots,X_r^-)\otimes v_0\,, \tag{2.111}$$

$$[(\Lambda+\rho)(H_\beta)-n]_{q_b} = 0\,, n\in\mathbb{N}, n<N_\beta\,, \tag{2.112}$$

Let us say that two elements $\Lambda, \Lambda' \in \mathscr{H}^*$ are *equivalent*, $\Lambda \cong \Lambda'$, if $\Lambda - \Lambda' = N\beta$, where β is any element of the dual integer root lattice, that is, $\beta = n_1\alpha_1^\vee + \cdots n_r\alpha_r^\vee$, $n_i \in \mathbb{Z}$, and N is such that $q_j = e^{2\pi i/N}$ for the shortest simple roots α_j whose duals enter the decomposition of β.

It is clear that if $\Lambda \cong \Lambda'$, then they obey or disobey (2.107b),(2.112) simultaneously. Thus the Verma modules V^Λ and $V^{\Lambda'}$ have the same structure and the corresponding irreducible factor modules will be equivalent as $\tilde{U}_q(\mathscr{G})$ modules. So the irreducible HWMs are described by their highest weights up to the above equivalence. Because of (2.112) it is also clear that the irreducible HWMs of $U_q(\mathscr{G})$ are finite-dimensional.

Actually it was proved in [175] that the maximal dimension of an irreducible finite-dimensional (not necessarily with highest weight) representation of $U_q(\mathscr{G})$ is equal to:

$$N^{\dim\mathscr{G}^+} = N^{|\Delta^+|}. \tag{2.113}$$

We consider the question of the irreducible representations as quotients of reducible Verma modules in the framework of embeddings between such modules. It is clear that the Verma module $V^{\Lambda'}$ is isomorphic to a submodule of V^Λ if $\Lambda \cong \Lambda'$ and $\Lambda - \Lambda' = N\beta$ for β, an element of the dual nonnegative integer root lattice, that is, $\beta = n_1\alpha_1^\vee + \cdots n_r\alpha_r^\vee$, $n_i \in \mathbb{Z}_+$. Thus to account for all other embeddings it is enough to consider the singular vectors in (2.111) with $k = 0, n \in \mathbb{N}, n < N$. It is clear that if (2.112) holds then $V^{\Lambda-n\beta}$ is isomorphic to a submodule of V^Λ. If (2.112) holds for several pairs $(n,\beta) = (m_i,\beta_i)$, $i = 1,\dots,k$, there are other Verma modules $V^{\Lambda-m_i\beta_i}$, all of which are isomorphic to submodules of V^Λ. Furthermore if (2.112) holds with $\beta \in \delta^+$ and $n \in -\mathbb{N}$ then V^Λ is a submodule of $V^{\Lambda+n\beta}$. Indeed, if $[(\Lambda+\rho)(H_\beta)+n] = 0$ then $[(\Lambda+n\beta+\rho)(H_\beta)-n] = 0$ because $\beta(H_\beta) = 2$ for all β.

What is more interesting and in contrast to the undeformed $q = 1$ case is that if V^Λ has a singular vector of type (2.111) with $k = 0, n = m \in \mathbb{N}, m < N$, then the embedded Verma module $V^{\Lambda-m\beta}$ has a singular vector of type (2.111) with $k = 0, n = N-m$. The embedded Verma module in $V^{\Lambda-n\beta}$ is easily seen to be $V^{\Lambda-N\beta}$. The latter is a submodule also of V^Λ, however, with a singular vector from (2.111) with $k = 1, n = 0$. The two embeddings coincide if $\beta = \alpha_i$ is a *simple* root. Indeed, the first embedding is a composition of two embeddings $V^\Lambda \to V^{\Lambda-m\beta} \to V^{\Lambda-N\beta}$; correspondingly if v_0', v_0'' are the highest-weight vectors of $V^{\Lambda-m\beta}$, $V^{\Lambda-N\beta}$, respectively, we have $\mathcal{P}^\beta_{N-m}\mathcal{P}^\beta_m \otimes v_0 \mapsto \mathcal{P}^\beta_{N-m} \otimes v_0' \mapsto 1 \otimes v_0''$; the second embedding is $V^\Lambda \to V^{\Lambda-N\beta}$; under this we have $\prod_{j=1}^r (X_j^-)^{n_j N} \otimes v_0 \mapsto 1 \otimes v_0''$, where $\beta = \sum n_j\alpha_j$. Thus if β is not a simple root we may have embedding of one and the same module in two different ways. (This is similar to the affine Kac–Moody case when β is an imaginary root (i. e., $(\beta,\beta) = 0$).)

2.7.2 The Example of $U_q(sl(2))$ at Roots of Unity

In this subsection we follow [198]. We consider $U_q(\mathcal{G})$ for $\mathcal{G} = sl(2)$; $r = 1, X_1^\pm = X^\pm$, $H_1 = H$, $\alpha_1 = \alpha = \alpha^\vee = 2\rho$. We take q as a primitive root of unity: $q = e^{2\pi i/N}$, $N \in \mathbb{N}+1$. We shall prove that all Verma modules V^Λ belong to multiplets of one of the two types described below.

The multiplets of the first type are in one-to-one correspondence with those equivalent classes for which

$$\Lambda(H) + n \neq 0\,, \forall n \in \mathbb{Z}\,, \tag{2.114}$$

for any representative. For a fixed class represented, say, by $\Lambda \in \mathcal{H}^*$, the corresponding multiplet consists of an infinite chain of embeddings

$$\cdots \to \tilde{V}_{-1} \to \tilde{V}_0 \to \tilde{V}_1 \to \cdots \tag{2.115}$$

where the Verma modules entering the multiplet are $\tilde{V}_k = V^{\Lambda-kN\alpha}, k \in \mathbb{Z}$; that is, they are in one-to-one correspondence with the elements of the class in consideration. Each embedding in (2.115) is realized by a singular vector $v_s = (X^-)^N \otimes v_0(\tilde{V}_k)$, where $v_0(V)$ denotes the highest-weight vector of the Verma module V. The factor modules $\tilde{L}_k = \tilde{V}_k/\tilde{V}_{k+1}$ are isomorphic $\tilde{L}_k \cong \tilde{L}_{k'} \cong \tilde{L}, \forall k, k' \in \mathbb{Z}$; moreover $\dim \tilde{L} = N$ and all states of $\tilde{L}$ are given by $(X^-)^m \otimes v_0$, $m = 0, \ldots, N-1$.

Thus, the highest weight of an irreducible HWM is determined only up to the equivalence defined above.

The multiplets of the second type are parametrized by a positive integer, say, m such that $m \leq N/2$. Fix such an m and choose an element $\Lambda \in \mathcal{H}^*$ such that

$$[\Lambda(H) + 1 - m] = 0\,, \quad m \in \mathbb{N}\,, \quad m \le N/2, \tag{2.116}$$

that is, $\Lambda' = \frac{m-1}{2}\alpha$ is an element of the class of Λ. If $m < N/2$ then V^Λ is part of an infinite chain of embeddings

$$\cdots \longrightarrow V'^m_{-1} \longrightarrow V^m_0 \longrightarrow V'^m_0 \longrightarrow V^m_1 \longrightarrow V'^m_1 \longrightarrow \cdots \tag{2.117}$$

where $V^m_k = V^{\Lambda-kN\alpha}\,, k \in \mathbb{Z}\,,\ V'^m_k = V^{\Lambda-m\alpha-kN\alpha}\,, k \in \mathbb{Z}$. (Thus, the classes which only have element Λ for which (2.116) holds with $N/2 < m < N$ are represented by the highest weights of V'^m_k.) The embeddings $V^m_k \longrightarrow V'^m_k$ are realized by $v_s = (X^-)^m \otimes v_0(V^m_k)$, while the embeddings $V'^m_k \longrightarrow V^m_{k+1}$ are realized by $v_s = (X^-)^{N-m} \otimes v_0(V'^m_k)$. The factor modules $L^m_k = V^m_k / V'^m_k$ are isomorphic: $L^m_k \cong L^m_{k'} \cong L^m\,, \forall k, k' \in \mathbb{Z}$; also the factor modules $L'^m_k = V'^m_k / V^m_{k+1}$ are isomorphic: $L'^m_k \cong L'^m_{k'} \cong L'^m\,, \forall k, k' \in \mathbb{Z}$; moreover $\dim L^m = m$, $\dim L'^m = N - m$, and all states of L^m (respectively L'^m) are given by $(X^-)^n \otimes v_0\,, n = 0, \ldots, m-1$ (resp. $n = 0, \ldots, N-m-1$).

If $N \in 2\mathbb{N}$ and $m = N/2$, then V^Λ is part of an infinite chain of embeddings

$$\cdots \longrightarrow V^{N/2}_{-1} \longrightarrow V^{N/2}_0 \longrightarrow V'^{N/2}_1 \longrightarrow \cdots \tag{2.118}$$

where $V^{N/2}_k = V^{\Lambda-kN\alpha/2}\,, k \in \mathbb{Z}$. Everything we said above for $L^m_k\,, L^m$ is valid here for $m = N/2$.

It is clear that all elements of Λ and thus all Verma modules V^Λ over $U_q(\mathcal{G})$ are accounted for. Thus we have proved:

Proposition 5. *Let $q^N = 1\,, N \in \mathbb{N}+1\,, \mathcal{G} = sl(2)$. (a) All Verma modules V^Λ over $U_q(sl(2))$ belong to multiplets of one of the two types described above. (b) There are exactly N inequivalent irreducible HWM of $\tilde{U}_q(sl(2))$ which have dimensions $1, 2, \ldots, N$.* ◊

The last conclusion was obtained by other methods in [30, 506, 529]. Note that the nonuniformity in N (denoted there by m) of the results of [529] is due to the fact that their q is a square root of ours.

2.7.3 Classification in the $U_q(sl(3,\mathbb{C}))$ Case

Our next detailed example is $U_q(sl(3,\mathbb{C}))$. Let $\mathcal{G} = sl(3,\mathbb{C})$. Let us denote:

$$\begin{aligned}
&X_3^\pm = \pm(q^{1/4} X_1^\pm X_2^\pm - q^{-1/4} X_2^\pm X_1^\pm)\,, \quad H_3 = H_1 + H_2, \\
&[H_i\,, X_3^\pm] = \pm X_3^\pm (= \pm\alpha_3(H_i) X_3^\pm)\,, \quad i = 1, 2\,, \\
&[X_3^+, X_3^-] = [H_3],
\end{aligned} \tag{2.119}$$

where H_i correspond to the roots α_i; $\alpha_3 = \alpha_1 + \alpha_2 = \rho$; α_1, α_2 are the simple roots with $(\alpha_1, \alpha_2) = -1$; $(\alpha_i, \alpha_i) = 2, i = 1, 2, 3$.

First we consider the case when q is not a root of unity. Then the Verma modules and the irreducible HWM over $U_q(sl(3, \mathbb{C}))$ are in one-to-one correspondence with their counterparts over $sl(3, \mathbb{C})$. We recall the classification of the Verma modules and of the irreducible HWM over (affine-) $sl(n, \mathbb{C})$ [40]. For $n = 3$ there are five types of multiplets of Verma modules.

The multiplets of the **first** type include Verma modules V^λ for which

$$\lambda(H_i) \notin \mathbb{Z}, \forall i = 1, 2, 3. \tag{2.120}$$

Each such multiplet is trivial containing only one irreducible Verma module V^λ.

The multiplets of the **second** type are parametrized by a positive integer, say, m. Fix such an m and choose an element $\lambda \in \mathcal{H}^*$ such that

$$\lambda(H_1) + 1 = m, \quad \lambda(H_i) \notin \mathbb{Z}, \quad i = 2, 3. \tag{2.121}$$

Then V^λ is part of the following multiplet:

$$V^\lambda \to V^{\lambda - m\alpha_1}, \tag{2.122}$$

where the Verma module $V^{\lambda - m\alpha_1}$ is irreducible. The embedding in (2.122) is realized by the singular vector $v_s = (X_1^-)^m \otimes v_0$. This multiplet is exactly like the only nontrivial multiplet in the case $sl(2, \mathbb{C})$ and (for q not a root of unity) $U_q(sl(2, \mathbb{C}))$. Note that one can exchange the roles of $\lambda(H_1)$ and $\lambda(H_2)$.

The multiplets of the **third** type are parametrized by a positive integer, say, m, and are characterized by elements $\lambda \in \mathcal{H}^*$ such that

$$\lambda(H_3) + 2 = m, \quad \lambda(H_i) \notin \mathbb{Z}, \quad i = 1, 2. \tag{2.123}$$

Then V^λ is part of a multiplet as in (2.122):

$$V^\lambda \to V^{\lambda - m\alpha_3}, \tag{2.124}$$

where $V^{\lambda - m\alpha_3}$ is irreducible. The embedding in (2.124) is realized by the singular vector in (2.37) for A_2.

The multiplets of the **fourth** type are parametrized by two positive integers, say, m_1, m_2, and are characterized by elements $\lambda \in \mathcal{H}^*$ such that

$$\lambda(H_i) + 1 = m_i, \quad i = 1, 2, \quad \Rightarrow \lambda(H_3) + 2 = m_1 + m_2. \tag{2.125}$$

The Verma module V^λ is part of the following multiplet (cf. [195] formula (38)):

$$\begin{array}{cccccc} & & V^{12} & & \longrightarrow & V^3 \\ & \nearrow & & & & \uparrow \\ & & \uparrow & & & \\ V^2 & \longrightarrow & & \longrightarrow & & V^{21} \\ & & \uparrow & & & \\ \uparrow & & & & \nearrow & \\ V & \longrightarrow & V^1 & & & \end{array} \tag{2.126}$$

where $V = V^\lambda$,

$$V^i = V^{\lambda - m_i \alpha_i}, \quad i = 1, 2, 3, \quad m_3 = m_1 + m_2, \tag{2.127a}$$

$$V^{ij} = V^{\lambda - m_i \alpha_i - m_3 \alpha_j}, \quad (ij) = (12), (21), \tag{2.127b}$$

and V^3 is irreducible. The embeddings $V \to V^1$, $V^2 \to V^{21}$, $V^{12} \to V^3$ are realized by singular vectors $(X_1^-)^p \otimes v_0$ with $p = m_1, m_3, m_2$, respectively. The embeddings $V \to V^2$, $V^1 \to V^{12}$, $V^{21} \to V^3$ are realized by singular vectors $(X_2^-)^p \otimes v_0$ with $p = m_2, m_3, m_1$, respectively. The embeddings $V^1 \to V^{21}$, $V^2 \to V^{12}$ are realized by singular vectors given by A_2 formula (2.37) with Λ replaced by $\Lambda - m_1\alpha_1$, $\Lambda - m_2\alpha_2$, respectively, and $m = m_2$, $m = m_1$, respectively. The sextet diagram (2.126) is commutative as in the $q = 1$ case.

The multiplets of the **fifth** type are parametrized by a positive integer, say, m, and are characterized by elements $\lambda \in \mathscr{H}^*$ such that

$$\lambda(H_1) + 1 = m_1 = m \in \mathbb{N}, \quad \lambda(H_2) + 1 = m_2 = 0. \tag{2.128}$$

Then V^λ is part of the following multiplet:

$$V^\lambda \to V^{\lambda - m\alpha_1} \to V^{\lambda - m\alpha_3}, \tag{2.129}$$

where $V^{\lambda - m\alpha_3}$ is irreducible. The embeddings $V^\lambda \to V^{\lambda - m\alpha_1}$, $V^{\lambda - m\alpha_1} \to V^{\lambda - m\alpha_3}$ are realized by singular vectors $(X_1^-)^m \otimes v_0$, $(X_2^-)^m \otimes v_0$, respectively. This case may be viewed as a reduction of the previous one for $m_2 = 0$. One can exchange the roles of $\lambda(H_1)$, m_1 and $\lambda(H_2)$, m_2.

Based on the above one can recover the classification of the irreducible HWM over $sl(n, \mathbb{C})$ [195] for $n = 3$ and consequently of the irreducible HWM over $U_q(sl(3, \mathbb{C}))$:

$$L_\lambda^{01}, \quad (\lambda + \rho)(H_i) \notin \mathbb{N}, \quad i = 1, 2, 3, \tag{2.130a}$$

$$L_m^{02}, \quad \lambda(H_1) + 1 = m_1 = m \in \mathbb{N}, \quad \lambda(H_2) + 1 = m_2 \notin \mathbb{N}, \tag{2.130b}$$

$$m_3 = m_1 + m_2 \notin \mathbb{N} \quad \text{except for} \quad m_2 = 0,$$

$$L^{03}_m\,,\quad \lambda(H_3)+2=m\in\mathbb{N},\quad \lambda(H_i)\notin\mathbb{Z}_+\,, i=1,2, \tag{2.130c}$$
$$L^{04}_{m_1m_2}\,,\quad \lambda(H_i)+1=m_i\in\mathbb{N}, i=1,2. \tag{2.130d}$$
$$L^{05}_{m_1m_3}\,,\quad (\lambda+\rho)(H_i)=m_i\in\mathbb{N}, i=1,3,$$
$$\lambda(H_2)+1=m_3-m_1<0. \tag{2.130e}$$

One should also take into account cases (2.130b,e) with the roles of $\lambda(H_1), m_1$ and $\lambda(H_2), m_2$ exchanged. Note that $L^{01}_\lambda \cong V^\lambda$ is present in all multiplets above – these are the Verma modules which are irreducible as noted; L^{02}_m is present in multiplets of the second type (this is the factor module of the reducible Verma module V^λ in (2.122)), of the fourth type (these are the factor modules of V^{12} and V^{21} in (2.126)), and of the fifth type (these are the factor modules of V^λ (the only case when $m_2=0$) and of $V^{\lambda-m\alpha_1}$ in (2.129)); L^{03}_m is present in multiplets of the third type (this is the factor module of the reducible Verma module V^λ in (2.124)); $L^{04}_{m_1m_2}$ and $L^{05}_{m_1m_3}$ are present in multiplets of the fourth type ($L^{04}_{m_1m_2}$ is the factor module of V in (2.126) while $L^{05}_{m_im_3}\,, i=1,2\,$, is the factor module of V^i in (2.126)). Moreover, $L^{04}_{m_1m_2}$ is the finite-dimensional HWM over $sl(3,\mathbb{C})$ or over $U_q(sl(3,\mathbb{C}))$.

Consider now the case when q is a root of unity and let N be the smallest positive integer such that $q^N=1$. The classification of the Verma modules is as follows [198]. There are five types of multiplets of such modules, the first four being direct counterparts of those for q, not a root of unity.

The multiplets of the **first** type include Verma modules V^λ for which

$$\lambda(H_i)\notin\mathbb{Z}\,,\quad \forall i=1,2,3. \tag{2.131}$$

For a fixed $\lambda\in\mathscr{H}^*$ the corresponding multiplet consists of the following diagram of embeddings:

$$\begin{array}{ccccccc}
 & & \vdots & & \vdots & & \\
 & & \uparrow & & \uparrow & & \\
\cdots\to & & V_{0,1} & \to & V_{1,1} & \to\cdots & \\
 & & \uparrow & & \uparrow & & \\
\cdots\to & & V_{0,0} & \to & V_{1,0} & \to\cdots & \\
 & & \uparrow & & \uparrow & & \\
 & & \vdots & & \vdots & &
\end{array} \tag{2.132}$$

where $V_{k,\ell} = V^{\lambda-kN\alpha_1-\ell N\alpha_2}$, $k,\ell \in \mathbb{Z}$. Each embedding in (2.132) is realized by a singular vector $v_s = (X_i^-)^N \otimes v_0$, for $i = 1$, $i = 2$, when the arrow depicting the embedding is horizontal and vertical, respectively. Because of the symmetry it is clear that the factor modules $L_{k,\ell} = V_{k,\ell}/I_{k,\ell}$, where $I_{k,\ell}$ is the maximal submodule of $V_{k,\ell}$, have the same structure $\forall k,\ell, \in \mathbb{Z}$. We shall denote by L^1 any of these representations.

In (2.132) and in all diagrams below we do not depict any embeddings outside the quadrangle $(V_{0,0}, V_{1,0}, V_{1,1}, V_{0,1})$ except the adjacent ones shown in (2.132).

The multiplets of the **second** type are parametrized by a positive integer, say, m such that $m \leq N/2$. Fix such an m and choose an element $\lambda \in \mathscr{H}^*$ such that

$$[\lambda(H_1) + 1 - m] = 0\,, \quad \lambda(H_i) + n \neq 0\,, \quad i = 2,3, \forall n \in \mathbb{Z}. \tag{2.133}$$

If $m_1 < N/2$, then V^λ is part of the following multiplet:

$$\begin{array}{ccccccc}
 & & \vdots & & \vdots & & \\
 & & \uparrow & & \uparrow & & \\
\cdots \to & V_{0,1} & \to & V^1_{0,1} & \to & V_{1,1} & \to \cdots \\
 & \uparrow & & \uparrow & & \uparrow & \\
\cdots \to & V_{0,0} & \to & V^1_{0,0} & \to & V_{1,0} & \to \cdots \\
 & \uparrow & & & & \uparrow & \\
 & \vdots & & & & \vdots &
\end{array} \tag{2.134}$$

where $V_{k,\ell}$ are as above and $V^1_{k,\ell} = V^{\lambda-m\alpha_1-kN\alpha_1-\ell N\alpha_2}$, $k,\ell \in \mathbb{Z}$. The embeddings $V_{k,\ell} \to V^1_{k,\ell}$ are realized by $v_s = (X_1^-)^m \otimes v_0(V_{k,\ell})$, while $V^1_{k,\ell} \to V_{k+1,\ell}$ are realized by $v_s = (X_1^-)^{N-m} \otimes v_0(V^1_{k,\ell})$. The factor modules $L_{k,\ell} = V_{k,\ell}/I_{k,\ell}$ and $L^1_{k,\ell} = V^1_{k,\ell}/I^1_{k,\ell}$ have the same structure $\forall k,\ell, \in \mathbb{Z}$. The representations $L_{k,\ell}$, $L^1_{k,\ell}$ shall be denoted by L^2_m, L^2_{N-m}, respectively. Thus there are $N-1$ essentially different irreducible HWMs with highest weights satisfying (78), namely, L^2_m for $m = 1,\dots,N-1$.

We do not consider separately the subcase obtained from this by exchanging the indices 1 and 2. The corresponding representations which are conjugate to L^2_m under the exchange $\alpha_1 \longrightarrow \alpha_2$ will be denoted by $\tilde{L}^2_{m_2}$.

The multiplets of the **third** type are parametrized by a positive integer, say, m_3 such that $m_3 \le N/2$. Fix such an m_3 and choose an element $\lambda \in \mathscr{H}^*$ such that

$$[\lambda(H_3) + 2 - m_3] = 0\,, \quad \lambda(H_i) + n \neq 0\,, \quad i = 1, 2, \forall n \in \mathbb{Z}. \tag{2.135}$$

The Verma module V^λ is part of the following multiplet:

$$\begin{array}{ccccc} & \vdots & & \vdots & \\ & \uparrow & & \uparrow & \\ \cdots \to & V_{0,1} & \to \qquad \to & V_{1,1} & \to \cdots \\ & & \nearrow & & \\ & \uparrow & V^3_{0,0} & \uparrow & \\ & & \nearrow & & \\ \cdots \to & V_{0,0} & \to \qquad \to & V_{1,0} & \to \cdots \\ & \uparrow & & \uparrow & \\ & \vdots & & \vdots & \end{array} \tag{2.136}$$

where $V_{k,\ell}$ are as above and $V^3_{k,\ell} = V^{\lambda - m_3\alpha_3 - kN\alpha_1 - \ell N\alpha_2}$, $k, \ell \in \mathbb{Z}$. The embeddings $V_{k,\ell} \to V^3_{k,\ell}$ and $V^3_{k,\ell} \to V_{k+1,\ell+1}$ are realized by the singular vector in (2.37) for A_2, $m = m_3$, $m = N - m_3$, respectively. Formula (2.37) (for A_2) is valid here for any $m \in \mathbb{N}$, if (2.135) holds (with m_3 replaced by m); however, if $m \ge N$, and $m = kN + t$, $k \in \mathbb{N}$, $t \in \mathbb{Z}_+$, $t < N$, it reduces to:

$$v_s^m = (X_1^-)^{kN}(X_2^-)^{kN} v_s^t. \tag{2.137}$$

Analogous to the previous case the representations $V_{k,\ell}/I_{k,\ell}$ $V^3_{k,\ell}/I^3_{k,\ell}$, $\forall k, \ell \in \mathbb{Z}$, shall be denoted by $L^3_{m_3}$ and $L^3_{N-m_3}$, respectively. Thus there are $N - 1$ essentially different irreducible HWMs with highest weights satisfying (2.135), namely, L^3_m for $m = 1, \dots, N-1$.

The multiplets of the **fourth** type are parametrized by two positive integers, say, m_1, m_2 such that $m_1 + m_2 < N$. Fix such m_1, m_2 and choose an element $\lambda \in \mathscr{H}^*$ such that

$$[\lambda(H_i) + 1 - m_i] = 0\,, i = 1, 2, \quad \Rightarrow [\lambda(H_3) + 2 - m_1 - m_2] = 0. \tag{2.138}$$

The Verma module V^λ is part of the following multiplet:

$$
\begin{array}{ccccccc}
\vdots & & & & & & \vdots \\
\uparrow & & & & & & \uparrow \\
\cdots \to V_{0,1} & \to & V^1_{0,1} & & \to & & V_{1,1} \to \cdots \\
 & & \uparrow & & & \nearrow & \\
\uparrow & & V^{12}_{0,0} & \to & V^3_{0,0} & & \uparrow \\
 & \nearrow & & & \uparrow & & \\
 & & \uparrow & & & & \\
V^2_{0,0} & \to & & \to & V^{21}_{0,0} & \to & V^2_{1,0} \\
 & & \uparrow & & & & \\
\uparrow & & & \nearrow & & & \uparrow \\
\cdots \to V_{0,0} & \to & V^1_{0,0} & & \to & & V_{1,0} \to \cdots \\
\uparrow & & & & & & \uparrow \\
\vdots & & & & & & \vdots
\end{array}
\tag{2.139}
$$

where $V_{k,\ell}$ is as before and

$$V^i_{k,\ell} = V^{\lambda - m_i\alpha_i - kN\alpha_1 - \ell N\alpha_2}, i = 1, 2, 3, m_3 = m_1 + m_2, k, \ell \in \mathbb{Z}, \tag{2.140a}$$

$$V^{ij}_{k,\ell} = V^{\lambda - m_i\alpha_i - m_3\alpha_j - kN\alpha_1 - \ell N\alpha_2}, (ij) = (12), (21), k, \ell \in \mathbb{Z}. \tag{2.140b}$$

We summarize the structure of the above multiplets as follows.

The embeddings $V_{k,\ell} \to V^1_{k,\ell}$, $V^1_{k,\ell} \to V_{k+1,\ell}$, $V^2_{k,\ell} \to V^{21}_{k,\ell}$, $V^{21}_{k,\ell} \to V^2_{k+1,\ell}$, $V^{12}_{k,\ell} \to V^3_{k,\ell}$, $V^3_{k,\ell} \to V^{12}_{k+1,\ell}$, $V_{k,\ell+1} \to V^1_{k,\ell+1}$, $V^1_{k,\ell+1} \to V_{k+1,\ell+1}$ are realized by singular vector $(X_1^-)^p \otimes v_0$, with $p = m_1, N - m_1, m_3, N - m_3, m_2, N - m_2, m_1, N - m_1$, respectively. The embeddings $V_{k,\ell} \to V^2_{k,\ell}$, $V^2_{k,\ell} \to V_{k,\ell+1}$, $V^1_{k,\ell} \to V^{12}_{k,\ell}$, $V^{12}_{k,\ell} \to V^1_{k,\ell+1}$, $V^{21}_{k,\ell} \to V^3_{k,\ell}$, $V^3_{k,\ell} \to V^{21}_{k,\ell+1}$, $V_{k+1,\ell} \to V^2_{k+1,\ell}$, $V^2_{k+1,\ell} \to V_{k+1,\ell+1}$ are realized by singular vector $(X_2^-)^p \otimes v_0$, with $p = m_2, N - m_2, m_3, N - m_3, m_1, N - m_1, m_2, N - m_2$, respectively. The embeddings $V^1_{k,\ell} \to V^{21}_{k,\ell}$, $V^2_{k,\ell} \to V^{12}_{k,\ell}$, $V^3_{k,\ell} \to V_{k+1,\ell+1}$ are realized by singular vectors given by formula (45), with λ replaced by $\lambda - m_1\alpha_1$, $\lambda - m_2\alpha_2$, $\lambda - m_3\alpha_3$ and $m = m_2$, $m = m_1$, $m = N - m_3$, respectively.

Note that the six HWMs $V_{k,\ell}, V^i_{k,\ell}, V^{ij}_{k,\ell}$ for fixed k, ℓ form the basic $sl(3,\mathbb{C})$ multiplet in the case $q = 1$ [195] or $U_q(sl(3,\mathbb{C}))$ multiplet (2.126) when q is not a root of unity. Let us say that the tip of this sextet is at $V_{k,\ell}$. This sextet shares one side with six sextets of the same type and orientation, and for the same k, ℓ their tips are at $V^{21}_{k-1,\ell-1}, V^{12}_{k,\ell-1}, V^{21}_{k,\ell}, V^{12}_{k,\ell}, V^{21}_{k-1,\ell}, V^{12}_{k-1,\ell-1}$. The role of (m_1, m_2) in these sextets is played by $(m_2, N - m_3)$ for $V^{12}_{*,*}$ and by $(N - m_3, m_1)$ for $V^{21}_{*,*}$. Moreover, this structure is periodic, and if we consider only such sextets then this multiplet looks like a honeycomb and resembles one of the multiplets of reducible Verma modules over the affine Lie algebra $sl(3,\mathbb{C})^{(1)}$, namely, the "maximal" multiplet in the sense that it represents the affine Weyl group W (cf. [195], Proposition 2 and the figure). However, in the affine case this honeycomb corresponding to the affine Weyl group has a distinguished point (corresponding to the unit element of W), that is, a Verma module which contains as submodules all other Verma modules in this multiplet (the irreducible subquotient of this distinguished Verma module is an integrable HWM, and all integrable HWM over $sl(3,\mathbb{C})^{(1)}$ can be obtained in this way).

Below we shall use also the fact that there are other sextets of HWMs, namely: $V^3_{k-1,\ell-1}, V^{12}_{k,\ell-1}, V^2_{k+1,\ell}, V_{k+1,\ell+1}, V^1_{k,\ell+1}, V^{21}_{k-1,\ell}$, for fixed k, ℓ and containing the sextet $V_{k,\ell}, V^i_{k,\ell}, V^{ij}_{k,\ell}$. Certainly these bigger sextets are more complicated.

Thus the structure of the representations $V_{k,\ell}, V^{12}_{k,\ell}, V^{21}_{k,\ell}$ is exactly the same; moreover, the range of their parameters is the same. The same holds for the representations $V^i_{k,\ell}$, $i = 1, 2, 3$. These are situated in the sextets at the site opposite to what we is called the tip. The values $(\lambda(H_1), \lambda(H_2))$, that is, the analogues of (m_1, m_2), are $(N - m_1, m_3)$, $(m_3, N - m_2)$, $(N - m_2, N - m_1)$ for $i = 1, 2, 3$, respectively, and they cover the same range. Moreover, this shows that the requirement $m_1 + m_2 < N$ is not a restriction. Indeed, the HWMs $V^i_{k,\ell}$ for one value of i exhaust all such cases.

From the above it is easy to see that there are the following essentially different irreducible HWMs with highest weights satisfying (2.138), namely, $L^4_{m_1 m_2}$ and $L'^4_{m_1 m_2}$ which will denote any of the factor modules $V_{k,\ell}/I_{k,\ell}$ and $V^3_{k,\ell}/I^3_{k,\ell}$, respectively.

The multiplets of the **fifth** type can be viewed as "analytic" continuation of the fourth type for $m_1 + m_2 = N$. Thus they are parametrized by a positive integer, say, m_1 such that $m_1 \leq N/2$. Fix such m_1 and choose an element $\lambda \in \mathscr{H}^*$ such that

$$[\lambda(H_i) + 1 - m_i] = 0, \quad m_2 = N - m_1. \tag{2.141}$$

The HWM V^λ is part of a multiplet containing the following HWM: $V_{k,\ell}$ and $V^i_{k,\ell}$, $i = 1, 2$ given by the same formulae as in the previous case with $m_2 = N - m_1$ and $m_3 = N$. It can be depicted using (2.139) and distorting it so that $V^3_{k,\ell}$ will coincide with $V_{k+1,\ell+1}$, $V^{12}_{k,\ell}$ with $V^1_{k,\ell+1}$, and $V^{21}_{k,\ell}$ with $V^2_{k+1,\ell}$. Thus the sextets with $V^{12}_{k,\ell}, V^{21}_{k,\ell}$ at the tips deteriorate into commutative triangles, and the latter representations do not have the structure of $V_{k,\ell}$. The singular vectors depicting the embeddings are, as in the previous case, however, taking into account the coincidences. It is easy to see that there are the following inequivalent irreducible HWMs with highest weights satisfying (86), namely,

$L^5_{m_1}$, $L^{51}_{m_1}$, $L^{52}_{m_1}$, which will denote the factor modules $V_{k,\ell}/I_{k,\ell}$, $V^1_{k,\ell}/I^1_{k,\ell}$, $V^2_{k,\ell}/I^2_{k,\ell}$. Note that $L^{51}_{m_1}$, $L^{52}_{m_2}$ are conjugate to each other under the exchange $\alpha_1 \longrightarrow \alpha_2$.

2.7.4 Cyclic Representations of $U_q(\mathscr{G})$

As we saw above when q is a root of unity the $2N$-th powers of the Cartan–Weyl generators form singular vectors of Verma modules. These are set to zero when we consider the irreducible factor modules. However, in a more general approach, when we consider non-highest-weight representations one may give nonzero values to these powers. These representations are called *cyclic representations* [170–172], the term referring to the fact that each root vector generates a multiplicative cyclic group.

2.7.4.1 $U_q(sl(2,\mathbb{C}))$

Let us start with the example of $U_q(sl(2,\mathbb{C}))$ considered in [529]. Let $q = e^{2\pi i/N}$. Their cyclic representation depends on three complex parameters a, b, μ such that $([p][\mu + 1 - p] + ab) \neq 0$ for $p = 1, \ldots, N-1$. It has the basis v_p, $p = 0, 1, \ldots, N-1$ and transforms under the generators of $U_q(sl(2,\mathbb{C}))$ as follows:

$$Hv_p = (\mu - 2p)v_p\,, \tag{2.142}$$

$$X^- v_p = ([p+1][\mu - p] + ab)^{1/2} v_{p+1}\,, p = 0, 1, \ldots, N-2\,, \tag{2.143a}$$

$$X^- v_{N-1} = a v_0\,, \tag{2.143b}$$

$$X^+ v_p = ([p][\mu + 1 - p] + ab)^{1/2} v_{p-1}\,, p = 1, \ldots, N-1\,, \tag{2.144a}$$

$$X^+ v_0 = b v_{N-1}\,, \tag{2.144b}$$

If $a \neq 0 \neq b$ this representation is not an HWM or lowest-weight module and is called cyclic because of formulae (2.143b) and (2.144b). If $a = 0, b \neq 0$, $\mu - 2m + 2 \notin \mathbb{Z}_+$, $(a \neq 0, b = 0, \mu \notin \mathbb{Z}_+)$, then it is a cyclic irreducible lowest (highest) weight module, with lowest (highest) weight $\lambda = (\mu - 2(m-1))\alpha/2$, $(\lambda = \mu\alpha/2)$.

If $a = b$ then $X^+ = (X^-)^t$. If $a = \bar{b}$ and μ real then $X^+ = (X^-)^+$. In this last case the representations with three real parameters correspond to representations obtained in [558].

Two such representations with parameters a, b, μ and a', b', μ' are isomorphic iff

$$\mu' = \mu + 2r, r \in \mathbb{Z}, a'b = ab', ab - a'b' = [2r][\mu + 2r + 1]. \tag{2.145}$$

2.7.4.2 $U_q(sl(n+1,\mathbb{C}))$

In this subsection we review the paper [171]. Let us consider $U_q = U_q(sl(n+1,\mathbb{C}))$, $n \geq 2$. In [171] is constructed (for generic q) an algebra map from U_q a

$\mathbb{C}(q)$ algebra $\mathscr{W}$ determined as follows. It is generated by x_{jk}, z_{jk}, $1 \le j \le k \le n$, and the inverses x_{jk}^{-1}, z_{jk}^{-1}, satisfying

$$[x_{jk}, x_{j'k'}] = [x_{jk}, z_{j'k'}] = [z_{jk}, z_{j'k'}] = 0, \quad \text{if} \quad (j,k) \neq (j',k'), \tag{2.146a}$$

$$z_{jk}x_{jk} = q x_{jk} z_{jk}. \tag{2.146b}$$

A $\mathbb{C}(q)$ linear involution * is defined by

$$(x_{jk})^* = x^{-1}_{k+1-jk}\,, (z_{jk})^* = z^{-1}_{k+1-jk}\,, \tag{2.147}$$

and $\mathbb{C}$ linear involution $\hat{}$ by

$$\hat{q} = q^{-1}, \quad \hat{x}_{jk} = x_{jk}\,, \quad \hat{z} = z^{-1}_{jk}\,, \tag{2.148}$$

the analogous involutions for $U_q(sl(n+1,\mathbb{C}))$ are defined by

$$(X_i^{\pm})^* = (X^{\mp}_{n+1-i}\,, \quad (H_i)^* = -H_{n+1-i}\,, \quad (K_i)^* = K^{-1}_{n+1-i}\,, \tag{2.149}$$

$$\hat{q} = q^{-1}, \quad \hat{X}_i^{\pm} = X_i^{\pm}\,, \quad \hat{H}_i = -H_i\,, \quad \hat{K}_i = K_i^{-1}. \tag{2.150}$$

For $r = (r_1, \dots, r_n) \in (\mathbb{C}^{\times})^n$ one defines $r^* = (r_n, \dots, r_1)$, $\hat{r} = (r_1^{-1}, \dots, r_n^{-1})$. The authors of [171] construct a family of $\mathbb{C}(q)$ homomorphisms

$$\rho_{r,s} : U_q(sl(n+1,\mathbb{C})) \to \mathscr{W}\,, \tag{2.151}$$

depending on $r, s \in (\mathbb{C}^{\times})^n$ by the formulae:

$$\rho_{r,s}(X_i^+) = \sum_{k=i}^{n} [\overline{z}_{ik}]\xi_{ik}\,, \tag{2.152}$$

$$\rho_{r,s}(X_i^-) = \rho_{s^*,r^*}(X^+_{n+1-i})^*\,,$$

$$\rho_{r,s}(K_i) = \frac{r_i}{s_i} z_{in}^2 z_{i+1n}^{-1}\,,$$

where

$$\xi_{ik} = x_{ik}x_{ik+1}\cdots x_{in}\,, \tag{2.153}$$

$$\overline{z}_{ik} = r_i z_{ik} z_{ik-1} z^{-1}_{i-1k-1} z^{-1}_{i+1k}. \tag{2.154}$$

Further let $N \ge 3$ be an odd positive integer and let $q = e^{2\pi i/N}$. Let $\Phi_N(x)$ denote the N-th cyclotomic polynomial so that $\Phi_N(q) = 0$. One sets

$$\mathscr{A} = \{f \in \mathbb{C}(q) | f \quad \text{is regular at} \quad \Phi_N(q) = 0\}. \tag{2.155}$$

Let $U_{\mathscr{A}}$ denote the $\mathscr{A}$-subalgebra of U_q generated by $X_i^{\pm}, K_i, i = 1, \dots, n$. Let further $U = U_{\mathscr{A}} \otimes_{\mathscr{A}} \mathbb{C}$. The algebra $\mathscr{W}_{\mathscr{A}}$ is defined analogously.

Consider an N-dimensional vector space with fixed basis $u_i, i = 0, \dots, N-1$:

$$V^1 = \oplus_{i=0}^{N-1} \mathbb{C} u_i . \tag{2.156}$$

One defines a representation σ of the Weyl algebra $\mathscr{W}_q^1$ with generators x, z by:

$$\sigma : \mathscr{W}_q^1 \to \mathrm{End}(V_q^1), \tag{2.157a}$$

$$\sigma(x)u_i = u_{i+1}, \quad (u_N = u_0), \quad \sigma(z)u_i = q^i u_i . \tag{2.157b}$$

Further let $m = n(n+1)/2 = \dim \mathscr{G}^+$ and $V = (V_q^1)^{\otimes m}$. Thus one obtains a representation $\sigma^{\otimes m} : \mathscr{W} \cong (\mathscr{W}_q^1)^{\otimes m} \to \mathrm{End}(V)$ by letting the generators x_{jk}, z_{jk} act on the (j,k)-component of V as $\sigma(x), \sigma(z)$ and as identity on the other components. Further one defines automorphisms $S_{\mathrm{n}}, T_{\mathrm{n}}$ of $\mathscr{W}$ for $\mathrm{n} = (\mathrm{n}_{jk}) \in (\mathbb{C}^{\times})^m$:

$$S_{\mathrm{n}}(x_{jk}) = \mathrm{n}_{jk} x_{jk}, \quad S_{\mathrm{n}}(z_{jk}) = z_{jk}, \tag{2.158a}$$

$$T_{\mathrm{n}}(x_{jk}) = x_{jk}, \quad T_{\mathrm{n}}(z_{jk}) = \mathrm{n}_{jk} z_{jk} . \tag{2.158b}$$

Now the representation of U is defined by the following composition of maps:

$$\pi : U \xrightarrow{\rho_{r,s}} \mathscr{W} \xrightarrow{S_g \circ T_h} \mathscr{W} \xrightarrow{\sigma^{\otimes m}} \mathrm{End}(V) \tag{2.159}$$

Besides $r, s \in (\mathbb{C}^{\times})^n$, the representation π contains $n(n+1)$ arbitrary parameters $g = (g_{jk}), h = (h_{jk}) \in (\mathbb{C}^{\times})^m$. Not all of these parameters are independent, and one can set $s_i = 1, i = 1, \dots, n$.

Further the authors of [171] show cyclicity of the representation and prove that it is irreducible for generic parameters r_i, g_{jk}, h_{jk}. For special values of the parameters, they obtain invariant subspaces.

2.7.4.3 $U_q(sl(n, \mathbb{C})^{(1)})$

Let $\hat{U}_q = U_q(sl(n, \mathbb{C})^{(1)}), n \geq 2$. Further we consider the cyclic representations of $\hat{U}_q$ following [170]. Let $q = e^{2\pi i/N}, N \geq 3$. Let V be an N-dimensional vector space. Let Y, Z be two linear operators on V satifying the relation $ZY = qYZ$. Denote by Y_i (respectively, Z_i) the operator on $\mathscr{W} = V^{\otimes n}$ which acts on the i-th component as Y (repectively, Z) and as identity on the other components. Set

$$\mathscr{W}^{(0)} = \left\{ w \in \mathscr{W} \,\middle|\, \left(\prod_{i=1}^{n} Z_i \right) w = w \right\} . \tag{2.160}$$

Let $a_0, \ldots, a_{n-1}$ and $x_0, \ldots, x_{n-1}$ be arbitrary nonzero numbers. An N^{n-1}-dimensional cyclic representation $\pi_{x,a}$ of $\tilde{U}_q$ on $\mathscr{W}^{(0)}_{x,a} = \mathscr{W}^{(0)}$ is defined as follows:

$$\pi_{x,a}(X_i^+) = \frac{x_i}{q^{1/2} - q^{-1/2}}(a_i Z_i^2 - a_i^{-1} Z_i^{-2}) Y_i Y_{i+1}^{-1}\,, \tag{2.161a}$$

$$\pi_{x,a}(X_i^-) = \frac{x_i^{-1}}{q^{1/2} - q^{-1/2}}(a_{i+1} Z_{i+1}^2 - a_{i+1}^{-1} Z_{i+1}^{-2}) Y_i^{-1} Y_{i+1}\,, \tag{2.161b}$$

$$\pi_{x,a}(K_i) = \frac{a_i}{a_{i+1}} Z_i Z_{i+1}^{-1}\,, \tag{2.161c}$$

where $a_0 = a_n$, $Y_0 = Y_n$, $Z_0 = Z_n$. Choose a basis v_k, $k = 0, \ldots, N-1$; $(v_N = v_0)$ of V on which Y, Z act by

$$Y v_k = q^k v_k\,, \quad Z v_k = v_{k-1}. \tag{2.162}$$

Then the basis vectors of $\mathscr{W}^{(0)}$ may be chosen as

$$w^{(0)}_{\mathbf{k}} = \sum_{p=0}^{N-1} v_{k_1+p} \otimes \cdots \otimes v_{k_n+p}\,, \tag{2.163}$$

where $\mathbf{k} = (k_1, \ldots, k_n)$, so that $\mathscr{W}^{(0)}_{k_1+1,\ldots,k_n+1} = \mathscr{W}^{(0)}_{k_1,\ldots,k_n}$.

2.8 Characters of Irreducible HWMs

2.8.1 Generalities

Let again $\mathscr{G}$ be any simple Lie algebra. We recall the decomposition (2.17). Following Dixmier [192] and Kac [372] let $E(\mathscr{H}^*)$ be the associative abelian algebra consisting of the series $\sum_{\mu\in\mathscr{H}^*} c_\mu e(\mu)$, where $c_\mu \in \mathbb{C}$ and $c_\mu = 0$ for μ outside the union of a finite number of sets of the form $D(\lambda) = \{\mu \in \mathscr{H}^* \,|\, \mu \leq \lambda\}$, using any ordering of $\mathscr{H}^*$; the *formal exponents* $e(\mu)$ have the properties $e(0) = 1$, $e(\mu)e(\nu) = e(\mu + \nu)$.

The *character* of V is defined by:

$$ch\ V = \sum_{\mu\in\Gamma_+} (\dim V_\mu) e(\lambda + \mu) = e(\lambda) \sum_{\mu\in\Gamma_+} (\dim V_\mu) e(\mu). \tag{2.164}$$

We recall [192] that for a Verma module $V = V^\Lambda$ we have $\dim V_\mu = P(\mu)$, where $P(\mu)$ is defined after (2.18). Analogously we use [192] to obtain:

$$ch\ V^\Lambda = e(\Lambda) \sum_{\mu\in\Gamma_+} P(\mu) e(\mu) = e(\Lambda) \prod_{\alpha\in\Delta^+} (1 - e(\alpha))^{-1}. \tag{2.165}$$

The *Weyl character formula* for the finite-dimensional irreducible lowest-weight representations over $\mathscr{G}$ has the form [192]:

$$ch\, L_\Lambda = ch\, V^\Lambda \sum_{w\in W} (-1)^{\ell(w)} e(w\cdot\Lambda - \Lambda) = \sum_{w\in W} (-1)^{\ell(w)} ch\, V^{w\cdot\Lambda}. \tag{2.166}$$

If q is not a root of unity, the above formula holds for the finite-dimensional irreducible HWM over $U_q(\mathscr{G})$ (this can be deduced from the results of [441, 532]). For other representations over $U_q(\mathscr{G})$ we have announced in [201] the results for $\mathscr{G} = sl(3,\mathbb{C})$; see next subsection.

2.8.2 $U_q(sl(3,\mathbb{C}))$

Consider $U_q(sl(3,\mathbb{C}))$ and let us denote $t_i \equiv e(-\alpha_i)\,, i = 1,2,$ then $e(-\alpha_3) = t_1t_2$. Then (2.165) can be rewritten as

$$ch\, V^\Lambda = e(\Lambda)/(1-t_1)(1-t_2)(1-t_1t_2). \tag{2.167}$$

In the case when q is not a root of unity the character formulae of the irreducible HWM over $U_q(sl(3,\mathbb{C}))$ are:

$$ch\, L^{01}_\Lambda = ch\, V^\Lambda\,, \tag{2.168a}$$
$$ch\, L^{02}_m = ch\, V^\Lambda(1-t_1^m)\,,\ ch\, L^{03}_m = ch\, V^\Lambda(1-(t_1t_2)^m) \tag{2.168b}$$
$$ch\, L^{05}_{m_1m_3} = ch\, V^\Lambda\,(1 - t_1^{m_1} - (t_1t_2)^{m_3} + t_1^{m_1}t_2^{m_3})\,, \tag{2.168c}$$

and the character formula for $L^{04}_{m_1m_2}$ is given by (2.166) which explicitly is (using the notation in (2.126)):

$$\begin{aligned} ch\, L^{04}_{m_1m_2} &= ch\, V^\Lambda - ch\, V^1 - ch\, V^2 + ch\, V^{12} + ch\, V^{21} - ch\, V^3 = \\ &= ch\, V^\Lambda\,(1 - t_1^{m_1} - t_2^{m_2} + \\ &\quad + t_1^{m_1}t_2^{m_3} + t_1^{m_3}t_2^{m_2} - (t_1t_2)^{m_3}) \end{aligned} \tag{2.169}$$

(The same formulae hold for the irreducible HWM over $sl(3,\mathbb{C})$.)

The proof of (2.168) is given in [201]. Actually there is nothing to prove for (2.168a) since V^Λ is irreducible in this case. Formula (2.168b) is just a rewriting of

$$ch\, L_\Lambda = ch\,(V^\Lambda/I^\Lambda) = ch\,(V^\Lambda/V^{\Lambda'}) = ch\, V^\Lambda - ch\, V^{\Lambda'}, \tag{2.170}$$

with $V^{\Lambda'} = V^{\Lambda - m\alpha_1}$ and $V^{\Lambda'} = V^{\Lambda - m\alpha_3}$. In the case of (2.168c) we use the explicit embedding structure of the Verma module V^Λ whose irreducible quotient is $L^{05}_{m_1m_3}$. This Verma module can be represented by V^2 on diagram (2.126), however, with Λ in (2.130e) replaced by $\Lambda' = \Lambda - m_2\alpha_2$. Thus we have:

$$ch\, I^\Lambda = ch\, V^{12} + ch\, V^{21} - ch\, V^3. \qquad (2.171)$$

Then (2.168c) follows from the combination of $ch\, L = ch(V^\Lambda/I^\Lambda)$ and (2.171). (Note that the Verma module V^Λ whose irreducible quotient is $L^{05}_{m_1m_3}$ could also be represented by V^2 on diagram (2.126), however, with Λ in (2.130e) replaced by $\Lambda' = \Lambda - m_1\alpha_1$.)

2.8.3 $U_q(sl(3,\mathbb{C}))$ at Roots of Unity

Let us consider now the case when q is a root of unity. The results (announced in [201]) on the characters of the irreducible HWM over $U_q(sl(3,\mathbb{C}))$
for q, a root of unity, can be summarized by the following:

Proposition 6. *Let $N \in \mathbb{N}+2$ be the smallest such number such that $q^N = 1$. Let L^1, L^2_m, L^3_m, $L^4_{m_1m_2}$, $L'^4_{m_1m_2}$, L^5_m, L^{51}_m, L^{52}_m be the representations of $U_q(sl(3,\mathbb{C}))$ defined as above. We have:*

$$ch\, L^1 = ch\, V^\Lambda(1-t_1^N)(1-t_2^N)(1-(t_1t_2)^N)\,, \qquad (2.172a)$$
$$ch\, L^2_m = ch\, V^\Lambda(1-t_1^m)(1-t_2^N)(1-(t_1t_2)^N)\,, \qquad (2.172b)$$
$$ch\, L^3_m = ch\, V^\Lambda(1-t_1^N)(1-t_2^N)(1-(t_1t_2)^m)\,, \qquad (2.172c)$$

$$ch\, L^4_{m_1m_2} = \sum_{w\in W}(-1)^{\ell(w)} ch\, V^{w\cdot\Lambda}\,, \qquad (2.173a)$$
$$= ch\, V_{k,\ell} - ch\, V^1_{k,\ell} - ch\, V^2_{k,\ell} + ch\, V^{12}_{k,\ell} + ch\, V^{21}_{k,\ell} - ch\, V^3_{k,\ell}\,, \qquad (2.173b)$$
$$ch\, L^5_m = ch\, L^4_{m,N-m}\,, \qquad (2.173c)$$

$$ch\, L'^4_{m_1m_2} = \sum_{w\in W}(-1)^{\ell(w)}(ch\, V^{w\cdot\Lambda} - ch\, V^{w\cdot\Lambda'})\,, \qquad (2.174)$$
$$\Lambda' = s_3\cdot\Lambda + N\alpha_3 = \Lambda - (m_3 - N)\alpha_3\,,$$

$$ch\, L^{51}_m = ch\, L^4_{N-m,N}\,, \qquad (2.175a)$$
$$ch\, L^{52}_m = ch\, L^4_{N,m}. \qquad (2.175b)$$

All states of L^1, L^2_m, L^3_m are given by:

$$(X_2^-)^{n_2}(X_3^-)^{n_3}(X_1^-)^{n_1}\otimes v_0\,, \qquad (2.176)$$
$n_i = 0,\dots,N-1,\quad i = 1,2,3,$ for L^1,
$n_1 = 0,\dots,m-1,\quad n_i = 0,\dots,N-1,\quad i = 2,3,$ for L^2_m,
$n_3 = 0,\dots,m-1,\quad n_i = 0,\dots,N-1,\quad i = 1,2,$ for L^3_m.

Further we have:

$$\dim L^1 = N^3\,, \tag{2.177a}$$
$$\dim L^2_m = \dim L^3_m = mN^2\,, \tag{2.177b}$$
$$\dim L^4_{m_1m_2} = m_1m_2(m_1+m_2)/2\,, \tag{2.177c}$$
$$\dim L^5_m = m(N-m)N/2\,, \tag{2.177d}$$

$$\begin{aligned}\dim L'^4_{m_1m_2} &= m_1m_2(m_1+m_2)/2 - \\ &\quad -(N-m_1)(N-m_2)(2N-m_1-m_2)/2 = \qquad (2.178)\\ &= (m_1+m_2-N)(2m_1m_2+N(2N-m_1-m_2))/2\,,\\ &\quad 1 < m_1, m_2 < N < m_1+m_2 < 2N, \qquad (2.179)\\ \dim L^{51}_m &= \dim L^4_{N-m,N} = \dim L'^4_{N-m,N} = \\ &= N(N-m)(2N-m)/2\,, \quad m \leq N/2\,, \qquad (2.180)\\ \dim L^{52}_m &= \dim L^4_{N,m} = \dim L'^4_{N,m} = \\ &= Nm(N+m)/2\,, \quad m < N/2. \qquad (2.181)\end{aligned}$$

The Proof of this proposition was given in [199–201] except (2.174), *which was given in [36] (communicated to us by V.G. Kac).* ◊

The most interesting case is (2.178) where we get representations which cannot occur classically though being parametrized as the finite-dimensional representations of $sl(3,\mathbb{C})$. These are called *irregular representations* or *modular representations*. Clearly, all representations for which either $m_1 = 1$ or $m_2 = 1$ remains classical. (This includes the (three-dimensional) fundamental representations, characterized by $(m_1, m_2) = (2,1), (1,2)$, which are not deformed for any q.)

For $U_q(sl(3,\mathbb{C}))$ the simplest irregular case is the one which classically is the (eight-dimensional) adjoint representation characterized by $(m_1, m_2) = (2,2)$. Indeed, for third root of unity, $N = 3$, the inequalities in (2.178) are satisfied and the dimension of the irreducible HWM is seven. The reason is that for third root of unity there is one additional singular vector which has to be taken into account besides $(X_1^-)^2 \otimes v_0$, $(X_2^-)^2 \otimes v_0$. Explicitly, we have:

$$v_s = v^{\alpha_3,1} = (X_1^-X_2^- - X_2^-X_1^-) \otimes v_0,\, q = e^{2\pi i/3},\, m_1 = m_2 = 2. \tag{2.182}$$

Thus in the irreducible HWM L_Λ with vacuum state $|\rangle$ such that $X_i^+|\rangle = 0$ we have:

$$(X_1^-)^2|\rangle = 0,\, (X_2^-)^2|\rangle = 0,\, (X_1^-X_2^- - X_2^-X_1^-)|\rangle = 0. \tag{2.183}$$

Table 2.1: Modular representations of $U_q(sl(3,\mathbb{C}))$

(m_1, m_2)	$d(m_1, m_2)$	N	m_3	$\dim L_\Lambda$	$\deg v_s$
(2,2)	8	3	4	7	1
(3,2)	15	4	5	12	1
(3,3)	27	4	6	26	2
(3,3)	27	5	6	19	1
(4,2)	24	5	6	18	1
(4,3)	42	5	7	39	2
(4,3)	42	6	7	27	1
(4,4)	64	5	8	63	3
(4,4)	64	6	8	56	2
(4,4)	64	7	8	37	1

Then the seven states in L_Λ are:

$$|\rangle, X_1^-|\rangle, X_2^-|\rangle, X_1^-X_2^-|\rangle, X_1^-X_2^-X_1^-|\rangle, X_2^-X_1^-X_2^-|\rangle, X_2^-X_1^-X_2^-X_1^-|\rangle. \tag{2.184}$$

Note that the additional state in the eight-dimensional regular case (i. e., the adjoint representation) is the state $X_2^-X_1^-|\rangle$. Here it is not an independent state since it coincides with the state $X_1^-X_2^-|\rangle$ because of the last equality in (2.183) which is due to the additional singular vector (2.182) not present in the regular case.

We present the low-dimensional irregular or modular representation L_Λ of $U_q(sl(3,\mathbb{C}))$ in table 2.1.

2.8.4 Conjectures

In this subsection we shall discuss several conjectures. Let $\mathcal{G}$ be any simple Lie algebra; let $q^N = 1$, $N \in \mathbb{N}+1$, $\Lambda \in \mathcal{H}^*$, $m_i \equiv \Lambda(H_i)+1 < N$, $i = 1,\dots,r$; let $\tilde\alpha$ be the highest root of Δ. Then we conjecture that (2.166) holds if

$$m_{\tilde\alpha} \equiv (\Lambda+\rho)(H_{\tilde\alpha}) \leq N. \tag{2.185}$$

The support for this conjecture is the following. If $m_{\tilde\alpha} = kN + n_{\tilde\alpha} > N$, $k, n_{\tilde\alpha} \in \mathbb{N}$, $n_{\tilde\alpha} < N$, then it is easy to see that there shall exist at least one $\beta' \in \Delta^+$ such that $m_{\beta'} \equiv (\Lambda + \rho)(H'_\beta) = k'N + n_{\beta'}$ with $k' \in \mathbb{Z}_+$, $n_{\beta'} \in \mathbb{N}$, $n_{\beta'} < N$ so that $n_{\tilde\alpha} < n_{\beta'}$. Then the singular vector given by formula (2.111) with $\beta = \tilde\alpha$, $k = 0$, and $n = n_{\tilde\alpha}$ shall not factorize including as a factor the singular vector given by (2.111) with $\beta = \beta'$, $k = 0$, and $n = n_{\beta'}$. Thus the embedding pattern of the submodules of V^Λ is not the same as of Verma modules $V(\Lambda)$ with Λ integral dominant.

In [36] it is conjectured by a different motivation that (2.166) holds when $m_{\tilde{\alpha}} < N$, that is, for the so-called regular representations. The latter can be extended (for $q = 1$) to the affine Lie algebra counterpart of $\mathscr{G}$ if $m_0 + m_{\tilde{\alpha}} \equiv k + g = N$, where k is the affine central charge, g is the *dual Coxeter number*, $m_0 \in \mathbb{N}$. This is natural in view of the connection (albeit in a partial case) with the affine Weyl group commented above (cf. also [195]).

A more general conjecture again involving the affine Weyl group was given in [442]. Let $\in \mathbb{N} + 1$. Let W be the affine Weyl group with simple reflections $s_0, \dots, s_r$. Let E be an $\mathbb{R}$–vector space with basis $\gamma_1, \dots, \gamma_r$. A positive definite inner product in $E \times E$ is defined by $(\gamma_i, \gamma_j) = a'_{ij}$, where $(a'_{ij})_{1\le i,j\le r}$ is the matrix inverse to $(a_{ij})_{1\le i,j\le r}$. Further, denote:

$$\mathscr{C}_N = \{x = \sum_{i=1}^{r} c_i\gamma_i \in E | c_i \in \mathbb{R}, \quad c_i \le -1 \quad \text{for} \quad i = 1, \dots, r,$$
$$\sum_{j=1}^{r} m_j c_j \ge 1 - N - g\}. \tag{2.186}$$

This is a simplex in E with $r+1$ walls given by $c_i = -1$ for $i = 1, \dots, r$ and $m_j c_j = 1 - N - g$. Denote by S_i, $i = 1, \dots, r$, and S_0^N the orthogonal reflections in E with respect to these walls. Then $s_i \mapsto S_i$, $i = 1, \dots, r$, and $s_0 \mapsto S_0^N$ defines an embedding $j_N : W \to \mathrm{Aff}(E)$. Further $\mathbb{Z}^r$ is identified with a lattice in E by $(z_1, \dots, z_r) \mapsto \sum_{i=1}^r z_i\gamma_i$. If $x \in E$, then $x \in j_N(w)(\Delta_N)$; for some $w \in W$; among such w there is a unique one, denoted by $w_{x,N}$ of maximal length. If w' is any element of W, one defines $x_{w',N} = w' w_{x,N}^{-1} \in E$.

Further, let W_0 be the finite subgroup of W generated by $s_1, \dots, s_r$ and let $j = j_N|_{W_0}$. Let $\mathscr{C} = \{x = \sum_{i=1}^r c_i\gamma_i \in E | c_i \in \mathbb{R}, c_i \le -1 \text{ for } i = 1, \dots, r\}$. If $x \in E$, then $x \in j(w)(\mathscr{C})$; for some $w \in W_0$; among such w there is a unique one, denoted by w_x of maximal length. If w' is any element of W_0, one defines $x'_w = w' w_x^{-1} \in E$.

Now the conjecture of Lusztig [442] uses the Kazhdan–Lusztig polynomials $P_{y,w}$ [381] and Bruhat order [192] for W and W_0. Let $\Lambda \in \Gamma$, then:

$$ch\, L_\Lambda = \sum_{\substack{w' \in W_0 \\ w' \le w}} (-1)^{\ell(ww')} P_{w',w}(1) ch\, V^{\Lambda_{w'}}, \ w = w_\Lambda, \tag{2.187}$$
$$q = 1 \text{ or } q \text{ not a root of } 1,$$

$$ch\, L_\Lambda = \sum_{\substack{w' \in W \\ w' \le w}} (-1)^{\ell(ww')} P_{w',w}(1) ch\, V^{\Lambda_{w',N}}, \ w = w_{\Lambda,N}, \tag{2.188}$$
$$q \text{ a primitive root of } 1.$$

For $q = 1$ (2.187) is the Kazhdan–Lusztig restatement of (2.166) [381]. For q which is not a root of unity both follow from results of [441, 531]. In formula (2.188), besides the mysterious connection with affine Lie algebras, there is also a mysterious connection with the representation theory of simple algebraic groups over an algebraically closed field of characteristic N [381, 442], or with the representation theory of modular Lie algebras in characteristic N (cf. [36]).

Let us consider the last formula in some detail. First let us note that the simplex $\mathscr{C}_N$ corresponds in our notation to the restrictions in (2.185), namely, $m_i \equiv \Lambda(H_i)+1 < N$, $i = 1, \dots, r$, $m_{\tilde{\alpha}} \equiv (\Lambda + \rho)(H_{\tilde{\alpha}}) \le N$. Thus if these restrictions hold, then (2.188) coincides with (2.166) and (2.185). Further we can convince ourselves that in the $U_q(sl(3, \mathbb{C}))$ case formulae (2.187) and (2.188) coincide with the corresponding results presented in Sections 2.8.2 and 2.8.3.

3 Positive-Energy Representations of Noncompact Quantum Algebras

Summary

We construct positive-energy representations of noncompact quantum algebras at roots of unity. We give the general setting, and then we consider in detail the examples of the q-deformed anti de Sitter algebra $\mathscr{A}_q = U_q(so(3,2))$ and q-deformed conformal algebra $\mathscr{C}_q = U_q(su(2,2))$. For $\mathscr{A}_q$ we discuss in detail the singleton representations, while for $\mathscr{C}_q$ we discuss in detail the massless representations. When the deformation parameter q is N-th root of unity, all irreducible representations are finite-dimensional. We give the dimensions of these representations and their character formulae. Generically, these dimensions are not classical, except in some special cases, including the deformations of the fundamental irreps of $so(3,2)$ and $su(2,2)$. We follow the papers [165, 212, 225, 231].

3.1 Preliminaries

Let G be a simple connected noncompact Lie group with unitary highest-weight representations [264], and let $\mathscr{G}_0$ be its Lie algebra. Thus, $\mathscr{G}_0$ is one of the following Lie algebras: $su(m,n)$, $so(n,2)$, $sp(2n,R)$, $so^*(2n)$, $E_{6(-14)}$, $E_{7(-25)}$. We consider q-deformations $U_q(\mathscr{G}_0)$ constructed by the procedure proposed in [204] and reviewed in Section 1.5. The positive-energy irreps of $U_q(\mathscr{G}_0)$ are realized as lowest-weight module M of $U_q(\mathscr{G})$, where $\mathscr{G}$ is the complexification of $\mathscr{G}_0$, together with a hermiticity condition necessary for the construction of a scalar product in M. We take lowest instead of the more often used highest-weight modules since we want the energy to be bounded from below. We use the standard deformation $U_q(\mathscr{G})$ [251, 360] given in terms of the Chevalley generators $X_i^\pm$ and H_i, $i = 1, \ldots, r = \text{rank}\,\mathscr{G}$ by the relations (1.19).

A lowest-weight module M^Λ is given by the lowest-weight $\Lambda \in \mathscr{H}^*$ ($\mathscr{H}^*$ is the dual of $\mathscr{H}$) and a lowest-weight vector v_0 so that $Xv_0 = 0$ if $X \in \mathscr{G}^-$ and $Hv_0 = \Lambda(H)v_0$ if $H \in \mathscr{H}$. In particular, we use the Verma modules V^Λ which are the lowest-weight modules such that $V^\Lambda = U_q(\mathscr{G}^+)v_0$. Thus the Poincaré–Birkhof–Witt theorem (cf., e. g., Section 2.5.1) tells us that the basis of V^Λ consists of monomial vectors

$$\Psi_{\{\bar{k}\}} = (Y_{i_1}^+)^{k_{i_1}} \ldots (Y_{i_n}^+)^{k_{i_n}} v_0 = \mathscr{P}_{\{\bar{k}\}} v_0, \quad k_{i_j} \in \mathbb{Z}_+, \tag{3.1}$$

where $Y_i^+ \in \mathscr{G}^+$, $i_1 < i_2 < \ldots < i_n$, in some fixed ordering of the basis. A $U_q(\mathscr{G}_0)$-invariant scalar product in V^Λ is given by:

$$\left(\Psi_{\{\bar{k}'\}}, \Psi_{\{\bar{k}\}}\right) = \left(\mathscr{P}_{\{\bar{k}'\}} v_0, \mathscr{P}_{\{\bar{k}\}} v_0\right) = \left(v_0, \omega\left(\mathscr{P}_{\{\bar{k}'\}}\right) \mathscr{P}_{\{\bar{k}\}} v_0\right), \tag{3.2}$$

with $(v_0, v_0) = 1$ and ω is the conjugation which singles out $\mathscr{G}_0$, which has the property that $\omega(X^\pm) \in \mathscr{G}^\mp$ if $X^\pm \in \mathscr{G}^\pm$.

DOI 10.1515/9783110427707-003

We use the information on Verma modules as given in Chapter 2. Specifically, we recall that when the deformation parameter q is a root of unity, the picture of the representations changes drastically. In this case all Verma modules V^Λ are reducible [198], and all irreducible representations are finite-dimensional [175]. Let q be a primitive N-th root of unity; that is, $q = e^{2\pi i/N}$, where $N \in \mathbb{N}$ and $N \geq 1 + n(\mathcal{G})$, where $n(\mathcal{G}) = 1$ for $\mathcal{G} = A_n, D_n, E_n$, $n(\mathcal{G}) = 2$ for $\mathcal{G} = B_n, C_n, F_4$, $n(\mathcal{G}) = 3$ for $\mathcal{G} = G_2$ ($n(\mathcal{G})$ is the ratio $(\alpha_L, \alpha_L)/(\alpha_S, \alpha_S)$, where α_L is a long root, and α_S a short root). The maximal dimension of any irreducible representation is equal to d_N for N odd [175]. There are singular vectors for all positive roots α [198]. Condition (2.2) also has more content now because if $(\Lambda - \rho)(H_\alpha) = -m \in \mathbb{Z}$, then (2.2) will be fulfilled for all $m + kN_\alpha$, $k \in \mathbb{Z}$, $N_\alpha = N/n(\mathcal{G})$ if $N \in n(\mathcal{G})\mathbb{N}$ and α is a long root and $N_\alpha = N$ in all other cases. In particular, there is an infinite series of positive integers m such that (2.2) is true [198]. For identical reasons, there is an infinite number of lowest weights Λ such that (2.2) is satisfied for the same set of positive integers $m = m_\alpha$. The structure of the corresponding finite-dimensional irreps is the same since it is fixed by these positive integers.

Some of the finite-dimensional irreducible representations can be unitary as we show in the examples in the next sections.

We also give an interpretation of the spectrum via character formulae.

3.2 Quantum Anti de Sitter Algebra

3.2.1 Representations

Here we follow mostly [212, 231]. The first example we consider is the quantum anti de Sitter algebra; that is, we take $\mathcal{G}_0 = so(3,2)$ and $\mathcal{G} = so(5,\mathbb{C})$. In this case $r = 2$ and the nonzero products between the simple roots are $(\alpha_1, \alpha_1) = 2$, $(\alpha_2, \alpha_2) = 4$, and $(\alpha_1, \alpha_2) = -2$; thus $a_{12} = -2$, $a_{21} = -1$. The non-simple positive roots are $\alpha_3 = \alpha_1 + \alpha_2$ and $\alpha_4 = 2\alpha_1 + \alpha_2$. The Cartan–Weyl basis for the nonsimple roots is given by [198, 576]:

$$X_3^\pm = \pm q^{\mp 1/2}(q^{1/2} X_1^\pm X_2^\pm - q^{-1/2} X_2^\pm X_1^\pm), \quad X_4^\pm = \pm \left(X_1^\pm X_3^\pm - X_3^\pm X_1^\pm \right)/[2]_q. \tag{3.3}$$

All commutation relations now follow from the above relations. We mention, in particular:

$$[X_3^+, X_3^-] = [H_3]_q, \quad H_3 = H_1 + 2H_2, \quad [X_4^+, X_4^-] = [H_4]_{q^2}, \quad H_4 = H_1 + H_2, \tag{3.4}$$

where the Cartan generators H_3, H_4 corresponding to the nonsimple roots α_3, α_4 are chosen as in [242].

We choose the generators of $U_q(so(3,2))$ as a real form of $U_q(so(5,\mathbb{C}))$ as follows [242]:

$$M_{21} = H_1/2, \quad M_{31} = (X_1^+ + X_1^-)/2,$$
$$M_{32} = i(X_1^+ - X_1^-)/2, \tag{3.5a}$$
$$M_{04} = (H_1 + H_2)/2, \quad M_{30} = i(X_3^+ + X_3^-)/2,$$
$$M_{34} = (X_3^- - X_3^+)/2, \tag{3.5b}$$
$$M_{10} = i(X_4^+ + X_4^- + X_2^+ + X_2^-)/2,$$
$$M_{20} = (X_4^+ - X_4^- - X_2^+ + X_2^-)/2, \tag{3.5c}$$
$$M_{41} = (X_2^+ - X_2^- + X_4^+ - X_4^-)/2,$$
$$M_{42} = i(X_2^+ + X_2^- - X_4^+ - X_4^-)/2. \tag{3.5d}$$

Clearly, for $q = 1$ the ten generators $M_{AB} = -M_{BA}$, $A, B = 0, 1, 2, 3, 4$, satisfy the $so(3, 2)$ commutation relations (with $\eta_{AB} = \mathrm{diag}(+ - - - +)$):

$$[M_{AB}, M_{CD}] = i(\eta_{BC}M_{AD} - \eta_{AC}M_{BD} - \eta_{BD}M_{AC} + \eta_{AD}M_{BC}), \quad q = 1.$$

The commutation relations for $U_q(so(3, 2))$ follow from (3.5) and the commutation relations of $U_q(so(5, \mathbb{C}))$. The Cartan subalgebras of $U_q(so(3, 2))$ and $U_q(so(5, \mathbb{C}))$ are generated by the same generators M_{21}, M_{04} or H_1, H_2. Note that the generators in (3.5a) and M_{04} are compact; the rest are noncompact. In particular, those in (3.5a) generate a $U_q(su(2))$ subalgebra, those in (3.5b) a $U_q(su(1, 1))$ subalgebra.

For $|q| = 1$ the generators in (3.5) are preserved by the following antilinear anti-involution ω of $U_q(so(5, \mathbb{C}))$ [231]:

$$\omega(H_j) = H_j, \quad j = 1, 2, \quad \omega(X_1^+) = X_1^-, \quad \omega(X_k^+) = -X_k^-, \quad k = 2, 3, 4. \tag{3.6}$$

The restriction $|q| = 1$ follows from requiring consistency between (3.3) and (3.6), which is necessary since the generators $X_3^\pm$, $X_4^\pm$ are given in terms of $X_1^\pm$, $X_2^\pm$. Thus in what follows we work with $|q| = 1$.

For the four positive roots of the root system of $so(5, \mathbb{C})$, one has from (2.2) (cf. [242]):

$$m_1 = -\Lambda(H_1) + 1 = 2s_0 + 1, \tag{3.7a}$$
$$m_2 = -\Lambda(H_2) + 1 = 1 - E_0 - s_0, \tag{3.7b}$$
$$m_3 = -\Lambda(H_3) + 3 = m_1 + 2m_2 = 3 - 2E_0, \tag{3.7c}$$
$$m_4 = -\Lambda(H_4) + 2 = m_1 + m_2 = 2 - E_0 + s_0, \tag{3.7d}$$

where the representations are labelled (as those of $so(3, 2)$) by the lowest value of the energy E_0 and by the spin $s_0 \in \mathbb{Z}_+/2$ of the state with this energy.

Let us recall the list of the positive-energy representations of $so(3, 2)$ (cf. [191, 267, 289, 302]):

$$\begin{aligned}
&\text{Rac: } D(E_0, s_0) = D(1/2, 0), \qquad \text{Di: } D(E_0, s_0) = D(1, 1/2),\\
&D(E_0 > 1/2, s_0 = 0), \quad D(E_0 > 1, s_0 = 1/2),\\
&D(E_0 \geq s_0 + 1, s_0 \geq 1).
\end{aligned} \tag{3.8}$$

The first two are the *singleton representations*, which were first discovered by Dirac in [191], and the last ones for $E_0 = s_0 + 1$ correspond to the spin-s_0 *massless representations of so(3,2)*.

Let us consider (3.7) for this list. We note that in all cases $m_1 \in \mathbb{N}$ (because $s_0 \in \mathbb{Z}_+/2$) and $m_2 \notin \mathbb{N}$ (because $m_2 \leq 1/2$). Next, we note that m_3 is a positive integer only for $E_0 = 1/2, 1$, in which case $m_3 = 2, 1$, respectively. Similarly, m_4 is a positive integer only for $E_0 - s_0 = 1$, and that integer is $m_4 = 1$. Accordingly, we find the following singular vectors of the Verma module over $U_q(so(3, 2))$ [231]:

$$v_1^a = (X_1^+)^{2s_0+1} v_0, \quad s_0 \in \mathbb{Z}_+/2, \tag{3.9a}$$

$$v_{31}^a = ([2s_0]_q X_3^+ - (1+q) X_2^+ X_1^+) v_0, \quad m_3 = 1, \tag{3.9b}$$

$$v_{32}^a = ((X_3^+)^2 - q^{1/2}[2]_q^2 X_2^+ X_4^+) v_0, \quad m_3 = 2, \tag{3.9c}$$

$$\begin{aligned} v_4^a = {} & ([2s_0]_q [2s_0 - 1]_q X_4^+ + q^{s_0}[1 - 2s_0]_q X_3^+ X_1^+ + \\ & + X_2^+ (X_1^+)^2) v_0, \quad m_4 = 1. \end{aligned} \tag{3.9d}$$

Note that (3.9b) for $s_0 = 0$ and (3.9d) for $s_0 = 0, 1/2$ are composite singular vectors being *descendants* of (3.9a). We take the basis of the Verma module (3.1) in terms of the Cartan–Weyl generators as:

$$\Psi_{\{\bar{k}\}} = (X_4^+)^{k_4} (X_3^+)^{k_3} (X_2^+)^{k_2} (X_1^+)^{k_1} v_0, \quad k_j \in \mathbb{Z}_+. \tag{3.10}$$

Further, we concentrate on the *singleton* representations. To obtain the irreducible factor-representations L_Λ with ground states denoted by $|E_0, s_0\rangle$, we have to impose the following null-state vanishing conditions (following from (3.9)):

$$\text{Rac:} \quad X_1^+ |1/2, 0\rangle = 0, \quad ((X_3^+)^2 - q^{1/2}[2]_q^2 X_2^+ X_4^+) |1/2, 0\rangle = 0; \tag{3.11}$$

$$\text{Di:} \quad (X_1^+)^2 |1, 1/2\rangle = 0, \quad (X_3^+ - (1+q) X_2^+ X_1^+) |1, 1/2\rangle = 0. \tag{3.12}$$

(For $q = 1$ formulae (3.9), (3.11), and (3.12) were obtained in [242].)

Now we give explicitly the basis of L_Λ. We consider the monomials as in (3.10), but on the vacuum $|E_0, s_0\rangle$. Condition (3.11) means that in (3.10) we have $k_1 = 0$ and $k_3 \leq 1$, since we replace $(X_3^+)^2$ by $X_2^+ X_4^+$ (one may replace also $X_2^+ X_4^+$ by $(X_3^+)^2$ as in [242]). Similarly, (3.12) means that in (3.10) we have $k_1 \leq 1$ and $k_3 = 0$, since we replace X_3^+ by $X_2^+ X_1^+$. Thus, we see that the basis of L_Λ consists, as in the classical case [242], of the following monomials [231]:

$$\text{Rac:}\quad (X_4^+)^j (X_3^+)^\varepsilon (X_2^+)^k |1/2, 0\rangle, \quad j,k = 0,1,\ldots, \quad \varepsilon = 0,1, \tag{3.13}$$

$$\text{Di:}\quad (X_4^+)^j (X_2^+)^k (X_1^+)^\varepsilon |1, 1/2\rangle, \quad j,k = 0,1,\ldots, \quad \varepsilon = 0,1. \tag{3.14}$$

Note that each weight has multiplicity one, which was the reason these representations were called singletons [289].

Now we shall calculate the norms of these states. First we calculate some norms valid for any Λ:

$$\|(X_2^+)^j (X_1^+)^k |\Lambda\rangle\|^2 = [j]_{q^2}! [k]_q! \left(\prod_{\ell=1}^{j} [\Lambda(H_2) - k - 1 + \ell]_{q^2} \right) \times \\ \times \prod_{s=1}^{k} [1 - \Lambda(H_1) - s]_q, \tag{3.15a}$$

$$\|(X_3^+)^j (X_1^+)^k |\Lambda\rangle\|^2 = [j]_q! [k]_q! \left(\prod_{\ell=1}^{j} [\Lambda(H_3) - 1 + \ell]_q \right) \times \\ \times \prod_{s=1}^{k} [1 - \Lambda(H_1) - s]_q, \tag{3.15b}$$

$$\|(X_3^+)^j (X_2^+)^k |\Lambda\rangle\|^2 = [j]_q! [k]_{q^2}! \left(\prod_{\ell=1}^{j} [\Lambda(H_3) + 2k - 1 + \ell]_q \right) \times \\ \times \prod_{s=1}^{k} [\Lambda(H_2) - 1 + s]_{q^2}, \tag{3.15c}$$

$$\|(X_4^+)^j (X_2^+)^k (X_1^+)^\varepsilon |\Lambda\rangle\|^2 = [j]_{q^2}! [k]_{q^2}! [-\Lambda(H_1)]_q^\varepsilon \times \\ \times \left(\prod_{\ell=1}^{j} [\Lambda(H_4) - 1 + \varepsilon + \ell]_{q^2} \right) \prod_{s=1}^{k} [\Lambda(H_2) - 1 - \varepsilon + s]_{q^2},$$

$$\|(X_4^+)^j (X_3^+)^\varepsilon (X_2^+)^k |\Lambda\rangle\|^2 = [j]_{q^2}! [k]_{q^2}! [\Lambda(H_3) + 2k]_q^\varepsilon \times \\ \times \left(\prod_{\ell=1}^{j} [\Lambda(H_4) - 1 + \varepsilon + \ell]_{q^2} \right) \prod_{s=1}^{k} [\Lambda(H_2) - 1 + s]_{q^2}.$$

In all cases we consider we have $\Lambda(H_1) = -2s_0$. Thus we get from (3.15a) with $j = 0$

$$\|(X_1^+)^k |E_0, s_0\rangle\|^2 = [k]_q! \prod_{\ell=1}^{k} [2s_0 + 1 - \ell]_q, \tag{3.16}$$

which vanishes if $k \geq 2s_0 + 1 = m_1$; the latter statement is clear also from the null-state condition. In the same way we see that (3.15a,b) vanish for $k \geq 2s_0 + 1$ and any j. To calculate the other norms we also use $\Lambda(H_2) = E_0 + s_0$ (then $\Lambda(H_3) = 2E_0$, $\Lambda(H_4) = E_0 - s_0$).

Finally, the norms of the basis states (3.13) and (3.14) are:

$$\|(X_4^+)^j(X_3^+)^\varepsilon(X_2^+)^k|1/2,0\rangle\|^2 = [2]_q^\varepsilon[j]_{q^2}![k]_{q^2}!\left(\prod_{\ell=1}^{j}[\ell-1/2+\varepsilon]_{q^2}\right)\times$$
$$\times\prod_{s=1}^{k}[s-1/2+\varepsilon]_{q^2}, \tag{3.17}$$

$$\|(X_4^+)^j(X_2^+)^k(X_1^+)^\varepsilon|1,1/2\rangle\|^2 = [j]_{q^2}![k]_{q^2}!\left(\prod_{\ell=1}^{j}[\ell-1/2+\varepsilon]_{q^2}\right)\times$$
$$\times\prod_{s=1}^{k}[s+1/2-\varepsilon]_{q^2}, \tag{3.18}$$

3.2.2 Roots of Unity Case

In this subsection we consider the case where the deformation parameter is a root of unity, namely, $q = e^{2\pi i/N}$, $N = 3, 4, \ldots$

Let us denote

$$\tilde{N} = \begin{cases} N & \text{for N odd} \\ N/2 & \text{for N even} \end{cases} \qquad N_j = \begin{cases} N & \text{for j} = 1,3 \\ \tilde{N} & \text{for j} = 2,4. \end{cases} \tag{3.19}$$

In this situation independently of the weight Λ there are singular vectors for all positive roots α_j, which are given by: $(X_j^+)^{kN}v_0$, $j = 1,3$, and $(X_j^+)^{k\tilde{N}}v_0$, $j = 2,4$, $k = 1,2,\ldots$ [198]. Thus we have to impose the following vanishing of null states in our representation spaces:

$$(X_j^+)^N|E_0,m_0\rangle = 0, \quad j = 1,3, \quad (X_j^+)^{\tilde{N}}|E_0,m_0\rangle = 0, \quad j = 2,4. \tag{3.20}$$

Taking into account condition (2.2) we see that if $m_j = (\rho-\Lambda)(H_j) \in \mathbb{Z}$, $j = 1,2,3,4$, there would be singular vectors of weights $(n_j' + kN_j)\alpha_j$, where $n_j' = \{m_j\}_{N_j}$, $\{x\}_p$ being the smallest positive integer equal to x (mod p), and $k = 0,1,\ldots$. Analogously, if $m_j \in 1/2+\mathbb{Z}$, $j = 2,4$, and N is odd, there would be singular vectors of weights $(n_j' + kN)\alpha_j$, $n_j' = \{m_j + N/2\}_N$, $k = 0,1,\ldots$. In particular, we have to impose:

$$(X_j^+)^{n_j'}|E_0,m_0\rangle = 0, \quad j = 1,2. \tag{3.21}$$

Further our representations will be characterized by the following positive integers:

$$n_1 = \{2s_0 + 1\}_N = \{m_1\}_N, \tag{3.22}$$

$$n_2 = \begin{cases} \{1 - E_0 - s_0\}_{\tilde{N}} = \{m_2\}_{\tilde{N}}, & \text{if} \quad \mathrm{E}_0 + \mathrm{s}_0 \in \mathbb{Z}, \\ \{1 - E_0 - s_0 + N/2\}_N = \{m_2 + N/2\}_N, & \text{if} \quad \mathrm{E}_0 + \mathrm{s}_0 \in 1/2 + \mathbb{Z}, \\ & N \text{ odd}, \\ \tilde{N}, & \text{otherwise}. \end{cases}$$

Note that $n_k \leq N_k$, $k = 1, 2$.

Let us recall that the finite-dimensional irreducible representations of $so(5, \mathbb{C})$ (or of other real form of $so(5, \mathbb{C})$ and of the corresponding quantum algebras when q is not a root of unity) are parametrized by two arbitrary positive integers, say, p_1, p_2, and the dimension of such a representation is given by:

$$d^c_{p_1,p_2} = \frac{1}{6} p_1\, p_2\, p_3\, p_4, \tag{3.23}$$

where $p_3 = p_1 + 2p_2$, $p_4 = p_1 + p_2$.

Now for N odd we divide our representations in classes depending on the values of $n_3 = n_1 + 2n_2$, $n_4 = n_1 + n_2$ and n_1, n_2:

$$a)\ n_3, n_4 \leq N, \tag{3.24a}$$

$$b)\ n_4 < N < n_3 < 2N, \tag{3.24b}$$

$$b')\ n_4 = N < n_3 \leq 2N, \qquad \text{or} \quad n_4 < n_3/2 = N,$$

$$c)\ n_1 < N < n_3, n_4 < 2N, \tag{3.24c}$$

$$c')\ n_1 = N < n_3, n_4 \leq 2N, \qquad \text{or} \quad n_1 < N < n_4 < n_3 = 2N,$$

$$d)\ n_2 < N < n_4 < 2N < n_3 < 3N, \tag{3.24d}$$

$$d')\ n_2 = N < n_4 \leq 2N < n_3 \leq 3N.$$

The same classification is valid for $U_q(so(5, \mathbb{C}))$, where (3.24a) is the regular case. This is a refinement of the classification of [231], the primed cases being separated out since, together with the regular case, these have the classical dimensions of the finite-dimensional irreps of $so(5, \mathbb{C})$; that is, a representation characterized by n_1, n_2 has dimension $d^c_{n_1,n_2}$. In particular, in case d') with $n_1 = n_2 = N$, we achieve the maximal possible dimension N^4 of an irrep of $U_q(so(5, \mathbb{C}))$ (cf. [175]). On the other hand, in the unprimed cases $b) - d)$, the dimension of a representation characterized by n_1, n_2 is strictly smaller than $d^c_{n_1,n_2}$. The representations $U_q(so(3, 2))$ inherit all the structure from their $U_q(so(5, \mathbb{C}))$ counterparts. Thus, the classification of the positive-energy representations of $U_q(so(3, 2))$ proceeds as follows.

Let us decompose: $2s_0 = 2S_0 + r_0 N$, $2S_0, r_0 \in \mathbb{Z}_+$, $2S_0 < N$. Then we have:

$$n_1 = 2S_0 + 1. \tag{3.25}$$

Now the formulae for n_2 depend on the combination $E_0 + s_0$.

Suppose first that $E_0 + s_0 \notin \mathbb{Z}/2$. Then we have:

$$n_2 = N, \quad n_3 = 2N + 2S_0 + 1 > 2N, \quad n_4 = N + 2S_0 + 1 > N, \quad \text{odd } N, \tag{3.26}$$

which is case (3.24d).

Next we consider the case $E_0 + s_0 \in \mathbb{Z}$. Taking into account the conditions of positive energy (3.8), we see that we have $E_0 \geq s_0 + 1$. Thus we set $E_0 = s_0 + 1 + p + kN$, where $p = 0, 1, \ldots, N-1$, $k \in \mathbb{Z}_+$. Let us also set $\kappa = 2S_0 + p$. Note that $0 \leq \kappa \leq 2N-2$. Then we have for N odd:

$$\begin{aligned}
n_2 &= N - \kappa,\\
n_3 &= 2N - \kappa - p + 1 \begin{cases} \leq N & \text{for} \quad \kappa + p > N,\\ > N \& \leq 2N & \text{for} \quad \kappa + p \leq N, \kappa > 0,\\ > 2N & \text{for} \quad \kappa = 0, \end{cases}\\
n_4 &= N - p + 1 \begin{cases} \leq N & \text{for} \quad p > 0,\\ > N & \text{for} \quad p = 0; \end{cases}\\
&\quad \kappa < N,
\end{aligned} \tag{3.27a}$$

$$\begin{aligned}
n_2 &= 2N - \kappa,\\
n_3 &= 4N - \kappa - p + 1 \begin{cases} > N \& \leq 2N & \text{for} \quad \kappa + p \geq 2N + 1,\\ > 2N & \text{for} \quad \kappa + p \leq 2N, \end{cases}\\
n_4 &= 2N - p + 1 > N,\\
&\quad \kappa \geq N\,.
\end{aligned} \tag{3.27b}$$

Thus we have case (3.24a) in (3.27a) when $\kappa + p \geq N + 1 \,\&\, p > 0$, case (3.24b) in (3.27a) when $\kappa + p \leq N \,\&\, p > 0$ ($\Rightarrow \kappa > 0$), case (3.24c) in (3.27a) when $p = 0 \,\&\, \kappa > 0$ and in (3.27b) when $\kappa + p \geq 2N + 1$, case (3.24d) in (3.27a) when $\kappa = 0$ ($\Rightarrow p = 0$), and in (3.27b) when $\kappa + p \leq 2N$.

Then we consider the case $E_0 + s_0 \in 1/2 + \mathbb{Z}$ for N odd. Taking into account the conditions of positive energy (3.8), we see that we have $E_0 \geq s_0 + 1/2$. Thus we set $E_0 = s_0 + 1/2 + p + kN$, where $p = 0, 1, \ldots, N-1$, $k \in \mathbb{Z}_+$. As above we set $\kappa = 2S_0 + p$ ($0 \leq \kappa \leq 2N-2$). We also denote $\hat{N} = (N+1)/2 \in \mathbb{N} + 1$. Then we have:

$$
\begin{aligned}
&n_2 = \hat{N} - \kappa, \\
&n_3 = N - \kappa - p + 2 \begin{cases} \leq N & \text{for} \quad \kappa + p \geq 2, \\ > N \& \leq 2N & \text{for} \quad \kappa + p \leq 1, \end{cases} \\
&n_4 = \hat{N} - p + 1 < N, \\
&\qquad\qquad\qquad\qquad \kappa < \hat{N}.
\end{aligned} \tag{3.28a}
$$

$$
\begin{aligned}
&n_2 = N + \hat{N} - \kappa, \\
&n_3 = 3N - \kappa - p + 2 \begin{cases} \leq N & \text{for} \quad \kappa + p \geq 2N + 2, \\ > N \& \leq 2N & \text{for} \quad N + 2 \leq \kappa + p \leq 2N + 1, \\ > 2N & \text{for} \quad \kappa + p \leq N + 1, \end{cases} \\
&n_4 = N + \hat{N} - p + 1 \begin{cases} \leq N & \text{for} \quad p > \hat{N}, \\ > N & \text{for} \quad p \leq \hat{N}, \end{cases} \\
&\qquad\qquad\qquad\qquad \hat{N} \leq \kappa \leq N + \hat{N}
\end{aligned} \tag{3.28b}
$$

$$
\begin{aligned}
&n_2 = 2N + \hat{N} - \kappa, \\
&n_3 = 5N - \kappa - p + 2 > 2N, \\
&n_4 = 2N + \hat{N} - p + 1 > N, \\
&\qquad\qquad\qquad\qquad \kappa \geq N + \hat{N}.
\end{aligned} \tag{3.28c}
$$

Thus we have case (3.24a) in (3.28a) when $\kappa+p \geq 2$ and in (3.28b) when $\kappa+p \geq 2N+2$ ($\Rightarrow p > \hat{N}$), case (3.24b) in (3.28a) when $\kappa + p \leq 1$ and in (3.28b) when $p > \hat{N}$ & $\kappa + p \leq 2N + 1$ ($\Rightarrow \kappa + p \geq N + 2$), case (3.24c) in (3.28b) when $p \leq \hat{N}$ & $\kappa + p \geq N + 2$ ($\Rightarrow \kappa + p \leq 2N + 1$), and case (3.24d) in (3.28b) when $\kappa + p \leq N + 1$ ($\Rightarrow p \leq \hat{N}$) and in (3.28c).

After the above analysis it remains to mention that the singleton irreps, $(E_0, s_0) = (1/2, 0), (1, 1/2)$, belong to case (3.24b) (cf. (3.28a) with $\kappa = 0, 1$, $p = 0$), while the massless irreps, $E_0 = s_0 + 1$, belong to case (3.24c).

This completes the classification of the positive-energy representations of $U_q(so(3,2))$ at odd roots of 1.

Further we treat in detail the singleton cases. In the case of the *Rac* besides (3.11) a new vanishing condition is:

$$(X_2^+)^{n_2} |1/2, 0\rangle = 0, \quad n_2 = [(N+1)/2]_{\text{int}}, \tag{3.29}$$

where $[x]_{\text{int}}$ is the biggest integer smaller or equal to x; note that this condition is (3.20) for N even and (3.21) for N odd. Further using (1.21) we find that the following states from (3.13) have positive norms [231]:

$$\|(X_4^+)^j (X_3^+)^\varepsilon (X_2^+)^k |1/2, 0\rangle\|^2 > 0, \quad \text{iff} \begin{cases} j, k \leq (N - 1 - 2\varepsilon)/2 & \text{for N odd} \\ j, k \leq (N-2)/2 & \text{for N even} \end{cases} \tag{3.30}$$

Due to factors in (3.17): $[j-1/2+\varepsilon]_{q^2}$, $[k-1/2+\varepsilon]_{q^2}$ for N odd, and $[j]_{q^2}$, $[k]_{q^2}$ for N even; all other states from (3.13) have zero norm and decouple from the irrep. Thus we calculate the dimension of the Rac irrep by counting the states in (3.30), which are $(N+1-2\varepsilon)^2/4$ for $\varepsilon = 0, 1$ and N odd, and $N^2/4$ for $\varepsilon = 0, 1$, and N even. Thus we get [231]:

$$\dim \text{Rac} = \begin{cases} \frac{N^2+1}{2}, & \text{for N odd} \\ \frac{N^2}{2}, & \text{for N even} \end{cases} \tag{3.31}$$

In the case of the *Di* besides (3.12) the new vanishing condition is:

$$(X_2^+)^{n_2}|1, 1/2\rangle = 0, \quad n_2 = [N/2]_{\text{int}}, \tag{3.32}$$

again this is (3.20) for N even and (3.21) for N odd. Then we find from (3.18) that the following states have positive norms [231]:

$$\|(X_4^+)^j(X_2^+)^k(X_1^+)^\varepsilon|1, 1/2\rangle\|^2 > 0, \quad \text{iff} \quad \begin{cases} j \le (N-1-2\varepsilon)/2 & \textit{and} \\ k \le (N-3+2\varepsilon)/2 & \text{for N odd} \\ j, k \le (N-2)/2 & \text{for N even} \end{cases} \tag{3.33}$$

and the counting of states gives [231]:

$$\dim \text{Di} = \begin{cases} \frac{N^2-1}{2}, & \text{for N odd} \\ \frac{N^2}{2}, & \text{for N even} \end{cases} \tag{3.34}$$

Thus the dimension of a singleton irrep for fixed N is strictly smaller than the minimal dimension of a (semi-) periodic irrep of $U_q(so(5, \mathbb{C}))$, which is N^2 [177]. The interesting thing is that the *sum* of the dimensions of the two singletons is exactly N^2. Thus we are led to the *conjecture* that passing from a minimal (semi-) periodic irrep of $U_q(so(5, \mathbb{C}))$ to a lowest-weight module of $U_q(so(5, \mathbb{C}))$ (by setting the corresponding Casimir values to zero), we obtain a reducile representation which is the direct sum of two irreps. The latter irreps when restricted to $U_q(so(3, 2))$ are the two singleton representations.

3.2.3 Character Formulae

When q is not a nontrivial root of 1, the spectrum of the singletons can be represented by the following character formulae (containing the same information as (3.13) and (3.14)):

$$ch\, L_{\text{Rac}} = e(\Lambda)(1 + t_3) \sum_{j=0}^{\infty} t_4^j \sum_{k=0}^{\infty} t_2^k, \tag{3.35}$$

$$ch\, L_{\mathrm{Di}} = e(\Lambda)(1+t_1)\sum_{j=0}^{\infty} t_4^j \sum_{k=0}^{\infty} t_2^k, \tag{3.36}$$

where $t_3 = e(\alpha_1+\alpha_2) = t_1 t_2$, $t_4 = e(2\alpha_1+\alpha_2) = t_1^2 t_2$. (For $q = 1$ these formulae were given in a slightly different, but equivalent, form in [242].) Now we note that the character formula for the Verma module with the same lowest weight here is:

$$ch\, V^\Lambda = e(\Lambda)/(1-t_1)(1-t_2)(1-t_3)(1-t_4). \tag{3.37}$$

Then we can rewrite the character formulae (3.35) and (3.36) as follows [242]:

$$ch\, L_{\mathrm{Rac}} = ch\, V^\Lambda(1 - t_1 - t_1^2 t_2^2 + t_1^3 t_2^2), \tag{3.38}$$

$$ch\, L_{\mathrm{Di}} = ch\, V^\Lambda(1 - t_1^2 - t_1 t_2 + t_1^3 t_2). \tag{3.39}$$

These formulae represent alternating sign summations over part of the Weyl group of $so(5,\mathbb{C})$, which was called *reduced Weyl group* in [196]).

Next we note that the spectrum given in (3.30) and (3.33) can be represented by the following character formulae for N odd:

$$ch\, L_{\mathrm{Rac}} = e(\Lambda)\left(\sum_{j=0}^{(N-1)/2} t_4^j \sum_{k=0}^{(N-1)/2} t_2^k + t_3 \sum_{j=0}^{(N-3)/2} t_4^j \sum_{k=0}^{(N-3)/2} t_2^k\right), \tag{3.40}$$

$$ch\, L_{\mathrm{Di}} = e(\Lambda)\left(\sum_{j=0}^{(N-1)/2} t_4^j \sum_{k=0}^{(N-3)/2} t_2^k + t_1 \sum_{j=0}^{(N-3)/2} t_4^j \sum_{k=0}^{(N-1)/2} t_2^k\right). \tag{3.41}$$

Let us denote by $L^c_{n_1,n_2}$ the finite-dimensional irreps of $so(5,\mathbb{C})$. The corresponding character formula, which is the classical Weyl character formula, is:

$$ch\, L^c_{n_1,n_2} = ch\, V^\Lambda(1 - t_1^{n_1} - t_2^{n_2} - t_3^{n_3} - t_4^{n_4} + t_1^{n_1} t_2^{n_4} + t_1^{n_3} t_2^{n_2} + t_1^{2n_4} t_2^{n_3}), \tag{3.42}$$

where the eight terms represent (alternating sign) summation over the (eight element) Weyl group of $so(5,\mathbb{C})$.

As we mentioned, the dimension of a unitary irrep of $U_q(so(3,2))$ characterized by n_1, n_2 is generically smaller than $d^c_{n_1,n_2}$. In particular, for the Rac when N is odd we have $(n_1,n_2) = (1,(N+1)/2)$. We have that $d^c_{1,(N+1)/2} = (N+1)(N+2)(N+3)/24 \geq dim_{\mathrm{Rac}} = (N^2+1)/2$. It is easy to notice that $\dim_{\mathrm{Rac}}$ may be represented as the difference of two dimensions:

$$\dim_{\mathrm{Rac}} = d^c_{1,(N+1)/2} - d^c_{1,(N-3)/2} \tag{3.43}$$

where the subtracted term corresponds to the weight $\Lambda' = \Lambda + 2\alpha_3$ with characterizing integers given by: $n'_j = (\rho - \Lambda')(H_j) = n_j - 2\alpha_3(H_j)$; that is, $(n'_1, n'_2) = (n_1, n_2 - 2)$. Correspondingly, the character formula for odd N is given by (cf. (3.40)):

$$\begin{aligned} ch\, L_{\mathrm{Rac}} &= ch\, L^c_{1,(N+1)/2} - ch\, L^c_{1,(N-3)/2} = \\ &= ch\, V^\Lambda \left(P_{1,(N+1)/2} - t_3^2 P_{1,(N-3)/2}\right), \end{aligned} \tag{3.44}$$

where we have introduced the notation: $ch\, L^c_{n_1,n_2} = ch\, V^\Lambda P_{n_1,n_2}$,

Note that the subtraction term vanishes only for $N = 3$, which is the only case when the quantum Rac dimension coincides with a classical dimension, here of one of the fundamental irreps of $so(3,2)$ with $d^c = 5$.

Analogously, for the Di when N is odd we have $(n_1, n_2) = (2, (N-1)/2)$. Here we have that $d^c_{2,(N-1)/2} = (N^2-1)(N+3)/12 \geq dim\mathrm{Di} = (N^2-1)/2$, and equality is possible only for $N = 3$; then the dimension is of the other fundamental irrep, $d^c = 4$. Here we have to subtract the character $ch\, L^c_{\Lambda'}$ with $\Lambda' = \Lambda + \alpha_3$, and $(n'_1, n'_2) = (n_1, n_2 - 1)$. We have for odd N:

$$\begin{aligned} ch\, L_{\mathrm{Di}} &= ch\, L^c_{2,(N-1)/2} - ch\, L^c_{2,(N-3)/2} = \\ &= ch\, V^\Lambda \left(P_{2,(N-1)/2} - t_3 P_{2,(N-3)/2}\right), \end{aligned} \tag{3.45}$$

$$\dim_{\mathrm{Di}} = d^c_{2,(N-1)/2} - d^c_{2,(N-3)/2}. \tag{3.46}$$

3.3 Conformal Quantum Algebra

3.3.1 Generic Case

The other example that we consider is the conformal algebra; that is, we take $\mathscr{G}_0 = su(2,2)$ and $\mathscr{G} = sl(4,\mathbb{C})$. In this case $r = 3$, and the nonzero products between the simple roots are $(\alpha_j, \alpha_j) = 2$, $j = 1,2,3$ and $(\alpha_1, \alpha_2) = (\alpha_2, \alpha_3) = -1$. The nonsimple positive roots are $\alpha_{12} = \alpha_1 + \alpha_2$, $\alpha_{23} = \alpha_2 + \alpha_3$, $\alpha_{13} = \alpha_1 + \alpha_2 + \alpha_3$. The Cartan–Weyl basis for the nonsimple roots is given by [202, 360]:

$$\begin{aligned} X^\pm_{jk} &= \pm q^{\mp 1/4}(q^{1/4} X^\pm_j X^\pm_k - q^{-1/4} X^\pm_k X^\pm_j), (jk) = (12), (23), \\ X^\pm_{13} &= \pm q^{\mp 1/4}(q^{1/4} X^\pm_1 X^\pm_{23} - q^{-1/4} X^\pm_{23} X^\pm_1) = \\ &= \pm q^{\mp 1/4}(q^{1/4} X^\pm_{12} X^\pm_3 - q^{-1/4} X^\pm_3 X^\pm_{12}). \end{aligned} \tag{3.47}$$

To single out $U_q(su(2,2))$ we use the following antilinear anti-involution [165]:

$$\omega(H) = H, \forall H \in \mathscr{H}, \quad \omega(X^\pm_{jk}) = \begin{cases} X^\mp_{jk}, & (jk) = (11), (33), \\ -X^\mp_{jk}, & \text{otherwise}. \end{cases} \tag{3.48}$$

For the six positive roots of the root system of $sl(4,\mathbb{C})$ one has from (2.2) that the Verma module V^Λ is reducible when:

$$m_1 = -\Lambda(H_1) + 1 = 2j_1 + 1, \tag{3.49a}$$
$$m_2 = -\Lambda(H_2) + 1 = 1 - d - j_1 - j_2, \tag{3.49b}$$
$$m_3 = -\Lambda(H_3) + 1 = 2j_2 + 1, \tag{3.49c}$$
$$m_{12} = -\Lambda(H_{12}) + 2 = m_1 + m_2 = 2 - d + j_1 - j_2\,, \tag{3.49d}$$
$$m_{23} = -\Lambda(H_{23}) + 2 = m_2 + m_3 = 2 - d - j_1 + j_2, \tag{3.49e}$$
$$m_{13} = -\Lambda(H_{13}) + 3 = m_1 + m_2 + m_3 = 3 - d + j_1 + j_2, \tag{3.49f}$$

where we use the classical labelling of the $su(2,2)$ representations: $2j_1, 2j_2$ are non-negative integers fixing finite-dimensional irreducible representations of the Lorentz subalgebra, and $d > 0$ is the energy (or conformal dimension). First we note that m_1 and m_3 are positive, since $2j_1$ and $2j_2$ are non-negative integers. The corresponding singular vectors are:

$$v_1 = \left(X_1^+\right)^{2j_1+1} v_0, \quad v_3 = \left(X_3^+\right)^{2j_2+1} v_0, \tag{3.50}$$

and these are present for all representations we discuss. Next, it is clear that depending on the value of d there may be other singular vectors. Since we are interested in the positive-energy irreps, we recall the list of these representations for $su(2,2)$ [449]:

$$\begin{aligned} &1)\ d > j_1 + j_2 + 2, && j_1j_2 \neq 0,\\ &2)\ d = j_1 + j_2 + 2\,, && j_1j_2 \neq 0,\\ &3)\ d > j_1 + j_2 + 1, && j_1j_2 = 0,\\ &4)\ d = j_1 + j_2 + 1, && j_1j_2 = 0, \end{aligned} \tag{3.51}$$

(omitting the one-dimensional representation with $d = j_1 = j_2 = 0$). In case 1) there are no additional singular vectors. If $d = j_1 + j_2 + 2$, which is case 2) and is also possible in case 3), then $m_{13} = 1$, and there is an additional singular vector:

$$\begin{aligned} v_{13}^{(1)} = \Big(&[2j_1][2j_2]X_1^+X_3^+X_2^+ - [2j_1][2j_2+1]X_1^+X_2^+X_3^+ - \\ &-[2j_1+1][2j_2]X_3^+X_2^+X_1^+ + [2j_1+1][2j_2+1]X_2^+X_1^+X_3^+\Big)v_0. \end{aligned} \tag{3.52}$$

Further, we concentrate on case 4), that is, to the *massless representations of so(4,2)* [165, 225, 449] for which $d = j_1 + j_2 + 1 \geq 1$, $j_1j_2 = 0$. For definiteness we choose first $j_2 = 0$. Then we see that in the case $j_1 \neq 0$, we have a singular vector corresponding to $m_{12} = 1$ [165, 225]:

$$v_{12} = \left([2j_1]X_{12}^+ - q^{j_1}X_2^+X_1^+\right)v_0, \quad d = j_1 + 1, j_2 = 0, \quad m_{12} = 1, \tag{3.53}$$

and another one which corresponds to $m_{13} = 2$ [165, 225], which, however, is a composite one and is not relevant. When $j_1 = 0$ there is still another composite singular vector corresponding to $m_{23} = 1$ [165, 225]. Furthermore, for $j_1 = 0$ the vector $v_{12} = X_2^+ X_1^+ v_0$ is also composite. Next we factor all invariant submodules built on these singular vectors. However, this factor representation is still reducible since it has an additional singular vector [225]:

$$v_f = \left(X_{13}^+ X_2^+ - q^{-1/2} X_{12}^+ X_{23}^+ \right) \widetilde{|\rangle}, \tag{3.54}$$

where $\widetilde{|\rangle}$ denotes the ground-state vector of this factor representation. [This is actually a *subsingular vector* of the Verma module V^Λ (cf. [215]).] Factoring out the submodule built on v_f, we obtain the irreducible lowest-weight representation L_Λ whose vacuum vector $|\rangle$ obeys [225]:

$$\left(X_1^+\right)^{2j_1+1} |\rangle = 0, \tag{3.55a}$$

$$X_3^+ |\rangle = 0, \tag{3.55b}$$

$$\left([2j_1] X_{12}^+ - q^{j_1} X_2^+ X_1^+ \right) |\rangle = 0, \tag{3.55c}$$

$$\left(X_{13}^+ X_2^+ - q^{-1/2} X_{12}^+ X_{23}^+ \right) |\rangle = 0. \tag{3.55d}$$

Now we can give explicitly the basis of L_Λ. We consider the monomials as in (3.1), but on the vacuum $|\rangle$. Taking into account all vanishing conditions we see that the basis of L_Λ consists of the following monomials [225]:

$$\begin{aligned}
\Phi^1_{\{k,\ell,n\}} &= (X_{13}^+)^k (X_{12}^+)^\ell (X_2^+)^n \,|\,\rangle\,, \quad k,\ell,n \in \mathbb{Z}_+\,, \\
\Phi^2_{\{k,\ell,n\}} &= (X_{13}^+)^k (X_{23}^+)^\ell (X_2^+)^n \,|\,\rangle\,, \quad k,n \in \mathbb{Z}_+\,,\ \ell \in \mathbb{N}, \\
\Phi^3_{\{k,\ell,n\}} &= (X_{13}^+)^k (X_{12}^+)^\ell (X_1^+)^n \,|\,\rangle\,, \quad k,\ell,n \in \mathbb{Z}_+\,,\ 1 \le n \le 2j_1\,,
\end{aligned} \tag{3.56}$$

the third case being absent for $j_1 = 0$. We note that the different vectors in (3.56) have different weights. Thus each weight has multiplicity one and is represented by a single vector just as the singletons of $so(3,2)$ (cf. the previous section).

The norms squared of the basis vectors $\|\Phi^a_{\{k,\ell,n\}}\|^2 \equiv (\Phi^a_{\{k,\ell,n\}}, \Phi^a_{\{k,\ell,n\}})$ are explicitly given by [225]:

$$\begin{aligned}
\|\Phi^1_{\{k,\ell,n\}}\|^2 &= [k]_q! [k+\ell]_q!\, [\ell+n]_q! [n+2j_1]_q! / [2j_1]_q! \\
\|\Phi^2_{\{k,\ell,n\}}\|^2 &= [k]_q! [k+\ell]_q!\, [\ell+n+2j_1]_q! [n]_q! / [2j_1]_q!, \\
\|\Phi^3_{\{k,\ell,n\}}\|^2 &= [k]_q! [k+\ell+n]_q!\, [\ell]_q! [2j_1]_q! / [2j_1-n]_q!\,.
\end{aligned} \tag{3.57}$$

When q is not a root of unity these norms can have both signs. They are positive only for $q = 1$, which is the well-known classical case of $su(2,2)$ [449]. Note, however, that such a basis is new also for the algebra $su(2,2)$. Unitarity can be achieved also when q is a nontrivial root of unity, which case we consider in the next subsection.

3.3.2 Roots of 1 Case

Let us now turn to the case of the deformation parameter q being a nontrivial root of unity, namely, $q = e^{2\pi i/N}$, $N = 2, 3, \ldots$.

Independently of the weight Λ there are singular vectors for all positive roots α, which are given by: $(X^+_\alpha)^{kN} v_0$, $k = 1, 2, \ldots$ [202]. Thus we have to impose the following vanishing of null states in our representation spaces:

$$(X^+_\alpha)^N |\rangle = 0. \tag{3.58}$$

Taking into account condition (2.2) we see that if $m_\alpha = (\rho - \Lambda)(H_\alpha) \in \mathbb{Z}$, there would be singular vectors of weights $(\{m_\alpha\}_N + kN)\alpha$, where $\{x\}_p$ is the smallest positive integer equal to x (mod p), and $k = 0, 1, \ldots$. In particular, we have to impose:

$$(X^+_j)^{\{m_j\}_N} |\rangle = 0, \quad j = 1, 2, 3. \tag{3.59}$$

Further our representations will be characterized by the following positive integers:

$$\begin{aligned} n_1 &= \{2j_1 + 1\}_N = \{m_1\}_N \\ n_2 &= \begin{cases} \{-d - j_1 - j_2 + 1\}_N = \{m_2\}_N, & \text{if } \mathrm{d} + \mathrm{j}_1 + \mathrm{j}_2 \in \mathbb{Z}, \\ N, & \text{if } \mathrm{d} + \mathrm{j}_1 + \mathrm{j}_2 \notin \mathbb{Z}, \end{cases} \\ n_3 &= \{2j_2 + 1\}_N = \{m_3\}_N. \end{aligned} \tag{3.60}$$

Note that $n_k \leq N$, $k = 1, 2, 3$.

Let us recall that the finite-dimensional irreducible representations of $sl(4, \mathbb{C})$ (or of $su(2, 2)$, or of $su(4)$, or of any other real form of $sl(4, \mathbb{C})$ and of the corresponding quantum algebras when q is not a root of unity) are parametrized by three arbitrary positive integers say, p_1, p_2, p_3, and the dimension of such a representation is given by:

$$d^C_{p_1,p_2,p_3} = \frac{1}{12} p_1\, p_2\, p_3\, p_{12}\, p_{23}\, p_{13}, \tag{3.61}$$

where $p_{12} = p_1 + p_2$, $p_{23} = p_2 + p_3$, $p_{13} = p_1 + p_2 + p_3$.

Now the representations are divided into classes [165] depending on the values of $n_{12} = n_1 + n_2$, $n_{23} = n_2 + n_3$, $n_{13} = n_1 + n_2 + n_3$ and n_k:

$$a)\ n_{jk} \leq N\,, \tag{3.62a}$$

$$b)\ n_{12},\, n_{23} < N < n_{13} \leq 2N\,, \tag{3.62b}$$

$$b')\ n_{12} < n_{23} = N < n_{13} \leq 2N\,, \quad \text{or} \quad n_{12} \longrightarrow n_{23}\,, \tag{3.62b$'$}$$

$$c)\ n_{12} \leq N < n_{23},\, n_{13} \leq 2N\,, \quad n_3 < N\,, \tag{3.62c}$$

$$c')\ n_{12} \leq N < n_{23},\, n_{13} \leq 2N\,, \quad n_3 = N\,, \tag{3.62c$'$}$$

$$d)\ n_{23} \le N < n_{12},\, n_{13} \le 2N\,, \qquad n_1 < N\,, \tag{3.62d}$$

$$d')\ n_{23} \le N < n_{12},\, n_{13} \le 2N\,, \qquad n_1 = N\,, \tag{3.62d$'$}$$

$$e)\ N < n_{12},\, n_{23},\, n_{13} \le 2N\,, \qquad n_2 + n_{13} < 3N\,, \tag{3.62e}$$

$$e')\ N < n_{12},\, n_{23} < 2N\,, \qquad n_2 = n_{13}/2 = N\,, \tag{3.62e$'$}$$

$$f)\ N < n_{12},\, n_{23} < 2N < n_{13} < 3N\,, \tag{3.62f}$$

$$f')\ n_1 = n_2 = N, \quad \text{or} \quad n_1 = n_3 = N, \quad \text{or} \quad n_2 = n_3 = N. \tag{3.62f$'$}$$

The same classification is valid for $U_q(sl(4,\mathbb{C}))$, where case (3.62a) is the so called regular case. This is a refinement of the classification of [165], the primed cases being separated out since together with the regular case these have the classical dimensions of the finite-dimensional irreps of $sl(4,\mathbb{C})$; that is, a representation characterized by n_1, n_2, n_3 has dimension $d^c_{n_1,n_2,n_3}$. In particular, in case $f')$ with $n_1 = n_2 = n_3 = N$ we achieve the maximal possible dimension N^6 of an irrep of $U_q(sl(4,\mathbb{C}))$ (cf. (2.113) and [175]). On the other hand, in the unprimed cases $b) - f)$, the dimension of a representation characterized by n_1, n_2, n_3 is strictly smaller than $d^c_{n_1,n_2,n_3}$.

The representations $U_q(su(2,2))$ inherit all the structure from their $U_q(sl(4,\mathbb{C}))$ counterparts. Thus, the classification of the positive-energy representations of $U_q(su(2,2))$ proceeds as follows.

Let us decompose: $2j_k = 2J_k + r_k N$, $2J_k, r_k \in \mathbb{Z}_+$, $2J_k < N$, $k = 1, 2$. Then we have:

$$n_1 = 2J_1 + 1, \qquad n_3 = 2J_2 + 1. \tag{3.62}$$

Let us consider now the conditions of positive energy (3.51). We see that in cases 1) and 3) we have to distinguish whether $d + j_1 + j_2$ is integer or not. If $d + j_1 + j_2 \notin \mathbb{N}$ then $n_2 = N$, $n_{12} = N + 2J_1 + 1 > N$, $n_{23} = N + 2J_2 + 1 > N$, $n_{13} = N + 2J_1 + 2J_2 + 2 > N$. Thus, depending on n_{13}, the possible cases are (3.62e,f).

Consider now the cases 1) and 3) of (3.51) with $d + j_1 + j_2 \in \mathbb{N}$. Then $d \ge j_1 + j_2 + 3$ and we set $d = p + j_1 + j_2 + 3 + kN$, where $p = 0, 1, \ldots, N-1$, $k \in \mathbb{Z}_+$. Let us also set $\kappa = 2J_1 + 2J_2 + 2 + p$. Note that $2 \le \kappa \le 3N - 1$. Then we have:

$$\begin{aligned} &n_2 = N - \kappa, n_{12} = N - 2J_2 - 1 - p < N,\\ &n_{23} = N - 2J_1 - 1 - p < N, n_{13} = N - p \le N,\\ &\quad \kappa < N \end{aligned} \tag{3.63a}$$

$$\begin{aligned} &n_2 = 2N - \kappa, n_{12} = 2N - 2J_2 - 1 - p,\\ &n_{23} = 2N - 2J_1 - 1 - p, N < n_{13} = 2N - p \le 2N,\\ &\quad N \le \kappa < 2N \end{aligned} \tag{3.63b}$$

$$\begin{aligned} &n_2 = 3N - \kappa, n_{12} = 3N - 2J_2 - 1 - p > N,\\ &n_{23} = 3N - 2J_1 - 1 - p > N, n_{13} = 3N - p > 2N,\\ &\quad 2N \le \kappa < 3N \end{aligned} \tag{3.63c}$$

Thus, all cases of (3.62) are possible: we have case (3.62a) in (3.63a) and (3.62f,f') in (3.63c), while (3.63b) contains all cases (3.62b,b'-e,e'), since both n_{12}, n_{23} can be bigger or smaller than N.

We pass now to case 2) of (3.51), $d = j_1 + j_2 + 2$, $j_1 j_2 \neq 0$, setting $\kappa' = 2J_1 + 2J_2 + 1$. Note that $1 \leq \kappa' \leq 2N - 1$. Then we have:

$$\begin{aligned} & n_2 = N - \kappa', n_{12} = N - 2J_2 \leq N, \\ & n_{23} = N - 2J_1 \leq N, n_{13} = N + 1 > N, \\ & \quad \kappa' < N \end{aligned} \tag{3.64a}$$

$$\begin{aligned} & n_2 = 2N - \kappa', n_{12} = 2N - 2J_2 > N, \\ & n_{23} = 2N - 2J_1 > N, n_{13} = 2N + 1 > 2N, \\ & \quad N \leq \kappa' < 2N \end{aligned} \tag{3.64b}$$

Thus, we have cases (3.62b,b') in (3.64a) and (3.62f,f') in (3.64b).

Finally we consider the massless case 4) of (3.51) $d = j_1 + j_2 + 1$, $j_1 j_2 = 0 = J_1 J_2$. We have:

$$\begin{aligned} & n_2 = N - 2J_1 - 2J_2, \\ & n_{12} = N + 1 - 2J_2 \quad \begin{cases} \leq N & \text{for} \quad \mathrm{J}_2 \neq 0, (\mathrm{J}_1 = 0) \\ > N & \text{for} \quad \mathrm{J}_2 = 0 \end{cases} \\ & n_{23} = N + 1 - 2J_1 \quad \begin{cases} \leq N & \text{for} \quad \mathrm{J}_1 \neq 0, (\mathrm{J}_2 = 0) \\ > N & \text{for} \quad \mathrm{J}_1 = 0 \end{cases} \\ & N < n_{13} = N + 2 \leq 2N. \end{aligned} \tag{3.65}$$

Thus, we have case (3.62c) if $0 < J_2 < (N-1)/2$, case (3.62c') if $J_2 = (N-1)/2$, case (3.62d) if $0 < J_1 < (N-1)/2$, case (3.62d') if $J_1 = (N-1)/2$, case (3.62e) if $J_1 = J_2 = 0$ and $N > 2$. case (3.62e') if $J_1 = J_2 = 0$ and $N = 2$.

This completes the classification of the positive-energy representations of $U_q(su(2,2))$ at roots of 1.

3.3.3 Massless Case

Further we treat in detail the massless case at roots of 1. Since $j_1 j_2 = 0$, let us choose for definiteness $j_2 = 0$. The additional vanishing conditions (3.59) besides (3.55) and (3.58) are:

$$(X_1^+)^{n_1} |\rangle = 0, \quad \text{if} \quad n_1 < 2j_1 + 1, N, \tag{3.66a}$$

$$(X_2^+)^{N-2J_1} |\rangle = 0, \quad \text{if} \quad J_1 > 0. \tag{3.66b}$$

To obtain the dimension $d(N, J_1)$ of these representations we first note that the norms given in (3.57) can be positive only in the following range of j_1 [165], [225]:

$$2rN \leq 2j_1 \leq (2r+1)N - 1, \quad \forall r \in \mathbb{Z}_+; \tag{3.67}$$

that is, in in terms of the decomposition $2j_1 = 2J_1 + r_1 N$ we consider only $r_1 = 2r \in 2\mathbb{Z}_+$.

For fixed j_1 in the above range, the basis of the massless unitary irreducible representation is given by [225]:

$$\Phi^1_{\{k,\ell,n\}}, \quad k,\ell,n \in \mathbb{Z}_+, \quad k+\ell, \ell+n \leq N-1, \\ n \leq N - 2J_1 - 1, \tag{3.68a}$$

$$\Phi^2_{\{k,\ell,n\}}, \quad k,n \in \mathbb{Z}_+, \quad \ell \in \mathbb{N}, \quad k+\ell \leq N-1, \\ \ell + n \leq N - 2J_1 - 1, \tag{3.68b}$$

$$\Phi^3_{\{k,\ell,n\}}, \quad k,\ell,n \in \mathbb{Z}_+, \quad k+\ell+n \leq N-1, \\ 1 \leq n \leq 2J_1. \tag{3.68c}$$

The norms of these vectors are given by (3.57) with j_1 replaced by J_1 and are strictly positive. Now we can find that the number of states in (3.68a), (3.68b) and (3.68c), respectively, is [225]:

$$\frac{1}{6}(N-2J_1)(2N^2 + N(4J_1+3) + 1 - 4J_1^2), \tag{3.69a}$$

$$\frac{1}{6}(N-2J_1)(N-2J_1-1)(2N+2J_1-1), \tag{3.69b}$$

$$\frac{1}{3}J_1(3N^2 - 6NJ_1 - 1 + 4J_1^2). \tag{3.69c}$$

The sum of these three numbers gives the dimension of the massless irreps (cf. [165],[225]):

$$d(N, J_1) = \frac{1}{3}\left[2N^3 - N(12J_1^2 - 1) + 3J_1(4J_1^2 - 1)\right]. \tag{3.70}$$

We recall that in the classical case the massless unitary representations are infinite-dimensional. However, we may compare our representations with the undeformed non-unitary finite-dimensional representations which have the same quantum numbers $(n_1, n_2, n_3) = (2J_1 + 1, N - 2J_1, 1)$. We note that the dimension of the former is generically smaller than the dimension of the latter, which is given by:

$$d^c_{2J_1+1,N-2J_1,1} = \frac{1}{12}(2J_1+1)(N-2J_1)(N+1)(N+1-2J_1)(N+2), \tag{3.71}$$

except when $N = 2, J_1 = 0$, and then $d(2,0) = d^c = 6$, or $N = 2J_1 + 1, J_1 > 0$, and then:

$$d_0 \equiv d(2J_1 + 1, J_1) = d^c = \frac{1}{3}(J_1 + 1)(2J_1 + 1)(2J_1 + 3) = \frac{1}{6}N(N + 1)(N + 2). \tag{3.72}$$

The irreps for $N = 2$ with $J_1 = 0, 1/2$ are deformations of two of the three fundamental representations of $su(2,2)$ with dimensions six and four, respectively, [165].

Finally, we note that one considers the remaining massless representations with $j_1 = 0$ and $j_2 \neq 0$ in the same way. Thus, in the dimension formulae one has to exchange all subscripts $1 \longrightarrow 3$. Also one may introduce the helicity $h = j_1 - j_2$, then all the formulae above may be written in terms of $|h|$. Thus, for the exceptional case $N = 2|h| + 1, h \neq 0$, we have (cf. (3.72)) [165]:

$$d_0 = \frac{1}{3}(|h| + 1)(2|h| + 1)(2|h| + 3) = \frac{1}{6}N(N + 1)(N + 2). \tag{3.73}$$

In particular, for $N = 2, J_2 = 1/2$ one obtains a deformation of the third fundamental representation of $su(2,2)$ with dimension four [165].

Thus the maximal possible dimension of a massless irrep for fixed N is d_0 for $N > 2$ and six for $N = 2$. Note that this maximal dimension is strictly smaller than the minimal dimension of a (semi-) periodic irrep of $U_q(sl(4,\mathbb{C}))$, which is N^3 [177].

3.3.4 Character Formulae

It is easy to see that the spectrum given in (3.56) can be represented by the following character formula [225]:

$$\begin{aligned} ch\, L = e(\Lambda)\Big(& \sum_{k=0}^{\infty}\sum_{\ell=0}^{\infty}\sum_{n=0}^{\infty} t_{13}^k t_{12}^\ell t_2^n + \\ & + \sum_{k=0}^{\infty}\sum_{\ell=1}^{\infty}\sum_{n=0}^{\infty} t_{13}^k t_{23}^\ell t_2^n + \\ & + \sum_{k=0}^{\infty}\sum_{\ell=0}^{\infty}\sum_{n=1}^{2j_1} t_{13}^k t_{12}^\ell t_1^n \Big) \end{aligned} \tag{3.74}$$

where $t_{12} = e(\alpha_{12}) = t_1 t_2$, $t_{23} = e(\alpha_{23}) = t_2 t_3$, $t_{13} = e(\alpha_{13}) = t_1 t_2 t_3$. Next we note that the character formula for the Verma module with the same lowest weight here is:

$$ch\, V^\Lambda = e(\Lambda)/(1 - t_1)(1 - t_2)(1 - t_3)(1 - t_{12})(1 - t_{23})(1 - t_{13}). \tag{3.75}$$

Now we can rewrite the character formula (3.74) as follows [225]:

$$\begin{aligned} ch\, L_\Lambda &= ch\, V^\Lambda Q(t_1, t_2, t_3) = \\ &= ch\, V^\Lambda \big(1 - t_1^{n_1} + t_1^{n_1} t_3 - t_3 - \end{aligned} \tag{3.76}$$

$$
\begin{aligned}
&-t_1t_2 + t_1^{n_1}t_2 - t_1^{n_1}t_2t_3^2 + t_1t_2t_3^2 - \\
&-t_1^{n_1}t_2^2t_3 + t_1^2t_2^2t_3 - t_1^2t_2^2t_3^2 + t_1^{n_1}t_2^2t_3^2), \\
&n_1 = 2j_1 + 1 \geq 1, \quad d = j_1 + 1, \quad j_2 = 0.
\end{aligned}
$$

This formula is valid for all $j_1 \in (1/2)\mathbb{Z}_+$, $j_2 = 0$. Note, however, that for $j_1 = 1/2$ the terms in the fourth row cancel each other, while for $j_1 = 0$ the terms in the third row cancel each other. To show that (3.76) concides with (3.74) amounts to the explicit straightforward division of the polynomials:

$$
\frac{Q(t_1, t_2, t_3)}{(1-t_1)(1-t_2)(1-t_3)(1-t_{12})(1-t_{23})(1-t_{13})}. \tag{3.77}
$$

The formula (3.76) represents an alternating sign summation over part of the Weyl group of $sl(4, \mathbb{C})$ (called reduced Weyl group in [209]) and may be obtained using [381, 382]. Note, however, that the ultimate formula is (3.74), which is obtained in a straightforward manner.

Analogously, the spectrum given in (3.68) can be represented by the following character formula:

$$
\begin{aligned}
ch\, L_\Lambda = e(\Lambda)\Bigg(&\sum_{k=0}^{N-1}\sum_{\ell=0}^{N-1-k}\sum_{n=0}^{\min(N-1-\ell, N-1-2J_1)} t_{13}^k t_{12}^\ell t_2^n + \\
+ &\sum_{k=0}^{N-1}\sum_{\ell=1}^{N-1-k}\sum_{n=0}^{N-1-\ell-2J_1} t_{13}^k t_{23}^\ell t_2^n + \\
+ &\sum_{k=0}^{N-1}\sum_{\ell=0}^{N-1-k}\sum_{n=1}^{\min(N-1-k-\ell, 2J_1)} t_{13}^k t_{12}^\ell t_1^n \Bigg)
\end{aligned} \tag{3.78}
$$

Finally, we can show that (3.78) may be represented as follows:

$$
\begin{aligned}
ch\, L_\Lambda &= ch\, L^c_{2J_1+1, N-2J_1, 1} - ch\, L^c_{2J_1, N-1-2J_1, 2} + ch\, L^c_{2J_1-1, N-1-2J_1, 1}, \\
&\quad J_1 \neq 0\,, \qquad (3.79a) \\
&= ch\, L^c_{1,N,1} - ch\, L^c_{1,N-2,1}, \\
&\quad J_1 = 0\,, \qquad (3.79b)
\end{aligned}
$$

where $L^c_{n_1,n_2,n_3}$, $n_1, n_2, n_3 \in \mathbb{N}$, denote the finite-dimensional irreducible (non-unitary) representation of $su(2,2)$ with character formula (cf. [195]):

$$
\begin{aligned}
ch\, L_{n_1,n_2,n_3} = ch\, V^\Lambda\Big(&1 - t_1^{n_1} - t_2^{n_2} - t_3^{n_3} + t_1^{n_1}t_3^{n_3} + t_1^{n_1}t_2^{n_{12}} + \\
&+ t_3^{n_3}t_2^{n_{23}} + t_1^{n_{12}}t_2^{n_2} + t_3^{n_{23}}t_2^{n_2} - t_1^{n_1}t_2^{n_{13}}t_3^{n_3} - \\
&- t_1^{n_{12}}t_2^{n_2}t_3^{n_{23}} - t_1^{n_{13}}t_2^{n_{23}}t_3^{n_3} - t_1^{n_1}t_2^{n_{12}}t_3^{n_{13}} - \\
&- (t_1t_2)^{n_{12}} - (t_2t_3)^{n_{23}} + t_1^{n_{12}}t_2^{n_2+n_{13}}t_3^{n_{23}} +
\end{aligned}
$$

$$+t_1^{n_1}(t_2t_3)^{n_{13}} + (t_1t_2)^{n_{12}}t_3^{n_{13}} + t_1^{n_{13}}(t_2t_3)^{n_{23}} +$$
$$+(t_1t_2)^{n_{13}}t_3^{n_3} - t_1^{n_{12}}t_2^{n_2+n_{13}}t_3^{n_{13}} - t_1^{n_{13}}t_2^{n_2+n_{13}}t_3^{n_{23}} -$$
$$-(t_1t_2t_3)^{n_{13}} + (t_1t_2t_3)^{n_{13}}t_2^{n_2}\Big) \tag{3.80}$$

and dimension $d^C_{n_1,n_2,n_3}$ (cf. (3.61a)) and in (3.79) we use the convention $chL^C_{n_1,n_2,n_3} = 0$ if any $n_k = 0$, which happens for $J_1 = 1/2$ or for $N = 2J_1 + 1$. A simple consequence of (3.79) is:

$$d(N, J_1) = \begin{cases} d^C_{2J_1+1,N-2J_1,1} - d^C_{2J_1,N-1-2J_1,2} + d^C_{2J_1-1,N-1-2J_1,1}, & J_1 \neq 0, \\ d^C_{1,N,1} - d^C_{1,N-2,1}, & J_1 = 0. \end{cases} \tag{3.81}$$

As we noted the dimensions of the massless representations are generically smaller than the corresponding classical dimensions (the first terms on the RHS of (3.81)).

4 Duality for Quantum Groups

Summary

We start this chapter by introducing matrix quantum groups. In the generic cases these are (one- or multiparameter) deformations of the classical Lie groups. Most of the matrix quantum groups are Hopf algebras though some are only bialgebras. They are in duality with the quantum algebras which are the corresponding deformations of the Lie algebras of the Lie groups under consideration. Actually, this duality is used to find unknown quantum algebras which are in duality with known matrix quantum groups. This was applied first in order to find the quantum algebras dual of the two-parameter matrix quantum group $GL_{p,q}(2)$ (deformation of the reductive Lie group $GL(2)$) [209]. The dual quantum algebra $U_{p,q}(gl(2))$ can be recast as a commutation algebra as the product two one-parameter deformations $U_{p/q}(sl(2)) \otimes U_{p/q}(\mathscr{Z})$) where we use the decomposition $gl(2) = sl(2) \oplus \mathscr{Z}$ (where $\mathscr{Z}$ is the centre of $gl(2)$). However, as a Hopf algebra $U_{p,q}(gl(2))$ cannot be split in this manner since the coalgebra action of $U_{p/q}(sl(2))$ involves also the generator K of $\mathscr{Z}$. Naturally, the splitting is recovered in the one-parameter case $p = q$. Further, the same approach was applied to the duality for multiparameter quantum $GL(n)$ for which the number of parameters is $n(n-1)/2 + 1$ [233]. Again the dual algebra may be split as commutation subalgebra as deformation of $sl(n)$ times the centre, but as Hopf algebra there is no splitting, unless there are $n-1$ relations between the parameters. Thus, there exists a Hopf algebra deformation of $U(sl(n))$ depending only on $(n^2 - 3n + 4)/2$ parameters. We present the duality for a Lorentz quantum group [234] and for the Jordanian matrix quantum group $GL_{g,h}(2)$ [39]. We present also the dualities for many exotic bialgebras following [49–52].

4.1 Matrix Quantum Groups

In the beginning we follow Manin [462]. The *quantum plane* [462] $R_q(n|0)$ or, rather the polynomial ring on it, is generated by coordinates x_i, $i = 1, \dots, n$, with commutation rules:

$$x_i x_j = q^{1/2} x_j x_i, \quad \text{for} \quad i < j. \tag{4.1}$$

The *Grassmannian quantum plane* [462] $R_q(0|n)$ is generated by coordinates ξ_i, $i = 1, \dots, n$, which satisfy:

$$\xi_i^2 = 0, \quad \xi_i \xi_j = -q^{-1/2} \xi_j \xi_i, \quad \text{for} \quad i < j. \tag{4.2}$$

Consider next $n \times n$ matrices M with noncommuting matrix elements, or *quantum matrices*, which perform linear transformations of $R_q(n|0)$ and $R_q(0|n)$:

$$\{x'_1, \dots, x'_n\} \in R_q(n|0), \quad x'_i = M_{ij} x_j, \tag{4.3a}$$

$$\{\xi'_1, \dots, \xi'_n\} \in R_q(0|n), \quad \xi'_i = M_{ij} \xi_j, \tag{4.3b}$$

DOI 10.1515/9783110427707-004

where one assumes that the elements of M commute with all x_i, ξ_i. Implementation of (4.3) gives the following restrictions upon the elements of M:

$$\begin{aligned} M_{ij}M_{i\ell} &= q^{1/2}M_{i\ell}M_{ij}, \quad \text{for} \quad j < \ell, && (4.4a)\\ M_{ij}M_{kj} &= q^{1/2}M_{kj}M_{ij}, \quad \text{for} \quad i < k, && (4.4b)\\ M_{i\ell}M_{kj} &= M_{kj}M_{i\ell}, \quad \text{for} \quad i < k, j < \ell, && (4.4c)\\ [M_{ij}, M_{k\ell}] &= (q^{1/2} - q^{-1/2})M_{i\ell}M_{kj}, \quad \text{for} \quad i < k, j < \ell. && (4.4d) \end{aligned}$$

Let us denote by $A_q(n)$ the bialgebra generated by the matrix elements M_{ij}, $i, j = 1, \ldots, n$, with the following comultiplication δ and counit ε:

$$\delta(M_{ij}) = \sum_{k=1}^{n} M_{ik} \otimes M_{kj}, \quad \text{or} \quad \delta(M) = M \hat{\otimes} M, \tag{4.5a}$$

$$\varepsilon(M_{ij}) = \delta_{ij}, \quad \text{or} \quad \varepsilon(M) = I_n, \tag{4.5b}$$

where $\hat{\otimes}$ denotes the tensor product of algebras and the usual product of matrices, I_n is the unit $n \times n$ matrix.

Note that the operations (4.5) do not depend on the deformation parameter; that is, they are *classical*.

Further, a *quantum determinant* is defined in the following way:

$$\begin{aligned} \mathscr{D} = \det\nolimits_q M &= \sum_{w \in S_n} \epsilon(w)\, M_{1,w(1)} \ldots M_{n,w(n)} = \\ &= \sum_{w \in S_n} \epsilon(w)\, M_{w(1),1} \ldots M_{w(n),n} \end{aligned} \tag{4.6}$$

where summations are over all permutations w of $\{1, \ldots, n\}$ and the *quantum signature* is:

$$\epsilon(w) = \prod_{\substack{j<k \\ w(j)>w(k)}} (-q^{1/2}) = (-q^{1/2})^{\ell(w)} \tag{4.7}$$

where $\ell(w)$ is the number of inversions in the permutation w. Note that

$$\delta(\det\nolimits_q M) = \det\nolimits_q M \otimes \det\nolimits_q M, \tag{4.8a}$$

$$\begin{aligned} \varepsilon(\det\nolimits_q M) &= \sum_{i_1,\ldots,i_n} \varepsilon(M_{1,i_1}) \ldots \varepsilon(M_{n,i_n})(-q^{1/2})^{\ell(i_1,\ldots,i_n)} = \\ &= \sum_{i_1,\ldots,i_n} \delta_{1,i_1} \ldots \delta_{n,i_n}(-q^{1/2})^{\ell(i_1,\ldots,i_n)} = 1, \end{aligned} \tag{4.8b}$$

It is easy to check that $\det_q M$ is central; that is, it commutes with the elements of M.

Further, if $\det_q M \neq 0$ one extends the algebra by an element $(\det_q M)^{-1}$ which obeys [462]:

$$\det_q M\,(\det_q M)^{-1} = (\det_q M)^{-1}\det_q M = 1_{\mathscr{A}} \tag{4.9}$$

Thus one can obtain the quantum groups $GL_q(n)$, $SL_q(n)$, respectively, as the Hopf algebras generated by the matrix elements $M_{ij}, i,j = 1,\dots,n$, such that $(\det_q M)^{-1}$ exists, $\det_q M = 1$ holds, respectively, [275, 462, 599]. The antipode is given by the formula:

$$\gamma(M) = M^{-1}, \quad \gamma(\det_{\mathrm{q}} M) = (\det_{\mathrm{q}} M)^{-1}. \tag{4.10}$$

(Woronowicz [599] calls these objects also *quantum pseudogroups*.)

The above notation is natural since for $q = 1$, and assuming that M_{ij} become complex numbers one obtains the standard commutative Hopf algebras of polynomial functions on the classical groups $GL(n)$, $SL(n)$ with comultiplication and counit given by (4.5) and the antipode given by (4.10) with $q = 1$. Of course in the $q = 1$ case one works usually with the groups $GL(n)$, $SL(n)$ themselves without reference to this related Hopf algebra (even when one considers tensor products of groups representations which are by default governed by the comultiplication structure).

The quantum group $SL_q(n)$ is in duality with the quantum algebra $U_q(sl(n))$. This duality is manifested in several forms. The first is through the R-matrices (cf. (1.31)). The R-matrix of $U_q(sl(n))$ in the fundamental representation has the form [272]:

$$R_n = q^{1/2}\sum_{i=1}^{n} E_{ii}\hat{\otimes}E_{ii} + \sum_{\substack{i,j=1\\ i\neq j}}^{n} E_{ii}\hat{\otimes}E_{jj} + (q^{1/2} - q^{-1/2})\sum_{\substack{i,j=1\\ i>j}}^{n} E_{ij}\hat{\otimes}E_{ji}\,, \tag{4.11}$$

Now one may check that the following relation holds:

$$R_n\, M_1\, M_2 = M_2\, M_1\, R_n\,, \tag{4.12}$$

where $M_1 = M\hat{\otimes}I_n$, $M_2 = I_n\hat{\otimes}M$.

Conversely, one may start with relation (4.12) imposing it on an arbitrary $n \times n$ matrix M; then one would obtain relation (4.4). This characterizes the approach of Faddeev–Reshetikhin–Takhtajan (FRT) [272] for which the starting point is formula (4.12) and the Yang–Baxter equation (1.58). Their motivation comes from the original context of the quantum inverse scattering method [269, 273, 274], where the matrix M played the role of *quantum monodromy matrix* (with operator-valued entries) of the auxiliary linear problem and the Yang–Baxter equation was a compatibility equation for equation (4.12). Following their approach Faddeev, Reshetikhin, and Takhtajan [272] have defined in a similar way the quantum groups $SO_q(n)$, $Sp_q(n)$.

Another manifestation of this duality is considered in Section 4.5.

Let us illustrate everything until now with the example of $A_q(2)$. For $n = 2$ from (4.1) we have:

$$x_1 x_2 = q^{1/2} x_2 x_1 , \tag{4.13a}$$

$$\xi_1^2 = \xi_2^2 = 0 , \quad \xi_1 \xi_2 = -q^{-1/2} \xi_2 \xi_1 . \tag{4.13b}$$

Writing the matrix M as:

$$M = \begin{pmatrix} a & b \\ c & d \end{pmatrix} , \tag{4.14}$$

we have from (4.4) and (4.6):

$$\begin{aligned} ab = q^{1/2} ba , \quad & ac = q^{1/2} ca , \quad bd = q^{1/2} db , \quad cd = q^{1/2} dc , \\ bc = cb , \quad & ad - da = (q^{1/2} - q^{-1/2}) bc \end{aligned} \tag{4.15}$$

$$\det_q M = ad - q^{1/2} bc = da - q^{-1/2} bc . \tag{4.16}$$

The left and right inverse matrix is given by

$$M^{-1} = (\det_q M)^{-1} \begin{pmatrix} d & -q^{-1/2} b \\ -q^{1/2} c & a \end{pmatrix} . \tag{4.17}$$

Further from (4.5a) we have:

$$\delta(M) = \begin{pmatrix} a \otimes a + b \otimes c & a \otimes b + b \otimes d \\ c \otimes a + d \otimes c & c \otimes b + d \otimes d \end{pmatrix} . \tag{4.18}$$

Next the R-matrix in this case is given by:

$$R_2 = \begin{pmatrix} q^{1/2} & 0 & 0 & 0 \\ 0 & 1 & 0 & 0 \\ 0 & q^{1/2} - q^{-1/2} & 1 & 0 \\ 0 & 0 & 0 & q^{1/2} \end{pmatrix} . \tag{4.19}$$

Using (4.19) it is easy to check (4.12) or to obtain (4.15) starting from (4.12).

There is a convenient enumeration of the matrix elements of R_2 given in [155]. Namely, it may be written as $R_{ijk\ell}$, $i, j, k, \ell = 1, 2$, so that the rows of (4.19) are enumerated from top to bottom by the pairs $(i, j) = (1, 1), (1, 2), (2, 1), (2, 2)$, and the columns of (4.19) are enumerated from left to right by the pairs $(k, \ell) = (1, 1), (1, 2), (2, 1), (2, 2)$. Then relation (4.12) may be rewritten as [155]:

$$R_{ijk\ell} M_{km} M_{\ell n} = M_{j\ell} M_{ik} R_{k\ell mn} , \tag{4.20}$$

with summation over repeated indices. The above enumeration may be written compactly also as [596]:

$$R_{ijk\ell} = \delta_{ik}\delta_{j\ell}(1 + (q^{1/2} - 1)\delta_{ij}) + \delta_{i\ell}\delta_{jk}(q^{1/2} - q^{-1/2})\theta(i - j)\,, \tag{4.21}$$

where

$$\theta(p) = \begin{cases} 1 & p > 0\,, \\ 0 & p \le 0. \end{cases} \tag{4.22}$$

Note that the example of $n = 2$ is actually representative of the general situation since for fixed i, j, k, ℓ formulae (4.4) are nothing else but (4.15) if we write (4.14) as:

$$M = \begin{pmatrix} a\ b \\ c\ d \end{pmatrix} = \begin{pmatrix} M_{ij} & M_{i\ell} \\ M_{kj} & M_{k\ell} \end{pmatrix} \tag{4.23}$$

((4.4a,b) should be used twice: (4.4a) also with i replaced by k, (4.4b) also with j replaced by ℓ.)

More general quantum groups, for example, multiparameter cases, are considered in Section 4.5.

4.1.1 Differential Calculus on Quantum Planes

Here we briefly review the noncommutative differential geometry and calculus initiated by Wess–Zumino (WZ) [596]. As noted by Manin [463] the WZ-calculus is different in spirit from that of Connes [152] but is compatible with the differential calculus of Woronowicz [599].

In the case of differential calculus on the quantum plane $R_q(n|0)$ with coordinates x_i (cf. (4.1)) the differentials

$$\xi_i = dx_i, \tag{4.24}$$

obey the relation

$$\xi_i\xi_j = -q^{-1/2}\xi_j\xi_i, \quad \text{for} \quad i < j, \tag{4.25}$$

while the derivatives

$$\partial_i = \frac{\partial}{\partial x_i}, \tag{4.26}$$

obey the relation

$$\partial_i\partial_j = q^{-1/2}\partial_j\partial_i, \quad \text{for} \quad i < j. \tag{4.27}$$

Since the differential calculus should be $GL_q(n,\mathbb{C})$-covariant all relations between variables, differentials, and derivatives are expressed through the R-matrix (cf. (4.11)). One may use for R_n the form (4.21) with $i,j,k,\ell = 1,\dots,n$. In [596] an R-matrix augmented by the permutation matrix is also used:

$$\hat{R} = PR, \quad \hat{R}_{ijk\ell} = R_{jik\ell}. \tag{4.28}$$

Thus in [596] were derived the following relations: between variables and differentials

$$x_i\xi_j = q^{1/2}R_{jik\ell}\xi_\ell x_k, \tag{4.29}$$

between variables and derivatives, considered as operators,

$$\partial_i x_j = \delta_{ij} + q^{1/2}R_{kji\ell}x_\ell\partial_k, \tag{4.30}$$

between derivatives and differentials

$$\partial_i\xi_j = q^{-1/2}(\hat{R})^{-1}_{jki\ell}\xi_\ell\partial_k. \tag{4.31}$$

Further the exterior differential

$$d = \sum \xi_i\partial_i \tag{4.32}$$

satisfies the Leibniz rule

$$d(fg) = (df)g + f(dg). \tag{4.33}$$

and has the usual properties

$$d^2 = 0 \tag{4.34}$$

$$dx_i - x_i d = \xi_i, \quad d\xi_i + \xi_i d = 0. \tag{4.35}$$

Only the commutation with derivatives is modified

$$d\partial_i = q\partial_i\, d \tag{4.36}$$

but this modification is compatible with (4.34), namely:

$$d^2 = d\sum \xi_i\partial_i = -\sum \xi_i\, d\,\partial_i = -q\sum \xi_i\partial_i d = -qd^2 \tag{4.37}$$

from which follows $d^2 = 0$, except in the case $q = -1$.

An $SO_q(n)$-invariant differential calculus was developed in [124]. Following the approach of Woronowicz [599], a differential calculus on quantum spheres was developed in [510], on $SO_q(n)$ and $SU_q(n)$ in [125], on classical simple quantum groups in [369] and on arbitrary quantum simple groups in [93]. We should mention also that there is much literature on q-difference operators (Eulerian calculus) related to quantum groups (cf., e. g., [283, 309, 466, 584]).

4.2 Duality between Hopf Algebras

Two bialgebras $\mathscr{U}$, $\mathscr{A}$ are said to be *in duality* [11] if there exists a doubly nondegenerate bilinear form

$$\langle\, ,\rangle : \mathscr{U} \times \mathscr{A} \to \mathbb{C}, \quad \langle\, ,\rangle : (u,a) \mapsto \langle u, a\rangle, \quad u \in \mathscr{U}, a \in \mathscr{A}, \tag{4.38}$$

such that for $u, v \in \mathscr{U}$, $a, b \in \mathscr{A}$:

$$\langle u, ab\rangle = \langle \delta_{\mathscr{U}}(u), a \otimes b\rangle, \quad \langle uv, a\rangle = \langle u \otimes v, \delta_{\mathscr{A}}(a)\rangle \tag{4.39a}$$

$$\langle 1_{\mathscr{U}}, a\rangle = \varepsilon_{\mathscr{A}}(a), \quad \langle u, 1_{\mathscr{A}}\rangle = \varepsilon_{\mathscr{U}}(u). \tag{4.39b}$$

Two Hopf algebras $\mathscr{U}$, $\mathscr{A}$ are said to be *in duality* [11] if they are in duality as bialgebras and if

$$\langle \gamma_{\mathscr{U}}(u), a\rangle = \langle u, \gamma_{\mathscr{A}}(a)\rangle. \tag{4.39c}$$

It is enough to define the pairing (4.38) between the generating elements of the two algebras. The pairing between any other elements of $\mathscr{U}$, $\mathscr{A}$ follows then from relations (4.39) and the standard bilinear form inherited by the tensor product. For example, suppose $\delta(u) = \sum_i u_i' \otimes u_i''$, then one has:

$$\langle u, ab\rangle = \langle \delta_{\mathscr{U}}(u), a \otimes b\rangle = \sum_i \langle u_i' \otimes u_i'', a \otimes b\rangle = \sum_i \langle u_i', a\rangle \langle u_i'', b\rangle. \tag{4.40}$$

4.3 Matrix Quantum Group $GL_{p,q}(2)$

In this subsection we review the two-parameter deformation of $GL(2)$ following [183].

Let $p, q \in \mathbb{C}\backslash\{0\}$. Consider next 2×2 matrices M with noncommuting matrix elements which perform linear transformations of $R_q(2|0)$ and $R_p(0|2)$; that is,

$$\{x_1', x_2'\} \in R_q(2|0), \quad x_i' = M_{ij} x_j, \tag{4.41a}$$

$$\{\xi_1', \xi_2'\} \in R_p(0|2), \quad \xi_i' = M_{ij} \xi_j, \tag{4.41b}$$

assuming that the elements of M commute with all x_i, ξ_i, and summation over repeated indices is understood. Let us write the matrix M as in (4.14):

$$M = \begin{pmatrix} a & b \\ c & d \end{pmatrix}$$

Then implementation of (4.41) gives that the matrix elements of M obey [183]:

$$\begin{aligned} &ab = p^{-1/2}ba\,, \quad ac = q^{-1/2}ca\,, \quad bd = q^{-1/2}db\,, \quad cd = p^{-1/2}dc\,, \\ &q^{1/2}bc = p^{1/2}cb\,, \quad ad - da = (p^{-1/2} - q^{1/2})bc \end{aligned} \tag{4.42}$$

Let us denote by $A_{p,q}(2)$ the bialgebra generated by the matrix elements a, b, c, d with the following comultiplication δ and counit ε (cf. also (4.5) for $n = 2$):

$$\delta(\begin{pmatrix} a & b \\ c & d \end{pmatrix}) = \begin{pmatrix} a\otimes a + b\otimes c & a\otimes b + b\otimes d \\ c\otimes a + d\otimes c & c\otimes b + d\otimes d \end{pmatrix}, \tag{4.43a}$$

$$\varepsilon(\begin{pmatrix} a & b \\ c & d \end{pmatrix}) = \begin{pmatrix} 1 & 0 \\ 0 & 1 \end{pmatrix}. \tag{4.43b}$$

Further, a quantum determinant $\det_{p,q} M \in A_{p,q}(2)$ here is defined as follows:

$$\mathscr{D} \equiv \det_{p,q} M = ad - p^{-1/2}bc = ad - q^{-1/2}cb = da - p^{1/2}cb = da - q^{1/2}bc. \tag{4.44}$$

and then we have (cf. (4.8)):

$$\delta(\mathscr{D}) = \mathscr{D} \otimes \mathscr{D}\,, \quad \varepsilon(\mathscr{D}) = 1. \tag{4.45}$$

The crucial difference with the one-parameter case which is obtained for $p = q$ (cf. Section 4.1) is that the quantum determinant is not central but satisfies the following relations [183]:

$$[\mathscr{D}, a] = [\mathscr{D}, d] = 0\,, \quad p^{1/2}\mathscr{D}b = q^{1/2}b\mathscr{D}\,, \quad q^{1/2}\mathscr{D}c = p^{1/2}c\mathscr{D}. \tag{4.46}$$

Further, if $\mathscr{D} \neq 0$ one extends the algebra by an element $\mathscr{D}^{-1}$ obeying

$$\mathscr{D}\mathscr{D}^{-1} = \mathscr{D}^{-1}\mathscr{D} = 1_{\mathscr{A}}\,, \tag{4.47a}$$

from which follows [183]:

$$\begin{aligned} &[\mathscr{D}^{-1}, a] = 0\,, \qquad [\mathscr{D}^{-1}, d] = 0 \\ &q^{1/2}\mathscr{D}^{-1}b = p^{1/2}b\mathscr{D}^{-1}\,, \; p^{1/2}\mathscr{D}^{-1}c = q^{1/2}c\mathscr{D}^{-1}. \end{aligned} \tag{4.47b}$$

Next one defines the left and right inverse matrix of M [183]:

$$M^{-1} = \mathscr{D}^{-1}\begin{pmatrix} d & -q^{1/2}b \\ -q^{-1/2}c & a \end{pmatrix} = \begin{pmatrix} d & -p^{1/2}b \\ -p^{-1/2}c & a \end{pmatrix}\mathscr{D}^{-1}. \tag{4.48}$$

Suppose that the bialgebra operations are defined on $\mathscr{D}^{-1}$. Then we have:

$$\delta(\mathscr{D}^{-1}) = \mathscr{D}^{-1} \otimes \mathscr{D}^{-1}, \quad \varepsilon(\mathscr{D}^{-1}) = 1. \tag{4.49}$$

The quantum group $GL_{p,q}(2)$ is defined as the Hopf algebra obtained from the bialgebra $A_{p,q}(2)$ extended by the element $\mathscr{D}^{-1}$ and with antipode given by the formula:

$$\gamma(M) = M^{-1}. \tag{4.50}$$

From the above definition we have:

$$\gamma(\mathscr{D}) = \mathscr{D}^{-1}, \quad \gamma(\mathscr{D}^{-1}) = \mathscr{D}. \tag{4.51}$$

For $p = q$ one obtains from $GL_{q,q}(2)$ the quantum groups $GL_q(2)$, respectively, $SL_q(2)$, if the condition $\mathscr{D} \neq 0$, respectively, $\mathscr{D} = 1_{\mathscr{A}}$, holds.

4.4 Duality for $GL_{p,q}(2)$

In this section we review the paper [209] where we have introduced (and applied to $A_{p,q}(2)$) a generalization of the approach which Sudbery [564] applied for $A_q(2) = A_{q,q}(2)$.

For $A_{p,q}(2)$ we use the basis given by all monomials $f = f_{k\ell mn} = a^k d^\ell b^m c^n$, where $k, \ell, m, n \in \mathbb{Z}_+$, and $f_{0000} = 1_{\mathscr{A}}$. We postulate the following pairings for $f = a^k d^\ell b^m c^n$:

$$\langle A, f \rangle = k\delta_{m0}\delta_{n0}, \tag{4.52a}$$

$$\langle B, f \rangle = \delta_{m1}\delta_{n0}, \tag{4.52b}$$

$$\langle C, f \rangle = \delta_{m0}\delta_{n1}, \tag{4.52c}$$

$$\langle D, f \rangle = \ell\delta_{m0}\delta_{n0}, \tag{4.52d}$$

Let us denote by $\mathscr{U}_{p,q}$ the bialgebra in duality with $A_{p,q}(2)$ and generated by A, B, C, D. Later we shall see that $\mathscr{U}_{p,q}$ has the structure of a Hopf algebra in duality with $GL_{p,q}(2)$.

The following relations hold as consequences from (4.52):

$$\langle A, \begin{pmatrix} a & b \\ c & d \end{pmatrix} \rangle = \begin{pmatrix} 1 & 0 \\ 0 & 0 \end{pmatrix}, \tag{4.53a}$$

$$\langle B, \begin{pmatrix} a & b \\ c & d \end{pmatrix} \rangle = \begin{pmatrix} 0 & 1 \\ 0 & 0 \end{pmatrix}, \tag{4.53b}$$

$$\langle C, \begin{pmatrix} a & b \\ c & d \end{pmatrix} \rangle = \begin{pmatrix} 0 & 0 \\ 1 & 0 \end{pmatrix}, \tag{4.53c}$$

$$\langle D, \begin{pmatrix} a & b \\ c & d \end{pmatrix} \rangle = \begin{pmatrix} 0 & 0 \\ 0 & 1 \end{pmatrix}, \tag{4.53d}$$

$$\langle Y, 1_{\mathscr{A}} \rangle = 0, \qquad Y = A, B, C, D, \tag{4.54}$$

$$\langle 1_{\mathscr{U}}, a^k d^\ell b^m c^n \rangle = \delta_{m0}\delta_{n0}. \tag{4.55}$$

We would like to find the commutation relations between the generators of $\mathscr{U}_{p,q}$. First we obtain that the action on $f = a^k d^\ell b^m c^n$ of the monomials in $\mathscr{U}_{p,q}$ which are quadratic in the generators is given by the following:

$$\langle BC, f \rangle = \delta_{m0}\delta_{n0} \sum_{j=0}^{k-1} (pq)^{(j-\ell)/2} + q^{-1/2}\delta_{m1}\delta_{n1}, \tag{4.56a}$$

$$\langle CB, f \rangle = \delta_{m0}\delta_{n0} \sum_{j=0}^{\ell-1} (pq)^{-j/2} + p^{1/2}\delta_{m1}\delta_{n1}, \tag{4.56b}$$

$$\langle AB, f \rangle = (k+1)\delta_{m1}\delta_{n0} = (k+1)\langle B, f \rangle, \tag{4.56c}$$

$$\langle BA, f \rangle = k\delta_{m1}\delta_{n0} = k\langle B, f \rangle, \tag{4.56d}$$

$$\langle AC, f \rangle = k\delta_{m0}\delta_{n1} = k\langle C, f \rangle, \tag{4.56e}$$

$$\langle CA, f \rangle = (k+1)\delta_{m0}\delta_{n1} = (k+1)\langle C, f \rangle, \tag{4.56f}$$

$$\langle DB, f \rangle = \ell\delta_{m1}\delta_{n0} = \ell\langle B, f \rangle, \tag{4.56g}$$

$$\langle BD, f \rangle = (\ell+1)\delta_{m1}\delta_{n0} = (\ell+1)\langle B, f \rangle, \tag{4.56h}$$

$$\langle DC, f \rangle = (\ell+1)\delta_{m0}\delta_{n1} = (\ell+1)\langle C, f \rangle, \tag{4.56i}$$

$$\langle CD, f \rangle = \ell\delta_{m0}\delta_{n1} = \ell\langle C, f \rangle, \tag{4.56j}$$

$$\langle AD, f \rangle = \langle DA, f \rangle = k\ell\delta_{m0}\delta_{n0} = k\ell\langle 1_{\mathscr{U}}, f \rangle. \tag{4.56k}$$

Then we have:

$$q^{1/2}\langle BC, a^k d^\ell b^m c^n \rangle - p^{-1/2}\langle CB, a^k d^\ell b^m c^n \rangle = \frac{(pq)^{(k-\ell)/2} - 1}{p^{1/2} - q^{-1/2}} \delta_{m0}\delta_{n0} \tag{4.57}$$

$$\langle [A, B], f \rangle = \langle B, f \rangle, \tag{4.58a}$$

$$\langle [A, C], f \rangle = -\langle C, f \rangle, \tag{4.58b}$$

$$\langle [D,B], f\rangle = -\langle B, f\rangle\,, \tag{4.58c}$$
$$\langle [D,C], f\rangle = \langle C, f\rangle\,, \tag{4.58d}$$
$$\langle [A,D], f\rangle = 0. \tag{4.58e}$$

We see that relations (4.56) depend on the element f; however, the commutation relations (4.58) do not. This is also true for (4.57); however, in order to see this we need the following formulae:

$$\langle A^s, a^k d^\ell b^m c^n\rangle = k^s \delta_{m0}\delta_{n0}\,, \quad s \in \mathbb{N}\,, \tag{4.59a}$$
$$\langle D^s, a^k d^\ell b^m c^n\rangle = \ell^s \delta_{m0}\delta_{n0}\,, \quad s \in \mathbb{N}\,, \tag{4.59b}$$
$$\langle r^A, a^k d^\ell b^m c^n\rangle = r^k \delta_{m0}\delta_{n0}\,, \quad r = p, q, \tag{4.59c}$$
$$\langle r^D, a^k d^\ell b^m c^n\rangle = r^\ell \delta_{m0}\delta_{n0}\,, \quad r = p, q, \tag{4.59d}$$

where we use the formal power series:

$$r^Y = 1_{\mathscr{U}} + \sum_{k=1}^{\infty} Y^k (\ln r)^k / k!.$$

Thus we obtain that the commutation relations in the algebra $\mathscr{U}_{p,q}$ are given by:

$$\begin{aligned} &q^{1/2} BC - p^{-1/2} CB = \frac{(pq)^{(A-D)/2} - 1_{\mathscr{U}}}{p^{1/2} - q^{-1/2}}\,, \\ &[A,B] = B\,, \quad [A,C] = -C\,, \\ &[D,B] = -B\,, \quad [D,C] = C\,, \quad [A,D] = 0. \end{aligned} \tag{4.60}$$

Note that the generator $K = A + D$ commutes with all other generators of $\mathscr{U}_{p,q}$. Let us denote by $\mathscr{Z}$ the algebra spanned by K.

Next we are looking for the analogue of the splitting $U_q(sl(2)) \otimes U_q(\mathscr{Z})$ which Sudbery [564] obtained in the one-parameter case. We try a similar change of basis:

$$H = A - D\,, \quad \tilde{X}^+ = q'^{-1/4} B q'^{-H/4}\,, \quad \tilde{X}^- = q'^{1/4} C q'^{-H/4}\,, \quad q' \equiv (pq)^{1/2}\,, \tag{4.61}$$

and we get that the generators $H, \tilde{X}^+, \tilde{X}^-$ satisfy commutation relations (1.19) with $\ell = 1$, $q_1 = q \to q'$, $H_1 = H$, $X_1^\pm = \tilde{X}^\pm$.

The factors $q'^{\pm 1/4}$ in (4.61) seem redundant, since factors $q'^{\pm\nu}$ for arbitrary $\nu \in \mathbb{C}$ will play the same role for the previous statement. Their significance becomes clear if we calculate the action of the new generators on $a^k d^\ell b^m c^n$, namely:

$$\langle H^s, a^k d^\ell b^m c^n\rangle = (k-\ell)^s \delta_{m0}\delta_{n0}\,, \tag{4.62a}$$
$$\langle q'^H, a^k d^\ell b^m c^n\rangle = q'^{k-\ell} \delta_{m0}\delta_{n0}\,, \tag{4.62b}$$
$$\langle K^s, a^k d^\ell b^m c^n\rangle = (k+\ell)^s \delta_{m0}\delta_{n0}\,, \tag{4.62c}$$

$$\langle q'^K, a^k d^\ell b^m c^n\rangle = q'^{k+\ell}\delta_{m0}\delta_{n0}, \qquad (4.62d)$$
$$\langle \tilde{X}^+, a^k d^\ell b^m c^n\rangle = q'^{(\ell-k)/4}\delta_{m1}\delta_{n0}, \qquad (4.62e)$$
$$\langle \tilde{X}^-, a^k d^\ell b^m c^n\rangle = q'^{(\ell-k)/4}\delta_{m0}\delta_{n1}. \qquad (4.62f)$$

Remark 4.1. Thus, the two parameters are glued together in the commutation relations and the action of the new basis of the algebra $\mathcal{U}_{p,q}$. This is in agreement with the general statement of Drinfeld [253] that the q-deformation of $U(sl(2))$ is unique. However, we shall see below that in the Hopf algebra relations the two parameters are not glued together, since in fact we are obtaining a deformation of $U(gl(2))$. ◊

We turn now to the bialgebra structure of $\mathcal{U}_{p,q}$. The comultiplication in the algebra $\mathcal{U}_{p,q}$ is given by:

$$\delta_{\mathcal{U}}(A) = A \otimes 1_{\mathcal{U}} + 1_{\mathcal{U}} \otimes A, \qquad (4.63a)$$
$$\delta_{\mathcal{U}}(B) = B \otimes p^{A/2}q^{-D/2} + 1_{\mathcal{U}} \otimes B, \qquad (4.63b)$$
$$\delta_{\mathcal{U}}(C) = C \otimes q^{A/2}p^{-D/2} + 1_{\mathcal{U}} \otimes C, \qquad (4.63c)$$
$$\delta_{\mathcal{U}}(D) = D \otimes 1_{\mathcal{U}} + 1_{\mathcal{U}} \otimes D, \qquad (4.63d)$$

or in the new basis by:

$$\delta_{\mathcal{U}}(H) = H \otimes 1_{\mathcal{U}} + 1_{\mathcal{U}} \otimes H, \qquad (4.64a)$$
$$\delta_{\mathcal{U}}(K) = K \otimes 1_{\mathcal{U}} + 1_{\mathcal{U}} \otimes K, \qquad (4.64b)$$
$$\delta_{\mathcal{U}}(\tilde{X}^+) = \tilde{X}^+ \otimes \left(\frac{p}{q}\right)^{K/4} q'^{H/4} + q'^{-H/4} \otimes \tilde{X}^+, \qquad (4.64c)$$
$$\delta_{\mathcal{U}}(\tilde{X}^-) = \tilde{X}^- \otimes \left(\frac{q}{p}\right)^{K/4} q'^{H/4} + q'^{-H/4} \otimes \tilde{X}^-. \qquad (4.64d)$$

For the Proof we use the duality property (4.39a), namely, we should have: $\langle Y, f\rangle = \langle \delta_{\mathcal{U}}(Y), f_1 \otimes f_2\rangle$, $Y = A, B, C, D$, for every splitting $f = f_1 f_2$.

The counit relations in $\mathcal{U}_{p,q}$ are given by:

$$\varepsilon_{\mathcal{U}}(Y) = 0, Y = A, B, C, D, H, K, \tilde{X}^\pm \qquad (4.65)$$

which follows from (4.54) and (4.61) and $\langle u, 1_{\mathcal{A}}\rangle = \varepsilon_{\mathcal{U}}(u)$ (cf. (4.39b)).

Let us assume now that $\mathcal{U}_{p,q}$ is a Hopf algebra in duality with $GL_{p,q}(2)$. This assumption would be correct if we can define consistently the action of the generators of $\mathcal{U}_{p,q}$ on $\mathcal{D}^{-1}$ and an antipode in $\mathcal{U}_{p,q}$. We are even in a better situation since the action on $\mathcal{D}^{-1}$ and the antipode map in $\mathcal{U}_{p,q}$ are uniquely obtained as a consequence of the assumed duality. Namely, we have that the action of $\mathcal{U}_{p,q}$ on $\mathcal{D}^{-1}$ is given by:

$$\langle 1_{\mathscr{U}}\,, \mathscr{D}^{-1}\rangle = 1\,, \tag{4.66a}$$

$$\langle \begin{pmatrix} A & B \\ C & D \end{pmatrix}, \mathscr{D}^{-1}\rangle = \begin{pmatrix} -1 & 0 \\ 0 & -1 \end{pmatrix}. \tag{4.66b}$$

To prove (4.66a) we use (4.39b) and (4.49): $\langle 1_{\mathscr{U}}\,, \mathscr{D}^{-1}\rangle = \varepsilon_{\mathscr{A}}(\mathscr{D}^{-1}) = 1$. For (4.66b) we use a corollary of (4.52):

$$\langle \begin{pmatrix} A & B \\ C & D \end{pmatrix}, \mathscr{D}\rangle = \begin{pmatrix} 1 & 0 \\ 0 & 1 \end{pmatrix}, \tag{4.66c}$$

and also (4.54) and (4.66a).

Next we obtain that the antipode map in $\mathscr{U}_{p,q}$ is given by:

$$\gamma_{\mathscr{U}} \begin{pmatrix} A & B \\ C & D \end{pmatrix} = \begin{pmatrix} -A & -Bp^{-A/2}q^{D/2} \\ -Cq^{-A/2}p^{D/2} & -D \end{pmatrix}. \tag{4.67}$$

Finally we can state the main result of [209]:

Theorem 4.1. *The Hopf algebra $\mathscr{U}_{p,q}$ in duality with $GL_{p,q}(2)$ by relations (4.52) is isomorphic to $U_{(pq)^{1/2}}(sl(2)) \otimes U_{p/q}(\mathscr{Z})$ as a commutation algebra, where $\mathscr{Z}$ is spanned by K, and $U_r(\mathscr{Z})$ is spanned by $K, r^{\pm K/4}$. The subalgebra $U_{p/q}(\mathscr{Z})$ is a Hopf subalgebra of $\mathscr{U}_{p,q}$, the commutation subalgebra generated by $H, \tilde{X}^{\pm}$ is not a Hopf subalgebra.* ◊

For $p = q$ the algebra in duality with $GL_q(2)$ is $U_q(sl(2)) \otimes U(\mathscr{Z})$ as a tensor product of Hopf subalgebras. For $q = 1$ the last statement reduces to the classical relation $U(gl(2)) = U(sl(2)) \otimes U(\mathscr{Z})$.

4.5 Duality for Multiparameter Quantum *GL*(*n*)

This section follows [233]. We show that the Hopf algebra $\mathscr{U}_{u\mathbf{q}}$ dual to the multiparameter matrix quantum group $GL_{u\mathbf{q}}(n)$ may be found applying the method of [209]; see also Section 4.4. Furthermore, we give the Cartan–Weyl basis of $\mathscr{U}_{u\mathbf{q}}$ and show that this is consistent with the duality. We show that as a commutation algebra $\mathscr{U}_{u\mathbf{q}} \cong U_u(sl(n,\mathbb{C})) \otimes U_u(\mathscr{Z})$, where $\mathscr{Z}$ is one-dimensional and $U_u(\mathscr{Z})$ is a central algebra in $\mathscr{U}_{u\mathbf{q}}$. However, as a coalgebra $\mathscr{U}_{u\mathbf{q}}$ cannot be split in this way and depends on all parameters.

4.5.1 Multiparameter Deformation of $GL(n)$

In [462] Manin has considered a family of quantum groups, deformations of the algebra of polynomial functions on $GL(n)$, depending on $n(n-1)/2$ parameters. Later, different multiparameter deformations were found in [183, 268, 277, 335, 403, 497, 500, 522, 542, 543, 565, 572]. The maximal number of parameters for $GL(n)$ is $N = n(n-1)/2+1$ [565]. Following [565] we denote these N parameters by u and q_{ij}, $1 \le i < j \le n$, and also for shortness by the pair $u, \tilde{q}$.

Let us consider an $n \times n$ quantum matrix M with noncommuting matrix elements a_{ij}, $1 \le i, j \le n$. The matrix quantum group $GL_{u\tilde{q}}(n)$ is generated by the matrix elements a_{ij} with the following commutation relations [565]:

$$\begin{aligned} a_{ij}a_{i\ell} &= p a_{i\ell}a_{ij}\,, \quad \text{for} \quad j<\ell\,, \\ a_{ij}a_{kj} &= q a_{kj}a_{ij}\,, \quad \text{for} \quad i<k\,, \\ p a_{i\ell}a_{kj} &= q a_{kj}a_{i\ell}\,, \quad \text{for} \quad i<k\,, j<\ell\,, \\ uq a_{k\ell}a_{ij} &= (up)^{-1}a_{ij}a_{k\ell} + \lambda a_{i\ell}a_{kj}\,, \quad \text{for} \quad i<k\,, j<\ell\,, \\ p &= q_{j\ell}/u^2\,, q = 1/q_{ik}\,, \lambda = u - 1/u. \end{aligned} \tag{4.68}$$

Considered as a bialgebra, it has the following comultiplication $\delta_{\mathcal{A}}$ and counit $\varepsilon_{\mathcal{A}}$:

$$\delta_{\mathcal{A}}(a_{ij}) = \sum_{k=1}^{n} a_{ik} \otimes a_{kj}\,, \quad \varepsilon_{\mathcal{A}}(a_{ij}) = \delta_{ij}. \tag{4.69}$$

This algebra has determinant $\mathcal{D}$ given by [542, 565]:

$$\begin{aligned} \mathcal{D} &= \textstyle\sum_{\rho\in S_n} \epsilon(\rho) a_{1,\rho(1)} \cdots a_{n,\rho(n)} = \\ &= \textstyle\sum_{\rho\in S_n} \epsilon'(\rho) a_{\rho(1),1} \cdots a_{\rho(n),n}\,, \end{aligned} \tag{4.70}$$

where summations are over all permutations ρ of $\{1, \dots, n\}$ and the quantum signatures are:

$$\begin{aligned} \epsilon(\rho) &= \prod_{\substack{j<k \\ \rho(j)>\rho(k)}} (-q_{\rho(k)\rho(j)}/u^2)\,, \\ \epsilon'(\rho) &= \prod_{\substack{j<k \\ \rho(j)>\rho(k)}} (-1/q_{\rho(k)\rho(j)}). \end{aligned} \tag{4.71}$$

The determinant obeys [542, 565]:

$$\delta_{\mathcal{A}}(\mathcal{D}) = \mathcal{D} \otimes \mathcal{D}\,, \quad \varepsilon_{\mathcal{A}}(\mathcal{D}) = 1. \tag{4.72}$$

The determinant is almost central; that is, it *q-commutes* with the elements a_{ij}:

$$a_{ik}\mathscr{D} = \frac{\prod_{j=1}^{k-1} q_{jk}^{-1} \prod_{\ell=k+1}^{n} q_{k\ell}/u^2}{\prod_{s=1}^{i-1} q_{si}^{-1} \prod_{t=i+1}^{n} q_{it}/u^2} \mathscr{D} a_{ik}. \tag{4.73}$$

Further, if $\mathscr{D} \neq 0$ one extends the algebra by an element $\mathscr{D}^{-1}$ which obeys:

$$\mathscr{D}\mathscr{D}^{-1} = \mathscr{D}^{-1}\mathscr{D} = 1_{\mathscr{A}}. \tag{4.74}$$

Note that for $q_{ij} = u$ for all i, j then the element $\mathscr{D}$ is central and it is possible that $\mathscr{D} = \mathscr{D}^{-1} = 1_{\mathscr{A}}$.

Next one defines the left and right *quantum cofactor matrices* A_{ij} and A'_{ij}:

$$A_{ij} = \sum_{\rho(i)=j} \frac{\epsilon(\rho\circ\sigma_i)}{\epsilon(\sigma_i)} a_{1,\rho(1)} \dots \widehat{a}_{ij} \dots a_{n,\rho(n)}\,, \tag{4.75a}$$

$$A'_{ij} = \sum_{\rho(j)=i} \frac{\epsilon(\rho\circ\sigma'_i)}{\epsilon(\sigma'_i)} a_{\rho(1),1} \dots \widehat{a}_{ij} \dots a_{\rho(n),n}\,, \tag{4.75b}$$

where σ_i and σ'_j denote the cyclic permutations:

$$\sigma_i = \{i, \dots, 1\}\,, \quad \sigma'_j = \{j, \dots, n\}\,, \tag{4.76}$$

and the notation $\hat{x}$ indicates that x is to be omitted. Now one can show that:

$$\sum_j a_{ij} A_{kj} = \sum_j A'_{ji} a_{jk} = \delta_{ik}\, \mathscr{D}\,, \tag{4.77}$$

and obtain the left and right inverse:

$$M^{-1} = \mathscr{D}^{-1} A' = A \mathscr{D}^{-1}. \tag{4.78}$$

Thus, one can introduce the antipode in $GL_{u\tilde{q}}(n)$ [542, 565]:

$$\gamma_{\mathscr{A}}(a_{ij}) = \mathscr{D}^{-1} A'_{ji} = A_{ji} \mathscr{D}^{-1}. \tag{4.79}$$

We are looking for the dual algebra to $GL_{u\tilde{q}}(n)$. As we have seen in Section 4.4 for this it is enough to define the pairing between the generating elements of the two algebras. However, we do not know the dual algebra completely. Then we need to know the action of the algebra $\mathscr{U}_{u\tilde{q}}$ dual to $GL_{u\tilde{q}}(n)$ on every element of $GL_{u\tilde{q}}(n)$. The basis of $GL_{u\tilde{q}}(n)$ consists of monomials

$$f = (a_{11})^{k_1} \dots (a_{nn})^{k_n} (a_{12})^{m_{12}} \dots (a_{n-1,n})^{m_{n-1,n}} (a_{n,n-1})^{n_{n,n-1}} \dots (a_{21})^{n_{21}}\,, \tag{4.80}$$

where $k_i, m_{ij}, n_{ij} \in \mathbb{Z}_+$ and we have used the so-called normal ordering of the element a_{ij}. Namely, we first put the elements a_{ii}; then we put the element a_{ij} with $i < j$ in lexicographic order; that is, if $i < k$ then a_{ij} $(i < j)$ is before $a_{k\ell}$ $(k < \ell)$ and a_{ti} $(t < i)$ is before a_{tk}; finally we put the elements a_{ij} with $i > j$ in antilexicographic order; that is, if $i > k$ then a_{ij} $(i > j)$ is before $a_{k\ell}$ $(k > \ell)$ and a_{ti} $(t > i)$ is before a_{tk}.

Similarly to the case $GL_{p,q}(2)$ we define the pairing only for the monomials in the normal order (4.80) as follows:

$$\begin{aligned}
\langle D_i\,, f\rangle &\equiv k_i\delta_{\mathbf{m}0}\delta_{\mathbf{n}0}\,, \quad 1 \le i \le n\,,\\
\langle E_{ij}\,, f\rangle &\equiv \delta_{m_{ij}1}\delta^{ij}_{\mathbf{m}0}\delta_{\mathbf{n}0}\,, \quad 1 \le i < j \le n\,,\\
\langle F_{ij}\,, f\rangle &\equiv \delta_{n_{ij}1}\delta_{\mathbf{m}0}\delta^{ij}_{\mathbf{n}0}\,, \quad 1 \le j < i \le n\,,\\
\langle 1_{\mathcal{U}}\,, f\rangle &\equiv \delta_{\mathbf{m}0}\delta_{\mathbf{n}0}\,,\\
\delta_{\mathbf{m}0} &= \prod_{1\le j<k\le n} \delta_{m_{jk}0}\,, \quad \delta_{\mathbf{n}0} = \prod_{1\le k<j\le n} \delta_{n_{jk}0}\,,\\
\delta^{ij}_{\mathbf{m}0} &= \prod_{\substack{1\le k<\ell\le n\\ (k,\ell)\ne(i,j)}} \delta_{m_{k\ell}0}\,, \quad \delta^{ij}_{\mathbf{n}0} = \prod_{\substack{1\le \ell<k\le n\\ (k,\ell)\ne(i,j)}} \delta_{n_{k\ell}0}\,.
\end{aligned} \tag{4.81}$$

If some monomial is not in normal order, then it should be brought to this order using commutation relations (4.68) and then (4.81) can be applied. Thus following [565] we can interpret formulae (4.81) as

$$\langle Y\,, f\rangle \equiv \varepsilon_{\mathcal{A}}\left(\frac{\partial f}{\partial y}\right)\,, \tag{4.82}$$

where y is a generating element of $GL_{u\mathbf{q}}(n)$ and differentiation is from the right. Actually, our interpretation is less restrictive than [565]: we differentiate as if y is an element of the classical $GL(n)$ and then postulate (4.81) for $GL_{u\mathbf{q}}(n)$. Our point is that in all cases this may bring only some differences in inessential numerical factors.

Note also that from (4.81) follows

$$\langle Y\,, 1_{\mathcal{A}}\rangle = 0\,, Y = D_i, E_{ij}, F_{ij}. \tag{4.83}$$

4.5.2 Commutation Relations of the Dual Algebra

To obtain the commutation relations between the generators D_i, E_{ij}, F_{ij}, we first need to evaluate the action of their bilinear products on the elements of $GL_{u\mathbf{q}}(n)$. We shall show that it is enough to do this for the Chevalley-like generators $D_i\,, 1 \le i \le n\,, E_i \equiv E_{i,i+1}\,, F_i \equiv F_{i+1,i}\,, 1 \le i \le n-1$. Then through them we shall express the rest of the generators E_{ij}, F_{ij}.

Using the defining relations and the second of relations (4.39a), we obtain:

$$\langle D_iD_j, f\rangle = \langle D_jD_i, f\rangle = k_ik_j\delta_{\mathbf{m}0}\delta_{\mathbf{n}0}\,, \quad 1 \le i, j \le n\,, \tag{4.84}$$

$$\begin{aligned}
\langle D_iE_j, f\rangle &= (k_i + \delta_{ij})\delta_{m_{j,j+1}1}\delta^j_{\mathbf{m}0}\delta_{\mathbf{n}0} = (k_i + \delta_{ij})\langle E_j, f\rangle\,, \quad \delta^j_{\mathbf{m}0} = \delta^{j,j+1}_{\mathbf{m}0}\,,\\
\langle E_jD_i, f\rangle &= (k_i + \delta_{i,j+1})\delta_{m_{j,j+1}1}\delta^j_{\mathbf{m}0}\delta_{\mathbf{n}0} = (k_i + \delta_{i,j+1})\langle E_j, f\rangle\,,
\end{aligned} \tag{4.85}$$

$$\begin{aligned}
\langle D_iF_j, f\rangle &= (k_i + \delta_{i,j+1})\delta_{n_{j+1,j}1}\delta_{\mathbf{m}0}\delta^j_{\mathbf{n}0} = (k_i + \delta_{i,j+1})\langle F_j, f\rangle\,,\\
&\delta^j_{\mathbf{n}0} = \delta^{j+1,j}_{\mathbf{n}0}\,,\\
\langle F_jD_i, f\rangle &= (k_i + \delta_{ij})\delta_{n_{j+1,j}1}\delta_{\mathbf{m}0}\delta^j_{\mathbf{n}0} = (k_i + \delta_{ij})\langle F_j, f\rangle\,,
\end{aligned} \tag{4.86}$$

$$\begin{aligned}
\langle E_iF_i, f\rangle &= u^{-2k_{i+1}}\frac{u^{2k_i} - 1}{u^2 - 1}\delta_{\mathbf{m}0}\delta_{\mathbf{n}0} + q^{-1}_{i,i+1}\delta_{m_{i,i+1}1}\delta_{n_{i+1,1}1}\delta^i_{\mathbf{m}0}\delta^i_{\mathbf{n}0}\,,\\
\langle F_iE_i, f\rangle &= \frac{u^{-2k_{i+1}} - 1}{u^{-2} - 1}\delta_{\mathbf{m}0}\delta_{\mathbf{n}0} + u^2q^{-1}_{i,i+1}\delta_{m_{i,i+1}1}\delta_{n_{i+1,i}1}\delta^i_{\mathbf{m}0}\delta^i_{\mathbf{n}0}
\end{aligned} \tag{4.87}$$

$$\begin{aligned}
\langle E_iF_j, f\rangle &= e_{ij}\delta_{m_{i,i+1}1}\delta_{n_{j+1,j}1}\delta^i_{\mathbf{m}0}\delta^j_{\mathbf{n}0}\,, \quad i \neq j\,,\\
e_{ij} &= \begin{cases} q_{i+1,j+1}/q_{i,j+1} & \text{for} \quad i < j\\ u^2/q_{i,i+1} & \text{for} \quad i = j+1\\ q_{j+1,i}/q_{j+1,i+1} & \text{for} \quad i > j+1 \end{cases}\\
\langle F_jE_i, f\rangle &= f_{ij}\delta_{m_{i,i+1}1}\delta_{n_{j+1,j}1}\delta^i_{\mathbf{m}0}\delta^j_{\mathbf{n}0}\,, \quad i \neq j\,,\\
f_{ij} &= \begin{cases} q_{i+1,j}/q_{ij} & \text{for} \quad i < j-1\\ 1/q_{i,i+1} & \text{for} \quad i = j-1\\ q_{ji}/q_{j,i+1} & \text{for} \quad i > j \end{cases}
\end{aligned} \tag{4.88}$$

$$\begin{aligned}
\langle E_iE_{i+1}, f\rangle &= q^{-1}_{i,i+1}\delta_{m_{i,i+1}1}\delta_{m_{i+1,i+2}1}\delta^{i,i+1}_{\mathbf{m}0}\delta_{\mathbf{n}0} + \delta_{m_{i,i+2}1}\delta^{i,i+2}_{\mathbf{m}0}\delta_{\mathbf{n}0}\,,\\
\langle E_iE_j, f\rangle &= e'_{ij}\delta_{m_{i,i+1}1}\delta_{m_{j,j+1}1}\delta^{ij}_{\mathbf{m}0}\delta_{\mathbf{n}0}, \quad i \neq j-1\,,\\
e'_{ij} &= \begin{cases} q_{i+1,j}/q_{ij} & \text{for} \quad i < j-1\\ (1+u^2)/q_{i,i+1} & \text{for} \quad i = j\\ q_{j+1,i+1}/q_{j,i+1} & \text{for} \quad i > j \end{cases}
\end{aligned} \tag{4.89}$$

$$\begin{aligned}
\langle F_{i+1}F_i, f\rangle &= q_{i+1,i+2}\delta_{n_{i+1,i}1}\delta_{n_{i+2,i+1}1}\delta^{i,i+1}_{\mathbf{n}0}\delta_{\mathbf{m}0} + \delta_{n_{i+2,i}1}\delta^{i+2,i}_{\mathbf{n}0}\delta_{\mathbf{m}0}\,,\\
\langle F_iF_j, f\rangle &= f'_{ij}\delta_{n_{i+1,i}1}\delta_{n_{j+1,j}1}\delta^{ij}_{\mathbf{n}0}\delta_{\mathbf{m}0}, \quad i \neq j+1\,,\\
f'_{ij} &= \begin{cases} q_{i,j+1}/q_{ij} & \text{for} \quad i < j\\ (1+u^{-2})q_{i,i+1} & \text{for} \quad i = j\\ q_{j+1,i+1}/q_{j+1,i} & \text{for} \quad i > j+1 \end{cases}
\end{aligned} \tag{4.90}$$

Thus, we have the following commutation relations:

$$[D_i, D_j] = 0 \tag{4.91a}$$

$$[D_i, E_j] = (\delta_{ij} - \delta_{i,j+1})E_j\,, \quad [D_i, F_j] = (-\delta_{ij} + \delta_{i,j+1})F_j \tag{4.91b}$$

$$uE_iF_i - u^{-1}F_iE_i = \lambda^{-1}\left(u^{2(D_i - D_{i+1})} - 1_{\mathcal{U}}\right) \tag{4.91c}$$

$$E_iF_j = g_{ij}F_jE_i\,, \quad i \neq j\,, \tag{4.91d}$$

$$g_{ij} = e_{ij}/f_{ij} = \begin{cases} q_{ij}q_{i+1,j+1}/q_{i,j+1}q_{i+1,j} & \text{for} \quad i < j-1 \\ q_{i,i+1}q_{i+1,i+2}/q_{i,i+2} & \text{for} \quad i = j-1 \\ u^2 q_{i-1,i+1}/q_{i-1,i}q_{i,i+1} & \text{for} \quad i = j+1 \\ q_{j,i+1}q_{j+1,i}/q_{ji}q_{j+1,i+1} & \text{for} \quad i > j+1 \end{cases}$$

$$E_iE_j = g_{ij}^{-1}E_jE_i\,, i < j-1 \tag{4.91e}$$

$$F_iF_j = g_{ij}^{-1}F_jF_i\,, i > j+1. \tag{4.91f}$$

It is convenient to use besides the generators D_i also the generators:

$$K = D_1 + \cdots + D_n\,, \quad H_i = D_i - D_{i+1}\,, \ 1 \le i \le n-1. \tag{4.92}$$

and we shall give many results for both sets. Let us note that the generator K commutes with all generators D_i, E_i, F_i:

$$[K, D_i] = 0\,, \quad [K, E_i] = 0\,, \quad [K, F_i] = 0\,, \tag{4.93}$$

while the generators H_i, E_i, F_i also form a commutation subalgebra, namely, instead of formulae (4.91a–c) we have:

$$[H_i, H_j] = 0 \tag{4.94a}$$

$$[H_i, E_j] = c_{ij}E_j\,, \quad [H_i, F_j] = -c_{ij}F_j \tag{4.94b}$$

$$c_{ij} = 2\delta_{ij} - \delta_{i,j+1} - \delta_{i+1,j}\,, \quad 1 \le i, j \le n-1$$

$$uE_iF_i - u^{-1}F_iE_i = \lambda^{-1}\left(u^{2H_i} - 1_{\mathcal{U}}\right) \tag{4.94c}$$

where the numbers c_{ij} form the Cartan matrix of the algebra $A_{n-1} = sl(n, \mathbb{C})$.

Similarly to (4.91) we derive the analogue of the Serre relations:

$$p_i^{\pm}E_i^2E_{i\pm1} - (u + u^{-1})E_iE_{i\pm1}E_i + (p_i^{\pm})^{-1}E_{i\pm1}E_i^2 = 0\,, \tag{4.95a}$$

$$p_i^{\pm}F_i^2F_{i\pm1} - (u + u^{-1})F_iF_{i\pm1}F_i + (p_i^{\pm})^{-1}F_{i\pm1}F_i^2 = 0\,, \tag{4.95b}$$

$$p_i^+ = q_{i,i+1}q_{i+1,i+2}/uq_{i,i+2}\,, \quad p_i^- = uq_{i-1,i+1}/q_{i-1,i}q_{i,i+1}$$

Now we shall express the rest of the generators E_{ij}, F_{ij} through the Chevalley-like ones. First let us rewrite relations (4.95) for sign "+" in a more suggestive way:

$$u^{-1}p_iE_i\left(E_iE_{i+1} - p_i^{-1}E_{i+1}E_i\right) - u\left(E_iE_{i+1} - p_i^{-1}E_{i+1}E_i\right)E_i = 0\,,$$
$$up_i^{-1}\left(F_{i+1}F_i - p_iF_iF_{i+1}\right)F_i - u^{-1}F_i\left(F_{i+1}F_i - p_iF_iF_{i+1}\right) = 0\,,$$
$$p_i = q_{i,i+1}q_{i+1,i+2}/q_{i,i+2}. \qquad (4.96)$$

Thus, we are prompted to define generators inductively analogously to one-parameter deformation (cf. (1.28)):

$$E_{ij} \equiv E_iE_{i+1,j} - p_{ij}^{-1}E_{i+1,j}E_i\,, \quad i<j \qquad (4.97)$$
$$F_{ij} \equiv F_{i,j+1}F_j - p_{ij}F_jF_{i,j+1}\,, \quad i>j$$
$$p_{ij} = q_{i,i+1}q_{i+1,j}/q_{ij}.$$

Thus we have two definitions for the generators E_{ij}, F_{ij} when $|i-j| \neq 1$ and we should check their consistency. The proof of this is inductive. We start with the case $|i-j| = 2$ where we have the desired consistency just using (4.89) and (4.90):

$$\langle E_{i,i+2}, f\rangle = \langle E_iE_{i+1} - p_i^{-1}E_{i+1}E_i, f\rangle = \delta_{m_{i,i+2}1}\delta_{\mathbf{m}0}^{i,i+2}\delta_{\mathbf{n}0} \qquad (4.98a)$$
$$\langle F_{i+2,i}, f\rangle = \langle F_{i+1}F_i - p_iF_iF_{i+1}, f\rangle = \delta_{n_{i,i+2}1}\delta_{\mathbf{n}0}^{i,i+2}\delta_{\mathbf{m}0} \qquad (4.98b)$$

Then we suppose that we have proved consistency for E_{ij}, F_{ij} when $1 < |i-j| < s$, and then we shall prove for E_{ij}, F_{ij} for $|i-j| = s$. Namely, using this supposition and (4.89) and (4.90) we find that:

$$\langle E_{ij}, f\rangle = \langle E_iE_{i+1,j} - p_{ij}^{-1}E_{i+1,j}E_i, f\rangle = \delta_{m_{ij}1}\delta_{\mathbf{m}0}^{ij}\delta_{\mathbf{n}0} \qquad (4.99a)$$
$$\langle F_{ij}, f\rangle = \langle F_{i,j+1}F_j - p_{ij}F_jF_{i,j+1}, f\rangle = \delta_{n_{ij}1}\delta_{\mathbf{n}0}^{ij}\delta_{\mathbf{m}0}. \qquad (4.99b)$$

For (4.99) we have used analogues of (4.89) and (4.90) for $E_iE_{i+1,j}$, $E_{i+1,j}E_i$, $F_{i,j+1}F_j$, $F_jF_{i,j+1}$.

4.5.3 Hopf Algebra Structure of the Dual Algebra

In this section we shall use the duality to derive the Hopf algebra structure of $\mathcal{U}_{u\tilde{q}}$. We start with the coproducts in $\mathcal{U}_{u\tilde{q}}$. Namely, we use repeatedly the first of relations (4.39a)

$$\langle Y, f\rangle = \langle \delta_{\mathcal{U}}(Y), f_1 \otimes f_2\rangle \qquad (4.100)$$

for every splitting $f = f_1f_2$. Thus we derive:

$$\delta_{\mathcal{U}}(D_i) = D_i \otimes 1_{\mathcal{U}} + 1_{\mathcal{U}} \otimes D_i\,, \qquad (4.101a)$$
$$\delta_{\mathcal{U}}(H_i) = H_i \otimes 1_{\mathcal{U}} + 1_{\mathcal{U}} \otimes H_i\,, \qquad (4.101b)$$
$$\delta_{\mathcal{U}}(K) = K \otimes 1_{\mathcal{U}} + 1_{\mathcal{U}} \otimes K. \qquad (4.101c)$$

Then we try the following Ansätze:

$$\delta_{\mathcal{U}}(E_i) = E_i \otimes \mathcal{P}_i + 1_{\mathcal{U}} \otimes E_i\,, \tag{4.102a}$$

$$\delta_{\mathcal{U}}(F_i) = F_i \otimes \mathcal{Q}_i + 1_{\mathcal{U}} \otimes F_i. \tag{4.102b}$$

We take in (4.100)$f_1 = a_{i,i+1}$, $f_2 = (a_{11})^{k_1} \dots (a_{nn})^{k_n}$ and using

$$\begin{aligned} f_1 f_2 &= a_{i,i+1}(a_{11})^{k_1} \dots (a_{nn})^{k_n} = \\ &= A_i (a_{11})^{k_1} \dots (a_{nn})^{k_n} a_{i,i+1} = A_i f_2 f_1\,, \\ A_i &= \left(\prod_{s=1}^{i-1} \left(\frac{q_{si}}{q_{s,i+1}}\right)^{k_s}\right) \left(\frac{u^2}{q_{i,i+1}}\right)^{k_i} \left(\frac{1}{q_{i,i+1}}\right)^{k_{i+1}} \left(\prod_{t=i+2}^{n} \left(\frac{q_{i+1,t}}{q_{it}}\right)^{k_t}\right), \end{aligned} \tag{4.103}$$

we obtain, on the one hand,

$$\langle E_i, f_1 f_2 \rangle = A_i \langle E_i, f_2 f_1 \rangle = A_i \tag{4.104}$$

while, on the other hand, using the Ansatz (4.102a), we have:

$$\langle E_i, f_1 f_2 \rangle = \langle E_i, f_1 \rangle \langle \mathcal{P}_i, f_2 \rangle = \langle \mathcal{P}_i, f_2 \rangle. \tag{4.105}$$

Comparing (4.104) with (4.105) we try

$$\mathcal{P}_i = \left(\prod_{s=1}^{i-1} \left(\frac{q_{si}}{q_{s,i+1}}\right)^{D_s}\right) \left(\frac{u^2}{q_{i,i+1}}\right)^{D_i} \left(\frac{1}{q_{i,i+1}}\right)^{D_{i+1}} \left(\prod_{t=i+2}^{n} \left(\frac{q_{i+1,t}}{q_{it}}\right)^{D_t}\right), \tag{4.106}$$

then we check that (4.102a) with this choice is consistent for all choices of f_1, f_2 in (4.100).

Analogously we proceed to obtain $\mathcal{Q}_i$: we take $f_1' = a_{i+1,i}$, f_2 as above to find:

$$f_1' f_2 = a_{i+1,i}(a_{11})^{k_1} \dots (a_{nn})^{k_n} = u^{2(k_i - k_{i+1})} A_i^{-1} f_2 f_1' \tag{4.107}$$

and thus we have:

$$\begin{aligned} \mathcal{Q}_i &= \left(\prod_{s=1}^{i-1} \left(\frac{q_{s,i+1}}{q_{si}}\right)^{D_s}\right) (q_{i,i+1})^{D_i} \left(\frac{q_{i,i+1}}{u^2}\right)^{D_{i+1}} \left(\prod_{t=i+2}^{n} \left(\frac{q_{it}}{q_{i+1,t}}\right)^{D_t}\right) = \\ &= u^{2H_i} \mathcal{P}_i^{-1}. \end{aligned} \tag{4.108}$$

The coproducts of the rest of the generators we obtain using (4.97) and the coproducts of the generators E_i, F_i, for example,

$$\delta_{\mathcal{U}}(E_{i,i+2}) = E_{i,i+2} \otimes \mathcal{P}_{i,i+2} + 1_{\mathcal{U}} \otimes E_{i,i+2} + (\lambda/u) E_{i+1} \otimes E_i \mathcal{P}_{i+1} \tag{4.109}$$

$$\mathscr{P}_{i,i+2} = \mathscr{P}_i\mathscr{P}_{i+1} = \left(\prod_{s=1}^{i-1}\left(\frac{q_{si}}{q_{s,i+2}}\right)^{D_s}\right)\left(\frac{u^2}{q_{i,i+2}}\right)^{D_i}\left(\frac{u^2}{q_{i,i+1}q_{i+1,i+2}}\right)^{D_{i+1}} \times$$
$$\times\left(\frac{1}{q_{i,i+2}}\right)^{D_{i+2}}\left(\prod_{t=i+3}^{n}\left(\frac{q_{i+2,t}}{q_{it}}\right)^{D_t}\right), \tag{4.110}$$

$$\delta_{\mathscr{U}}(F_{i+2,i}) = F_{i+2,i}\otimes\mathscr{Q}_{i+2,i} + 1_{\mathscr{U}}\otimes F_{i+2,i} - u\lambda F_i\otimes F_{i+1}\mathscr{Q}_i \tag{4.111}$$

$$\mathscr{Q}_{i+2,i} = \mathscr{Q}_i\mathscr{Q}_{i+1} = u^{2(H_i+H_{i+1})}(\mathscr{P}_i\mathscr{P}_{i+1})^{-1} = u^{2H_{i,i+2}}\mathscr{P}_{i,i+2}^{-1}, \tag{4.112}$$

$$H_{i,i+2} \equiv H_i + H_{i+1} = D_i - D_{i+2}, \tag{4.113}$$

where we have used:

$$\mathscr{P}_iE_j = \begin{cases} u^2E_i\mathscr{P}_i & \text{for} \quad i=j \\ g_{ij}^{-1}E_j\mathscr{P}_i & \text{for} \quad i\neq j \end{cases}, \quad \mathscr{P}_iF_j = \begin{cases} u^{-2}F_i\mathscr{P}_i & \text{for} \quad i=j \\ g_{ij}F_j\mathscr{P}_i & \text{for} \quad i\neq j \end{cases} \tag{4.114}$$
$$\mathscr{Q}_iE_j = \begin{cases} u^2E_i\mathscr{Q}_i & \text{for} \quad i=j \\ u^{2c_{ij}}g_{ij}E_j\mathscr{Q}_i & \text{for} \quad i\neq j \end{cases}, \quad \mathscr{Q}_iF_j = \begin{cases} u^{-2}F_i\mathscr{Q}_i & \text{for} \quad i=j \\ u^{-2c_{ij}}g_{ij}^{-1}F_j\mathscr{P}_i & \text{for} \quad i\neq j. \end{cases}$$

The counit relations in $\mathscr{U}_{u\tilde{q}}$ are given by:

$$\varepsilon_{\mathscr{U}}(Y) = 0\,, Y = D_i, E_{ij}, F_{ij}, K, H_i\,, \tag{4.115}$$

which follows easily using (4.39b), (4.83), and (4.92):

$$\varepsilon_{\mathscr{U}}(Y) = \langle Y, 1_{\mathscr{A}}\rangle = 0. \tag{4.116}$$

Finally, the antipode map in $\mathscr{U} = \mathscr{U}_{u\tilde{q}}$ is given by:

$$\gamma_{\mathscr{U}}(D_i) = -D_i\,, \quad \gamma_{\mathscr{U}}(H_i) = -H_i\,, \quad \gamma_{\mathscr{U}}(K) = -K\,, \tag{4.117a}$$
$$\gamma_{\mathscr{U}}(E_i) = -E_i\mathscr{P}_i^{-1}, \quad \gamma_{\mathscr{U}}(F_i) = -F_i\mathscr{Q}_i^{-1}. \tag{4.117b}$$

This follows from (4.101),(4.102), and (4.115) with elementary application of one of the basic axioms of Hopf algebras [11]:

$$m\circ(\mathrm{id}_{\mathscr{U}}\otimes\gamma_{\mathscr{U}})\circ\delta_{\mathscr{U}} = i\circ\varepsilon_{\mathscr{U}}\,, \tag{4.118}$$

where both sides are maps $\mathscr{U}\to\mathscr{U}$, m is the usual product in the algebra: $m(Y\otimes Z) = YZ\,, Y, Z\in\mathscr{U}$ and i is the natural embedding of $\mathbb{C}$ into $\mathscr{U}$: $i(c) = c1_{\mathscr{U}}\,, c\in\mathbb{C}$. To obtain (4.117) we just apply both sides of (4.118) to D_i, H_i, K, E_i, F_i. For (4.117c), we also use $\gamma_{\mathscr{U}}(\mathscr{P}_i) = \mathscr{P}_i^{-1}$, $\gamma_{\mathscr{U}}(\mathscr{Q}_i) = \mathscr{Q}_i^{-1}$, which follow from (4.117a). The antipode map for the rest of the generators E_{ij}, F_{ij}, we obtain using (4.97) and (4.117).

4.5.4 Drinfeld–Jimbo Form of the Dual Algebra

In this section we show how to transform the algebra $\mathscr{U}_{u\tilde{q}}$ to a Drinfeld–Jimbo form. (It could be transformed also to the algebra given in [565] in terms only of the Chevalley generators.) We first note that if we set all parameters equal $q_{ij} = u$ for all i, j and make the change

$$E_i = X_i^+ u^{H_i/2}, \quad F_i = X_i^- u^{H_i/2}, \tag{4.119}$$

then the generators $H_i, X_i^\pm, 1 \le i \le n-1$ obey the commutation rules and Serre relations of the standard Drinfeld–Jimbo deformation $U_u(sl(n, \mathbb{C}))$.

Then we note that if $q_{ij} = u$ for all i, j, then we have $\mathscr{P}_i = u^{H_i} = \mathscr{Q}_i$. Thus, we are prompted to try for the analogue of the transformation (4.119) in the multiparametric case the following:

$$E_i = X_i^+ \mathscr{P}_i^{1/2}, \quad F_i = X_i^- \mathscr{Q}_i^{1/2}. \tag{4.120}$$

Indeed we have:

$$[H_i, X_j^+] = [H_i, E_j \mathscr{P}_i^{-1/2}] = c_{ij} X_j^+ \tag{4.121a}$$

$$[H_i, X_j^-] = [H_i, F_j \mathscr{Q}_i^{-1/2}] = -c_{ij} X_j^- \tag{4.121b}$$

$$\begin{aligned}[X_i^+, X_i^-] &= [E_i \mathscr{P}_i^{-1/2}, F_i \mathscr{Q}_i^{-1/2}] = \left(uE_iF_i - u^{-1}F_iE_i\right) u^{-H_i} = \\ &= \lambda^{-1}\left(u^{H_i} - u^{-H_i}\right) \equiv [H_i]_u\,,\end{aligned} \tag{4.121c}$$

where we have used (4.120), (4.94b–c), (4.106), (4.108), and (4.114);

$$\begin{aligned}[X_i^+, X_j^-] &= [E_i \mathscr{P}_i^{-1/2}, F_j \mathscr{Q}_j^{-1/2}] = \\ &= E_iF_j \left(\frac{\mathscr{P}_j}{g_{ij}\mathscr{P}_i}\right)^{1/2} u^{-H_j} - F_jE_i \left(\frac{\mathscr{P}_j}{g_{ji}\mathscr{P}_i}\right)^{1/2} u^{-H_j} u^{\delta_{j,i\pm1}} = \\ &= F_jE_i \left(\frac{\mathscr{P}_j}{g_{ji}\mathscr{P}_i}\right)^{1/2} u^{-H_j}\left(g_{ij}^{1/2} g_{ji}^{1/2} - u^{\delta_{j,i\pm1}}\right) = 0\,, \text{for } i \neq j\,,\end{aligned} \tag{4.122}$$

where we have used (4.120), (4.106), (4.108), (4.114), and

$$g_{ij}g_{ji} = \begin{cases} u^2 & \text{for} \quad j = i \pm 1 \\ 1 & \text{otherwise.} \end{cases} \tag{4.123}$$

Next we have:

$$(X_i^+)^2 X_{i\pm1}^+ - [2]_u X_i^+ X_{i\pm1}^+ X_i^+ + X_{i\pm1}^+ (X_i^+)^2 = \\ = u^{-1} g_{i,i\pm1} E_i^2 E_{i\pm1} - [2]_u E_i E_{i\pm1} E_i + u^{-1} g_{i\pm1,i} E_{i\pm1} E_i^2 = 0\,, \tag{4.124}$$

where we have used (4.120) and (4.114), the facts that $g_{i,i\pm1}/u = p_i^\pm$, $g_{i\pm1,i}/u = (p_i^\pm)^{-1}$, and (4.95a):

$$X_i^+ X_j^+ = E_i \mathscr{P}_i^{-1/2} E_j \mathscr{P}_j^{-1/2} = g_{ij}^{1/2} E_i E_j \mathscr{P}_i^{-1/2} \mathscr{P}_j^{-1/2} = \\ = g_{ij}^{-1/2} E_j E_i \mathscr{P}_i^{-1/2} \mathscr{P}_j^{-1/2} = g_{ij}^{-1/2} g_{ji}^{-1/2} E_j \mathscr{P}_j^{-1/2} E_i \mathscr{P}_i^{-1/2} = X_j^+ X_i^+\,, \tag{4.125}$$

where $i < j-1$, and we have used (4.120), (4.114), (4.91e), and (4.123). Formulae (4.94a), (4.121), (4.122), (4.124), and (4.125) and the analogues of (4.124) and (4.125) for sign "-" are the defining relations of the one-parameter deformation $U_u(sl(n,\mathbb{C}))$ in terms of the Chevalley generators H_i, $X_i^\pm$, $i = 1, \dots, n-1$.

Thus as a **commutation algebra** we have $\mathscr{U}_{u\tilde{q}} \cong U_u(sl(n,\mathbb{C})) \otimes U_u(\mathscr{Z})$, where $U_u(\mathscr{Z})$ is spanned by $K, u^{\pm K/2}$. This splitting is preserved also by the counit and the antipode (cf. (4.115) and (4.117b)) for the generators H_i and K, while for $X_i^\pm$ we have:

$$\begin{aligned} \varepsilon_{\mathscr{U}}(X_i^+) &= \varepsilon_{\mathscr{U}}(E_i)\varepsilon_{\mathscr{U}}(\mathscr{P}_i^{-1/2}) = 0\,, \\ \varepsilon_{\mathscr{U}}(X_i^-) &= \varepsilon_{\mathscr{U}}(F_i)\varepsilon_{\mathscr{U}}(\mathscr{Q}_i^{-1/2}) = 0 \\ \gamma_{\mathscr{U}}(X_i^+) &= \gamma_{\mathscr{U}}(\mathscr{P}_i^{-1/2})\gamma_{\mathscr{U}}(E_i) = -\mathscr{P}_i^{1/2} E_i \mathscr{P}_i^{-1} = \\ &\quad -u E_i \mathscr{P}_i^{-1/2} = -u X_i^+ \\ \gamma_{\mathscr{U}}(X_i^-) &= \gamma_{\mathscr{U}}(\mathscr{Q}_i^{-1/2})\gamma_{\mathscr{U}}(F_i) = -\mathscr{Q}_i^{1/2} F_i \mathscr{P}_i^{-1} = \\ &\quad -u^{-1} F_i \mathscr{Q}_i^{-1/2} = -u^{-1} X_i^-\,, \end{aligned} \tag{4.126}$$

where we have used (4.114). The splitting is also preserved by the coproducts of H_i, K (cf. (4.101b)).

However, for the coproducts of the Chevalley generators $X_i^\pm$ we have:

$$\delta_{\mathscr{U}}(X_i^+) = \delta_{\mathscr{U}}(E_i)\delta_{\mathscr{U}}(\mathscr{P}_i^{-1/2}) = (E_i \otimes \mathscr{P}_i + 1_{\mathscr{U}} \otimes E_i)\left(\mathscr{P}_i^{-1/2} \otimes \mathscr{P}_i^{-1/2}\right) = \\ = X_i^+ \otimes \mathscr{P}_i^{1/2} + \mathscr{P}_i^{-1/2} \otimes X_i^+\,, \tag{4.127a}$$

$$\delta_{\mathscr{U}}(X_i^-) = \delta_{\mathscr{U}}(F_i)\delta_{\mathscr{U}}(\mathscr{Q}_i^{-1/2}) = (F_i \otimes \mathscr{Q}_i + 1_{\mathscr{U}} \otimes F_i)\left(\mathscr{Q}_i^{-1/2} \otimes \mathscr{Q}_i^{-1/2}\right) = \\ = X_i^- \otimes \mathscr{Q}_i^{1/2} + \mathscr{Q}_i^{-1/2} \otimes X_i^-\,. \tag{4.127b}$$

Thus, as a coalgebra $\mathscr{U}_{u\tilde{q}}$ cannot be split as above, and furthermore it depends on all parameters. Only if we set $q_{ij} = u$ for all i, j then $\mathscr{P}_i = u^{H_i} = \mathscr{Q}_i$ and (4.127) become the standard coproducts of the Chevalley generators $X_i^\pm$ of $U_u(sl(n,\mathbb{C}))$.

4.5.5 Special Cases of Hopf Algebra Splitting

In this section we consider the special case when some of the parameters coincide, so that the central generator K would decouple as in the one-parameter deformation. For this we first need to express the operators $\mathscr{P}_i$ (and through them $\mathscr{Q}_i$) in terms of the generators H_i and K. For this we first express the generators D_i through H_i and K:

$$D_i = \frac{1}{n}\left(K - \sum_{j=1}^{i-1} jH_j + \sum_{j=i}^{n-1}(n-j)H_j\right) = \hat{K} + \hat{H}_i \tag{4.128}$$

$$\hat{K} \equiv \frac{1}{n}\left(K - \sum_{j=1}^{n-1} jH_j\right), \qquad \hat{H}_i \equiv \sum_{j=i}^{n-1} H_j, \qquad (\hat{H}_n \equiv 0).$$

Now we substitute (4.128) in (4.106) to obtain:

$$\mathscr{P}_i = (\tilde{q}_i)^{\hat{K}} \left(\prod_{s=1}^{i-1}\left(\frac{q_{si}}{q_{s,i+1}}\right)^{\hat{H}_s}\right)\left(\frac{u^2}{q_{i,i+1}}\right)^{\hat{H}_i}\left(\frac{1}{q_{i,i+1}}\right)^{\hat{H}_{i+1}} \prod_{t=i+2}^{n-1}\left(\frac{q_{i+1,t}}{q_{it}}\right)^{\hat{H}_t}$$

$$\tilde{q}_i \equiv \left(\prod_{s=1}^{i-1}\frac{q_{si}}{q_{s,i+1}}\right)\frac{u^2}{q_{i,i+1}^2}\prod_{t=i+2}^{n}\frac{q_{i+1,t}}{q_{it}}. \tag{4.129}$$

From the above expression it is clear that in order for K to decouple from the system the $n-1$ constants $\tilde{q}_i$ should become equal to unity. This brings $n-1$ conditions on the parameters q_{ij}. It seems natural to use these conditions to fix the $n-1$ next-to-main-diagonal parameters $q_{i,i+1}$, and indeed, a natural choice for this exists, namely, we may set:

$$\begin{aligned} q^0_{i,i+1} &\equiv u^{i(n-i)} \prod_{\substack{1\le s\le i,\quad i+1\le t\le n \\ s<t-1}} q_{st}^{-1} = \\ &= u\widetilde{\prod_{s=1}^{i}\prod_{t=i+1}^{n}}\frac{u}{q_{st}}, \qquad 1 \le i \le n-1, \end{aligned} \tag{4.130}$$

where the tilde over the double product means that the case $s = i = t-1$ should be omitted. Then we obtain:

$$(\tilde{q}_i)_{q_{i,i+1}=q^0_{i,i+1}} = 1, \qquad 1 \le i \le n-1 \tag{4.131}$$

and substituting this in the operators $\mathscr{P}_i$ we get in terms of $\hat{H}_i$ and in terms of H_i:

$$\begin{aligned} \tilde{\mathscr{P}}_i \equiv (\mathscr{P}_i)_{q_{i,i+1}=q^0_{i,i+1}} &= \left(\prod_{s=1}^{i-2}\left(\frac{q_{si}}{q_{s,i+1}}\right)^{\hat{H}_s}\right)\left(\frac{u}{q_{i-1,i+1}}\widetilde{\prod_{s=1}^{i-1}\prod_{t=i}^{n}}\frac{u}{q_{st}}\right)^{\hat{H}_{i-1}} \times \\ &\times u^{\hat{H}_i-\hat{H}_{i+1}}\left(\widetilde{\prod_{s=1}^{i}\prod_{t=i+1}^{n}}\frac{q_{st}}{u}\right)^{\hat{H}_i+\hat{H}_{i+1}} \times \end{aligned} \tag{4.132}$$

$$\times \left(\frac{u}{q_{i,i+2}} \overbrace{\prod_{s=1}^{i+1} \prod_{t=i+2}^{n}} \frac{u}{q_{st}} \right)^{\hat{H}_{i+2}} \prod_{t=i+3}^{n-1} \left(\frac{q_{i+1,t}}{q_{it}} \right)^{\hat{H}_t} =$$

$$= \left(\prod_{j=1}^{i-2} \left(\prod_{s=1}^{j} \frac{q_{si}}{q_{s,i+1}} \right)^{H_j} \right) \left(\left(\prod_{s=1}^{i-1} \frac{u^2}{q^2_{s,i+1}} \right) \prod_{s=1}^{i-1} \prod_{t=i+2}^{n} \frac{u}{q_{st}} \right)^{H_{i-1}} \times$$

$$\times \left(u \left(\prod_{s=1}^{i-1} \frac{u}{q_{s,i+1}} \right) \prod_{t=i+2}^{n} \frac{q_{it}}{u} \right)^{H_i} \times$$

$$\times \left(\left(\prod_{t=i+2}^{n} \frac{q^2_{it}}{u^2} \right) \prod_{s=1}^{i-1} \prod_{t=i+2}^{n} \frac{q_{st}}{u} \right)^{H_{i+1}} \left(\prod_{j=i+2}^{n-1} \left(\prod_{t=j+1}^{n} \frac{q_{it}}{q_{i+1,t}} \right)^{H_j} \right).$$

Thus, for the particular choice $q_{i,i+1} = q^0_{i,i+1}$ we have the splitting $\mathscr{U}_{u\tilde{q}} \cong U_{u,\tilde{\tilde{q}}}(sl(n,\mathbb{C})) \otimes U_u(\mathscr{Z})$ as tensor product of two Hopf subalgebras. Here by $U_{u,\tilde{\tilde{q}}}(sl(n,\mathbb{C}))$ we denote the Hopf algebra which is a deformation of $U(sl(n,\mathbb{C}))$ and is of Drinfeld–Jimbo form with deformation parameter u as commutation algebra, while as a coalgebra it depends on all remaining $(n^2 - 3n + 4)/2$parameters $\tilde{\tilde{q}} = \{q_{ij} \mid j - i > 1\}$ (cf. (4.127) and (4.132)).

4.6 Duality for a Lorentz Quantum Group

This section follows [234]. We find the dual algebra $\mathscr{L}_q^*$ to the matrix Lorentz quantum group $\mathscr{L}_q$ of Podles–Woronowicz [511] and Watamura et al. [123]. In fact, we start with a larger matrix quantum group $\widetilde{\mathscr{L}}_q$ and we find first its dual algebra $\widetilde{\mathscr{L}}_q^*$. As in the previous sections we start by postulating the pairings between the generating elements of the two algebras. We find that the algebra $\mathscr{L}_q^*$ is split in two mutually commuting subalgebras as in the classical case; that is, we can write $\mathscr{L}_q^* \cong U_q(sl(2,\mathbb{C})) \otimes U_q(sl(2,\mathbb{C}))$. We give also the coalgebra structure which, however, does not preserve this splitting.

4.6.1 Matrix Lorentz Quantum Group

In this section we recall the matrix Lorentz quantum group introduced in [123, 511]. It is more convenient to start with a larger matrix quantum group denoted by $\widetilde{\mathscr{L}}_q$ and generated by the elements $\alpha, \beta, \gamma, \delta, \bar{\alpha}, \bar{\beta}, \bar{\gamma}, \bar{\delta}$ with the following commutation relations ($q \in \mathbb{R}$, $\lambda = q - q^{-1}$):

$$\alpha\beta = q\beta\alpha, \quad \alpha\gamma = q\gamma\alpha, \quad \beta\delta = q\delta\beta, \quad \gamma\delta = q\delta\gamma,$$
$$\alpha\delta - \delta\alpha = \lambda\beta\gamma, \qquad \beta\gamma = \gamma\beta, \tag{4.133}$$

$$\begin{aligned} \bar{\beta}\bar{\alpha} &= q\bar{\alpha}\bar{\beta}, \quad \bar{\gamma}\bar{\alpha} = q\bar{\alpha}\bar{\gamma}, & \bar{\delta}\bar{\beta} &= q\bar{\beta}\bar{\delta}, \quad \bar{\delta}\bar{\gamma} = q\bar{\gamma}\bar{\delta}, \\ \bar{\delta}\bar{\alpha} - \bar{\alpha}\bar{\delta} &= \lambda\bar{\beta}\bar{\gamma}, & \bar{\beta}\bar{\gamma} &= \bar{\gamma}\bar{\beta}, \end{aligned} \tag{4.134}$$

$$\begin{aligned} \alpha\bar{\alpha} &= \bar{\alpha}\alpha - q\lambda\bar{\gamma}\gamma, & \alpha\bar{\beta} &= q^{-1}\bar{\beta}\alpha - \lambda\bar{\delta}\gamma, \\ \alpha\bar{\gamma} &= q\bar{\gamma}\alpha, & \alpha\bar{\delta} &= \bar{\delta}\alpha, \\ \beta\bar{\alpha} &= q^{-1}\bar{\alpha}\beta - \lambda\bar{\gamma}\delta, & \beta\bar{\beta} &= \bar{\beta}\beta + q\lambda(\bar{\alpha}\alpha - \bar{\delta}\delta - q\lambda\bar{\gamma}\gamma), \\ \beta\bar{\gamma} &= \bar{\gamma}\beta, & \beta\bar{\delta} &= q\bar{\delta}\beta + q^2\lambda\bar{\gamma}\alpha, \\ \gamma\bar{\alpha} &= q\bar{\alpha}\gamma, & \gamma\bar{\beta} &= \bar{\beta}\gamma, \\ \gamma\bar{\gamma} &= \bar{\gamma}\gamma, & \gamma\bar{\delta} &= q^{-1}\bar{\delta}\gamma, \\ \delta\bar{\alpha} &= \bar{\alpha}\delta, & \delta\bar{\beta} &= q\bar{\beta}\delta + q^2\lambda\bar{\alpha}\gamma, \\ \delta\bar{\gamma} &= q^{-1}\bar{\gamma}\delta, & \delta\bar{\delta} &= \bar{\delta}\delta + q\lambda\bar{\gamma}\gamma. \end{aligned} \tag{4.135}$$

Note that relations (4.134) may be obtained from (4.134) by the anti-involution

$$\alpha \mapsto \bar{\alpha}, \beta \mapsto \bar{\beta}, \gamma \mapsto \bar{\gamma}, \delta \mapsto \bar{\delta}, q \mapsto q. \tag{4.136}$$

We note that there are two central elements in the algebra $\widetilde{\mathscr{L}}_q$:

$$\begin{aligned} \mathscr{D} &= \alpha\delta - q\beta\gamma = \delta\alpha - q^{-1}\beta\gamma, \\ \mathscr{D}^* &= \bar{\alpha}\bar{\delta} - q^{-1}\bar{\beta}\bar{\gamma} = \bar{\delta}\bar{\alpha} - q\bar{\beta}\bar{\gamma}, \end{aligned} \tag{4.137}$$

which are conjugated under (4.136).

Considered as a bialgebra, $\widetilde{\mathscr{L}}_q$ has the following comultiplication $\Delta_{\widetilde{\mathscr{L}}_q}$ and counit $\varepsilon_{\widetilde{\mathscr{L}}_q}$ given on its generating elements:

$$\Delta_{\widetilde{\mathscr{L}}_q}\left(\begin{pmatrix} \alpha & \beta \\ \gamma & \delta \end{pmatrix}\right) = \begin{pmatrix} \alpha\otimes\alpha + \beta\otimes\gamma & \alpha\otimes\beta + \beta\otimes\delta \\ \gamma\otimes\alpha + \delta\otimes\gamma & \gamma\otimes\beta + \delta\otimes\delta \end{pmatrix} \tag{4.138a}$$

$$\Delta_{\widetilde{\mathscr{L}}_q}\left(\begin{pmatrix} \bar{\alpha} & \bar{\beta} \\ \bar{\gamma} & \bar{\delta} \end{pmatrix}\right) = \begin{pmatrix} \bar{\alpha}\otimes\bar{\alpha} + \bar{\beta}\otimes\bar{\gamma} & \bar{\alpha}\otimes\bar{\beta} + \bar{\beta}\otimes\bar{\delta} \\ \bar{\gamma}\otimes\bar{\alpha} + \bar{\delta}\otimes\bar{\gamma} & \bar{\gamma}\otimes\bar{\beta} + \bar{\delta}\otimes\bar{\delta} \end{pmatrix},$$

$$\varepsilon_{\widetilde{\mathscr{L}}_q}\left(\begin{pmatrix} \alpha & \beta \\ \gamma & \delta \end{pmatrix}\right) = \begin{pmatrix} 1 & 0 \\ 0 & 1 \end{pmatrix}, \tag{4.138b}$$

$$\varepsilon_{\widetilde{\mathscr{L}}_q}\left(\begin{pmatrix} \bar{\alpha} & \bar{\beta} \\ \bar{\gamma} & \bar{\delta} \end{pmatrix}\right) = \begin{pmatrix} 1 & 0 \\ 0 & 1 \end{pmatrix},$$

where for convenience we have used matrix notation.

Note that the bialgebra $\widetilde{\mathscr{L}}_q$ contains two conjugated (by (4.136)) sub-bialgebras, denoted as in [462] by $A_q(2)$ and $A_{q^{-1}}(2)$, and generated by $\alpha, \beta, \gamma, \delta$ and $\bar{\alpha}, \bar{\beta}, \bar{\gamma}, \bar{\delta}$, respectively (cf. (4.134),(4.134), and (4.138)).

If we impose the restrictions $\mathscr{D} \neq 0_{\widetilde{\mathscr{L}}_q} \neq \mathscr{D}^*$ then we may extend the algebra with two new central elements $\mathscr{D}^{-1}$ and $\mathscr{D}^{*-1}$ such that in the extended algebra, which we denote by $\widehat{\mathscr{L}}_q$, we have $\mathscr{D}\mathscr{D}^{-1} = 1_{\widehat{\mathscr{L}}_q} = \mathscr{D}^{-1}\mathscr{D}$ and $\mathscr{D}^*\mathscr{D}^{*-1} = 1_{\widehat{\mathscr{L}}_q} = \mathscr{D}^{*-1}\mathscr{D}^*$. The algebra $\widehat{\mathscr{L}}_q$ is a Hopf algebra since we can define an antipode $S_{\widehat{\mathscr{L}}_q}$ by:

$$S_{\widehat{\mathscr{L}}_q}\left(\begin{pmatrix} \alpha & \beta \\ \gamma & \delta \end{pmatrix}\right) = \mathscr{D}^{-1}\begin{pmatrix} \delta & -q^{-1}\beta \\ -q\gamma & \alpha \end{pmatrix}, \tag{4.139}$$
$$S_{\widehat{\mathscr{L}}_q}\left(\begin{pmatrix} \bar{\alpha} & \bar{\beta} \\ \bar{\gamma} & \bar{\delta} \end{pmatrix}\right) = \mathscr{D}^{*-1}\begin{pmatrix} \bar{\delta} & -q\bar{\beta} \\ -q^{-1}\bar{\gamma} & \bar{\alpha} \end{pmatrix},$$

from which follows:

$$S_{\widehat{\mathscr{L}}_q}(\mathscr{D}) = \mathscr{D}^{-1}, \quad S_{\widehat{\mathscr{L}}_q}(\mathscr{D}^{-1}) = \mathscr{D}, \quad S_{\widehat{\mathscr{L}}_q}(\mathscr{D}^*) = \mathscr{D}^{*-1}, \quad S_{\widehat{\mathscr{L}}_q}(\mathscr{D}^{*-1}) = \mathscr{D}^*. \tag{4.140}$$

If we impose the restrictions $\mathscr{D} = \mathscr{D}^* = 1_{\widetilde{\mathscr{L}}_q}$, we obtain the matrix Lorentz quantum group $\mathscr{L}_q$ introduced in [123, 511]. It is a Hopf algebra with coalgebra relations given by (4.138) and (4.139) with $\mathscr{D}^{-1} = \mathscr{D}^{*-1} = 1_{\widetilde{\mathscr{L}}_q}$. Note that $\mathscr{L}_q$ contains two conjugated (by (4.78)) Hopf subalgebras $SL_q(2)$ and $SL_{q^{-1}}(2)$, which are obtained from $A_q(2)$ and $A_{q^{-1}}(2)$, mentioned above, respectively.

4.6.2 Dual Algebras to the Algebras $\mathscr{L}_q$ and $\widetilde{\mathscr{L}}_q$

We are looking for the dual algebras to $\mathscr{L}_q$ and $\widetilde{\mathscr{L}}_q$. Following our procedure we first need to fix a basis in $\widetilde{\mathscr{L}}_q$. We choose the following basis in $\widetilde{\mathscr{L}}_q$:

$$f = \alpha^k \bar{\alpha}^{\bar{k}} \delta^\ell \bar{\delta}^{\bar{\ell}} \beta^m \bar{\beta}^{\bar{m}} \gamma^n \bar{\gamma}^{\bar{n}}, \quad k, \bar{k}, \ell, \bar{\ell}, m, \bar{m}, n, \bar{n} \in \mathbb{Z}_+. \tag{4.141}$$

The basis of the matrix Lorentz quantum group given in [511] may be obtained from the above after some rearrangement and by replacing:

$$\alpha^k \delta^\ell \to \begin{cases} \alpha^{k-\ell} & \text{for} \quad k \geq \ell \\ \delta^{\ell-k} & \text{for} \quad k < \ell \end{cases}, \qquad \bar{\delta}^{\bar{\ell}} \bar{\alpha}^{\bar{k}} \to \begin{cases} \bar{\alpha}^{\bar{k}-\bar{\ell}} & \text{for} \quad \bar{k} \geq \bar{\ell} \\ \bar{\delta}^{\bar{\ell}-\bar{k}} & \text{for} \quad \bar{k} < \bar{\ell} \end{cases}. \tag{4.142}$$

Let us denote by $A, B, C, D, \bar{A}, \bar{B}, \bar{C}, \bar{D}$ the generators of the dual algebra $\widetilde{\mathscr{L}}_q^*$ of $\widetilde{\mathscr{L}}_q$. For the action of $\widetilde{\mathscr{L}}_q^*$ on $\widetilde{\mathscr{L}}_q$ we set (as in Section 4.4):

$$\langle A, f \rangle \equiv k\delta_{m0}\delta_{n0}\delta_{\bar{m}0}\delta_{\bar{n}0}, \tag{4.143a}$$
$$\langle B, f \rangle \equiv \delta_{m1}\delta_{n0}\delta_{\bar{m}0}\delta_{\bar{n}0}, \tag{4.143b}$$
$$\langle C, f \rangle \equiv \delta_{m0}\delta_{n1}\delta_{\bar{m}0}\delta_{\bar{n}0}, \tag{4.143c}$$

$$\langle D, f\rangle \equiv \ell\delta_{m0}\delta_{n0}\delta_{\bar{m}0}\delta_{\bar{n}0}, \tag{4.143d}$$
$$\langle \bar{A}, f\rangle \equiv \bar{k}\delta_{m0}\delta_{n0}\delta_{\bar{m}0}\delta_{\bar{n}0}, \tag{4.143e}$$
$$\langle \bar{B}, f\rangle \equiv \delta_{m0}\delta_{n0}\delta_{\bar{m}1}\delta_{\bar{n}0}, \tag{4.143f}$$
$$\langle \bar{C}, f\rangle \equiv \delta_{m0}\delta_{n0}\delta_{\bar{m}0}\delta_{\bar{n}1}, \tag{4.143g}$$
$$\langle \bar{D}, f\rangle \equiv \bar{\ell}\delta_{m0}\delta_{n0}\delta_{\bar{m}0}\delta_{\bar{n}0}, \tag{4.143h}$$
$$\langle 1_{\widetilde{\mathscr{L}}_q^*}, f\rangle \equiv \delta_{m0}\delta_{n0}\delta_{\bar{m}0}\delta_{\bar{n}0}. \tag{4.143i}$$

If some monomial is not in normal order (4.141), then it should be brought to this order using commutation relations (4.134),(4.134), and (4.135), and then (4.143) can be applied.

We note also corollaries of (4.143):

$$\begin{aligned}
&\langle \begin{pmatrix} A & B \\ C & D \end{pmatrix}, \mathscr{D}\rangle = \begin{pmatrix} 1 & 0 \\ 0 & 1 \end{pmatrix},\\
&\langle \begin{pmatrix} A & B \\ C & D \end{pmatrix}, \mathscr{D}^*\rangle = \begin{pmatrix} 0 & 0 \\ 0 & 0 \end{pmatrix},\\
&\langle \begin{pmatrix} \bar{A} & \bar{B} \\ \bar{C} & \bar{D} \end{pmatrix}, \mathscr{D}\rangle = \begin{pmatrix} 0 & 0 \\ 0 & 0 \end{pmatrix},\\
&\langle \begin{pmatrix} \bar{A} & \bar{B} \\ \bar{C} & \bar{D} \end{pmatrix}, \mathscr{D}^*\rangle = \begin{pmatrix} 1 & 0 \\ 0 & 1 \end{pmatrix},
\end{aligned} \tag{4.144}$$

$$\langle Y, 1_{\widetilde{\mathscr{L}}_q}\rangle = 0,\ Y = A, B, C, D, \bar{A}, \bar{B}, \bar{C}, \bar{D}. \tag{4.145}$$

Next we would like to derive the commutation relations between the generators of $\widetilde{\mathscr{L}}_q^*$. For the bilinear products we obtain using (4.39):

$$\begin{aligned}
\langle BC, f\rangle &= \delta_{m0}\delta_{n0}\delta_{\bar{m}0}\delta_{\bar{n}0}\sum_{j=0}^{k-1} q^{2(\ell-j)} + q\delta_{m1}\delta_{n1}\delta_{\bar{m}0}\delta_{\bar{n}0},\\
\langle CB, f\rangle &= \delta_{m0}\delta_{n0}\delta_{\bar{m}0}\delta_{\bar{n}0}\sum_{j=0}^{\ell-1} q^{2j} + q^{-1}\delta_{m1}\delta_{n1}\delta_{\bar{m}0}\delta_{\bar{n}0},\\
\langle AB, f\rangle &= (k+1)\delta_{m1}\delta_{n0}\delta_{\bar{m}0}\delta_{\bar{n}0} = (k+1)\langle B, f\rangle,\\
\langle BA, f\rangle &= k\delta_{m1}\delta_{n0}\delta_{\bar{m}0}\delta_{\bar{n}0} = k\langle B, f\rangle,\\
\langle AC, f\rangle &= k\delta_{m0}\delta_{n1}\delta_{\bar{m}0}\delta_{\bar{n}0} = k\langle C, f\rangle,\\
\langle CA, f\rangle &= (k+1)\delta_{m0}\delta_{n1}\delta_{\bar{m}0}\delta_{\bar{n}0} = (k+1)\langle C, f\rangle,\\
\langle DB, f\rangle &= \ell\delta_{m1}\delta_{n0}\delta_{\bar{m}0}\delta_{\bar{n}0} = \ell\langle B, f\rangle,\\
\langle BD, f\rangle &= (\ell+1)\delta_{m1}\delta_{n0}\delta_{\bar{m}0}\delta_{\bar{n}0} = (\ell+1)\langle B, f\rangle,\\
\langle DC, f\rangle &= (\ell+1)\delta_{m0}\delta_{n1}\delta_{\bar{m}0}\delta_{\bar{n}0} = (\ell+1)\langle C, f\rangle,
\end{aligned}$$

$$\begin{aligned}
\langle CD, f\rangle &= \ell\delta_{m0}\delta_{n1}\delta_{\bar m0}\delta_{\bar n0} = \ell\langle C, f\rangle,\\
\langle AD, f\rangle &= \langle DA, f\rangle = k\ell\delta_{m0}\delta_{n0}\delta_{\bar m0}\delta_{\bar n0} =\\
&= k\ell\langle 1_{\mathscr{L}_q^*}, f\rangle,
\end{aligned}\tag{4.146}$$

$$\begin{aligned}
\langle \bar B\bar C, f\rangle &= \delta_{m0}\delta_{n0}\delta_{\bar m0}\delta_{\bar n0}\sum_{j=0}^{\bar k-1} q^{2(j-\bar\ell)} + q^{-1}\delta_{m0}\delta_{n0}\delta_{\bar m1}\delta_{\bar n1},\\
\langle \bar C\bar B, f\rangle &= \delta_{m0}\delta_{n0}\delta_{\bar m0}\delta_{\bar n0}\sum_{j=0}^{\bar\ell-1} q^{-2j} + q\delta_{m0}\delta_{n0}\delta_{\bar m1}\delta_{\bar n1},\\
\langle \bar A\bar B, f\rangle &= (\bar k+1)\delta_{m0}\delta_{n0}\delta_{\bar m1}\delta_{\bar n0} = (\bar k+1)\langle \bar B, f\rangle,\\
\langle \bar B\bar A, f\rangle &= \bar k\delta_{m0}\delta_{n0}\delta_{\bar m1}\delta_{\bar n0} = \bar k\langle \bar B, f\rangle,\\
\langle \bar A\bar C, f\rangle &= \bar k\delta_{m0}\delta_{n0}\delta_{\bar m0}\delta_{\bar n1} = \bar k\langle \bar C, f\rangle,\\
\langle \bar C\bar A, f\rangle &= (\bar k+1)\delta_{m0}\delta_{n0}\delta_{\bar m0}\delta_{\bar n1} = (\bar k+1)\langle \bar C, f\rangle,\\
\langle \bar D\bar B, f\rangle &= \bar\ell\delta_{m0}\delta_{n0}\delta_{\bar m1}\delta_{\bar n0} = \bar\ell\langle \bar B, f\rangle,\\
\langle \bar B\bar D, f\rangle &= (\bar\ell+1)\delta_{m0}\delta_{n0}\delta_{\bar m1}\delta_{\bar n0} = (\bar\ell+1)\langle \bar B, f\rangle,\\
\langle \bar D\bar C, f\rangle &= (\bar\ell+1)\delta_{m0}\delta_{n0}\delta_{\bar m0}\delta_{\bar n1} = (\bar\ell+1)\langle \bar C, f\rangle,\\
\langle \bar C\bar D, f\rangle &= \bar\ell\delta_{m0}\delta_{n0}\delta_{\bar m0}\delta_{\bar n1} = \bar\ell\langle \bar C, f\rangle,\\
\langle \bar A\bar D, f\rangle &= \langle \bar D\bar A, f\rangle = \bar k\bar\ell\delta_{m0}\delta_{n0}\delta_{\bar m0}\delta_{\bar n0} =\\
&= \bar k\bar\ell\langle 1_{\mathscr{L}_q^*}, f\rangle,
\end{aligned}\tag{4.147}$$

$$\begin{aligned}
\langle A\bar A, f\rangle &= \langle \bar A A, f\rangle = k\bar k\langle 1_{\mathscr{L}_q^*}, f\rangle,\\
\langle A\bar B, f\rangle &= \langle \bar B A, f\rangle = k\langle \bar B, f\rangle,\\
\langle A\bar C, f\rangle &= \langle \bar C A, f\rangle = k\langle \bar C, f\rangle,\\
\langle A\bar D, f\rangle &= \langle \bar D A, f\rangle = k\bar\ell\langle 1_{\mathscr{L}_q^*}, f\rangle,\\
\langle B\bar A, f\rangle &= \langle \bar A B, f\rangle = \bar k\langle B, f\rangle,\\
\langle B\bar B, f\rangle &= \langle \bar B B, f\rangle = \delta_{m1}\delta_{n0}\delta_{\bar m1}\delta_{\bar n0},\\
\langle B\bar C, f\rangle &= \langle \bar C B, f\rangle = \delta_{m1}\delta_{n0}\delta_{\bar m0}\delta_{\bar n1},\\
\langle B\bar D, f\rangle &= \langle \bar D B, f\rangle = \bar\ell\langle B, f\rangle,\\
\langle C\bar A, f\rangle &= \langle \bar A C, f\rangle = \bar k\langle C, f\rangle,\\
\langle C\bar B, f\rangle &= \langle \bar B C, f\rangle = \delta_{m0}\delta_{n1}\delta_{\bar m1}\delta_{\bar n0},\\
\langle C\bar C, f\rangle &= \langle \bar C C, f\rangle = \delta_{m0}\delta_{n1}\delta_{\bar m0}\delta_{\bar n1},\\
\langle C\bar D, f\rangle &= \langle \bar D C, f\rangle = \bar\ell\langle C, f\rangle,\\
\langle D\bar A, f\rangle &= \langle \bar A D, f\rangle = \ell\bar k\langle 1_{\mathscr{L}_q^*}, f\rangle,\\
\langle D\bar B, f\rangle &= \langle \bar B D, f\rangle = \ell\langle \bar B, f\rangle,
\end{aligned}$$

$$\langle D\bar{C}, f\rangle = \langle \bar{C}D, f\rangle = \ell\langle \bar{C}, f\rangle,$$
$$\langle D\bar{D}, f\rangle = \langle \bar{D}D, f\rangle = \ell\bar{\ell}\langle 1_{\mathscr{L}_q^*}, f\rangle. \tag{4.148}$$

We note that the bilinear products involving the generators A, B, C, D may be obtained on elements $f = \alpha^k\delta^\ell\beta^m\gamma^n$. Thus, these relations may be taken from Section 4.3 setting $p = q$ and replacing $q \to q^2$. Analogously one may obtain the bilinear products involving the generators $\bar{A}, \bar{B}, \bar{C}, \bar{D}$ on elements $f = \bar{\alpha}^{\bar{k}}\bar{\delta}^{\bar{\ell}}\bar{\beta}^{\bar{m}}\bar{\gamma}^{\bar{n}}$, and these may also be taken from Section 4.3.

Using (4.146), (4.147), and (4.148) we obtain for the commutation relations:

$$[A, B] = B, \quad [A, C] = -C, \quad [D, B] = -B,$$
$$[D, C] = C, \quad [A, D] = 0,$$
$$q^{-1}BC - qCB = \lambda^{-1}\left(1_{\mathscr{L}_q^*} - q^{2(D-A)}\right), \tag{4.149}$$
$$[\bar{A}, \bar{B}] = \bar{B}, \quad [\bar{A}, \bar{C}] = -\bar{C}, \quad [\bar{D}, \bar{B}] = -\bar{B},$$
$$[\bar{D}, \bar{C}] = \bar{C}, \quad [\bar{A}, \bar{D}] = 0,$$
$$q\bar{B}\bar{C} - q^{-1}\bar{C}\bar{B} = \lambda^{-1}\left(q^{2(\bar{A}-\bar{D})} - 1_{\mathscr{L}_q^*}\right), \tag{4.150}$$
$$[Y, Z] = 0, \quad Y = A, B, C, D, \quad Z = \bar{A}, \bar{B}, \bar{C}, \bar{D}. \tag{4.151}$$

Thus, the algebra $\mathscr{L}_q^*$ is split in two mutually commuting algebras, which we denote by $\mathscr{L}_{q1}^*$ and $\mathscr{L}_{q2}^*$, generated by A, B, C, D and $\bar{A}, \bar{B}, \bar{C}, \bar{D}$, respectively. Moreover, there are two central elements $K_1 = (A+D)/2$ and $K_2 = (\bar{A}+\bar{D})/2$. Furthermore, these two algebras may be brought to a Drinfeld–Jimbo [251, 360] form by the following substitutions:

$$A - D = H_1, \quad B = X_1^+ q^{-H_1/2}, \quad C = X_1^- q^{-H_1/2}, \tag{4.152a}$$
$$\bar{A} - \bar{D} = H_2, \quad \bar{B} = X_2^+ q^{H_2/2}, \quad \bar{C} = X_2^- q^{H_2/2}. \tag{4.152b}$$

Indeed, it is easy to check that:

$$[H_j, X_j^\pm] = \pm 2X_j^\pm, \quad [X_j^+, X_j^-] = \lambda^{-1}(q_j^H - q_j^{-H}) = [H_j]_q, \quad j = 1, 2, \tag{4.153}$$

which are the commutation rules of the standard Drinfeld–Jimbo deformation $U_q(sl(2, \mathbb{C}))$.

Thus, one result here is that the algebra $\mathscr{L}_q^*$, which is dual to the algebra $\mathscr{L}_q$, has the following form with respect to the commutation relations:

$$\mathscr{L}_q^* \cong \mathscr{L}_{q1}^* \otimes \mathscr{L}_{q2}^*, \tag{4.154a}$$
$$\mathscr{L}_{qj}^* \cong U_q(sl(2, \mathbb{C}))_j \otimes U_q(\mathscr{Z}_j), \quad j = 1, 2, \tag{4.154b}$$

where $U_q(sl(2,\mathbb{C}))_j$ is generated by $H_j, X_j^{\pm}$ and has commutation relations given in (4.153), $\mathscr{Z}_j$ is generated by K_j, $U_q(\mathscr{Z}_j)$ is central in $\widetilde{\mathscr{L}}_q^*$ and is generated by $K_j, q^{\pm K_j/2}$, $j = 1, 2$.

Further, we note that the above results also mean that the algebra $\widetilde{\mathscr{L}}_{q1}^*, \widetilde{\mathscr{L}}_{q2}^*$ is dual in the sense (4.38) to the algebra $A_q(2)$, $A_{q^{-1}}(2)$, respectively. It is easy also to see that $U_q(sl(2,\mathbb{C}))_1$, $U_q(sl(2,\mathbb{C}))_2$, respectively, is dual in the sense (4.38) to the algebra $SL_q(2)$, $SL_{q^{-1}}(2)$, respectively. Thus, we can state another result here, namely, that the algebra $\mathscr{L}_q^*$, which is dual to the matrix Lorentz quantum group $\mathscr{L}_q$, has the following form:

$$\mathscr{L}_q^* \cong U_q(sl(2,\mathbb{C}))_1 \otimes U_q(sl(2,\mathbb{C}))_2. \tag{4.155}$$

However, the splittings (4.154) and (4.155) are not preserved by the coalgebra operations as we shall see in the next subsection.

4.6.3 Coalgebra Structure of the Dual Algebras

Here we shall use the duality to derive the coalgebra structure of $\mathscr{L}_q^*$ and $\widetilde{\mathscr{L}}_q^*$. We start with the coproducts in $\widetilde{\mathscr{L}}_q^*$. Namely, we use repeatedly the first of relations (4.39a)

$$\langle Y, f\rangle = \langle \Delta_{\widetilde{\mathscr{L}}_q^*}(Y), f_1 \otimes f_2\rangle \tag{4.156}$$

for every splitting $f = f_1 f_2$. Thus, we derive:

$$\Delta_{\widetilde{\mathscr{L}}_q^*}(Y) = Y \otimes 1_{\widetilde{\mathscr{L}}_q^*} + 1_{\widetilde{\mathscr{L}}_q^*} \otimes Y, \quad Y = A, D, \bar{A}, \bar{D}, \tag{4.157a}$$

$$\Delta_{\widetilde{\mathscr{L}}_q^*}(Y) = Y \otimes q^{D-A+\bar{A}-\bar{D}} + 1_{\widetilde{\mathscr{L}}_q^*} \otimes Y, \quad Y = C, \bar{C}. \tag{4.157b}$$

Already here we see that because of (4.157b) the splitting mentioned in the previous section is not preserved. The coproducts of the generators $B, \bar{B}$ are even more complicated. We try the following Ansätze:

$$\Delta_{\widetilde{\mathscr{L}}_q^*}(B) = B \otimes q^{D-A-\bar{A}+\bar{D}} + \sum_{s\in\mathbb{Z}_+} b_s \bar{B}^s \otimes B^{s+1}, \tag{4.158a}$$

$$\Delta_{\widetilde{\mathscr{L}}_q^*}(\bar{B}) = \bar{B} \otimes q^{A-D+\bar{A}-\bar{D}} + \sum_{s\in\mathbb{N}} \tilde{b}_s \bar{B}^{s+1} \otimes B^s. \tag{4.158b}$$

We have to find the coefficients b_s, $\tilde{b}_s$. First, we consider the following pairing: $\langle B^t, \bar{\beta}^{\bar{m}}\beta^m\rangle$. To evaluate this pairing we have to bring $\bar{\beta}^{\bar{m}}\beta^m$ to normal form using the formula for $\bar{\beta}\beta$ in (4.135). It is clear that this pairing will be nonzero iff $m = \bar{m} + t$ since only in this case there will be terms in normal order which are proportional to β^t and do not contain $\bar{\beta}$. In this case we have:

$$\bar{\beta}^m \beta^{m+t} = \sum_{j=0}^{m} P_{mtj}(\alpha\bar{\alpha}, \delta\bar{\delta}, q)\beta^{m+t-j}\bar{\beta}^{m-j}, \tag{4.159}$$

where $P_{mtj}(x, y, q)$ is a homogeneous polynomial of degree j in the first two variables; that is, $P_{mtj}(\mu x, \mu y, q) = \mu^j P_{mtj}(x, y, q)$. Clearly, $P_{mt0}(x, y, q) = 1$. Then for the pairing in question we have:

$$\langle B^t, \bar{\beta}^m \beta^{m+t}\rangle = P_{mtm}(1, 1, q)\langle B^t, \beta^t\rangle = [t]! P_{mtm}(1, 1, q). \tag{4.160}$$

Thus, for splittings of $\bar{\beta}^m \beta^{m+1} = \bar{\beta}^j \bar{\beta}^{m-j}\beta^{m+1}$ for $0 \le j \le m$ we have:

$$\begin{aligned}
\langle B, \bar{\beta}^m \beta^{m+1}\rangle &= \langle \Delta_{\mathscr{L}_q^*}(B), \bar{\beta}^j \otimes \bar{\beta}^{m-j}\beta^{m+1}\rangle = \\
&= \sum_{s\in\mathbb{Z}_+} b_s \langle \bar{B}^s \otimes B^{s+1}, \bar{\beta}^j \otimes \bar{\beta}^{m-j}\beta^{m+1}\rangle = \\
&= \sum_{s\in\mathbb{Z}_+} b_s \langle \bar{B}^s, \bar{\beta}^j\rangle\langle B^{s+1}, \bar{\beta}^{m-j}\beta^{m+1}\rangle = \\
&= [j]! b_j \langle B^{j+1}, \bar{\beta}^{m-j}\beta^{m+1}\rangle = \\
&= [j]! b_j [j+1]! P_{m-j,j+1,m-j}(1, 1, q),
\end{aligned} \tag{4.161}$$

while, on the other hand, using (4.160) for $t = 1$ we have

$$\langle B, \bar{\beta}^m \beta^{m+1}\rangle = P_{m1m}(1, 1, q). \tag{4.162}$$

Thus, we find that

$$b_j = P_{m1m}(1, 1, q)/[j]![j+1]! P_{m-j,j+1,m-j}(1, 1, q) = P_{j1j}(1, 1, q)/[j]![j+1]!, \tag{4.163}$$

where we have used the fact that b_j does not depend on m.

Next we shall use also the pairing $\langle \bar{B}^t, \bar{\beta}^{\bar{m}} \beta^m\rangle$ which is nonzero iff $\bar{m} = m + t$. Let us write the analogue of (4.159):

$$\bar{\beta}^{m+t} \beta^{m} = \sum_{j=0}^{m} Q_{mtj}(\alpha\bar{\alpha}, \delta\bar{\delta}, q)\beta^{m-j}\bar{\beta}^{m+t-j}, \tag{4.164}$$

where $Q_{mtj}(x, y, q)$ is a homogeneous polynomial of degree j in the first two variables. Clearly, $Q_{mt0}(x, y, q) = 1$. Then for the analogue of (4.160) we have:

$$\langle \bar{B}^t, \bar{\beta}^{m+t} \beta^{m}\rangle = Q_{mtm}(1, 1, q)\langle \bar{B}^t, \bar{\beta}^t\rangle = [t]!\, Q_{mtm}(1, 1, q). \tag{4.165}$$

Thus, for splittings of $\bar{\beta}^{m+1} \beta^{m} = \bar{\beta}^{m+1}\beta^{m-j}\beta^{j}$ for $1 \le j \le m$ we have:

$$\langle \bar{B}, \bar{\beta}^{m+1} \beta^{m}\rangle = [j]!\, \tilde{b}_j\, [j+1]!\, Q_{m-j,j+1,m-j}(1, 1, q), \tag{4.166}$$

while, on the other hand, using (4.165) for $t = 1$ we have

$$\langle \bar{B}, \bar{\beta}^{m+1} \beta^m \rangle = Q_{m1m}(1, 1, q). \tag{4.167}$$

Thus, we find as in (4.163):

$$\tilde{b}_j = Q_{j1j}(1, 1, q)/[j]!\,[j+1]!. \tag{4.168}$$

Thus, finding the coefficients b_j and $\tilde{b}_j$ is reduced to the knowledge of the functions $P_{j1j}(1, 1, q)$ and $Q_{j1j}(1, 1, q)$. The latter may be found from (4.159) and (4.164) after tedious calculations for any fixed j.

The counit relations in $\widetilde{\mathscr{L}}_q^*$ are given by:

$$\varepsilon_{\widetilde{\mathscr{L}}_q^*}(Y) = 0, \quad Y = A, B, C, D, \bar{A}, \bar{B}, \bar{C}, \bar{D}, \tag{4.169}$$

which follows easily using (4.39b) and (4.83):

$$\varepsilon_{\widetilde{\mathscr{L}}_q^*}(Y) = \langle Y, 1_{\widetilde{\mathscr{L}}_q} \rangle = 0. \tag{4.170}$$

The coproduct and counit operations in the algebra $\mathscr{L}_q^*$, which is dual to the matrix Lorentz quantum group $\mathscr{L}_q$, are given by the same formulae (4.157),(4.158), and (4.169) as for $\widetilde{\mathscr{L}}_q^*$. For the antipode map in $\mathscr{L}_q^*$ we have:

$$S_{\mathscr{L}_q^*}(Y) = -Y, \quad Y = A, D, \bar{A}, \bar{D}, \tag{4.171a}$$

$$S_{\mathscr{L}_q^*}(Y) = -Yq^{A-D-\bar{A}+\bar{D}}, \quad Y = C, \bar{C}, \tag{4.171b}$$

This follows from (4.101) and (4.169) with elementary application of one of the basic axioms of Hopf algebras (1.6). To obtain (4.171) we just apply both sides of (1.6) to $A, D, \bar{A}, \bar{D}, C, \bar{C}$. The antipode map for the generators $B, \bar{B}$ may be obtained in the same way using (4.158).

4.7 Duality for the Jordanian Matrix Quantum Group $GL_{g,h}(2)$

The group $GL(2)$ admits two distinct quantum group deformations with central quantum determinant: $GL_q(2)$ [251] and the Jordanian $GL_h(2)$ [183, 608]. These are the only such possible deformations (up to isomorphism) [416]. Both may be viewed as special cases of two-parameter deformations: $GL_{p,q}(2)$ [183] and $GL_{g,h}(2)$ [13]. In the initial years of the development of quantum group theory, mostly $GL_q(2)$ and $GL_{p,q}(2)$ were considered. Later started research on $SL_h(2)$ and its dual quantum algebra $U_h(sl(2))$ [499]. In particular, aspects of differential calculus [13], and differential geometry [376] were developed, the universal R-matrix for $U_h(sl(2))$ was given in [68, 384, 592], representations of $U_h(sl(2))$ were constructed in [5, 16, 218, 587].

In this section (following mostly [39]) we give the Hopf algebra $\mathcal{U}_{g,h}$ dual to the Jordanian matrix quantum group $GL_{g,h}(2)$. As an algebra it depends on the sum $\tilde{g} = (g+h)/2$ of the two parameters and is split in two subalgebras: $\mathcal{U}'_{g,h}$ (with three generators) and $U(\mathscr{Z})$ (with one generator). The subalgebra $U(\mathscr{Z})$ is a central Hopf subalgebra of $\mathcal{U}_{g,h}$. The subalgebra $\mathcal{U}'_{g,h}$ is not a Hopf subalgebra, and its coalgebra structure depends on both parameters. We discuss also two interesting one-parameter special cases: $g = h$ and $g = -h$. The subalgebra $\mathcal{U}'_{h,h}$ is a Hopf algebra and coincides with the algebra introduced by Ohn as the dual of $SL_h(2)$. The subalgebra $\mathcal{U}'_{-h,h}$ is isomorphic to $U(sl(2))$ as an algebra but has a nontrivial coalgebra structure and again is not a Hopf subalgebra of $\mathcal{U}_{-h,h}$.

4.7.1 Jordanian Matrix Quantum Group $GL_{g,h}(2)$

Here we recall the Jordanian two-parameter deformation $GL_{g,h}(2)$ of $GL(2)$ introduced in [13] (and denoted $GL_{h,h'}$). One starts with a unital associative algebra generated by four elements a, b, c, d of a quantum matrix $M = \begin{pmatrix} a & b \\ c & d \end{pmatrix}$ with the following relations ($g, h \in \mathbb{C}$):

$$\begin{gathered}
[a,c] = gc^2\,, \qquad [d,c] = hc^2\,, \qquad [a,d] = gdc - hac \\
[a,b] = h(\mathscr{D} - a^2)\,, \qquad [d,b] = g(\mathscr{D} - d^2)\,, \qquad [b,c] = gdc + hac - ghc^2 \\
\mathscr{D} = ad - bc + hac = ad - cb - gdc + ghc^2
\end{gathered} \tag{4.172}$$

where $\mathscr{D}$ is a multiplicative quantum determinant which is not central (unless $g = h$):

$$[a,\mathscr{D}] = [\mathscr{D},d] = (g-h)\mathscr{D}c\,, \qquad [b,\mathscr{D}] = (g-h)(\mathscr{D}d - a\mathscr{D})\,, \qquad [c,\mathscr{D}] = 0 \tag{4.173}$$

Relations (4.172) are obtained by applying either the Faddeev–Reshetikhin–Takhtajan method [272], namely, by solving the monodromy equation:

$$R M_1 M_2 = M_2 M_1 R$$

($M_1 = M \hat{\otimes} I_2$, $M_2 = I_2 \hat{\otimes} M$), with R-matrix:

$$R = \begin{pmatrix} 1 & -h & h & gh \\ 0 & 1 & 0 & -g \\ 0 & 0 & 1 & g \\ 0 & 0 & 0 & 1 \end{pmatrix} \tag{4.174}$$

or the method of Manin [461] using M as transformation matrix of the appropriate quantum planes [13].

The above algebra is turned into a bialgebra $A_{g,h}(2)$ with the standard $GL(2)$ coproduct δ and counit ε:

$$\delta\left(\begin{pmatrix} a & b \\ c & d \end{pmatrix}\right) = \begin{pmatrix} a\otimes a + b\otimes c & a\otimes b + b\otimes d \\ c\otimes a + d\otimes c & c\otimes b + d\otimes d \end{pmatrix} \tag{4.175}$$

$$\varepsilon\left(\begin{pmatrix} a & b \\ c & d \end{pmatrix}\right) = \begin{pmatrix} 1 & 0 \\ 0 & 1 \end{pmatrix}. \tag{4.176}$$

From (4.175) and (4.176), respectively, it follows:

$$\delta(\mathcal{D}) = \mathcal{D}\otimes\mathcal{D}, \qquad \varepsilon(\mathcal{D}) = 1. \tag{4.177}$$

Further, we shall suppose that $\mathcal{D}$ is invertible; that is, there is an element $\mathcal{D}^{-1}$ which obeys:

$$\mathcal{D}\mathcal{D}^{-1} = \mathcal{D}^{-1}\mathcal{D} = 1_{\mathcal{A}}, \qquad (\mathcal{D}^{-1}) = \mathcal{D}^{-1}\otimes\mathcal{D}^{-1}, \qquad \varepsilon(\mathcal{D}^{-1}) = 1. \tag{4.178}$$

(Alternatively one may say that the algebra is extended with the element $\mathcal{D}^{-1}$.) In this case one defines the left and right inverse matrix of M [13]:

$$\begin{aligned} M^{-1} &= \mathcal{D}^{-1}\begin{pmatrix} d+gc & -b+g(d-a)+g^2c \\ -c & a-gc \end{pmatrix} = \\ &= \begin{pmatrix} d+hc & -b+h(d-a)+h^2c \\ -c & a-hc \end{pmatrix}\mathcal{D}^{-1}. \end{aligned} \tag{4.179}$$

The quantum group $GL_{g,h}(2)$ is defined as the Hopf algebra obtained from the bialgebra $A_{g,h}(2)$ when $\mathcal{D}^{-1}$ exists and with antipode given by the formula:

$$\gamma(M) = M^{-1} \quad \Rightarrow \quad \gamma(\mathcal{D}) = \mathcal{D}^{-1}, \quad \gamma(\mathcal{D}^{-1}) = \mathcal{D} \tag{4.180}$$

For $g = h$ one obtains from $GL_{g,h}(2)$ the matrix quantum group $GL_h(2) = GL_{h,h}(2)$, and, if the condition $\mathcal{D} = 1_{\mathcal{A}}$ holds, the matrix quantum group $SL_h(2)$. Analogously, for $g = h = 0$ one obtains from $GL_{g,h}(2)$ the algebra of functions over the classical groups $GL(2)$ and $SL(2)$, respectively.

4.7.2 The Dual of $GL_{g,h}(2)$

Following our general method we first need to fix a PBW basis of $GL_{g,h}(2)$. We first choose the following PBW basis:

$$a^k\, d^\ell\, c^n\, b^m, \qquad k,\ell,m,n \in \mathbb{Z}_+, \tag{4.181}$$

the reasoning being that the elements a, d, c generate a subalgebra (though not a Hopf subalgebra) of $GL_{g,h}(2)$ (cf. the first line of (4.172)). Further simplification results if we make the following change of generating elements and parameters:

$$\begin{aligned} \tilde{a} &= \tfrac{1}{2}(a+d)\,, & \tilde{d} &= \tfrac{1}{2}(a-d) \\ \tilde{g} &= \tfrac{1}{2}(g+h)\,, & \tilde{h} &= \tfrac{1}{2}(g-h). \end{aligned} \tag{4.182}$$

With these generating elements and parameters the algebra relations become:

$$\begin{aligned} c\tilde{a} &= \tilde{a}c - \tilde{g}c^2\,, \qquad c\tilde{d} = \tilde{d}c - \tilde{h}c^2\,, \qquad \tilde{d}\tilde{a} = \tilde{a}\tilde{d} - \tilde{g}\tilde{d}c + \tilde{h}\tilde{a}c \\ b\tilde{a} &= \tilde{a}b + \tilde{g}cb - 2\tilde{h}\tilde{a}\tilde{d} + 2\tilde{g}\tilde{d}^2 + (\tilde{g}^2 - \tilde{h}^2)\tilde{a}c + \tilde{g}(\tilde{h}^2 - \tilde{g}^2)c^2 \\ b\tilde{d} &= \tilde{d}b - \tilde{h}cb + 2\tilde{g}\tilde{a}\tilde{d} - 2\tilde{h}\tilde{d}^2 + (\tilde{h}^2 - \tilde{g}^2)\tilde{d}c + \tilde{h}(\tilde{g}^2 - \tilde{h}^2)c^2 \\ bc &= cb + 2\tilde{g}\tilde{a}c - 2\tilde{h}\tilde{d}c + (\tilde{h}^2 - \tilde{g}^2)c^2 \\ \mathscr{D} &= \tilde{a}^2 - \tilde{d}^2 - cb + (\tilde{g}^2 - \tilde{h}^2)c^2 - \tilde{g}\tilde{a}c + \tilde{h}\tilde{d}c. \end{aligned} \tag{4.183}$$

Note that these relations are written in anticipation of the PBW basis:

$$f = f_{k,\ell,m,n} = \tilde{a}^k\,\tilde{d}^\ell\,c^n\,b^m\,, \qquad k,\ell,m,n \in \mathbb{Z}_+. \tag{4.184}$$

Note also that the relations in the subalgebras generated by a, d, c and $\tilde{a}, \tilde{d}, c$ are isomorphic under the change: $a \mapsto \tilde{a},\ d \mapsto \tilde{d},\ c \mapsto c,\ g \mapsto \tilde{g},\ h \mapsto \tilde{h}$ (cf. the first lines in (4.172) and (4.183)).

The coalgebra relations become:

$$\begin{aligned} &\delta\left(\begin{pmatrix} \tilde{a} & b \\ c & \tilde{d} \end{pmatrix}\right) = \\ &= \begin{pmatrix} \tilde{a}\otimes\tilde{a} + \tilde{d}\otimes\tilde{d} + \tfrac{1}{2}b\otimes c + \tfrac{1}{2}c\otimes b & \tilde{a}\otimes b + \tilde{d}\otimes b + b\otimes\tilde{a} - b\otimes\tilde{d} \\ c\otimes\tilde{a} + c\otimes\tilde{d} + \tilde{a}\otimes c - \tilde{d}\otimes c & \tilde{a}\otimes\tilde{d} + \tilde{d}\otimes\tilde{a} + \tfrac{1}{2}b\otimes c - \tfrac{1}{2}c\otimes b \end{pmatrix} \\ &\varepsilon\left(\begin{pmatrix} \tilde{a} & b \\ c & \tilde{d} \end{pmatrix}\right) = \begin{pmatrix} 1 & 0 \\ 0 & 0 \end{pmatrix}, \end{aligned} \tag{4.185}$$

$$\begin{aligned} \gamma\left(\begin{pmatrix} \tilde{a} & b \\ c & \tilde{d} \end{pmatrix}\right) &= \mathscr{D}^{-1}\begin{pmatrix} \tilde{a} - \tilde{d} + (\tilde{g}+\tilde{h})c & -b - 2(\tilde{g}+\tilde{h})\tilde{d} + (\tilde{g}+\tilde{h})^2 c \\ -c & \tilde{a} + \tilde{d} - (\tilde{g}+\tilde{h})c \end{pmatrix} \\ &= \begin{pmatrix} \tilde{a} - \tilde{d} + (\tilde{g}-\tilde{h})c & -b + 2(\tilde{h}-\tilde{g})\tilde{d} + (\tilde{g}-\tilde{h})^2 c \\ -c & \tilde{a} + \tilde{d} + (\tilde{h}-\tilde{g})c \end{pmatrix}\mathscr{D}^{-1}. \end{aligned}$$

Let us denote by $\mathscr{U}_{g,h} = U_{g,h}(gl(2))$ the unknown yet dual algebra of $GL_{g,h}(2)$ and by A, B, C, D the four generators of $\mathscr{U}_{g,h}$. Following [209] and Section 4.4 we define the pairing $\langle Z, f\rangle$, $Z = A, B, C, D$, f is from (4.184), as the classical tangent vector at the identity:

$$\langle Z, f \rangle \equiv \varepsilon\left(\frac{\partial f}{\partial y}\right), \qquad (Z, y) = (A, \tilde{a}),\ (B, b),\ (C, c),\ (D, \tilde{d}). \tag{4.186}$$

From this we get the explicit expressions:

$$\langle A, f \rangle = \varepsilon\left(\frac{\partial f}{\partial \tilde{a}}\right) = k\delta_{\ell 0}\delta_{m0}\delta_{n0} \tag{4.187a}$$

$$\langle B, f \rangle = \varepsilon\left(\frac{\partial f}{\partial b}\right) = \delta_{\ell 0}\delta_{m1}\delta_{n0} \tag{4.187b}$$

$$\langle C, f \rangle = \varepsilon\left(\frac{\partial f}{\partial c}\right) = \delta_{\ell 0}\delta_{m0}\delta_{n1} \tag{4.187c}$$

$$\langle D, f \rangle = \varepsilon\left(\frac{\partial f}{\partial \tilde{d}}\right) = \delta_{\ell 1}\delta_{m0}\delta_{n0} \tag{4.187d}$$

4.7.3 Algebra Structure of the Dual

First we find the commutation relations between the generators of $\mathcal{U}_{g,h}$. Below we shall need expressions like $e^{\nu B}$, which we define as formal power series $e^{\nu B} = 1_{\mathcal{U}} + \sum_{p\in\mathbb{N}} \frac{\nu^p}{p!} B^p$. We have:

Proposition 1. *The commutation relations of the generators A, B, C, D introduced by* (4.187) *are:*

$$[B, C] = D \tag{4.188a}$$

$$[D, B] = \tfrac{1}{\tilde{g}}(e^{2\tilde{g}B} - 1_{\mathcal{U}}) \tag{4.188b}$$

$$[D, C] = -2C + \tilde{g}D^2 - \tilde{g}A \tag{4.188c}$$

$$[A, B] = 0, \qquad [A, C] = 0, \qquad [A, D] = 0 \tag{4.188d}$$

Proof. Using the assumed duality the above relations are shown by calculating their pairings with the basis monomials $f = \tilde{a}^k \tilde{d}^\ell c^n b^m$ of the dual algebra. In particular, The pairing of any quadratic monomial of the unknown dual algebra with $f = \tilde{a}^k \tilde{d}^\ell c^n b^m$ is given by the duality properties (4.39):

$$\begin{aligned}\langle WZ, f \rangle &= \langle W \otimes Z, \delta_{\mathcal{A}}(f) \rangle = \langle W \otimes Z, \sum_j f_j' \otimes f_j'' \rangle = \\ &= \sum_j \langle W, f_j' \rangle \langle Z, f_j'' \rangle\end{aligned} \tag{4.189}$$

where f_j', f_j'' are elements of the basis (4.184), and so further a direct application of (4.187) is used. Thus, we have:

$$\begin{aligned}\langle BC, f \rangle = {} & \tfrac{1}{2}\delta_{\ell 1}\delta_{m0}\delta_{n0} + \tilde{h}\delta_{\ell 1}\delta_{m1}\delta_{n0} + \\ & + \delta_{\ell 0}\delta_{n0}\theta_{m2}\tfrac{1}{2}(\tilde{g}^2 - \tilde{h}^2)\tilde{g}^{m-2} + \delta_{\ell 0}\delta_{m1}\delta_{n1}\end{aligned}$$

$$
\begin{aligned}
\langle CB, f\rangle &= -\tfrac{1}{2}\delta_{\ell 1}\delta_{m0}\delta_{n0} + \\
&\quad + \tilde{h}\delta_{\ell 1}\delta_{m1}\delta_{n0} + \delta_{\ell 0}\delta_{n0}\theta_{m2}\tfrac{1}{2}(\tilde{g}^2 - \tilde{h}^2)\tilde{g}^{m-2} + \delta_{\ell 0}\delta_{m1}\delta_{n1} \\
\langle DB, f\rangle &= \delta_{\ell 0}\delta_{n0}(\delta_{m1} + \theta_{m2}2^{m-1}\tilde{g}^{m-2}(\tilde{g} - \tilde{h})) \qquad (4.190)\\
\langle BD, f\rangle &= -\delta_{\ell 0}\delta_{n0}(\delta_{m1} + \theta_{m2}2^{m-1}\tilde{g}^{m-2}(\tilde{g} + \tilde{h})) \\
\langle DC, f\rangle &= -\delta_{\ell 0}\delta_{m0}\delta_{n1} + (\tilde{h} + \tilde{g})\delta_{\ell 2}\delta_{m0}\delta_{n0} + k\tilde{g}\delta_{\ell 1}\delta_{m0}\delta_{n0} + \delta_{\ell 1}\delta_{m0}\delta_{n1} \\
\langle CD, f\rangle &= \delta_{\ell 0}\delta_{m0}\delta_{n1} + (\tilde{h} - \tilde{g})\delta_{\ell 2}\delta_{m0}\delta_{n0} + k\tilde{g}\delta_{\ell 1}\delta_{m0}\delta_{n0} + \delta_{\ell 1}\delta_{m0}\delta_{n1} \\
\theta_{rs} &\equiv \begin{cases} 1 & r \geq s \\ 0 & r < s. \end{cases}
\end{aligned}
$$

From these follow the pairing of f with the commutators:

$$
\begin{aligned}
\langle [B, C], f\rangle &= \delta_{\ell 1}\delta_{m0}\delta_{n0}, \qquad (4.191a)\\
\langle [D, B], f\rangle &= \delta_{\ell 0}\theta_{m1}\delta_{n0}2^m\tilde{g}^{m-1}, \qquad (4.191b)\\
\langle [D, C], f\rangle &= -2\delta_{\ell 0}\delta_{m0}\delta_{n1} + 2\tilde{g}\delta_{\ell 2}\delta_{m0}\delta_{n0} \qquad (4.191c)
\end{aligned}
$$

Note that quadratic relations (4.190) depend on both parameters, while the commutation relations (4.191), which follow from (4.190), depend only on the parameter $\tilde{g}$.

Now in order to establish (4.188a) it is enough to compare the RHS of (4.191a) and (4.187d).

Further, for relation (4.188b) we use (4.191b) and

$$\langle B^p, f\rangle = p!\delta_{\ell 0}\delta_{mp}\delta_{n0} \qquad (4.192)$$

(proved by induction) and its consequence:

$$
\begin{aligned}
\langle (e^{2\tilde{g}B} - 1_{\mathcal{U}}), f\rangle &= \sum_{p\in\mathbb{N}} \frac{(2\tilde{g})^p}{p!}\langle B^p, f\rangle = \sum_{p\in\mathbb{N}} \frac{(2\tilde{g})^p}{p!} p!\delta_{\ell 0}\delta_{mp}\delta_{n0} = \\
&= (2\tilde{g})^m\delta_{\ell 0}\theta_{m1}\delta_{n0} \qquad (4.193)
\end{aligned}
$$

To establish (4.188c) we compare the RHS of (4.191c) with the appropriate linear combination of the right-hand sides of three equations, namely, (4.187a), and (4.187c) and

$$\langle D^2, f\rangle = 2\delta_{\ell 2}\delta_{m0}\delta_{n0} + k\delta_{\ell 0}\delta_{m0}\delta_{n0}. \qquad (4.194)$$

Finally, to establish (4.188d) we use:

$$[A, B], f\rangle = \langle [A, C], f\rangle = \langle [A, D], f\rangle = 0 \qquad (4.195)$$

which are straightforward. This finishes the proof. ∎

Note that the commutation relations (4.188) depend only on the parameter $\tilde{g}$ and that the generator A is central. This is similar to the situation of the dual algebra $\mathcal{U}_{p,q}$ of the standard matrix quantum group $GL_{p,q}$ the commutation relations of which depend only on the combination $q' = \sqrt{pq}$ and also one generator is central (Section 4.4). Here the central generator appears as a central extension, but this is fictitious since this may be corrected by a change of basis, namely, by replacing the generator C by a generator $\tilde{C}$:

$$C = \tilde{C} - \tfrac{\tilde{g}}{2}A \tag{4.196}$$

With this only (4.188c) changes to:

$$[D, \tilde{C}] = -2\tilde{C} + \tilde{g}D^2 \tag{4.197}$$

Besides this change we shall make a change of generating elements of $\mathcal{U}_{g,h}$ in order to bring the commutation relations to a form closer to the algebra of [499]. Thus, we make the following substitutions:

$$D = e^{\mu B}\, H\, e^{\nu B} \tag{4.198a}$$

$$C = e^{\mu' B}\, Y\, e^{\nu' B} - \tfrac{\tilde{g}}{2}\sinh(\tilde{g}B)e^{(\mu'+\nu')B} - \tfrac{\tilde{g}}{2}A. \tag{4.198b}$$

Substituting (4.198) into (4.188a) we get the desired result $[B, Y] = H$ if we choose: $\mu' = \mu$, $\nu' = \nu$. Substituting (4.198) into (4.188b) we get the desired result $[H, B] = \frac{2}{\tilde{g}}\sinh(\tilde{g}B)$ if we choose: $\mu + \nu = \tilde{g}$. Thus with conditions:

$$\mu + \nu = \tilde{g}, \qquad \mu' = \mu, \qquad \nu' = \nu$$

we obtain the following commutation relations instead of (4.188):

$$[B, Y] = H \tag{4.199a}$$

$$[H, B] = \tfrac{2}{\tilde{g}}\sinh(\tilde{g}B) \tag{4.199b}$$

$$\begin{aligned}[H, Y] &= -Y\cosh(\tilde{g}B) - \cosh(\tilde{g}B)Y = \\ &= -2Y\cosh(\tilde{g}B) - \tilde{g}H\sinh(\tilde{g}B) + \\ &\quad + \tilde{g}\sinh(\tilde{g}B)\cosh(\tilde{g}B)\end{aligned} \tag{4.199c}$$

$$[A, B] = 0, \qquad [A, Y] = 0, \qquad [A, H] = 0 \tag{4.199d}$$

Note that relations (4.199a,b,c) coincide with those of the one-parameter algebra of [499], though the coalgebra structure is different as we shall see below. We can use this coincidence to derive the Casimir operator of $\mathcal{U}_{g,h}$:

$$\hat{\mathcal{C}}_2 = f_1(A)\,\mathcal{C}_2 + f_2(A) \tag{4.200}$$

$$\mathcal{C}_2 = \tfrac{1}{2}(H^2 + \sinh^2(\tilde{g}B)) + \tfrac{1}{\tilde{g}}(Y\sinh(\tilde{g}B) + \sinh(\tilde{g}B)Y),$$

where $f_1(A)$, $f_2(A)$ are arbitrary polynomials in the central generator A. To derive (4.200), it is enough to check that $[\mathscr{C}_2, Z] = 0$ for $Z = B, Y, H$. For the latter one may also notice [74] that $\mathscr{C}_2$ is the Casimir of the one-parameter algebra of [499].

Finally we also write a subalgebra $\widetilde{\mathscr{U}}_{g,h}$ of $\mathscr{U}_{g,h}$ with the basis: $A,\ K = e^{\tilde{g}B} = K^+$, $K^{-1} = e^{-\tilde{g}B} = K^-,\ Y,\ H$, so that in terms of A, K, K^{-1}, Y, H no exponents of generators appear in the algebra and coalgebra relations. Thus instead of (4.199) we have:

$$\begin{aligned}
[K^{\pm}, Y] &= \pm\tilde{g}HK^{\pm} \pm \tfrac{\tilde{g}}{2}(1_{\mathscr{U}} - (K^{\pm})^2) \\
[H, K^{\pm}] &= (K^{\pm})^2 - 1_{\mathscr{U}} \\
[H, Y] &= -Y(K + K^{-1}) + \tfrac{\tilde{g}}{2}H(K^{-1} - K) + \\
&\quad + \tfrac{\tilde{g}}{4}(K^2 - (K^-)^2) \\
KK^{-1} &= K^{-1}K = 1_{\mathscr{U}} \\
[A, K] &= [A, K^{-1}] = 0, \quad [A, Y] = 0, \quad [A, H] = 0.
\end{aligned} \tag{4.201}$$

4.7.4 Coalgebra Structure of the Dual

We turn now to the coalgebra structure of $\mathscr{U}_{g,h}$. We have:

Proposition 2.

(i) *The comultiplication in the algebra $\mathscr{U}_{g,h}$ is given by:*

$$\delta_{\mathscr{U}}(A) = A \otimes 1_{\mathscr{U}} + 1_{\mathscr{U}} \otimes A \tag{4.202a}$$

$$\delta_{\mathscr{U}}(B) = B \otimes 1_{\mathscr{U}} + 1_{\mathscr{U}} \otimes B \tag{4.202b}$$

$$\begin{aligned}
\delta_{\mathscr{U}}(Y) &= Y \otimes e^{-\tilde{g}B} + e^{\tilde{g}B} \otimes Y - \\
&\quad - \tfrac{\tilde{h}^2}{\tilde{g}} \sinh(\tilde{g}B) \otimes A^2 e^{-\tilde{g}B} + \tilde{h}H \otimes Ae^{-\tilde{g}B}
\end{aligned} \tag{4.202c}$$

$$\delta_{\mathscr{U}}(H) = H \otimes e^{-\tilde{g}B} + e^{\tilde{g}B} \otimes H - \tfrac{2\tilde{h}}{\tilde{g}} \sinh(\tilde{g}B) \otimes Ae^{-\tilde{g}B} \tag{4.202d}$$

(ii) *The counit relations in $\mathscr{U}_{g,h}$ are given by:*

$$\varepsilon_{\mathscr{U}}(Z) = 0, \qquad Z = A, B, Y, H \tag{4.203}$$

(iii) *The antipode in the algebra $\mathscr{U}_{g,h}$ is given by:*

$$\begin{aligned}
\gamma_{\mathscr{U}}(A) &= -A \\
\gamma_{\mathscr{U}}(B) &= -B \\
\gamma_{\mathscr{U}}(Y) &= -e^{-\tilde{g}B}Ye^{\tilde{g}B} + \tfrac{\tilde{h}^2}{\tilde{g}} \sinh(\tilde{g}B)A^2 + \tilde{h}e^{-\tilde{g}B}HAe^{\tilde{g}B} \\
\gamma_{\mathscr{U}}(H) &= -e^{-\tilde{g}B}He^{\tilde{g}B} - \tfrac{2\tilde{h}}{\tilde{g}} \sinh(\tilde{g}B)A.
\end{aligned} \tag{4.204}$$

We omit the easy proof given in [39]. ◊

Corollary 1. For later reference we mention also the coproduct and antipode of the intermediate generator $\tilde{C}$ and the antipode of the initial generator D:

$$\begin{aligned}\delta_{\mathcal{U}}(\tilde{C}) &= \tilde{C}\otimes 1_{\mathcal{U}} + e^{2\tilde{g}B}\otimes\hat{\mathcal{C}} - \\ &\quad -\tfrac{\tilde{h}^2}{2\tilde{g}}(e^{2\tilde{g}B}-1_{\mathcal{U}})\otimes A^2 + \tilde{h}D\otimes A \end{aligned} \tag{4.205a}$$

$$\gamma_{\mathcal{U}}(\tilde{C}) = -e^{-2\tilde{g}B}\tilde{C} + \tfrac{\tilde{h}^2}{2\tilde{g}}(1_{\mathcal{U}} - e^{-2\tilde{g}B})A^2 + \tilde{h}e^{-2\tilde{g}B}DA \tag{4.205b}$$

$$\gamma_{\mathcal{U}}(D) = -e^{-2\tilde{g}B}D + \tfrac{\tilde{h}}{\tilde{g}}(e^{-2\tilde{g}B}-1_{\mathcal{U}})A \tag{4.205c}$$

Corollary 2. The coalgebra structure in the subalgebra $\widetilde{\mathcal{U}}_{g,h}$ is given as follows:

(i) comultiplication:

$$\begin{aligned}\delta_{\mathcal{U}}(A) &= A\otimes 1_{\mathcal{U}} + 1_{\mathcal{U}}\otimes A \\ \delta_{\mathcal{U}}(K^{\pm}) &= K^{\pm}\otimes K^{\pm} \\ \delta_{\mathcal{U}}(Y) &= Y\otimes K^{-1} + K\otimes Y - \\ &\quad -\tfrac{\tilde{h}^2}{2\tilde{g}}(K-K^{-1})\otimes A^2K^{-1} + \tilde{h}H\otimes AK^{-1} \\ \delta_{\mathcal{U}}(H) &= H\otimes K^{-1} + K\otimes H + \tfrac{\tilde{h}}{\tilde{g}}(K^{-1}-K)\otimes AK^{-1}\end{aligned} \tag{4.206}$$

(ii) counit:

$$\varepsilon_{\mathcal{U}}(Z) = 0, \quad Z = A, Y, H, \qquad \varepsilon_{\mathcal{U}}(Z) = 1, \quad Z = K, K^{-1} \tag{4.207}$$

(iii) antipode:

$$\begin{aligned}\gamma_{\mathcal{U}}(A) &= -A \\ \gamma_{\mathcal{U}}(K^{\pm}) &= K^{\mp} \\ \gamma_{\mathcal{U}}(Y) &= -K^{-1}YK + \tfrac{\tilde{h}^2}{2\tilde{g}}(K-K^{-1})A^2 + \tilde{h}K^{-1}HAK \\ \gamma_{\mathcal{U}}(H) &= -K^{-1}HK + \tfrac{\tilde{h}}{\tilde{g}}(K^{-1}-K)A\end{aligned} \tag{4.208}$$

The result of this section can be summarized as follows:

Theorem 4.2. *The Hopf algebra $\mathcal{U}_{g,h}$ dual to $GL_{g,h}(2)$ is generated by A, B, Y, H (or A, K, K^{-1}, Y, H) (cf. relations (4.187) and (4.198)). It is given by relations (4.199), (4.202), (4.203), and (4.204) (respectively, (4.201), (4.206), (4.207), and (4.208)). As an algebra it depends only on one parameter $\tilde{g} = (g+h)/2$ and is split in two subalgebras: $\mathcal{U}'_{g,h}$ (respectively, $\widetilde{\mathcal{U}'}_{g,h}$) generated by B, Y, H (respectively, K, K^{-1}, Y, H) and $U(\mathcal{Z})$, where the algebra $\mathcal{Z}$ is spanned by A. The subalgebra $U(\mathcal{Z})$ is central in $\mathcal{U}_{g,h}$ and is also a Hopf subalgebra of $\mathcal{U}_{g,h}$. The subalgebra $\mathcal{U}'_{g,h}$ (respectively, $\widetilde{\mathcal{U}'}_{g,h}$) is not a Hopf subalgebra.* ◊

4.7.5 One-Parameter Cases

4.7.5.1 Case g=h

The one-parameter matrix quantum group $GL_{\tilde{g}}(2)$ [183, 608] is obtained from $GL_{g,h}(2)$ by setting $g = h = \tilde{g}$. Thus the dual algebra $\mathcal{U}_{\tilde{g}} \equiv \mathcal{U}_{\tilde{g},\tilde{g}}$ of $GL_{\tilde{g}}(2)$ is obtained by setting $\tilde{h} = \frac{1}{2}(g - h) = 0$ in (4.199), (4.202), (4.203), and (4.204). Since the commutation relations (4.199) and the counit relations (4.203) do not depend on $\tilde{h}$, they remain unchanged for $\mathcal{U}_{\tilde{g}}$. The coproduct and antipode relations of $\mathcal{U}_{\tilde{g}}$ are:

$$\begin{aligned}
\delta_{\mathcal{U}}(B) &= B \otimes 1_{\mathcal{U}} + 1_{\mathcal{U}} \otimes B \\
\delta_{\mathcal{U}}(Y) &= Y \otimes e^{-\tilde{g}B} + e^{\tilde{g}B} \otimes Y \\
\delta_{\mathcal{U}}(H) &= H \otimes e^{-\tilde{g}B} + e^{\tilde{g}B} \otimes H \\
\delta_{\mathcal{U}}(A) &= A \otimes 1_{\mathcal{U}} + 1_{\mathcal{U}} \otimes A
\end{aligned} \tag{4.209}$$

$$\begin{aligned}
\gamma_{\mathcal{U}}(B) &= -B \\
\gamma_{\mathcal{U}}(Y) &= -e^{-\tilde{g}B} Y e^{\tilde{g}B} \\
\gamma_{\mathcal{U}}(H) &= -e^{-\tilde{g}B} H e^{\tilde{g}B} \\
\gamma_{\mathcal{U}}(A) &= -A
\end{aligned} \tag{4.210}$$

We see that the one-parameter Hopf algebra $\mathcal{U}_{\tilde{g}}$ is split in two Hopf subalgebras $\mathcal{U}'_{\tilde{g}} \equiv \mathcal{U}'_{\tilde{g},\tilde{g}}$ and $U(\mathcal{Z})$, and we may write:

$$\mathcal{U}_{\tilde{g}} = \mathcal{U}'_{\tilde{g}} \otimes U(\mathcal{Z}) \tag{4.211}$$

Now we compare the algebra $\mathcal{U}'_{\tilde{g}}$ with the algebra of [499]. We see that after the identification $B \mapsto X$, $\tilde{g} \mapsto -h$, the algebra $\mathcal{U}'_{\tilde{g}}$ coincides with the algebra of Ohn. We also note that the algebra $\mathcal{U}'_{\tilde{g}}$ in the basis $B, \tilde{C}, D$ coincides for $\tilde{h} = 0$ with the version given in [68] after the identification: $(B, \tilde{C}, D; \tilde{g}) \mapsto (A_+, A_-, A; z)$, and by using the opposite coalgebra structure.

4.7.5.2 Case g = −h

Here we consider another one-parameter case: $g = -h = \tilde{h}$; that is, $\tilde{g} = 0$. From (4.199), (4.202), and (4.204), we obtain:

$$\begin{aligned}
[B, Y] &= H \\
[H, B] &= 2B \\
[H, Y] &= -2Y \\
[A, B] &= 0, \quad [A, Y] = 0, \quad [A, H] = 0 \\
\delta_{\mathcal{U}}(A) &= A \otimes 1_{\mathcal{U}} + 1_{\mathcal{U}} \otimes A \\
\delta_{\mathcal{U}}(B) &= B \otimes 1_{\mathcal{U}} + 1_{\mathcal{U}} \otimes B
\end{aligned} \tag{4.212}$$

$$\begin{aligned}
\delta_{\mathcal{U}}(Y) &= Y \otimes 1_{\mathcal{U}} + 1_{\mathcal{U}} \otimes Y - \tilde{h}^2 B \otimes A^2 + \tilde{h} H \otimes A \\
\delta_{\mathcal{U}}(H) &= H \otimes 1_{\mathcal{U}} + 1_{\mathcal{U}} \otimes H - 2\tilde{h} B \otimes A
\end{aligned} \tag{4.213}$$

$$\begin{aligned}
\gamma_{\mathcal{U}}(A) &= -A \\
\gamma_{\mathcal{U}}(B) &= -B \\
\gamma_{\mathcal{U}}(Y) &= -Y + \tilde{h}^2 BA^2 + \tilde{h} HA \\
\gamma_{\mathcal{U}}(H) &= -H - 2\tilde{h} BA .
\end{aligned} \tag{4.214}$$

Thus, for $\tilde{g} = 0$ the interesting feature is that the subalgebra $\mathcal{U}'_{\tilde{h},-\tilde{h}}$ is isomorphic to the undeformed $U(sl(2))$ with $sl(2)$ spanned by B, Y, H. However, as in the general case, the coalgebra sector is not classical, and the generators B, Y, H do not close a co-subalgebra.

4.7.6 Application of a Nonlinear Map

In [5] a nonlinear map was proposed under which the one-parameter Ohn's algebra was brought to undeformed $sl(2)$ form, though, the coalgebra structure becomes even more complicated (cf. [16, 587]). Since our two-parameter dual is like Ohn's algebra in the algebra sector, we can also apply the map of [5]. We give the map in our notation, namely, following (28) and (33) of [5] we set:

$$\begin{aligned}
I_+ &= \frac{2}{\tilde{g}} \tanh\left(\frac{\tilde{g} B}{2}\right) = -\frac{2}{\tilde{g}} \left(1_{\mathcal{U}} + 2\sum_{\ell=1}^{\infty} (-K)^{\ell}\right) \left(= \frac{2}{\tilde{g}} \left(\frac{K - 1_{\mathcal{U}}}{K + 1_{\mathcal{U}}}\right)\right) \\
I_- &= \cosh\left(\frac{\tilde{g} B}{2}\right) Y \cosh\left(\frac{\tilde{g} B}{2}\right) = \\
&= \tfrac{1}{4} \left(K^{1/2} + K^{-1/2}\right) Y \left(K^{1/2} + K^{-1/2}\right).
\end{aligned} \tag{4.215}$$

Then we have, as in [5] for the case $U_h(sl(2))$ (note though that we do not rescale H), the classical $gl(2)$ commutation relations and Casimir:

$$[H, I_\pm] = \pm 2 I_\pm , \qquad [I_+, I_-] = H , \qquad [A, I_\pm] = [A, H] = 0 \tag{4.216}$$

$$\hat{\mathcal{C}}_2^c = f_1(A)\, \mathcal{C}_2^c + f_2(A) , \qquad \mathcal{C}_2^c = I_+ I_- + I_- I_+ + \tfrac{1}{2} H^2 . \tag{4.217}$$

Of course, our aim is to write the coproducts. Actually, for I^+ we use (4.5) of [587] (since I^+ is expressed through B which has the (parameter-independent) classical coproduct (4.202b) as in the one-parameter case), which in our notation gives:

$$\delta_{\mathcal{U}}(I_+) = I_+ \otimes 1_{\mathcal{U}} + 1_{\mathcal{U}} \otimes I_+ + \sum_{n=1}^{\infty} \left(-\frac{\tilde{g}^2}{4}\right)^n \left(I_+^{n+1} \otimes I_+^n + I_+^n \otimes I_+^{n+1}\right) . \tag{4.218}$$

For the coproduct of H we need the inverse of (4.215a) (cf. (3.1) of [16]):

$$K^{\pm 1} = e^{\pm \tilde{g} B} = 1_{\mathscr{U}} + 2 \sum_{\ell=1}^{\infty} \left(\pm \frac{\tilde{g}}{2} I_+ \right)^{\ell} \left(= \frac{1_{\mathscr{U}} \pm \frac{\tilde{g}}{2} I_+}{1_{\mathscr{U}} \mp \frac{\tilde{g}}{2} I_+} \right). \tag{4.219}$$

Then we have using (4.202d):

$$\begin{aligned} \delta_{\mathscr{U}}(H) &= H \otimes 1_{\mathscr{U}} + 1_{\mathscr{U}} \otimes H + \\ &+ 2 \sum_{n=1}^{\infty} \left(H \otimes \left(-\frac{\tilde{g}}{2} I_+ \right)^n + \left(\frac{\tilde{g}}{2} I_+ \right)^n \otimes H \right) - \\ &- 2 \tilde{h} I_+ \sum_{k=0}^{\infty} \left(\frac{\tilde{g}}{2} I_+ \right)^{2k} \otimes A \left(1_{\mathscr{U}} + 2 \sum_{\ell=1}^{\infty} \left(-\frac{\tilde{g}}{2} I_+ \right)^{\ell} \right), \end{aligned} \tag{4.220}$$

For the coproduct of I_- we use (4.202c) and

$$\begin{aligned} \delta_{\mathscr{U}}(I_-) &= \delta_{\mathscr{U}} \left(\cosh\left(\frac{\tilde{g} B}{2} \right) \right) \delta_{\mathscr{U}}(Y) \, \delta_{\mathscr{U}} \left(\cosh\left(\frac{\tilde{g} B}{2} \right) \right), \\ \delta_{\mathscr{U}} \left(\cosh\left(\frac{\tilde{g} B}{2} \right) \right) &= \cosh\left(\frac{\tilde{g} B}{2} \right) \otimes \cosh\left(\frac{\tilde{g} B}{2} \right) + \\ &+ \sinh\left(\frac{\tilde{g} B}{2} \right) \otimes \sinh\left(\frac{\tilde{g} B}{2} \right) \end{aligned} \tag{4.221}$$

to obtain:

$$\begin{aligned} \delta_{\mathscr{U}}(I_-) &= I_- \otimes \sum_{\ell=0}^{\infty} (\ell+1) \left(-\frac{\tilde{g}}{2} I_+ \right)^{\ell} + \sum_{\ell=0}^{\infty} (\ell+1) \left(\frac{\tilde{g}}{2} I_+ \right)^{\ell} \otimes I_- - \\ &- \frac{\tilde{g}}{2} (I_+ I_- + I_+ I_-) \otimes \sum_{\ell=1}^{\infty} \ell \left(-\frac{\tilde{g}}{2} I_+ \right)^{\ell} + \\ &+ \frac{\tilde{g}}{2} \sum_{\ell=1}^{\infty} \ell \left(\frac{\tilde{g}}{2} I_+ \right)^{\ell} \otimes (I_+ I_- + I_+ I_-) + \\ &+ \frac{\tilde{g}^2}{4} I_+ I_- I_+ \otimes \sum_{\ell=2}^{\infty} (\ell - 1) \left(-\frac{\tilde{g}}{2} I_+ \right)^{\ell} + \\ &+ \frac{\tilde{g}^2}{4} \sum_{\ell=2}^{\infty} (\ell - 1) \left(\frac{\tilde{g}}{2} I_+ \right)^{\ell} \otimes I_+ I_- I_+ - \\ &- \tilde{h}^2 (I_+ \otimes A^2) \left\{ \sum_{k=0}^{\infty} (k+1) \left(\frac{\tilde{g}}{2} I_+ \right)^{2k} \otimes 1_{\mathscr{U}} + \right. \\ &+ \sum_{k=0}^{\infty} \left(\frac{\tilde{g}}{2} I_+ \right)^{2k} \otimes \sum_{\ell=1}^{\infty} \left(-\frac{\tilde{g}}{2} I_+ \right)^{\ell} + \\ &+ \left. \sum_{k=0}^{\infty} (k+1) \left(-\frac{\tilde{g}}{2} I_+ \right)^{k} \otimes \sum_{\ell=1}^{\infty} \ell \left(-\frac{\tilde{g}}{2} I_+ \right)^{\ell} \right\} + \end{aligned}$$

$$
\begin{aligned}
&+\tilde{h}(1_{\mathscr{U}} \otimes A)\Big\{(H \otimes 1_{\mathscr{U}}) \times \\
&\times\Big(\sum_{k=0}^{\infty}\left(\frac{\tilde{g}}{2}I_+\right)^{2k} \otimes 1_{\mathscr{U}} + 1_{\mathscr{U}} \otimes \sum_{\ell=1}^{\infty}(\ell+1)\left(-\frac{\tilde{g}}{2}I_+\right)^{\ell} + \\
&+2\sum_{k=1}^{\infty}\left(-\frac{\tilde{g}}{2}I_+\right)^{k} \otimes \sum_{\ell=1}^{\infty}\ell\left(-\frac{\tilde{g}}{2}I_+\right)^{\ell}\Big) - \\
&-2\Big(\sum_{k=1}^{\infty}k\left(-\frac{\tilde{g}}{2}I_+\right)^{2k} \otimes 1_{\mathscr{U}} + \\
&+\sum_{k=1}^{\infty}k\left(-\frac{\tilde{g}}{2}I_+\right)^{k} \otimes \sum_{\ell=1}^{\infty}\ell\left(-\frac{\tilde{g}}{2}I_+\right)^{\ell}\Big)\Big\}.
\end{aligned}
\tag{4.222}
$$

In the special case $\tilde{h} = 0$ the coproducts of H and I_- coincide with the one-parameter formulae of [16] (cf. (3.2) and (5.3)), respectively (with $\tilde{g} \mapsto -h$). In the special case $\tilde{g} = 0$ the nonlinear map becomes an identity and naturally the coproducts of I_+, I_-, H, coincide with those of B, Y, H, respectively (cf. (4.213b,c,d)).

4.8 Duality for Exotic Bialgebras

4.8.1 Exotic Bialgebras: General Setting

This section follows [49, 50]. For some time it was not clear how many distinct quantum group deformations are admissible for the group $GL(2)$ and the supergroup $GL(1|1)$. For the group $GL(2)$ there were the well-known standard $GL_{pq}(2)$ [183] and non-standard (Jordanian) $GL_{gh}(2)$ [13] two-parameter deformations. For the supergroup $GL(1|1)$ there were the standard $GL_{pq}(1|1)$ [119, 166, 341] and the hybrid (standard–nonstandard) $GL_{qh}(1|1)$ [295] two-parameter deformations. Then, in [48] it was shown that the list of these four deformations is exhaustive (refuting a long-standing claim of [416]) for the existence of a hybrid (standard–nonstandard) two-parameter deformation of $GL(2)$); see also [144]. In particular, it was shown that these four deformations match the distinct triangular 4×4 R-matrices from the classification of [339], which are deformations of the trivial R-matrix (corresponding to undeformed $GL(2)$).

The matching mentioned above was done by applying the FRT formalism [272] to these R-matrices. This analysis revealed altogether *three triangular* R-matrices and *two nontriangular* R-matrices which are *not* deformations of the trivial R-matrix. These new matrix bialgebras, which we called *exotic*, are very interesting and deserve further study. One of the first problems when dealing with such matrix bialgebras is to find the bialgebras with which they are in duality, since some of the structural characteristics are more transparent for the duals. The bialgebras in duality are also the interesting objects with respect to the development of the representation theory.

This is the first problem we solve here. We then find also the quantum planes corresponding to these bialgebras by the Wess–Zumino R-matrix method [596]. For the latter we find the minimal polynomials pol(·) in one variable such that $\mathrm{pol}(\widehat{R}) = 0$ is the lowest-order polynomial identity satisfied by the singly permuted R-matrix $\widehat{R} \equiv PR$ (P is the permutation matrix). These minimal polynomials indeed separate the three cases of $R_{H2,3}$ [339]. (Recall that the corresponding minimal polynomial in the Jordanian case is only quadratic.) We find also the quantum planes by Manin's method [461].

4.8.2 Exotic Bialgebras: Triangular Case 1

In this subsection we consider the matrix bialgebra, denoted here by $\mathcal{A}_1$, which is obtained by applying the RTT relations of [272]:

$$R\,T_1\,T_2 = T_2\,T_1\,R, \tag{4.223}$$

where $T_1 = T \otimes \mathbf{1}_2$, $T_2 = \mathbf{1}_2 \otimes T$, for the case when $R = R_1$:

$$R_1 = \begin{pmatrix} 1 & h & -h & h_3 \\ & 1 & 0 & -h \\ & & 1 & h \\ & & & 1 \end{pmatrix}, \qquad h_3 \neq -h^2. \tag{4.224}$$

This R-matrix, together with the condition on the parameters, is one of the three special cases of the R-matrix denoted by $R_{H2,3}$ in [339]. The algebraic relations of $\mathcal{A}_1$ obtained in this way are given by formulae (5.11) of [48], namely:

$$\begin{aligned} &c^2 = 0, \qquad ca = ac = 0, \qquad dc = cd = 0, \\ &da = ad, \qquad cb = bc, \qquad a^2 = d^2 \\ &ab = ba + h(a^2 + bc - ad), \qquad db = bd - h(a^2 + bc - ad). \end{aligned} \tag{4.225}$$

Note that the constant h_3 does not enter the above relations.

Note that this bialgebra is not a Hopf algebra. Indeed, suppose that it is and there is an antipode γ, then we use one of the Hopf algebra axioms:

$$m \circ (\mathrm{id} \otimes \gamma) \circ \delta = i \circ \varepsilon \tag{4.226}$$

as maps $\mathcal{A} \to \mathcal{A}$, where m is the usual product in the algebra: $m(Y \otimes Z) = YZ, Y, Z \in \mathcal{A}$ and i is the natural embedding of the number field F into $\mathcal{A}$: $i(c) = \mu 1_{\mathcal{A}}, \mu \in F$. Applying this to the element d we would have:

$$c\,\gamma(b) + d\,\gamma(d) = 1_{\mathcal{A}}$$

which leads to contradiction after multiplying from the left by c (one would get $0 = c$).

The algebra $\mathscr{A}_1$ has the following PBW basis:

$$b^n a^k d^\ell\,, \quad b^n c\,, \qquad n,k \in \mathbb{Z}_+\,, \quad \ell = 0,1. \tag{4.227}$$

The last line of (4.225) strongly suggests the substitution:

$$\tilde{a} \;=\; \tfrac{1}{2}(a+d), \quad \tilde{d} \;=\; \tfrac{1}{2}(a-d), \tag{4.228}$$

so that the new algebraic relations and PBW basis are:

$$\begin{aligned} &c^2 \;=\; 0, \quad \tilde{a}c \;=\; c\tilde{a} \;=\; \tilde{d}c \;=\; c\tilde{d} \;=\; \tilde{a}\tilde{d} \;=\; \tilde{d}\tilde{a} \;=\; 0, \quad cb \;=\; bc, \\ &\tilde{a}b \;=\; b\tilde{a}, \quad \tilde{d}b \;=\; b\tilde{d} + 2h\tilde{d}^2 + hbc \end{aligned} \tag{4.229}$$

$$b^n \tilde{a}^k, \quad b^n \tilde{d}^\ell, \quad b^n c, \qquad n,k \in \mathbb{Z}_+\,, \quad \ell \in \mathbb{N}. \tag{4.230}$$

The coalgebra relations become:

$$\delta \begin{pmatrix} \tilde{a} \\ b \\ c \\ \tilde{d} \end{pmatrix} = \begin{pmatrix} \tilde{a}\otimes\tilde{a} + \tilde{d}\otimes\tilde{d} + \frac{1}{2}\, b\otimes c + \frac{1}{2}\, c\otimes b \\ \tilde{a}\otimes b + \tilde{d}\otimes b + b\otimes\tilde{a} - b\otimes\tilde{d} \\ c\otimes\tilde{a} + c\otimes\tilde{d} + \tilde{a}\otimes c - \tilde{d}\otimes c \\ \tilde{a}\otimes\tilde{d} + \tilde{d}\otimes\tilde{a} + \frac{1}{2}\, b\otimes c - \frac{1}{2}\, c\otimes b \end{pmatrix} \tag{4.231}$$

$$\varepsilon\begin{pmatrix} \tilde{a} & b \\ c & \tilde{d} \end{pmatrix} = \begin{pmatrix} 1 & 0 \\ 0 & 0 \end{pmatrix}. \tag{4.232}$$

4.8.2.1 Duality

Let us denote by $\mathscr{U}_1$the unknown yet dual algebra of $\mathscr{A}_1$, and by $\tilde{A}, B, C, \tilde{D}$the four generators of $\mathscr{U}_1$. We would like as in Section 4.4 and [209] to define the pairing $\langle Z, f\rangle$, $Z = \tilde{A}, B, C, \tilde{D}$, $f$is from (4.230), as the classical tangent vector at the identity (4.186) $\langle Z, f\rangle = \varepsilon\left(\frac{\partial f}{\partial y}\right)$; however, here this would work only for the pairs: $(Z, y) = (\tilde{A}, \tilde{a})$, (B, b), $(\tilde{D}, \tilde{d})$, but not for (C, c). The reason is that classically some of the relations in (4.229) are constraints and we have to differentiate internally with respect to the manifold described by these constraints. In particular, if a constraint is given by setting $g = 0$, where g is some function of $\tilde{a}, b, c, \tilde{d}$, then any differentiation $\mathscr{D}$ should respect:

$$(\mathscr{D} g f)_{g=0} \;=\; 0, \tag{4.233}$$

where f is any polynomial function of $\tilde{a}, b, c, \tilde{d}$. Thus, we are lead to define:

$$\langle C, f\rangle \equiv \varepsilon\left(E\,\frac{\partial}{\partial c} f\right) \tag{4.234}$$

where:

$$E = \hat{E}(-\tilde{a}, \frac{\partial}{\partial \tilde{a}}), \quad \hat{E}(x,y) \equiv \sum_{k=0}^{\infty} \frac{x^k y^k}{k!} \tag{4.235}$$

From the above definitions we get:

$$\langle \tilde{A}, f \rangle = \varepsilon\left(\frac{\partial f}{\partial \tilde{a}}\right) = \delta_{n0} \begin{cases} k & \text{for } f = b^n \tilde{a}^k \\ 0 & \text{for } f = b^n \tilde{d}^\ell \\ 0 & \text{for } f = b^n c \end{cases} \tag{4.236a}$$

$$\langle B, f \rangle = \varepsilon\left(\frac{\partial f}{\partial b}\right) = \delta_{n1} \begin{cases} 1 & \text{for } f = b^n \tilde{a}^k \\ 0 & \text{for } f = b^n \tilde{d}^\ell \\ 0 & \text{for } f = b^n c \end{cases} \tag{4.236b}$$

$$\langle C, f \rangle = \varepsilon\left(E\,\frac{\partial f}{\partial c}\right) = \delta_{n0} \begin{cases} 0 & \text{for } f = b^n \tilde{a}^k \\ 0 & \text{for } f = b^n \tilde{d}^\ell \\ 1 & \text{for } f = b^n c \end{cases} \tag{4.236c}$$

$$\langle \tilde{D}, f \rangle = \varepsilon\left(\frac{\partial f}{\partial \tilde{d}}\right) = \delta_{\ell 1}\delta_{n0} \begin{cases} 0 & \text{for } f = b^n \tilde{a}^k \\ 1 & \text{for } f = b^n \tilde{d}^\ell \\ 0 & \text{for } f = b^n c \end{cases} \tag{4.236d}$$

$$\langle E, f\rangle = \begin{cases} 1 & \text{for } \quad f = 1_{\mathcal{A}} \\ 0 & \text{otherwise.} \end{cases} \tag{4.236e}$$

We have included above also the auxiliary generator E since it will appear in the coproduct relations (cf. below). Note that if we have taken the definition (4.186) for (C, c), the result in (4.236) would superficially be the same.

Now we can find the relations between the generators of $\mathcal{U}_1$. We have:

Proposition 3. *The generators* $\tilde{A}, B, C, \tilde{D}, E$ *introduced above obey the following relations:*

$$\begin{aligned} &[\tilde{D}, C] = -2C, \quad [B, C] = \tilde{D}, \quad [B, C]_+ = \tilde{D}^2, \\ &[\tilde{D}, B] = 2B\tilde{D}^2, \quad [\tilde{D}, B]_+ = 0, \\ &\tilde{D}^3 = \tilde{D}, \quad C^2 = 0, \\ &[\tilde{A}, B] = 0, \quad [\tilde{A}, C] = 0, \quad [\tilde{A}, \tilde{D}] = 0, \\ &EZ = ZE = 0, \quad Z = \tilde{A}, B, C, \tilde{D}. \end{aligned} \tag{4.237}$$

For the proof we refer to [49]. ◊

We note that the algebraic relations (4.237) for $\mathcal{U}_1$ do not depend on the constant h present in the relations (4.229) of the dual algebra $\mathcal{A}_1$. Later, we shall see that the established duality reduces also the algebra $\mathcal{A}_1$ so that it also does not depend on h.

4.8.2.2 Coalgebra Structure of the Dual

We turn now to the coalgebra structure of $\mathcal{U}_1$. We have:

Proposition 4.

(i) *The comultiplication in the algebra $\mathcal{U}_1$ is given by:*

$$\delta(\tilde{A}) = \tilde{A} \otimes 1_{\mathcal{U}} + 1_{\mathcal{U}} \otimes \tilde{A}, \tag{4.238a}$$
$$\delta(B) = B \otimes 1_{\mathcal{U}} + 1_{\mathcal{U}} \otimes B, \tag{4.238b}$$
$$\delta(C) = C \otimes E + E \otimes C, \tag{4.238c}$$
$$\delta(\tilde{D}) = \tilde{D} \otimes E + E \otimes \tilde{D}, \tag{4.238d}$$
$$\delta(E) = E \otimes E. \tag{4.238e}$$

(ii) *The counit relations in $\mathcal{U}_1$ are given by:*

$$\varepsilon_{\mathcal{U}}(Z) = 0, \qquad Z = \tilde{A}, B, C, \tilde{D} \tag{4.239a}$$
$$\varepsilon_{\mathcal{U}}(E) = 1, \tag{4.239b}$$

where we have included also the auxiliary operator E.

For the proof we refer to [49]. ◊

There is no antipode for the bialgebra $\mathcal{U}_1$. Indeed, suppose that there was such. Then by applying the Hopf algebra axiom (4.226) to the generator E, we would get:

$$E\,\gamma(E) = 1_{\mathcal{U}},$$

which would lead to contradiction after multiplication from the left with $Z = \tilde{A}, B, C, \tilde{D}$ (we would get $0 = Z$).

4.8.2.3 Reduction of the Bialgebra

We noticed that the algebraic relations (4.237) of $\mathcal{U}_1$ do not depend on the constant h from relations (4.229) of $\mathcal{A}_1$. The coproduct relations (4.238) also do not depend on h. We now clarify the reason for this. First we note that $\mathcal{A}_1$ has the following two-sided ideals and coideals:

$$I = \mathcal{A}_1 b\tilde{d} \oplus \mathcal{A}_1 \tilde{d}^2 \oplus \mathcal{A}_1 bc \tag{4.240a}$$
$$I_2 = \mathcal{A}_1 \tilde{d}^2 \oplus \mathcal{A}_1 bc \tag{4.240b}$$
$$I_1 = \mathcal{A}_1 bc \tag{4.240c}$$

so that

$$I_1 \subset I_2 \subset I \subset \mathscr{A}_1 \tag{4.241}$$

Furthermore the pairing of all these ideals with the dual algebra $\mathscr{U}_1$ vanish; thus we can set them consistently equal to zero. Thus, the basis of $\mathscr{A}_1$ is reduced to the following monomials:

$$b^n \tilde{a}^k \,, \; n, k \in \mathbb{Z}_+ \,, \quad \tilde{d}, \quad c \tag{4.242}$$

Actually, it were only these monomials that appeared in the proof of the dual relations (4.237). The algebraic relations of the reduced algebra become rather trivial:

$$\begin{aligned} &\tilde{a}c = c\tilde{a} = \tilde{d}c = c\tilde{d} = \tilde{a}\tilde{d} = \tilde{d}\tilde{a} = cb = bc = \tilde{d}b = b\tilde{d} = 0, \\ &c^2 = 0, \quad \tilde{a}b = b\tilde{a}, \end{aligned} \tag{4.243}$$

while the coalgebra relations remain unchanged and nontrivial. It is remarkable that the dual algebra has much richer structure in both the algebraic and coalgebraic sectors.

4.8.2.4 Consistency with the FRT Approach

For the application of the FRT approach to duality we need the 4×4 R-matrix, which for the algebra $\mathscr{A}_1$ is given by (4.224). In the duality relations enter actually the matrices $R_1^{\pm}$:

$$\begin{aligned} R_1^+ &\equiv P R_1 P = R_1(-h) = \begin{pmatrix} 1 & -h & h & h_3 \\ & 1 & 0 & h \\ & & 1 & -h \\ & & & 1 \end{pmatrix} \\ R_1^- &\equiv R_1^{-1} = \begin{pmatrix} 1 & -h & h & -h_3 - 2h^2 \\ & 1 & 0 & h \\ & & 1 & -h \\ & & & 1 \end{pmatrix}. \end{aligned} \tag{4.244}$$

where P is the *permutation matrix*:

$$P \equiv \begin{pmatrix} 1&0&0&0 \\ 0&0&1&0 \\ 0&1&0&0 \\ 0&0&0&1 \end{pmatrix}. \tag{4.245}$$

These R-matrices encode (part of) the duality between $\mathscr{U}_1$ and $\mathscr{A}_1$ by formula (2.1) of [272] taken for $k = 1$ and written in our setting:

$$\langle L^{\pm}, T \rangle = R_1^{\pm}, \tag{4.246}$$

where $L^{\pm}$ are 2×2 matrices whose elements are functions of the generators of $\mathscr{U}_1$, T is the 2×2 matrix formed by the generators of $\mathscr{A}_1$. In order to make formula (4.246) explicit we have to adopt some convention on the indices. We choose to write it as:

$$\langle L_{ik}^{\pm}, T_{\ell j} \rangle = (R_1^{\pm})_{ijk\ell}, \quad i, j, k, \ell = 1, 2, \tag{4.247}$$

where the enumeration of the R-matrices is done as in [155], namely, the rows are enumerated from top to bottom by the pairs $(i, j) = (1, 1), (1, 2), (2, 1), (2, 2)$, and the columns are enumerated from left to right by the pairs $(k, \ell) = (1, 1), (1, 2), (2, 1), (2, 2)$.

Using all this and rewriting the result in terms of the new basis (4.229) of $\mathscr{A}_1$ we have:

$$\left\langle L_{11}^{\pm}, \begin{pmatrix} \tilde{a} & b \\ c & \tilde{d} \end{pmatrix} \right\rangle = \left\langle L_{22}^{\pm}, \begin{pmatrix} \tilde{a} & b \\ c & \tilde{d} \end{pmatrix} \right\rangle = \begin{pmatrix} 1 & -h \\ 0 & 0 \end{pmatrix} \tag{4.248}$$

$$\left\langle L_{12}^{\pm}, \begin{pmatrix} \tilde{a} & b \\ c & \tilde{d} \end{pmatrix} \right\rangle = \begin{pmatrix} h & h_{\pm} \\ 0 & 0 \end{pmatrix}, \tag{4.249}$$

where $h_+ = h_3$ and $h_- = -h_3 - 2h^2$. Note that the elements $L_{21}^{\pm}$ have zero products with all generators so we can set them to zero. Next we calculate the pairings with arbitrary elements of $\mathscr{A}_1$ for which we use the fact that the coproducts of the $L_{jk}^{\pm}$ generators are canonically given by [272]:

$$\delta\left(L_{ik}^{\pm}\right) = \sum_{j=1}^{2} L_{ij}^{\pm} \otimes L_{jk}^{\pm}. \tag{4.250}$$

Using this we obtain:

$$\left\langle L_{11}^{\pm}, b^n \tilde{a}^k \right\rangle = \left\langle L_{22}^{\pm}, b^n \tilde{a}^k \right\rangle = (-h)^n \tag{4.251}$$

$$\left\langle L_{12}^{\pm}, b^n \tilde{a}^k \right\rangle = (-1)^n h^{n-1}((k+n)h^2 - n(h_{\pm} + h^2)) \tag{4.252}$$

All other pairings are zero.

Computing the above pairings with the defining relations (4.236) we conclude that these L operators are expressed in terms of the generators of the dual algebra $\mathscr{U}_1$ as follows:

$$L_{11}^{\pm} = L_{22}^{\pm} = e^{-hB} \tag{4.253a}$$
$$L_{12}^{\pm} = ((h_{\pm} + h^2)B + h\tilde{A})e^{-hB} \tag{4.253b}$$

where expressions like $e^{\nu B}$ are defined as formal power series $e^{\nu B} = 1_{\mathcal{U}} + \sum_{p\in\mathbb{Z}_+} \frac{\nu^p}{p!} B^p$. Formulae (4.253) are compatible with the coproducts (4.238a,b) of the generators $\tilde{A}, B$. However, as we see this approach does not say anything about the generators $C, \tilde{D}$.

4.8.3 Exotic Bialgebras: Triangular Case 2

In this section we consider the bialgebra, denoted here by $\mathcal{A}_2$, which is obtained by applying the basic relations (4.223) for the case when $R = R_2$:

$$R_2 = \begin{pmatrix} 1 & h_1 & h_2 & h_3 \\ & 1 & 0 & h_2 \\ & & 1 & h_1 \\ & & & 1 \end{pmatrix}, \qquad h_1 + h_2 \neq 0 \tag{4.254}$$

This R-matrix together with the condition on the parameters is the second of the special cases (mentioned in the Introduction) of the R-matrix denoted by $R_{H2,3}$ in [339]. Its algebraic relations thus obtained are given by formulae (5.9) of [48], namely:

$$\begin{aligned} &c^2 = 0, \quad ca = ac = 0, \quad dc = cd = 0, \\ &da = ad, \quad cb = bc, \quad a^2 = d^2 = ad + bc, \\ &ab = bd = ba + (h_1 - h_2)bc, \quad db = bd + (h_2 - h_1)bc. \end{aligned} \tag{4.255}$$

Note that the constant h_3 does not enter the above relations.

The coalgebra relations are the same as for $\mathcal{A}_1$. Also the demonstration that this bialgebra is not a Hopf algebra is done as for $\mathcal{A}_1$. The PBW basis in this case is:

$$b^n a^k, \quad a^\ell d, \quad c, \qquad n, k \in \mathbb{Z}_+, \quad \ell = 0, 1. \tag{4.256}$$

Also in this case we make the change of basis (4.228) to obtain:

$$\begin{aligned} &c^2 = 0, \quad \tilde{a}c = c\tilde{a} = \tilde{d}c = c\tilde{d} = \tilde{a}\tilde{d} = \tilde{d}\tilde{a} = 0, \\ &\tilde{a}b = b\tilde{a}, \quad bc = cb = 2\tilde{d}^2, \quad \tilde{d}^3 = 0 \\ &\tilde{d}b = -b\tilde{d} = (h_1 - h_2)\tilde{d}^2. \end{aligned} \tag{4.257}$$

The PBW basis becomes:

$$b^n \tilde{a}^k, \quad \tilde{d}^\ell, \quad c, \qquad n, k \in \mathbb{Z}_+, \quad \ell = 1, 2. \tag{4.258}$$

Thus, this bialgebra looks “smaller” than $\mathcal{A}_1$ – compare with (4.258). It has also a smaller structure of two-sided ideals and coideals:

$$I_2 = \mathscr{A}_2 \tilde{d}^2 \oplus \mathscr{A}_2 bc \tag{4.259a}$$

$$I_1 = \mathscr{A}_2 bc \tag{4.259b}$$

so that

$$I_1 \subset I_2 \subset \mathscr{A}_2 \tag{4.260}$$

– compare with (4.240, 4.241).

4.8.3.1 Algebra and Coalgebra Structure of the Dual

In view of the similarities between the algebras $\mathscr{A}_1$ and $\mathscr{A}_2$ it is natural do use the same generators $\tilde{A}, B, C, \tilde{D}, E$ for the dual $\mathscr{U}_2$. It is not surprising that we get the same algebraic and coalgebraic relations. We have:

Proposition 5. *The generators $\tilde{A}, B, C, \tilde{D}, E$ of the bialgebra $\mathscr{U}_2$ obey the same algebraic and coalgebraic relations as for the algebra $\mathscr{U}_1$ given in Propositions 3 and 4.* ◊

Proof. The proof is based on the fact that the bialgebras $\mathscr{A}_1$ and $\mathscr{A}_2$ differ in the relations involving the (co)ideals I_k, which have no bearing on the relations of $\mathscr{U}_1$. Thus, we need only to show that all bilinears built from the generators $\tilde{A}, B, C, \tilde{D}, E$ have zero pairings with the ideals I_k (cf. (4.259, 4.260)), which is easy to demonstrate. ∎

As a corollary also here the basis and algebraic relations of $\mathscr{A}_2$ reduce to (4.242) and (4.243).

Thus, we have shown the following important conclusion:

Proposition 6. *The bialgebras $\mathscr{A}_1$ and $\mathscr{A}_2$ considered as bialgebras in duality with the bialgebras $\mathscr{U}_1$, $\mathscr{U}_2$, respectively, coincide.* ◊

We recall that the notion of duality we use does not coincide with the FRT definition of duality. The latter is more stringent as we shall see in the next subsection.

4.8.3.2 Consistency with the FRT Approach

The 4×4 R-matrix needed for the FRT approach is given in (4.254). The matrices $R_2^{\pm}$ entering the duality relations are:

$$R_2^+ \equiv P R_2 P = \begin{pmatrix} 1 & h_2 & h_1 & h_3 \\ & 1 & 0 & h_1 \\ & & 1 & h_2 \\ & & & 1 \end{pmatrix} \tag{4.261a}$$

$$R_2^- \equiv R_2^{-1} = \begin{pmatrix} 1 & -h_1 & -h_2 & 2h_1h_2 - h_3 \\ & 1 & 0 & -h_2 \\ & & 1 & -h_1 \\ & & & 1 \end{pmatrix} \tag{4.261b}$$

Using the above and relations (4.247) (with $R_1 \to R_2$) we obtain:

$$\begin{aligned} \left\langle L_{11}^+, \begin{pmatrix} \tilde{a} & b \\ c & \tilde{d} \end{pmatrix} \right\rangle &= \left\langle L_{22}^+, \begin{pmatrix} \tilde{a} & b \\ c & \tilde{d} \end{pmatrix} \right\rangle = \begin{pmatrix} 1 & h_2 \\ 0 & 0 \end{pmatrix} \\ \left\langle L_{12}^+, \begin{pmatrix} \tilde{a} & b \\ c & \tilde{d} \end{pmatrix} \right\rangle &= \begin{pmatrix} h_1 & h_3 \\ 0 & 0 \end{pmatrix} \\ \left\langle L_{11}^-, \begin{pmatrix} \tilde{a} & b \\ c & \tilde{d} \end{pmatrix} \right\rangle &= \left\langle L_{22}^-, \begin{pmatrix} \tilde{a} & b \\ c & \tilde{d} \end{pmatrix} \right\rangle = \begin{pmatrix} 1 & -h_1 \\ 0 & 0 \end{pmatrix} \\ \left\langle L_{12}^-, \begin{pmatrix} \tilde{a} & b \\ c & \tilde{d} \end{pmatrix} \right\rangle &= \begin{pmatrix} -h_2 & -h_3 + 2h_1h_2 \\ 0 & 0 \end{pmatrix} \end{aligned} \tag{4.262}$$

Iterating this we obtain:

$$\begin{aligned} \left\langle L_{11}^+, b^n \tilde{a}^k \right\rangle &= \left\langle L_{22}^+, b^n \tilde{a}^k \right\rangle = h_2^n \\ \left\langle L_{12}^+, b^n \tilde{a}^k \right\rangle &= h_2^{n-1}((k+n)h_1h_2 + n(h_3 - h_1h_2)) \\ \left\langle L_{11}^-, b^n \tilde{a}^k \right\rangle &= \left\langle L_{22}^-, b^n \tilde{a}^k \right\rangle = (-h_1)^n \\ \left\langle L_{12}^-, b^n \tilde{a}^k \right\rangle &= (-h_1)^{n-1}((k+n)h_1h_2 + n(-h_3 + h_1h_2)) \end{aligned} \tag{4.263}$$

From the above follow:

$$L_{11}^+ = L_{22}^+ = e^{h_2 B} \tag{4.264a}$$

$$L_{12}^+ = ((h_3 - h_1h_2)B + h_1\tilde{A})e^{h_2 B} \tag{4.264b}$$

$$L_{11}^- = L_{22}^+ = e^{-h_1 B} \tag{4.264c}$$

$$L_{12}^- = ((-h_3 + h_1h_2)B - h_2\tilde{A})e^{-h_1 B} \tag{4.264d}$$

This is compatible with the coproducts for the operators $\tilde{A}, B$.

Thus, we see that the L operators in this case are different from those of $\mathscr{U}_1$ (cf. (4.253)). Thus, the FRT approach is more stringent than the notion of duality we use since it distinguishes the two pairs of bialgebras. However, this difference is not as drastic as the difference between the algebraic relations (4.229) and (4.257) of $\mathscr{A}_1$ and $\mathscr{A}_2$, respectively, since (4.253) is just a special case of (4.264) obtained for $h_1 = -h_2 = h$.

On the other hand the FRT approach is incomplete in the cases at hand since it gives info only about part of the generators, namely, $\tilde{A}$ and B, and says nothing about the generators $C, \tilde{D}$.

4.8.4 Exotic Bialgebras: Triangular Case 3

In this section we consider the bialgebra which we denote here by $\mathscr{A}_3$. It is obtained by applying the basic relations (4.223) for the case when $R = R_3$:

$$R_3 = \begin{pmatrix} 1 & 0 & 0 & 1 \\ & -1 & 0 & 0 \\ & & -1 & 0 \\ & & & 1 \end{pmatrix} \tag{4.265}$$

This R-matrix is denoted by $R_{S0,2}$ in [339]. The algebraic relations of $\mathscr{A}_3$ are given by formulae (5.13) of [48], namely:

$$\begin{aligned} &c^2 = 0, \qquad ca = ac = 0, \qquad dc = cd = 0, \\ &da = ad, \qquad cb = bc, \qquad a^2 = d^2 \\ &ab + ba = 0, \qquad db + bd = 0 \end{aligned} \tag{4.266}$$

The coalgebra relations and the demonstration that this bialgebra is not a Hopf algebra are as for $\mathscr{A}_1$, $\mathscr{A}_2$.

Also in this case we make the change of basis (4.228) to obtain:

$$\begin{aligned} &c^2 = 0, \quad \tilde{a}c = c\tilde{a} = \tilde{d}c = c\tilde{d} = \tilde{a}\tilde{d} = \tilde{d}\tilde{a} = 0, \quad cb = bc, \\ &\tilde{a}b + b\tilde{a} = 0, \quad \tilde{d}b + b\tilde{d} = 0. \end{aligned} \tag{4.267}$$

The algebra $\mathscr{A}_3$ has the same PBW bases (4.227) and (4.230) as the algebra $\mathscr{A}_1$. It has also the same (co)ideals as $\mathscr{A}_1$ (cf. (4.240, 4.241)).

4.8.4.1 Algebra and Coalgebra Structure of the Dual

In view of the similarities between the algebras $\mathscr{A}_1$ and $\mathscr{A}_3$ it is natural do use the same generators $\tilde{A}, B, C, \tilde{D}, E$ for the dual $\mathscr{U}_3$. It is not surprising that we get the same algebraic relations between generators $\tilde{A}, B, C, \tilde{D}, E$. However, unlike the bialgebras $\mathscr{A}_1$, $\mathscr{A}_2$ the coalgebraic relations and the relation with the FRT formalism here are different, and it is even necessary to introduce two new auxilliary operators $F_\pm$ defined as:

$$\langle F_\pm, f \rangle \equiv \varepsilon\left(\hat{E}(\pm 1, \frac{\partial}{\partial \tilde{d}})f\right) = \varepsilon\left(\exp(\pm\frac{\partial}{\partial \tilde{d}})f\right). \tag{4.268}$$

Explicitly we have:

$$\langle F_+, f \rangle = \begin{cases} 1 & \text{for } f = \tilde{d}^{\ell} \\ 1 & \text{for } f = 1_{\mathscr{A}} \\ 0 & \text{otherwise} \end{cases} \tag{4.269a}$$

$$\langle F_-, f \rangle = \begin{cases} (-1)^{\ell} & \text{for } f = \tilde{d}^{\ell} \\ 1 & \text{for } f = 1_{\mathscr{A}} \\ 0 & \text{otherwise.} \end{cases} \tag{4.269b}$$

We have for the algebraic and coalgebraic structure of $\mathscr{U}_3$:

Proposition 7. *The generators $\tilde{A}, B, C, \tilde{D}, E, F_{\pm}$ obey the following algebraic relations:*

$$\begin{aligned}
[\tilde{D}, C] &= -2C, \quad [B, C] = \tilde{D} \quad [B, C]_+ = \tilde{D}^2 \\
[\tilde{D}, B] &= 2B\tilde{D}^2, \quad [\tilde{D}, B]_+ = 0 \\
\tilde{D}^3 &= \tilde{D}, \quad C^2 = 0 \\
[\tilde{A}, B] &= 0, \quad [\tilde{A}, C] = 0, \quad [\tilde{A}, \tilde{D}] = 0, \\
EZ &= ZE = 0, \quad Z = \tilde{A}, B, C, \tilde{D} \\
F_+^2 &= F_-^2 = 1_{\mathscr{U}}, \quad [F_+, F_-] = 0 \\
[\tilde{A}, F_{\pm}] &= 0, \quad BF_{\pm} \pm F_{\mp}B = 0 \\
[C, F_{\pm}]_+ &= 0, \quad [\tilde{D}, F_{\pm}] = 0 \\
EF_{\pm} &= F_{\pm}E = E.
\end{aligned} \tag{4.270}$$

For the proof we refer to [49]. ◊

Proposition 8.

(i) *The comultiplication in the algebra $\mathscr{U}_3$ is given by:*

$$\begin{aligned}
\delta(\tilde{A}) &= \tilde{A} \otimes 1_{\mathscr{U}} + 1_{\mathscr{U}} \otimes \tilde{A}, \\
\delta(B) &= B \otimes 1_{\mathscr{U}} + F_+F_- \otimes B, \\
\delta(C) &= C \otimes E + E \otimes C, \\
\delta(\tilde{D}) &= \tilde{D} \otimes E + E \otimes \tilde{D}, \\
\delta(E) &= E \otimes E \\
\delta(F_{\pm}) &= F_{\pm} \otimes F_{\pm}.
\end{aligned} \tag{4.271}$$

(ii) *The counit relations in $\mathscr{U}_3$ are given by:*

$$\begin{aligned}
\varepsilon_{\mathscr{U}}(Z) &= 0, \qquad Z = \tilde{A}, B, C, \tilde{D} \\
\varepsilon_{\mathscr{U}}(Z) &= 1, \qquad Z = E, F_{\pm}
\end{aligned} \tag{4.272}$$

For the proof we refer to [49]. ◊

There is no antipode for the bialgebra $\mathcal{U}_3$ – this is proved exactly as for $\mathcal{U}_1$.

As in the case of $\mathcal{U}_1 \longrightarrow \mathcal{A}_1$ (and $\mathcal{U}_2 \longrightarrow \mathcal{A}_2$) duality, one may reduce the basis of $\mathcal{A}_3$ from the $\mathcal{U}_3 \longrightarrow \mathcal{A}_3$ duality, but only with the ideal $I_1 = \mathcal{A}_3\, bc$ (since $\tilde{d}^2$ is not annihilated by $F_\pm$). Thus, the basis of $\mathcal{A}_3$ is reduced to the following monomials:

$$b^n \tilde{a}^k\,, \quad b^n \tilde{d}^\ell\,, \quad c\,, \qquad n, k \in \mathbb{Z}_+\,, \quad \ell \in \mathbb{N}. \tag{4.273}$$

The algebraic relations of the reduced algebra become:

$$\begin{aligned} &c^2 = 0, \quad \tilde{a}c = c\tilde{a} = \tilde{d}c = c\tilde{d} = \tilde{a}\tilde{d} = \tilde{d}\tilde{a} = cb = bc = 0,\\ &\tilde{a}b + b\tilde{a} = 0, \quad \tilde{d}b + b\tilde{d} = 0. \end{aligned} \tag{4.274}$$

4.8.4.2 Consistency with the FRT Approach

The 4×4 R-matrix needed for the FRT approach is given in (4.265). The matrices $R_3^\pm$ entering the duality relations are:

$$R_3^+ \equiv P R_3 P = R_3 \tag{4.275a}$$

$$R_3^- \equiv R_3^{-1} = \begin{pmatrix} 1 & 0 & 0 & -1 \\ & -1 & 0 & 0 \\ & & -1 & 0 \\ & & & 1 \end{pmatrix}. \tag{4.275b}$$

Using the above and relations (4.247) (with $R_1 \to R_3$) we obtain:

$$\begin{aligned} \left\langle L_{11}^\pm, \begin{pmatrix} \tilde{a} & b \\ c & \tilde{d} \end{pmatrix} \right\rangle &= \begin{pmatrix} 0 & 0 \\ 0 & 1 \end{pmatrix} \\ \left\langle L_{22}^\pm, \begin{pmatrix} \tilde{a} & b \\ c & \tilde{d} \end{pmatrix} \right\rangle &= \begin{pmatrix} 0 & 0 \\ 0 & -1 \end{pmatrix} \\ \left\langle L_{12}^\pm, \begin{pmatrix} \tilde{a} & b \\ c & \tilde{d} \end{pmatrix} \right\rangle &= \begin{pmatrix} 0 & \pm 1 \\ 0 & 0 \end{pmatrix} \end{aligned} \tag{4.276}$$

Iterating these relations for arbitrary elements of the basis of $\mathcal{A}_3$ we can show that the L generators are given in terms of some of the other generators in the following way:

$$L_{11}^\pm = F_+\,, \qquad L_{22}^\pm = F_-\,, \qquad L_{12}^\pm = \pm B F_- \tag{4.277}$$

Formulae (4.277) are compatible with the coproducts in (4.271) of the generators $B, F_\pm$. However, as we see this approach does not say anything about the basic generators $\tilde{A}, C, \tilde{D}$.

4.8.5 Higher-Order R-matrix Relations and Quantum Planes

In order to address the question of the quantum planes corresponding to the exotic bialgebras we have to know the relations which the R-matrices fulfil. As we know the R-matrices producing deformations of the $GL(2)$ and $GL(1|1)$ fulfil second-order relations. However, in the cases at hand we have higher-order relations.

We start with the R-matrix $R_{H2,3}$ of [339]:

$$R = \begin{pmatrix} 1 & h_1 & h_2 & h_3 \\ & 1 & 0 & h_2 \\ & & 1 & h_1 \\ & & & 1 \end{pmatrix} \tag{4.278}$$

We need actually the singly permuted R-matrix:

$$\hat{R} \equiv PR = \begin{pmatrix} 1 & h_1 & h_2 & h_3 \\ 0 & 0 & 1 & h_1 \\ 0 & 1 & 0 & h_2 \\ 0 & 0 & 0 & 1 \end{pmatrix} \tag{4.279}$$

Explicit calculation shows now that we have:

$$(\hat{R} - \mathbf{1})(\hat{R} + \mathbf{1}) = 0, \qquad h_1 = -h_2 = h,\ h_3 = -h^2, \tag{4.280a}$$

$$(\hat{R} - \mathbf{1})^2(\hat{R} + \mathbf{1}) = 0, \qquad h_1 = -h_2 = h,\ h_3 \neq -h^2,$$

$$\hat{R} = PR_1, \tag{4.280b}$$

$$(\hat{R} - \mathbf{1})^3(\hat{R} + \mathbf{1}) = 0, \qquad h_1 + h_2 \neq 0, \quad \hat{R} = PR_2 \tag{4.280c}$$

where $\mathbf{1}$ is the 4×4 unit matrix. Thus the minimal polynomials are:

$$\mathrm{pol}(\hat{R}) = \begin{cases} (\hat{R} - \mathbf{1})(\hat{R} + \mathbf{1}) & \text{for } h_1 = -h_2 = h,\ h_3 = -h^2, \\ (\hat{R} - \mathbf{1})^2(\hat{R} + \mathbf{1}) & \text{for } h_1 = -h_2 = h,\ h_3 \neq -h^2 \\ (\hat{R} - \mathbf{1})^3(\hat{R} + \mathbf{1}) & \text{for } h_1 + h_2 \neq 0 \end{cases} \tag{4.281}$$

Remark 4.2. We recall that (4.280a) is the Jordanian subcase which produces the $GL_{h,h}(2)$ deformation of $GL(2)$. Thus, the three subcases of Hietarinta's R-matrix $R_{H2,3}$ are distiguished not only and not so much by the algebras they produce but intrinsically by their minimal polynomials. ◊

To derive the corresponding quantum planes we shall apply the Wess–Zumino formalism [596]. The commutation relations between the coordinates z^i and differentials $\bar{z}^i$, $i = 1, 2$, are given as follows:

$$z^i z^j = \mathscr{P}_{ijk\ell}\, z^k z^\ell \tag{4.282}$$

$$\zeta^i \zeta^j = -\mathscr{Q}_{ijk\ell}\, \zeta^k \zeta^\ell \tag{4.283}$$

$$z^i \zeta^j = \mathscr{Q}_{ijk\ell}\, \zeta^k z^\ell \tag{4.284}$$

where the operators $\mathscr{P}$, $\mathscr{Q}$ are functions of $\hat{R}$ and must satisfy:

$$(\mathscr{P} - \mathbf{1})\,(\mathscr{Q} + \mathbf{1}) = 0. \tag{4.285}$$

In the well-studied deformations of $GL(2)$ there are quadratic minimal polynomials, and there are only two choices for the operators $\mathscr{P}$, $\mathscr{Q}$ (cf. e. g., (4.280a)). Here we have more choices. In particular, for the case (4.280b) we have four choices:

$$(\mathscr{P} - \mathbf{1},\ \mathscr{Q} + \mathbf{1}) = \begin{cases} (\hat{R} - \mathbf{1},\ \hat{R}^2 - \mathbf{1}) \\ (\hat{R} + \mathbf{1},\ (\hat{R} - \mathbf{1})^2) \\ (\hat{R}^2 - \mathbf{1},\ \hat{R} - \mathbf{1}) \\ ((\hat{R} - \mathbf{1})^2,\ \hat{R} + \mathbf{1}), \end{cases} \tag{4.286}$$

while in the case (4.280c) we have six choices:

$$(\mathscr{P} - \mathbf{1},\ \mathscr{Q} + \mathbf{1}) = \begin{cases} (\hat{R} - \mathbf{1},\ (\hat{R}^2 - \mathbf{1})(\hat{R} - \mathbf{1})) \\ (\hat{R} + \mathbf{1},\ (\hat{R} - \mathbf{1})^3) \\ (\hat{R}^2 - \mathbf{1},\ (\hat{R} - \mathbf{1})^2) \\ ((\hat{R} - \mathbf{1})^2,\ \hat{R}^2 - \mathbf{1}) \\ ((\hat{R}^2 - \mathbf{1})(\hat{R} - \mathbf{1}),\ \hat{R} - \mathbf{1}) \\ ((\hat{R} - \mathbf{1})^3,\ \hat{R} + \mathbf{1}). \end{cases} \tag{4.287}$$

Our choice will be the last possibility of both (4.286) and (4.287); that is, we shall use $\mathscr{P} - \mathbf{1} = (\hat{R} - \mathbf{1})^a$ with $a = 2, 3$, respectively, and $\mathscr{Q} = \hat{R}$ in all cases. With this choices and denoting $(x, y) = (z^1, z^2)$ we obtain from (4.282), respectively,

$$xy - yx = hy^2, \quad h_1 = -h_2 = h, \quad \mathscr{P} - \mathbf{1} = (\hat{R} - \mathbf{1})^2, \tag{4.288}$$

$$xy - yx = \tfrac{1}{2}(h_1 - h_2)y^2, \quad h_1 \neq -h_2, \quad \mathscr{P} - \mathbf{1} = (\hat{R} - \mathbf{1})^3. \tag{4.289}$$

We note that the quantum planes corresponding to the bialgebras $\mathscr{A}_1$ and $\mathscr{A}_2$ are not essentially different. Furthermore, the quantum plane (4.288) is the same as for the Jordanian subcase if we choose $\mathscr{P} - \mathbf{1} = \hat{R} - \mathbf{1}$.

Denoting $(\xi, \eta) = (\zeta^1, \zeta^2)$ we obtain from (4.283) with $\mathscr{Q} = \hat{R}$:

$$\xi^2 + \frac{h_1 - h_2}{2}\xi\eta = 0 \tag{4.290a}$$

$$\eta^2 = 0 \tag{4.290b}$$

$$\xi\eta = -\eta\xi \tag{4.290c}$$

Of course, for $\hat{R} = PR_1$ (4.290a) simplifies to

$$\xi^2 + h\,\xi\eta = 0, \tag{4.291}$$

which is valid also for the Jordanian subcase.

Finally, for the coordinates–differentials relations we obtain from (4.284) with $\mathscr{Q} = \hat{R}$ again for all subcases:

$$x\xi = \xi x + h_1\xi y + h_2\eta x + h_3\eta y \tag{4.292a}$$

$$x\eta = \eta x + h_1\eta y \tag{4.292b}$$

$$y\xi = \xi y + h_2\eta y \tag{4.292c}$$

$$y\eta = \eta y. \tag{4.292d}$$

Finally we derive the quantum plane relations for the case of the R_3 matrix. It is easy to see that (4.280b) holds also in this case; that is, for

$$\hat{R}_3 \equiv PR_3 = \begin{pmatrix} 1 & 0 & 0 & 1 \\ 0 & 0 & -1 & 0 \\ 0 & -1 & 0 & 0 \\ 0 & 0 & 0 & 1 \end{pmatrix}. \tag{4.293}$$

Using (4.282–4.284) with $\mathscr{P} - \mathbf{1} = (\hat{R}_3 - \mathbf{1})^2$, $\mathscr{Q} = \hat{R}_3$, we obtain, respectively:

$$xy = -yx \tag{4.294}$$

$$\xi^2 = 0\,, \eta^2 = 0\,, \xi\eta = \eta\xi \tag{4.295}$$

$$x\xi = \xi x + \eta y\,, x\eta = -\eta x\,, y\xi = -\xi y\,, y\eta = \eta y. \tag{4.296}$$

Finally, we note that a check of consistency of this formalism is to implement Manin's approach to quantum planes [461]. Namely, one takes quantum matrix T (cf. (4.14)) as transformation matrix of the two-dimensional quantum planes. This means that if we define:

$$z'^i = T_{ij}\, z^j, \qquad \zeta'^i = T_{ij}\, \zeta^j, \tag{4.297}$$

then $(x', y') = (z'^1, z'^2)$ and $(\xi', \eta') = (\zeta'^1, \zeta'^2)$ should satisfy the same relations as (x, y) and (ξ, η). The latter statement may be used to recover the algebraic relations of the bialgebras. Namely, suppose, that relations (4.288), (4.290b,c), (4.291), (4.292), or relations (4.289), (4.290), (4.292), or relations (4.294), (4.295), and (4.296), hold for both (x, y) and (ξ, η) and (x', y') and (ξ', η'); then substitute the expressions for (x', y') and (ξ', η') in the these relations, under the assumption that a, b, c, d commute with (x, y) and (ξ, η); then the coefficients of the independent bilinears that may be built from (x, y) and (ξ, η) will reproduce the algebraic relations of the bialgebras $\mathscr{A}_1$, $\mathscr{A}_2$, $\mathscr{A}_3$, respectively.

4.8.6 Exotic Bialgebras: Nontriangular Case *S*03

In this subsection and the next we find the exotic matrix bialgebras which correspond to the two nontriangular nonsingular 4×4 R-matrices of [339], namely, $R_{S0,3}$ and $R_{S1,4}$, which are not deformations of the trivial R-matrix. We study three bialgebras denoted by $S03$, $S14$, $S14o$, the latter two cases corresponding to $R_{S1,4}$ for deformation parameter $q^2 \neq 1$ and $q^2 = 1$, respectively.

Again we consider matrix bialgebras which are unital associative algebras generated by four elements a, b, c, d. The coproduct and counit relations are the classical ones (4.5).

Here it shall be convenient to make the following change of generators:

$$\tilde{a} = \tfrac{1}{2}(a + d), \quad \tilde{d} = \tfrac{1}{2}(a - d), \quad \tilde{b} = \tfrac{1}{2}(b + c), \quad \tilde{c} = \tfrac{1}{2}(b - c). \tag{4.298}$$

For the new generators we have instead of (4.5):

$$\delta\begin{pmatrix} \tilde{a} & \tilde{b} \\ \tilde{c} & \tilde{d} \end{pmatrix} = \begin{pmatrix} \tilde{a}\otimes\tilde{a} + \tilde{b}\otimes\tilde{b} - \tilde{c}\otimes\tilde{c} + \tilde{d}\otimes\tilde{d} & \tilde{a}\otimes\tilde{b} + \tilde{b}\otimes\tilde{a} - \tilde{c}\otimes\tilde{d} + \tilde{d}\otimes\tilde{c} \\ \tilde{a}\otimes\tilde{c} + \tilde{c}\otimes\tilde{a} - \tilde{b}\otimes\tilde{d} + \tilde{d}\otimes\tilde{b} & \tilde{a}\otimes\tilde{d} + \tilde{d}\otimes\tilde{a} - \tilde{b}\otimes\tilde{c} + \tilde{c}\otimes\tilde{b} \end{pmatrix} \tag{4.299}$$

$$\varepsilon\begin{pmatrix} \tilde{a} & \tilde{b} \\ \tilde{c} & \tilde{d} \end{pmatrix} = \begin{pmatrix} 1 & 0 \\ 0 & 0 \end{pmatrix} \tag{4.300}$$

Here we consider the matrix bialgebra $S03$, which we obtain by applying the RTT relations of [272]:

$$R\, T_1\, T_2 = T_2\, T_1\, R\,, \tag{4.301}$$

where $T_1 = T \otimes \mathbf{1}_2$, $T_2 = \mathbf{1}_2 \otimes T$, for the case when $R = R_{S0,3}$, [339]:

$$R_{S0,3} \equiv \begin{pmatrix} 1 & 0 & 0 & 1 \\ 0 & 1 & 1 & 0 \\ 0 & 1 & -1 & 0 \\ -1 & 0 & 0 & 1 \end{pmatrix} \tag{4.302}$$

The relations which follow from (4.301) and (4.302) are:

$$\begin{aligned} b^2 + c^2 &= 0, & a^2 - d^2 &= 0, \\ cd &= ba, & dc &= -ab, \\ bd &= ca, & db &= -ac, \\ da &= ad, & cb &= -bc. \end{aligned} \tag{4.303}$$

In terms of the generators $\tilde{a}, \tilde{b}, \tilde{c}, \tilde{d}$ we have:

$$\begin{aligned} \tilde{b}^2 = \tilde{c}^2 &= 0, & \tilde{a}\tilde{d} = \tilde{d}\tilde{a} &= 0, \\ \tilde{a}\tilde{b} &= 0, & \tilde{b}\tilde{d} &= 0, \\ \tilde{d}\tilde{c} &= 0, & \tilde{c}\tilde{a} &= 0. \end{aligned} \tag{4.304}$$

In view of the above relations we conclude that this bialgebra has no PBW basis. Indeed, the ordering following from (4.304) is cyclic:

$$\tilde{a} > \tilde{c} > \tilde{d} > \tilde{b} > \tilde{a}. \tag{4.305}$$

Thus, the basis consists of building blocks like $\tilde{a}^k \hat{\mathscr{C}} \tilde{d}^\ell \tilde{b}$ and cyclic. Explicitly the basis can be described by the following monomials:

$$\tilde{a}^{k_1} \tilde{c} \tilde{d}^{\ell_1} \tilde{b} \cdots \tilde{a}^{k_n} \tilde{c} \tilde{d}^{\ell_n} \tilde{b} \tilde{a}^{k_{n+1}}, \qquad n, k_i, \ell_i \in \mathbb{Z}_+ \tag{4.306a}$$

$$\tilde{d}^{\ell_1} \tilde{b} \tilde{a}^{k_1} \tilde{c} \cdots \tilde{d}^{\ell_n} \tilde{b} \tilde{a}^{k_n}, \qquad n, k_i, \ell_i \in \mathbb{Z}_+ \tag{4.306b}$$

$$\tilde{a}^{k_1} \tilde{c} \tilde{d}^{\ell_1} \tilde{b} \cdots \tilde{a}^{k_n} \tilde{c} \tilde{d}^{\ell_n}, \qquad n, k_i, \ell_i \in \mathbb{Z}_+ \tag{4.306c}$$

$$\tilde{d}^{\ell_1} \tilde{b} \tilde{a}^{k_1} \tilde{c} \cdots \tilde{d}^{\ell_n} \tilde{b} \tilde{a}^{k_n} \tilde{c} \tilde{d}^{\ell_{n+1}}, \qquad n, k_i, \ell_i \in \mathbb{Z}_+ \tag{4.306d}$$

We shall call the elements of the basis "words". The one-letter words are the generators $\tilde{a}, \tilde{b}, \tilde{c}, \tilde{d}$; they are obtained from (4.306a), (4.306b), (4.306c), and (4.306d), respectively, for $n = 0, k_1 = 1$, $n = 1, k_1 = \ell_1 = 0$, $n = 1, k_1 = \ell_1 = 0$, $n = 0, \ell_1 = 1$, respectively. The unit element $1_{\mathscr{A}}$ is obtained from (4.306b) or (4.306c) for $n = 0$.

4.8.6.1 Dual Algebra

Let us denote by $s03$ the unknown yet dual algebra of $S03$, and by $\tilde{A}, \tilde{B}, \tilde{C}, \tilde{D}$ the four generators of $s03$. Like in [209] we define the pairing $\langle Z, f \rangle$, $Z = \tilde{A}, \tilde{B}, \tilde{C}, \tilde{D}$, f is from (4.306), as the classical tangent vector at the identity (4.186) $\langle Z, f \rangle = \varepsilon\left(\frac{\partial f}{\partial y}\right)$, where $(Z, y) = (\tilde{A}, \tilde{a}), (\tilde{B}, \tilde{b}), (\tilde{C}, \tilde{c}), (\tilde{D}, \tilde{d})$. Explicitly, we get:

$$\langle \tilde{A}, f \rangle = \varepsilon\left(\frac{\partial f}{\partial \tilde{a}}\right) = \begin{cases} k & \text{for } f = \tilde{a}^k \\ 0 & \text{otherwise} \end{cases} \tag{4.307a}$$

$$\langle \tilde{B}, f \rangle = \varepsilon\left(\frac{\partial f}{\partial \tilde{b}}\right) = \begin{cases} 1 & \text{for } f = \tilde{b}\tilde{a}^k \\ 0 & \text{otherwise} \end{cases} \tag{4.307b}$$

$$\langle \tilde{C}, f \rangle = \varepsilon\left(\frac{\partial f}{\partial \tilde{c}}\right) = \begin{cases} 1 & \text{for } f = \tilde{a}^k\tilde{c} \\ 0 & \text{otherwise} \end{cases} \tag{4.307c}$$

$$\langle \tilde{D}, f \rangle = \varepsilon\left(\frac{\partial f}{\partial \tilde{d}}\right) = \begin{cases} 1 & \text{for } f = \tilde{d} \\ 0 & \text{otherwise} \end{cases} \tag{4.307d}$$

Using the above we obtain:

Proposition 9. *The generators $\tilde{A}, \tilde{B}, \tilde{C}, \tilde{D}$ introduced above obey the following relations:*

$$\begin{aligned}
&[\tilde{A}, Z] = 0, \quad Z = \tilde{B}, \tilde{C}, \\
&\tilde{A}\tilde{D} = \tilde{D}\tilde{A} = \tilde{D}^3 = \tilde{B}^2\tilde{D} = \tilde{D}\tilde{B}^2 = \tilde{D}, \\
&\tilde{D}\tilde{B} = -\tilde{B}\tilde{D} = \tilde{C}\tilde{D}^2 = \tilde{D}^2\tilde{C}, \\
&[\tilde{B}, \tilde{C}] = -2\tilde{D}, \quad \{\tilde{C}, \tilde{D}\} = 0, \\
&\tilde{B}^2 + \tilde{C}^2 = 0, \quad \tilde{B}^3 = \tilde{B}, \\
&\tilde{C}^3 = -\tilde{C}, \quad \tilde{B}^2\tilde{A} = \tilde{A}.
\end{aligned} \tag{4.308}$$

$$\begin{aligned}
\delta_{\mathcal{U}}(\tilde{A}) &= \tilde{A} \otimes 1_{\mathcal{U}} + 1_{\mathcal{U}} \otimes \tilde{A} \\
\delta_{\mathcal{U}}(\tilde{B}) &= \tilde{B} \otimes 1_{\mathcal{U}} + (1_{\mathcal{U}} - \tilde{B}^2) \otimes \tilde{B} \\
\delta_{\mathcal{U}}(\tilde{C}) &= \tilde{C} \otimes (1_{\mathcal{U}} - \tilde{B}^2) + 1_{\mathcal{U}} \otimes \tilde{C} \\
\delta_{\mathcal{U}}(\tilde{D}) &= \tilde{D} \otimes (1_{\mathcal{U}} - \tilde{B}^2) + (1_{\mathcal{U}} - \tilde{B}^2) \otimes \tilde{D} \\
\varepsilon_{\mathcal{U}}(Z) &= 0, \qquad Z = \tilde{A}, \tilde{B}, \tilde{C}, \tilde{D}.
\end{aligned} \tag{4.309}$$

$\tilde{A}$, $\tilde{B}^2 = -\tilde{C}^2$ and $\tilde{D}^2$ are Casimir operators. The bialgebra s03 is not a Hopf algebra. For the Proof we refer to [50]. ◊

Corollary: The algebra generated by the generator $\tilde{A}$ is a sub-bialgebra of $s03$. The algebra $s03'$ generated by the generators $\tilde{B}, \tilde{C}, \tilde{D}$ is a nine-dimensional subbialgebra of $s03$ with PBW basis:

$$1_{\mathcal{U}}, \ \tilde{B}, \ \tilde{C}, \ \tilde{D}, \ \tilde{B}\tilde{C}, \ \tilde{B}\tilde{D}, \ \tilde{D}\tilde{C}, \ \tilde{B}^2, \ \tilde{D}^2. \tag{4.310}$$

Proof. The statement follows immediately from relations (4.308, 4.309). We comment only the PBW basis of the subalgebra $s03'$. Indeed, a priori it has a PBW basis:

$$\tilde{B}^k\,\tilde{D}^\ell\,\tilde{C}^m,\qquad k,\ell\le 2,\ m\le 1 \tag{4.311}$$

the restrictions following from (4.308). Furthermore it is easy to see that there are no cubic (and consequently higher order) elements of the basis. For some of the cubic elements this is clear from (4.308). For the rest we have:

$$\begin{aligned}\tilde{B}\tilde{D}\tilde{C} &= -\tilde{D}^2\tilde{C}^2 = \tilde{D}^2\tilde{B}^2 = \tilde{D}^2 \\ \tilde{B}^2\tilde{C} &= -\tilde{C}^3 = \tilde{C} \\ \tilde{B}\tilde{D}^2 &= -\tilde{C}\tilde{D}^3 = \tilde{D}\tilde{C}\end{aligned} \tag{4.312}$$

also using (4.308). Thus, the basis is given by (4.310) that the algebra is indeed nine-dimensional. ◊

Remark 4.3. The algebra $s03$ is not the direct sum of the two subalgebras described in the preceding corollary since both subalgebras have nontrivial action on each other; for example, $\tilde{B}^2\tilde{A} = \tilde{A}$, $\tilde{A}\tilde{D} = \tilde{D}$. The algebra $s03'$ is a nine-dimensional associative algebra over the central algebra generated by $\tilde{A}$. ◊

4.8.6.2 Regular Representation

We start with the study of the left regular representation (LRR) of the subalgebra $s03'$. For this we need the left multiplication table:

	1	$\tilde{B}$	$\tilde{C}$	$\tilde{B}^2$	$\tilde{B}\tilde{C}$	...
$\tilde{B}$	$\tilde{B}$	$\tilde{B}^2$	$\tilde{B}\tilde{C}$	$\tilde{B}$	$\tilde{C}$	...
$\tilde{C}$	$\tilde{C}$	$\tilde{B}\tilde{C}+2\tilde{D}$	$-\tilde{B}^2$	$\tilde{C}$	$-\tilde{B}+2\tilde{D}\tilde{C}$	...
$\tilde{D}$	$\tilde{D}$	$-\tilde{B}\tilde{D}$	$\tilde{D}\tilde{C}$	$\tilde{D}$	$-\tilde{D}^2$	...

	...	$\tilde{D}$	$\tilde{D}^2$	$\tilde{B}\tilde{D}$	$\tilde{D}\tilde{C}$
$\tilde{B}$	...	$\tilde{B}\tilde{D}$	$\tilde{D}\tilde{C}$	$\tilde{D}$	$\tilde{D}^2$
$\tilde{C}$	...	$-\tilde{D}\tilde{C}$	$-\tilde{B}\tilde{D}$	$\tilde{D}^2$	$\tilde{D}$
$\tilde{D}$	...	$\tilde{D}^2$	$\tilde{D}$	$-\tilde{D}\tilde{C}$	$-\tilde{B}\tilde{D}$

The LRR hence contains the subrepresentation generated as a vector space by $\{\tilde{D}, \tilde{D}^2, \tilde{B}\tilde{D}, \tilde{D}\tilde{C}\}$, which decomposes into two two-dimensional irreps:

$$v_0^1 = \tilde{D} + \tilde{D}^2 \,, \quad v_1^1 = \tilde{B}\tilde{D} + \tilde{D}\tilde{C} \,, \tag{4.313}$$
$$\tilde{B}\begin{pmatrix} v_0^1 \\ v_1^1 \end{pmatrix} = \begin{pmatrix} v_1^1 \\ v_0^1 \end{pmatrix}, \quad \tilde{C}\begin{pmatrix} v_0^1 \\ v_1^1 \end{pmatrix} = \begin{pmatrix} -v_1^1 \\ v_0^1 \end{pmatrix}, \quad \tilde{D}\begin{pmatrix} v_0^1 \\ v_1^1 \end{pmatrix} = \begin{pmatrix} v_0^1 \\ -v_1^1 \end{pmatrix}$$
$$v_0^2 = \tilde{B}\tilde{D} - \tilde{D}\tilde{C} \,, \quad v_1^2 = \tilde{D} - \tilde{D}^2 \,, \tag{4.314}$$
$$\tilde{B}\begin{pmatrix} v_0^2 \\ v_1^2 \end{pmatrix} = \begin{pmatrix} v_1^2 \\ v_0^2 \end{pmatrix}, \quad \tilde{C}\begin{pmatrix} v_0^2 \\ v_1^2 \end{pmatrix} = \begin{pmatrix} -v_1^2 \\ v_0^2 \end{pmatrix}, \quad \tilde{D}\begin{pmatrix} v_0^2 \\ v_1^2 \end{pmatrix} = \begin{pmatrix} v_0^2 \\ -v_1^2 \end{pmatrix}.$$

These two irreps are isomorphic by the map $(v_0^1, v_1^1) \to (v_0^2, v_1^2)$. On both of them the Casimirs $\tilde{B}^2, \tilde{D}^2$ take the value 1. (Also the Casimir $\tilde{A}$ of $s03$ has the value 1.)

The LRR contains also the trivial one-dimensional representation generated by the vector $v = \tilde{B}^2 - 1_{\mathcal{U}}$. On this vector all Casimirs and moreover all generators of $s03$ take the value 0.

The quotient of the LRR by the above three submodules has the following multiplication table:

	1	$\tilde{B}$	$\tilde{C}$	$\tilde{B}\tilde{C}$
$\tilde{B}$	$\tilde{B}$	$\tilde{B}^2$	$\tilde{B}\tilde{C}$	$\tilde{C}$
$\tilde{C}$	$\tilde{C}$	$\tilde{B}\tilde{C}$	$-\tilde{B}^2$	$-\tilde{B}$
$\tilde{D}$	0	0	0	0

Thus the quotient decomposes into a direct sum of four one-dimensional representations, generated as vector spaces by

$$v_{\epsilon,\epsilon'} = \tilde{B} + \epsilon 1_{\mathcal{U}} - i\epsilon\epsilon'\tilde{C} - i\epsilon'\tilde{B}\tilde{C} \,, \qquad \epsilon, \epsilon' = \pm. \tag{4.315}$$

On the latter vectors we have the following action:

$$\tilde{B}v_{\epsilon,\epsilon'} = \epsilon v_{\epsilon,\epsilon'} \,, \qquad \tilde{C}v_{\epsilon,\epsilon'} = i\epsilon' v_{\epsilon,\epsilon'} \,, \qquad \tilde{D}v_{\epsilon,\epsilon'} = 0 \,. \tag{4.316}$$

Obviously, on all of them the Casimirs $\tilde{B}^2, \tilde{D}^2$ take the values 1, 0, respectively. However, these four representations are not isomorphic to each other.

To summarize, there are seven irreps of $s03'$ which are obtained from the LRR:
- one-dimensional trivial (all generators act by zero)
- two-dimensional with both Casimirs $\tilde{B}^2, \tilde{D}^2$ having value 1
- four one-dimensional with Casimir values 1, 0 for $\tilde{B}^2, \tilde{D}^2$, respectively

Turning to the algebra $s03$ we note that it inherits the representation structure of its subalgebra $s03'$. On the representations (4.313,4.314) the Casimir $\tilde{A}$ has the value 1, while on the trivial irrep generated by $v = \tilde{B}^2 - 1_{\mathcal{U}}$ the Casimir $\tilde{A}$ has the value 0.

However, on the one-dimensional irreps generated by (4.315) the Casimir $\tilde{A}$ has no fixed value. Thus, the list of the irreps of $s03$ arising from the LRR is:
- one-dimensional trivial
- two-dimensional with all Casimirs $\tilde{A}, \tilde{B}^2, \tilde{D}^2$ having value 1
- four one-dimensional with Casimir values $\mu, 1, 0$ for $\tilde{A}, \tilde{B}^2, \tilde{D}^2$, respectively, $\mu \in \mathbb{C}$

Finally, we note that we could have studied also the right regular representation of $s03$. The list of irreps would be the same as the above one.

4.8.6.3 Weight Representations

Here we consider *weight representations*. These are representations which are built from the action of the algebra on a *weight vector* with respect to one of the generators. We start with a weight vector v_0 such that:

$$\tilde{D}\, v_0 = \lambda\, v_0 \tag{4.317}$$

where $\lambda \in \mathbb{C}$ is the weight. As we shall see the cases $\lambda \neq 0$ and $\lambda = 0$ are very different.

We start with $\lambda \neq 0$. In that case from from $\tilde{D}^3 = \tilde{D}$ follows that $\lambda^2 = 1$, while from $\tilde{B}^2\tilde{D} = \tilde{D}$ follows that $\tilde{B}^2 v_0 = v_0$. Further, from (4.308d) follows that $\tilde{C}\, v_0 = -\lambda\, \tilde{B}\, v_0$. Thus, acting with the elements of $s03$ on v_0 we obtain a two-dimensional representation, for example:

$$v_0\,, \tilde{B}\, v_0\,, \tag{4.318}$$

(and we could have chosen $v_0\,, \tilde{C}\, v_0$ as its basis). This representation is irreducible. The action is given as follows:

	v_0	$\tilde{B}v_0$
$\tilde{B}$	$\tilde{B}v_0$	v_0
$\tilde{C}$	$-\lambda\tilde{B}v_0$	λv_0
$\tilde{D}$	λv_0	$-\lambda\tilde{B}v_0$

Both Casimirs $\tilde{B}^2, \tilde{D}^2$ take the value 1.

Let now $\lambda = 0$. In this case acting with the elements of $s03$ on v_0 we obtain a five-dimensional representation:

$$v_0\,, \tilde{B}\, v_0\,, \tilde{C}\, v_0\,, \tilde{B}\tilde{C}\, v_0\,, \tilde{B}^2\, v_0. \tag{4.319}$$

This representation is reducible. It has a one-dimensional subrepresentation spanned by the vector $w = v\, v_0 = (\tilde{B}^2 - 1_{\mathcal{U}})v_0$. This is the trivial representation since all generators act by zero on it. After we factor out this representation the factor representation splits into four one-dimensional representations spanned by the following vectors $w_{\epsilon,\epsilon'} = v_{\epsilon,\epsilon'} v_0$, where $v_{\epsilon,\epsilon'}$ is from (4.315) and the action of the generators is as given in (4.316). Thus, these irreps are as those obtained from the LRR.

To summarize, there are six irreps of $s03'$ which are obtained as weight irreps of the generator $\tilde{D}$:

- one-dimensional trivial
- one two-dimensional with both Casimirs $\tilde{B}^2, \tilde{D}^2$ having value 1
- four one-dimensional with Casimir values 1, 0 for $\tilde{B}^2, \tilde{D}^2$, respectively

Turning to the algebra $s03$, we note that it inherits the representation structure of its subalgebra $s03'$; however, the value of the Casimir $\tilde{A}$ is not fixed except on the trivial irrep. Thus, the list of the irreps of $s03$ which are obtained as weight irreps of the generator $\tilde{D}$ is:

- one-dimensional trivial
- one two-dimensional with Casimir values $\mu, 1, 1$ for $\tilde{A}, \tilde{B}^2, \tilde{D}^2$, respectively, $\mu \in \mathbb{C}$
- four one-dimensional with Casimir values $\mu, 1, 0$ for $\tilde{A}, \tilde{B}^2, \tilde{D}^2$, respectively, $\mu \in \mathbb{C}$

Finally, we note that it is not possible to construct weight representations w.r.t. generator $\tilde{B}$ (or $\tilde{C}$).

4.8.6.4 Representations of *s03* on *S03*

Here we shall study the representations of $s03$ obtained by the use of its right regular representation (RRR) on the dual bialgebra $S03$. The RRR is defined as follows:

$$\begin{aligned} &\pi_R(Z)f \equiv f_{(1)}\langle Z, f_{(2)}\rangle\,, \qquad Z \in s03\,,\ Z \neq 1_{\mathcal{U}}\,, \quad f \in S03\,, \\ &\pi_R(1_{\mathcal{U}})f \equiv f\,, \qquad f \in S03\,, \end{aligned} \tag{4.320}$$

where we use Sweedler's notation for the coproduct: $\delta(f) = f_{(1)} \otimes f_{(2)}$. (Note that we cannot use the left regular action since that would be given by the formula: $\pi_L(Z)f = \langle \gamma_{\mathcal{U}}(Z), f_{(1)}\rangle f_{(2)}$ and we do not have an antipode.) More explicitly, for the generators of $s03$ we have:

$$\begin{aligned} \pi_R(\tilde{A})\begin{pmatrix} \tilde{a} & \tilde{b} \\ \tilde{c} & \tilde{d} \end{pmatrix} &= \begin{pmatrix} \tilde{a} & \tilde{b} \\ \tilde{c} & \tilde{d} \end{pmatrix} \\ \pi_R(\tilde{B})\begin{pmatrix} \tilde{a} & \tilde{b} \\ \tilde{c} & \tilde{d} \end{pmatrix} &= \begin{pmatrix} \tilde{b} & \tilde{a} \\ \tilde{d} & \tilde{c} \end{pmatrix} \\ \pi_R(\tilde{C})\begin{pmatrix} \tilde{a} & \tilde{b} \\ \tilde{c} & \tilde{d} \end{pmatrix} &= \begin{pmatrix} -\tilde{c} & \tilde{d} \\ \tilde{a} & -\tilde{b} \end{pmatrix} \end{aligned} \tag{4.321}$$

$$\pi_R(\tilde D)\begin{pmatrix}\tilde a & \tilde b\\ \tilde c & \tilde d\end{pmatrix} = \begin{pmatrix}\tilde d & -\tilde c\\ -\tilde b & \tilde a\end{pmatrix}$$
$$\pi_R(Z)\,1_{\mathscr A} = 1_{\mathscr A}\,\langle Z, 1_{\mathscr A}\rangle = 1_{\mathscr A}\,\varepsilon_{\mathscr U}(Z) = 0\,, \qquad Z = \tilde A, \tilde B, \tilde C, \tilde D.$$

For the action on the elements (words) of $SO3$ we use a corollary of (4.320):

$$\pi_R(Z)fg = \pi_R(\delta_{\mathscr U}(Z))(f \otimes g), \tag{4.322}$$

where f, g are arbitrary words from (4.306). Further we shall need the notion of the "length" $\ell(f)$ of the word f. It is defined naturally as the number of the letters of f; in addition we set $\ell(1_{\mathscr A}) = 0$. Now we obtain from (4.322):

$$\pi_R(\tilde A)f = \ell(f)f \tag{4.323a}$$
$$\pi_R(\tilde B)f\cdot g = (\pi_R(\tilde B)f)\cdot g \tag{4.323b}$$
$$\pi_R(\tilde C)f\cdot g = f\cdot(\pi_R(\tilde C)g) \tag{4.323c}$$
$$\pi_R(\tilde D)f = 0\,, \qquad \text{if } \ell(f) > 1 \tag{4.323d}$$

From (4.323b,c) it is obvious that the only nonzero action of $\tilde B, \tilde C$ actually is:

$$\pi_R(\tilde B)\begin{pmatrix}\tilde a & \tilde b\\ \tilde c & \tilde d\end{pmatrix}\cdot f = \begin{pmatrix}\tilde b & \tilde a\\ \tilde d & \tilde c\end{pmatrix}\cdot f \tag{4.324a}$$
$$\pi_R(\tilde C)f\cdot\begin{pmatrix}\tilde a & \tilde b\\ \tilde c & \tilde d\end{pmatrix} = f\cdot\begin{pmatrix}-\tilde c & \tilde d\\ \tilde a & -\tilde b\end{pmatrix} \tag{4.324b}$$

From (4.323a) it is obvious that we can classify the irreps by the value μ_A of the Casimir $\tilde A$ which runs over the non-negative integers. For fixed μ_A the basis of the corresponding representations is spanned by the words f such that $\ell(f) = \mu_A$. Thus, we have:

- $\mu_A = 0$

  This is the one-dimensional trivial representation spanned by the unit element $1_{\mathscr A}$ on which all generators of $s03$ have zero action.

- $\mu_A = 1$

  This representation is four-dimensional spanned by the four generators $\tilde a, \tilde b, \tilde c, \tilde d$ of $SO3$. It is reducible and decomposes in two two-dimensional irreps with basis vectors:

$$v_0^1 = \tilde a + \tilde d = a\,, \quad v_1^1 = \tilde b + \tilde c = b\,, \tag{4.325a}$$
$$v_0^2 = \tilde b - \tilde c = c\,, \quad v_1^2 = \tilde a - \tilde d = d. \tag{4.325b}$$

The RRR of $\tilde{B}, \tilde{C}, \tilde{D}$ on these vectors is as (4.313,4.314):

$$\pi_R(\tilde{B})\begin{pmatrix} v_0^k \\ v_1^k \end{pmatrix} = \begin{pmatrix} v_1^k \\ v_0^k \end{pmatrix}, \qquad \pi_R(\tilde{C})\begin{pmatrix} v_0^k \\ v_1^k \end{pmatrix} = \begin{pmatrix} -v_1^k \\ v_0^k \end{pmatrix},$$

$$\pi_R(\tilde{D})\begin{pmatrix} v_0^k \\ v_1^k \end{pmatrix} = \begin{pmatrix} v_0^k \\ -v_1^k \end{pmatrix} \tag{4.326}$$

These two irreps are isomorphic by the map $(v_0^1, v_1^1) \to (v_0^2, v_1^2)$. On both of them the Casimirs $\tilde{B}^2, \tilde{D}^2$ take the value 1.

– $\mu_A = 2$

This representation is eight-dimensional spanned by $\tilde{a}^2, \tilde{a}\tilde{c}, \tilde{b}\tilde{a}, \tilde{b}\tilde{c}, \tilde{c}\tilde{b}, \tilde{c}\tilde{d}, \tilde{d}^2, \tilde{d}\tilde{b}$. It is reducible and decomposes in eight one-dimensional irreps with basis vectors:

$$v^1_{\epsilon,\epsilon'} = (\tilde{a} + \epsilon\tilde{b})(\tilde{a} + i\epsilon'\tilde{c}) \tag{4.327a}$$

$$v^2_{\epsilon,\epsilon'} = (\tilde{d} + \epsilon\tilde{c})(\tilde{d} + i\epsilon'\tilde{b}) \tag{4.327b}$$

$$\epsilon, \epsilon' = \pm$$

The RRR of $\tilde{B}, \tilde{C}, \tilde{D}$ on these vectors is as (4.316):

$$\pi_R(\tilde{B})v^k_{\epsilon,\epsilon'} = \epsilon v^k_{\epsilon,\epsilon'}, \qquad \pi_R(\tilde{C})v^k_{\epsilon,\epsilon'} = i\epsilon' v^k_{\epsilon,\epsilon'}, \qquad \pi_R(\tilde{D})v^k_{\epsilon,\epsilon'} = 0\,. \tag{4.328}$$

The irrep with vector $v^1_{\epsilon,\epsilon'}$ is isomorphic to the irrep with vector $v^2_{\epsilon,\epsilon'}$. Thus, there are only four distinct irreps parametrized by ϵ, ϵ'. On all of them the Casimirs $\tilde{B}^2, \tilde{D}^2$ take the value 1,0, respectively.

– $\mu_A = N > 2$

These representations are reducible and decompose in one-dimensional irreps with basis vectors:

$$v^1_{\epsilon,\epsilon'} = (\tilde{a} + \epsilon\tilde{b}) \cdot f_1 \cdot (\tilde{a} + i\epsilon'\tilde{c}) \tag{4.329a}$$

$$v^2_{\epsilon,\epsilon'} = (\tilde{d} + \epsilon\tilde{c}) \cdot f_2 \cdot (\tilde{d} + i\epsilon'\tilde{b}) \tag{4.329b}$$

$$v^3_{\epsilon,\epsilon'} = (\tilde{a} + \epsilon\tilde{b}) \cdot f_3 \cdot (\tilde{d} + i\epsilon'\tilde{b}) \tag{4.329c}$$

$$v^4_{\epsilon,\epsilon'} = (\tilde{d} + \epsilon\tilde{c}) \cdot f_4 \cdot (\tilde{a} + i\epsilon'\tilde{c}) \tag{4.329d}$$

$$\epsilon, \epsilon' = \pm\,, \qquad \ell(f_k) = N - 2$$

The RRR of $\tilde{B}, \tilde{C}, \tilde{D}$ on these vectors is as exactly as (4.316). The irrep with vector $v^k_{\epsilon,\epsilon'}$ is isomorphic to the irrep with vector $v^n_{\epsilon,\epsilon'}$. Thus, there are only four distinct irreps as in the case above. On all of them the Casimirs $\tilde{B}^2, \tilde{D}^2$ take the value 1,0, respectively.

To summarize the list of irreps of $s03'$ is the same as given in Section 4.8.6.2. The list of irreps of $s03$ here is smaller since the Casimir $\tilde{A}$ can take only non-negative integer values. Thus, the list of the irreps of $s03$ using the dual bialgebra $S03$ as carrier space is:

- one-dimensional trivial
- two-dimensional with all Casimirs $\tilde{A}, \tilde{B}^2, \tilde{D}^2$ having value 1
- four one-dimensional with Casimir values $\mu, 1, 0$ for $\tilde{A}, \tilde{B}^2, \tilde{D}^2$, respectively, $\mu \in \mathbb{N} + 1$

The difference in the two lists is natural since here more structure (the co-product) is involved. Speaking more loosely the irreps here may be looked upon as “integrals” of the irreps obtained in Section 4.8.6.2.

4.8.7 Exotic Bialgebras: Nontriangular Case *S*14

In this subsection we consider the matrix bialgebra $S14$. We obtain it by applying the RTT relation (4.223) for the case $R = R_{S1,4}$, when $q^2 \neq 1$ where:

$$R_{S1,4} \equiv \begin{pmatrix} 0 & 0 & 0 & q \\ 0 & 0 & 1 & 0 \\ 0 & 1 & 0 & 0 \\ q & 0 & 0 & 0 \end{pmatrix} \tag{4.330}$$

This R-matrix is given in [339].

The relations which follow from (4.223) and (4.330) when $q^2 \neq 1$ are:

$$\begin{aligned} b^2 - c^2 &= 0\,, & a^2 - d^2 &= 0 \\ ab = ba &= 0\,, & ac = ca &= 0 \\ bd = db &= 0\,, & cd = dc &= 0. \end{aligned} \tag{4.331}$$

In terms of the generators $\tilde{a}, \tilde{b}, \tilde{c}, \tilde{d}$

$$\begin{aligned} \tilde{b}\tilde{c} + \tilde{c}\tilde{b} &= 0 & \tilde{a}\tilde{d} + \tilde{d}\tilde{a} &= 0 \\ \tilde{a}\tilde{b} = \tilde{b}\tilde{a} &= 0 & \tilde{a}\tilde{c} = \tilde{c}\tilde{a} &= 0 \\ \tilde{b}\tilde{d} = \tilde{d}\tilde{b} &= 0 & \tilde{c}\tilde{d} = \tilde{d}\tilde{c} &= 0. \end{aligned} \tag{4.332}$$

From the above relations it is clear that the PBW basis of $S14$ is:

$$\tilde{a}^k \tilde{d}^\ell\,, \quad \tilde{b}^k \tilde{c}^\ell. \tag{4.333}$$

4.8.7.1 Dual Algebra

Let us denote by $s14$ the unknown yet dual algebra of $S14$, and by $\tilde{A}, \tilde{B}, \tilde{C}, \tilde{D}$ the four generators of $s14$. We define the pairing as (4.186):$\langle Z, f\rangle$, $Z = \tilde{A}, \tilde{B}, \tilde{C}, \tilde{D}$, $f$is from (4.333). Explicitly, we obtain:

$$\langle \tilde{A}, f \rangle = \varepsilon\left(\frac{\partial f}{\partial \tilde{a}}\right) = \begin{cases} k\delta_{\ell 0} & f = \tilde{a}^k \tilde{d}^\ell \\ 0 & f = \tilde{b}^k \tilde{c}^\ell \end{cases} \tag{4.334a}$$

$$\langle \tilde{B}, f \rangle = \varepsilon\left(\frac{\partial f}{\partial \tilde{b}}\right) = \begin{cases} 0 & f = \tilde{a}^k \tilde{d}^\ell \\ \delta_{k1}\delta_{\ell 0} & f = \tilde{b}^k \tilde{c}^\ell \end{cases} \tag{4.334b}$$

$$\langle \tilde{C}, f \rangle = \varepsilon\left(\frac{\partial f}{\partial \tilde{c}}\right) = \begin{cases} 0 & f = \tilde{a}^k \tilde{d}^\ell \\ \delta_{k0}\delta_{\ell 1} & f = \tilde{b}^k \tilde{c}^\ell \end{cases} \tag{4.334c}$$

$$\langle \tilde{D}, f \rangle = \varepsilon\left(\frac{\partial f}{\partial \tilde{d}}\right) = \begin{cases} \delta_{\ell 1} & f = \tilde{a}^k \tilde{d}^\ell \\ 0 & f = \tilde{b}^k \tilde{c}^\ell \end{cases} \tag{4.334d}$$

We shall need (as in Section 4.8.2) the auxiliary operator E:

$$\langle E, f\rangle = \begin{cases} 1 & \text{for } f = 1_{\mathcal{A}} \\ 0 & \text{otherwise} \end{cases} \tag{4.335}$$

Using the above we obtain:

Proposition 10. *The generators $\tilde{A}, \tilde{B}, \tilde{C}, \tilde{D}$ introduced above obey the following relations:*

$$\begin{aligned} &\tilde{C} = \tilde{D}\tilde{B} = -\tilde{B}\tilde{D}, \quad [\tilde{A}, \tilde{D}] = 0 \\ &\tilde{A}\tilde{B} = \tilde{B}\tilde{A} = \tilde{D}^2\tilde{B} = \tilde{B}^3 = \tilde{B}, \\ &EZ = ZE = 0\,, \quad Z = \tilde{A}, \tilde{B}, \tilde{D}\,, \end{aligned} \tag{4.336}$$

$$\begin{aligned} \delta_{\mathcal{U}}(\tilde{A}) &= \tilde{A} \otimes 1_{\mathcal{U}} + 1_{\mathcal{U}} \otimes \tilde{A} \\ \delta_{\mathcal{U}}(\tilde{B}) &= \tilde{B} \otimes E + E \otimes \tilde{B} \\ \delta_{\mathcal{U}}(\tilde{D}) &= \tilde{D} \otimes K + 1_{\mathcal{U}} \otimes \tilde{D}\,, \qquad K \equiv (-1)^{\tilde{A}} \\ \delta(E) &= E \otimes E \end{aligned} \tag{4.337}$$

$$\varepsilon_{\mathcal{U}}(Z) = 0\,, \quad Z = \tilde{A}, \tilde{B}, \tilde{D}, \quad \varepsilon_{\mathcal{U}}(E) = 1 \tag{4.338}$$

$\tilde{A}$, $\tilde{B}^2$ and $\tilde{D}^2$ are Casimir operators. The bialgebra s14 is not a Hopf algebra.
For the Proof we refer to [50]. ◊

Corollary 3. The algebra generated by the generator $\tilde{A}$ is a sub-bialgebra of $s14$. The algebra $s14'$ generated by $\tilde{B}, \tilde{D}$ is a subalgebra of $s14$, but is not a sub-bialgebra (cf. (4.337b,c)). It has the following PBW basis:

$$\tilde{B},\ \tilde{B}^2,\ \tilde{D}\tilde{B},\ \tilde{D}\tilde{B}^2,\ \tilde{D}^\ell,\ \ell = 0, 1, 2, \ldots \tag{4.339}$$

where we use the convention $\tilde{D}^0 = 1_{\mathcal{U}}$. ◊

4.8.7.2 Regular Representation

We start with the study of the right regular representation of the subalgebra $s14'$. For this we use the right multiplication table:

	$\tilde{B}$	$\tilde{B}^2$	$\tilde{D}\tilde{B}$	$\tilde{D}\tilde{B}^2$	$\tilde{D}^{2k}$	$\tilde{D}^{2k+1}$
$\tilde{B}$	$\tilde{B}^2$	$\tilde{B}$	$\tilde{D}\tilde{B}^2$	$\tilde{D}\tilde{B}$	$\tilde{B}$	$\tilde{D}\tilde{B}$
$\tilde{D}$	$-\tilde{D}\tilde{B}$	$\tilde{D}\tilde{B}^2$	$-\tilde{B}$	$\tilde{B}^2$	$\tilde{D}^{2k+1}$	$\tilde{D}^{2k+2}$

From the above table follows that there is a four-dimensional subspace spanned by $\tilde{B}, \tilde{B}^2, \tilde{D}\tilde{B}, \tilde{D}\tilde{B}^2$. It is reducible and decomposes into four one-dimensional representations spanned by:

$$v_{\epsilon,\epsilon'} = \tilde{B} + \epsilon\tilde{B}^2 - \epsilon'\tilde{D}\tilde{B} + \epsilon\epsilon'\tilde{D}\tilde{B}^2 \tag{4.340}$$

The action of $\tilde{B}, \tilde{D}$ on these vectors is:

$$\tilde{B}v_{\epsilon,\epsilon'} = \epsilon v_{\epsilon,\epsilon'}\,, \qquad \tilde{D}v_{\epsilon,\epsilon'} = \epsilon' v_{\epsilon,\epsilon'} \tag{4.341}$$

The value of the Casimirs $\tilde{B}^2, \tilde{D}^2$ on these vectors is 1.

The quotient of the RRR by the above submodules has the following multiplication table:

	$\tilde{D}^{2k}$	$\tilde{D}^{2k+1}$
$\tilde{B}$	0	0
$\tilde{D}$	$\tilde{D}^{2k+1}$	$\tilde{D}^{2k+2}$

This representation is reducible. It contains an infinite set of nested submodules $V^n \supset V^{n+1}$, $n = 0, 1, \ldots$, where V^n is spanned by $\tilde{D}^{n+\ell}$, $\ell = 0, 1, \ldots$. Correspondingly there is an infinite set of one-dimensional irreducible factor-modules $F^n \equiv V^n/V^{n+1}$ (generated by $\tilde{D}^n$), which are all isomorphic to the trivial representation since the generators $\tilde{B}, \tilde{D}$ act as zero on them. Thus there are five irreps arising from the RRR of $s14'$:

- one-dimensional trivial
- four one-dimensional with both Casimirs $\tilde{B}^2, \tilde{D}^2$ having value 1

Turning to the algebra $s14$ we note that it inherits the representation structure of its subalgebra $s14'$. On the representations (4.340) the Casimir $\tilde{A}$ has the value 1. However, on the one-dimensional irreps F^n the Casimir $\tilde{A}$ has no fixed value. Thus, the list of the irreps arising from the RRR of $s14$ is:
- one-dimensional with Casimir values μ, 0, 0 for $\tilde{A}, \tilde{B}^2, \tilde{D}^2$, respectively, $\mu \in \mathbb{C}$
- four one-dimensional with all Casimirs $\tilde{A}, \tilde{B}^2, \tilde{D}^2$ having value 1

4.8.7.3 Weight Representations

Here we study weight representations, first w.r.t. $\tilde{D}$, as in (4.317). The resulting representation of $s14'$ is three-dimensional:

$$v_0\,,\tilde{B}\,v_0\,,\tilde{B}^2\,v_0. \tag{4.342}$$

It is reducible and contains one one-dimensional and one two-dimensional irrep:
- one-dimensional $\lambda \in \mathbb{C}$:

$$w_0 = (\tilde{B}^2 - 1_{\mathcal{U}})v_0\,, \tag{4.343}$$

$$\tilde{B}w_0 = 0\,, \qquad \tilde{D}w_0 = \lambda w_0\,, \tag{4.344}$$

- two-dimensional with $\lambda = \pm 1$:

$$\{v_0, v_1 = \tilde{B}\,v_0\} \tag{4.345}$$

$$\tilde{B}\begin{pmatrix} v_0 \\ v_1 \end{pmatrix} = \begin{pmatrix} v_1 \\ v_0 \end{pmatrix}, \qquad \tilde{D}\begin{pmatrix} v_0 \\ v_1 \end{pmatrix} = \lambda\begin{pmatrix} v_0 \\ -v_1 \end{pmatrix} \tag{4.346}$$

Turning to the algebra $s14$ we note that it inherits the representation structure of its subalgebra $s14'$. On the one-dimensional irrep (4.343) the Casimir $\tilde{A}$ has no fixed value since $\tilde{B}$ is trivial, and $[\tilde{A}, \tilde{D}] = 0$. On the two-dimensional irrep (4.345) the Casimir $\tilde{A}$ has the value 1 since $\tilde{A}\tilde{B} = \tilde{B}$.

Thus, there are the following irreps of $s14$ which are obtained as weight irreps of the generator $\tilde{D}$:
- one-dimensional with Casimir values μ, 0, λ^2 for $\tilde{A}, \tilde{B}^2, \tilde{D}^2$, respectively, $\mu, \lambda \in \mathbb{C}$
- two two-dimensional with all Casimirs $\tilde{A}, \tilde{B}^2, \tilde{D}^2$ having the value 1

Next we consider weight representations w.r.t. $\tilde{B}$:

$$\tilde{B}\,v_0 = \nu\,v_0\,, \tag{4.347}$$

with $\nu \in \mathbb{C}$. From $\tilde{B}^3 = \tilde{B}$ follows that $\nu = 0, \pm 1$. Acting with the generators we obtain the following representation vectors: $v_\ell = \tilde{D}^\ell\,v_0$. We have that $\tilde{D}v_\ell = v_{\ell+1}$.

Further we consider first the case $v^2 = 1$. Then we apply the relation $\tilde{D}^2\tilde{B} = \tilde{B}$ to v_ℓ and we get:

$$\tilde{D}^2\tilde{B}v_\ell = (-1)^\ell v v_{\ell+2} = \tilde{B}v_\ell = (-1)^\ell v v_\ell$$

from which follows that we have to identify $v_{\ell+2}$ with v_ℓ. Thus the representation is given as follows:

$$\{v_0, v_1 = \tilde{D}v_0\} \tag{4.348}$$

$$\tilde{B}\begin{pmatrix} v_0 \\ v_1 \end{pmatrix} = v\begin{pmatrix} v_0 \\ -v_1 \end{pmatrix}, \qquad \tilde{D}\begin{pmatrix} v_0 \\ v_1 \end{pmatrix} = \begin{pmatrix} v_1 \\ v_0 \end{pmatrix}$$

On this irrep both Casimirs $\tilde{B}^2, \tilde{D}^2$ have value 1 ($v^2 = 1$).

Further we consider the case $v = 0$. This representation is reducible. It contains an infinite set of nested submodules $V^n \supset V^{n+1}$, $n = 0, 1, \ldots$, where V^n is spanned by $\tilde{D}^{n+\ell}v_0$, $\ell = 0, 1, \ldots$. Correspondingly there is an infinite set of one-dimensional irreducible factor-modules $F^n \equiv V^n/V^{n+1}$ (generated by $\tilde{D}^n v_0$), which are all isomorphic to the trivial representation since the generators $\tilde{B}, \tilde{D}$ act as zero on them.

Turning to the algebra $s14$ we note that it inherits the representation structure of its subalgebra $s14'$, with the value of the Casimir $\tilde{A}$ being not fixed if $\tilde{B}$ acts trivially, and being 1, if $\tilde{B}$ acts non trivially.

Thus, there are the following irreps of $s14$ which are obtained as weight irreps of the generator $\tilde{B}$:

- one-dimensional with Casimir values $\mu, 0, 0$ for $\tilde{A}, \tilde{B}^2, \tilde{D}^2$, respectively, $\mu \in \mathbb{C}$
- two two-dimensional with all Casimirs $\tilde{A}, \tilde{B}^2, \tilde{D}^2$ having the value 1

4.8.7.4 Representations of *s*14 on *S*14

Here we shall study the representations of $s14$ obtained by the use of its right regular representation (RRR) on the dual bialgebra $S14$. The RRR is defined as in (4.320). For the generators of $s14$ we have:

$$\pi_R(\tilde{A})\begin{pmatrix} \tilde{a} & \tilde{b} \\ \tilde{c} & \tilde{d} \end{pmatrix} = \begin{pmatrix} \tilde{a} & \tilde{b} \\ \tilde{c} & \tilde{d} \end{pmatrix} \tag{4.349}$$

$$\pi_R(\tilde{B})\begin{pmatrix} \tilde{a} & \tilde{b} \\ \tilde{c} & \tilde{d} \end{pmatrix} = \begin{pmatrix} \tilde{b} & \tilde{a} \\ \tilde{d} & \tilde{c} \end{pmatrix}$$

$$\pi_R(\tilde{D})\begin{pmatrix} \tilde{a} & \tilde{b} \\ \tilde{c} & \tilde{d} \end{pmatrix} = \begin{pmatrix} \tilde{d} & -\tilde{c} \\ -\tilde{b} & \tilde{a} \end{pmatrix}$$

$$\pi_R(E)\begin{pmatrix} \tilde{a} & \tilde{b} \\ \tilde{c} & \tilde{d} \end{pmatrix} = \begin{pmatrix} 0 & 0 \\ 0 & 0 \end{pmatrix}$$

$$\pi_R(Z)\,1_{\mathcal{A}} = 1_{\mathcal{A}}\,\langle Z, 1_{\mathcal{A}}\rangle = 1_{\mathcal{A}}\,\varepsilon_{\mathcal{U}}(Z) = \begin{cases} 0\,, & Z = \tilde{A}, \tilde{B}, \tilde{D} \\ 1\,, & Z = E \end{cases}$$

For the action on the basis of S14 we use formula (4.322). We obtain:

$$\begin{aligned}
&\pi_R(A)\,\tilde{a}^n\tilde{d}^k = (n+k)\tilde{a}^n\tilde{d}^k\,, \qquad \pi_R(A)\,\tilde{b}^n\tilde{c}^k = (n+k)\tilde{b}^n\tilde{c}^k \\
&\pi_R(B)\,\tilde{a}^n\tilde{d}^k = \delta_{k0}\delta_{n1}\tilde{b} + \delta_{n0}\delta_{k1}\tilde{c}\,, \qquad \pi_R(B)\,\tilde{b}^n\tilde{c}^k = \delta_{k0}\delta_{n1}\tilde{a} + \delta_{n0}\delta_{k1}\tilde{d} \\
&\pi_R(D)\,\tilde{a}^k\tilde{d}^\ell = (-1)^{\ell+1}\ell\tilde{a}^{k+1}\tilde{d}^{\ell-1} + (-1)^\ell k\tilde{a}^{k-1}\tilde{d}^{\ell+1} \\
&\pi_R(D)\,\tilde{b}^k\tilde{c}^\ell = (-1)^\ell \ell\tilde{b}^{k+1}\tilde{c}^{\ell-1} + (-1)^{\ell+1}k\tilde{b}^{k-1}\tilde{c}^{\ell+1}
\end{aligned} \tag{4.350}$$

We see that similarly to Section 4.8.6.4 the Casimir $\tilde{A}$ acts as the length of the elements of S14, that is, (4.321) holds. Thus, also here we classify the irreps by the value μ_A of the Casimir $\tilde{A}$ which runs over the non-negative integers. For fixed μ_A the basis of the corresponding representations is spanned by the elements f such that $\ell(f) = \mu_A$. The dimension of each such representation is:

$$\dim(\mu_A) = \begin{cases} 2(\mu_A+1) & \text{for } \mu_A \geq 1 \\ 1 & \text{for } \mu_A = 0 \end{cases} \tag{4.351}$$

The classification goes as follows:

– $\mu_A = 0$

This is the one-dimensional trivial representation spanned by $1_{\mathcal{A}}$.

– $\mu_A = 1$

This representation is four-dimensional spanned by the four generators $\tilde{a}, \tilde{b}, \tilde{c}, \tilde{d}$ of S14. It decomposes in two two-dimensional isomorphic to each other irreps with basis vectors as in (4.325) – this is due to the fact that the action (4.349b,c) is the same as the action (4.321). The value of the Casimirs $\tilde{B}^2, \tilde{D}^2$ is 1.

– Each representation for fixed $\mu_A \geq 2$ is reducible and decomposes in two isomorphic representations: one built on the basis $\tilde{a}^k\tilde{d}^\ell$, and the other built on the basis $\tilde{b}^k\tilde{c}^\ell$, each of dimension $\mu_A + 1$. Thus, for $\mu_A \geq 2$ we shall consider only the representations built on the basis $\tilde{a}^k\tilde{d}^\ell$. These representations are also reducible and they all decompose in one-dimensional irreps. Further, the action of $\tilde{B}$ is zero, thus, we speak only about the action of $\tilde{D}$.

– $\mu_A = 2n$, $n = 1, 2, \dots$

For fixed n the representation decomposes into $2n + 1$ one-dimensional irreps. On one of these, which is spanned by the element:

$$w_0 = \sum_{k=0}^{n} \binom{n}{k} \tilde{a}^{2n-2k}\tilde{d}^{2k}\,, \tag{4.352}$$

the generator $\tilde{D}$ acts by zero. The rest of the irreps are enumerated by the parameters: $\pm$, τ, where $\tau = 2, 4, \dots, 2n = \mu_A$, and are spanned by the vectors:

$$
\begin{aligned}
u_\tau^\pm &= u_0 \pm \tau u_1 \,, \\
u_0 &= \sum_{k=0}^{n} \alpha_k \, \tilde{a}^{2n-2k} \tilde{d}^{2k} \,, \quad \alpha_0 = 1 \,, \\
u_1 &= \sum_{k=0}^{n-1} \beta_k \, \tilde{a}^{2n-2k-1} \tilde{d}^{2k+1} \,, \quad \beta_0 = 1 \,,
\end{aligned}
\tag{4.353}
$$

on which $\tilde{D}$ acts by:

$$\pi_R(\tilde{D}) \, u_\tau^\pm = \pm \tau \, u_\tau^\pm \tag{4.354}$$

which follows from:

$$\pi_R(\tilde{D}) \begin{pmatrix} u_0 \\ u_1 \end{pmatrix} = \begin{pmatrix} \tau^2 \, u_1 \\ u_0 \end{pmatrix} \tag{4.355}$$

Note that the value of the Casimir $\tilde{D}^2$ is equal to τ^2. The coefficients $\alpha_k \,, \beta_k$ depend on τ and are fixed from the two recursive equations which follow from (4.355):

$$
\begin{aligned}
\tau^2 \beta_k &= 2(n-k)\alpha_k - 2(k+1)\alpha_{k+1} \,, \quad k = 0, \dots, n-1 \\
\alpha_k &= (2k+1)\beta_k - (2n-2k+1)\beta_{k-1} \,, \quad k = 0, \dots, n,
\end{aligned}
\tag{4.356}
$$

where we set $\beta_{-1} \equiv 0, \beta_n \equiv 0$.

- $\mu_A = 2n+1, \, n = 1, 2, \dots$
  For fixed n the representation is $(2n+2)$-dimensional and decomposes into $2n+2$ irreps which are enumerated by two parameters: $\pm$, τ, where $\tau = 1, 3, 5, \dots, 2n+1 = \mu_A$, and are spanned by the vectors:

$$
\begin{aligned}
w_\tau^\pm &= w_0 \pm \tau w_1 \,, \\
w_0 &= \sum_{k=0}^{n} \alpha'_k \, \tilde{a}^{2n-2k+1} \tilde{d}^{2k} \,, \quad \alpha'_0 = 1 \,, \\
w_1 &= \sum_{k=0}^{n} \beta'_k \, \tilde{a}^{2n-2k} \tilde{d}^{2k+1} \,, \quad \beta'_0 = 1 \,,
\end{aligned}
\tag{4.357}
$$

on which $\tilde{D}$ acts by (4.354). Note that the value of the Casimir $\tilde{D}^2$ is equal to τ^2. The coefficients $\alpha'_k \,, \beta'_k$ are fixed from the two recursive equations which follow from (4.354):

$$
\begin{aligned}
\tau^2 \beta'_k &= (2n-2k+1)\alpha'_k - 2(k+1)\alpha'_{k+1} \,, \quad k = 0, \dots, n; \\
\alpha'_k &= (2k+1)\beta'_k - 2(n-k+1)\beta'_{k-1} \,, \quad k = 0, \dots, n,
\end{aligned}
\tag{4.358}
$$

where we set $\alpha'_{n+1} \equiv 0, \beta'_{-1} \equiv 0$.

To summarize the list of irreps of $s14$ on $S14$ is:

- one-dimensional trivial
- two two-dimensional with all Casimirs $\tilde{A}, \tilde{B}^2, \tilde{D}^2$ having the value 1
- one-dimensional enumerated by $n = 1, 2, \ldots$, which for fixed n have Casimir values $2n, 0, 0$ for $\tilde{A}, \tilde{B}^2, \tilde{D}^2$, respectively
- pairs of one-dimensional irreps enumerated by $n = 1, 2, \ldots\ \tau = 2, 4, \ldots \ldots, 2n$, which have Casimir values $2n, 0, \tau^2$ for $\tilde{A}, \tilde{B}^2, \tilde{D}^2$, respectively.
- pairs of one-dimensional irreps enumerated by $n = 1, 2, \ldots;\ \tau = 1, 3, \ldots \ldots, (2n+1)$, which have Casimir values $2n+1, 0, \tau^2$ for $\tilde{A}, \tilde{B}^2, \tilde{D}^2$, respectively

Finally, we note in the irreps of $s14$ on $S14$ all Casimirs can take only non-negative integer values.

4.8.8 Exotic Bialgebras: Nontriangular Case *S14o*

In this section we consider the matrix bialgebra $S14o$. We obtain it by applying the RTT relations (4.223) for the case $R = R_{S1,4}$ (cf. (4.330)), when $q^2 = 1$. We shall consider the case $q = 1$ (the case $q = -1$ is equivalent, cf. below). For $q = 1$ the relations following from (4.223) and (4.330) are:

$$a^2 = d^2\,, \quad b^2 = c^2 = 0\,, \quad ab = ba = ac = ca = bd = db = cd = dc = 0 \tag{4.359}$$

or in terms of the generators $\tilde{a}, \tilde{b}, \tilde{c}, \tilde{d}$:

$$\tilde{b}\tilde{a} = \tilde{a}\tilde{b}\,, \quad \tilde{c}\tilde{a} = -\tilde{a}\tilde{c}\,, \quad \tilde{d}\tilde{a} = -\tilde{a}\tilde{d}\,, \quad \tilde{c}\tilde{b} = -\tilde{b}\tilde{c}\,, \quad \tilde{d}\tilde{b} = -\tilde{b}\tilde{d}\,, \quad \tilde{d}\tilde{c} = \tilde{c}\tilde{d} \tag{4.360}$$

(The case $q = -1$ is obtained from the above through the exchange $\tilde{b} \leftrightarrow \tilde{c}$.)

From the above relations it is clear that we can choose any ordering of the PBW basis. For definiteness we choose for the PBW basis of $S14o$:

$$\tilde{a}^k \tilde{b}^\ell \tilde{c}^m \tilde{d}^n \tag{4.361}$$

4.8.8.1 Dual Algebra

Let us denote by $s14o$the unknown yet dual algebra of $S14o$, and by $\tilde{A}, \tilde{B}, \tilde{C}, \tilde{D}$ the four generators of $s14o$. We define the pairing $\langle Z, f\rangle$, $Z = \tilde{A}, \tilde{B}, \tilde{C}, \tilde{D}$, $f$is from (4.361), as in (4.186). Explicitly, we obtain:

$$\langle \tilde{A}\,,\, f\,\rangle = \varepsilon\left(\frac{\partial f}{\partial \tilde{a}}\right) = \begin{cases} k & \text{for } f = \tilde{a}^k \\ 0 & \text{otherwise} \end{cases} \tag{4.362a}$$

$$\langle \tilde{B}, f \rangle = \varepsilon\left(\frac{\partial f}{\partial \tilde{b}}\right) = \begin{cases} 1 & \text{for } f = \tilde{a}^k \tilde{b} \\ 0 & \text{otherwise} \end{cases} \tag{4.362b}$$

$$\langle \tilde{C}, f \rangle = \varepsilon\left(\frac{\partial f}{\partial \tilde{c}}\right) = \begin{cases} 1 & \text{for } f = \tilde{a}^k \tilde{c} \\ 0 & \text{otherwise} \end{cases} \tag{4.362c}$$

$$\langle \tilde{D}, f \rangle = \varepsilon\left(\frac{\partial f}{\partial \tilde{d}}\right) = \begin{cases} 1 & \text{for } f = \tilde{a}^k \tilde{d} \\ 0 & \text{otherwise} \end{cases} \tag{4.362d}$$

Using the above we obtain:

Proposition 11. *The generators $\tilde{A}, \tilde{B}, \tilde{C}, \tilde{D}$ introduced above obey the following relations:*

$$\begin{aligned} &[\tilde{A}, Z] = 0\,, \qquad Z = \tilde{B}, \tilde{C}, \tilde{D} \\ &[\tilde{B}, \tilde{C}] = -2\tilde{D}\,, \quad [\tilde{B}, \tilde{D}] = -2\tilde{C}\,, \quad [\tilde{C}, \tilde{D}] = -2\tilde{B} \end{aligned} \tag{4.363}$$

$$\begin{aligned} \delta_{\mathcal{U}}(\tilde{A}) &= \tilde{A} \otimes 1_{\mathcal{U}} + 1_{\mathcal{U}} \otimes \tilde{A} \\ \delta_{\mathcal{U}}(\tilde{B}) &= \tilde{B} \otimes 1_{\mathcal{U}} + 1_{\mathcal{U}} \otimes \tilde{B} \end{aligned} \tag{4.364}$$

$$\begin{aligned} \delta_{\mathcal{U}}(\tilde{C}) &= \tilde{C} \otimes K + 1_{\mathcal{U}} \otimes \tilde{C}\,, \qquad K = (-1)^{\tilde{A}} \\ \delta_{\mathcal{U}}(\tilde{D}) &= \tilde{D} \otimes K + 1_{\mathcal{U}} \otimes \tilde{D} \end{aligned} \tag{4.365}$$

$$\varepsilon_{\mathcal{U}}(Z) = 0\,, \qquad Z = \tilde{A}, \tilde{B}, \tilde{C}, \tilde{D} \tag{4.366}$$

$$\gamma_{\mathcal{U}}(\tilde{A}) = -\tilde{A}\,, \ \gamma_{\mathcal{U}}(\tilde{B}) = -\tilde{B}\,, \ \gamma_{\mathcal{U}}(\tilde{C}) = -\tilde{C}K\,, \ \gamma_{\mathcal{U}}(\tilde{D}) = -\tilde{D}K. \tag{4.367}$$

For the Proof we refer to [50]. ◊

Corollary: The auxiliary generator $K = (-1)^{\tilde{A}}$ is central and $K^{-1} = K$. Its coalgebra relations are:

$$\delta_{\mathcal{U}}(K) = K \otimes K\,, \quad \varepsilon_{\mathcal{U}}(K) = 1\,, \quad \gamma_{\mathcal{U}}(K) = K \ \lozenge \tag{4.368}$$

Corollary: The algebra generated by the generator $\tilde{A}$ is a Hopf subalgebra of $s14o$. The algebra $s14o'$ generated by $\tilde{B}, \tilde{C}, \tilde{D}$ is a subalgebra of $s14o$, but is not a Hopf subalgebra because of the operator K in the coalgebra structure. The algebras $s14o, s14o'$ are isomorphic to $U(gl(2))$, $U(sl(2))$, respectively. The latter is seen from the following:

$$\begin{aligned} &X^{\pm} \equiv \tfrac{1}{2}(\tilde{D} \mp \tilde{C}) \\ &[\tilde{B}, X^{\pm}] = \pm 2X^{\pm}, \quad [X^+, X^-] = \tilde{B}. \end{aligned} \tag{4.369}$$

Indeed the last line presents the standard $sl(2)$ commutation relations. However, the generators $X^{\pm}$ inherit the K dependence in the coalgebra operations:

$$\begin{aligned}\delta_{\mathcal{U}}(X^{\pm}) &= X^{\pm} \otimes K + 1_{\mathcal{U}} \otimes X^{\pm} \\ \varepsilon_{\mathcal{U}}(X^{\pm}) &= 0 \\ \gamma_{\mathcal{U}}(X^{\pm}) &= -X^{\pm}K\end{aligned} \tag{4.370}$$

The algebra $s14o$ is a *graded algebra*:

$$\deg X^{\pm} = \pm 1, \qquad \deg \tilde{A} = \deg \tilde{B} = 0, \qquad (\Longrightarrow \deg K = 0) \ \Diamond \tag{4.371}$$

Based on the above corollary we are able to make the following important observation. The algebra $s14o$ may be identified with a special case of the Hopf algebra $\mathcal{U}_{p,q}$ which was found in [209] as the dual of $GL_{p,q}(2)$ (see Section 4.4 here). To make direct contact with [209], we need to replace there $(p^{1/2}, q^{1/2}) \longrightarrow (p, q)$, then to set $q = p^{-1}$, and at the end to set $p = -1$. (The necessity to set values in such order is clear from, e. g., the formula for the coproduct in (5.21) of [209].) The generators from [209] $K, p^K, H, X^{\pm}$ correspond to $\tilde{A}, K, \tilde{B}, X^{\pm}$ in the notation at hand.

More than this. It turns out that the corresponding algebras in duality, namely, $S14o$ and $GL_{p,q}(2)$ may be identified setting q, p as above. To make this evident we make the following change of generators:

$$\hat{a} = \tilde{a} + \tilde{b}, \quad \hat{b} = \tilde{d} - \tilde{c}, \quad \hat{c} = \tilde{c} + \tilde{d}, \quad \hat{d} = \tilde{a} - \tilde{b}. \tag{4.372}$$

For these generators the commutation relations are:

$$\hat{b}\hat{a} = -\hat{a}\hat{b}, \quad \hat{c}\hat{a} = -\hat{a}\hat{c}, \quad \hat{d}\hat{a} = \hat{a}\hat{d}, \quad \hat{c}\hat{b} = \hat{b}\hat{c}, \quad \hat{b}\hat{d} = -\hat{d}\hat{b}, \quad \hat{c}\hat{d} = -\hat{d}\hat{c} \tag{4.373}$$

that is, exactly those of $GL_{p,q}(2)$ (cf. [183]) for $p = q = -1$. Furthermore the coproduct and and counit are as for $GL_{p,q}(2)$ or $GL(2)$, that is, as in (4.393). For the antipode we have to suppose that the determinant $ad - p^{-1}bc$ from [183], which here becomes (cf. $p = -1$):

$$\omega = \hat{a}\hat{d} + \hat{b}\hat{c}, \tag{4.374}$$

is invertible, or, that $\omega \neq 0$ and we extend the algebra by an element ω^{-1} so that:

$$\begin{aligned}&\omega\omega^{-1} = \omega^{-1}\omega = 1_{\mathcal{A}}, \qquad \delta_{\mathcal{U}}(\omega^{\pm 1}) = \omega^{\pm 1} \otimes \omega^{\pm 1}, \\ &\varepsilon_{\mathcal{U}}(\omega^{\pm 1}) = 1, \, \gamma_{\mathcal{U}}(\omega^{\pm 1}) = \omega^{\mp 1}\end{aligned} \tag{4.375}$$

Then the antipode is given by:

$$\gamma_{\mathscr{U}}\begin{pmatrix}\hat{a} & \hat{b}\\ \hat{c} & \hat{d}\end{pmatrix} = \omega^{-1}\begin{pmatrix}\hat{d} & \hat{b}\\ \hat{c} & \hat{a}\end{pmatrix} \tag{4.376}$$

or in a more compact notation:

$$\gamma_{\mathscr{U}}(M) = M^{-1} \tag{4.377}$$

This relation between $s14o, S14o$ and $\mathscr{U}_{p,q}$, $GL_{p,q}(2)$ was not anticipated since the corresponding R-matrices $R_{S1,4}$ and $R_{S2,1}$ are listed in [339] as different and furthermore nonequivalent. It turns out that this is indeed the case, except in the case we have stumbled upon. To show this we first recall:

$$R_{S2,1} = \begin{pmatrix}1 & 0 & 0 & 0\\ 0 & p & 1-pq & 0\\ 0 & 0 & q & 0\\ 0 & 0 & 0 & 1\end{pmatrix} \tag{4.378}$$

which for $q = p^{-1} = -1$ becomes:

$$R_0 \equiv (R_{S2,1})_{q=p^{-1}=-1} = \begin{pmatrix}1 & 0 & 0 & 0\\ 0 & -1 & 0 & 0\\ 0 & 0 & -1 & 0\\ 0 & 0 & 0 & 1\end{pmatrix} \tag{4.379}$$

Further, we need:

$$R_{\pm} \equiv (R_{S1,4})_{q=\pm 1} = \begin{pmatrix}0 & 0 & 0 & \pm 1\\ 0 & 0 & 1 & 0\\ 0 & 1 & 0 & 0\\ \pm 1 & 0 & 0 & 0\end{pmatrix} \tag{4.380}$$

Now we can show that $R_{\pm}$ can be transformed by “gauge transformations” to R_0, namely, we have:

$$R_0 = (U_{\pm} \otimes U_{\pm})\, R_{\pm}\, (U_{\pm} \otimes U_{\pm})^{-1} \tag{4.381a}$$

$$U_{+} = \frac{1}{\sqrt{2}}\begin{pmatrix}1 & 1\\ 1 & -1\end{pmatrix}, \qquad U_{-} = \frac{1}{\sqrt{2}}\begin{pmatrix}1 & i\\ i & 1\end{pmatrix} \tag{4.381b}$$

In accord with this we have:

$$\hat{T} \equiv \begin{pmatrix} \hat{a} & \hat{b} \\ \hat{c} & \hat{d} \end{pmatrix}, \qquad T \equiv \begin{pmatrix} a & b \\ c & d \end{pmatrix}, \qquad \hat{T} = U_+ \, T \, (U_+)^{-1} \Rightarrow$$

$$\begin{aligned} \hat{a} &= \tfrac{1}{2}(a + b + c + d), & \hat{b} &= \tfrac{1}{2}(a - b + c - d), \\ \hat{c} &= \tfrac{1}{2}(a + b - c - d), & \hat{d} &= \tfrac{1}{2}(a - b - c + d), \end{aligned} \tag{4.382}$$

which is equivalent to substituting (4.298) in (4.372).

The use of U_- would lead to different relations between hatted and unhatted generators, which, however, would not affect the algebra relations. Indeed:

$$\hat{T}' \equiv \begin{pmatrix} \hat{a}' & \hat{b}' \\ \hat{c}' & \hat{d}' \end{pmatrix}, \quad T' \equiv \begin{pmatrix} a' & b' \\ c' & d' \end{pmatrix}, \qquad \hat{T}' = U_- \, T' \, (U_-)^{-1}$$

$$\begin{aligned} \hat{a}' &= \tfrac{1}{2}(a' - ib' + ic' + d'), & \hat{b}' &= \tfrac{1}{2}(-ia' + b' + c' + id'), \\ \hat{c}' &= \tfrac{1}{2}(ia' + b' + c' - id'), & \hat{d}' &= \tfrac{1}{2}(a' + ib' - ic' + d') \end{aligned} \tag{4.383}$$

But this becomes equivalent to (4.382) with the changes:

$$(\hat{a}', i\hat{b}', -i\hat{c}', \hat{d}') \mapsto (\hat{a}, \hat{b}, \hat{c}, \hat{d}), \qquad (a', -ib', ic', d') \mapsto (a, b, c, d) \tag{4.384}$$

while the (inverse) changes (4.384) do not affect (4.373) and (4.359).

Representations of *s14o* on *S14o*

The regular representation of $s14o$ ($s14o'$) on itself and its weight representations are the same as those of $U(gl(2))$ and $(U(sl(2)))$ due to (4.369). The situation is different for the representations of $s14o$ on $S14o$ since these involve the coalgebra structure. However, in treating the representations of $s14o$ on $S14o$ we can use the the relation between $s14o$, $S14o$ and $\mathcal{U}_{p,q}$, $GL_{p,q}(2)$ that we established in the previous subsection. Then we employ the construction for the induced representations of $\mathcal{U}_{p,q}$ on $GL_{p,q}(2)$ from [229] and Section 5.1 below, to which we refer.

4.8.9 Exotic Bialgebras: Higher Dimensions

In the previous sections were exposed the studies [49–52] of our initial collaboration on the algebraic structures coming from 4x4 R-matrices (solutions of the Yang–Baxter equation) that are not deformations of classical ones (i. e., the identity up to signs). More recently, our follow-up collaboration (with Boucif Abdesselam replacing our deceased friend and coauthor Daniel Arnaudon) constructed $N^2 \times N^2$ unitary braid matrices $\widehat{R}$ for $N > 2$ generalizing the class known for $N = 2$ [7, 8].

Here we follow [9] to study of the bialgebras that arise from these higher dimensional unitary braid matrices with the simplest possible case $N = 3$ in order to get the necessary expertise. However, even this case is complicated enough.

4.8.9.1 Preliminaries

Our starting point is the following 9×9 braid matrix from [7]:

$$\widehat{R}(\theta) = \begin{vmatrix} a_+ & 0 & 0 & 0 & 0 & 0 & 0 & 0 & a_- \\ 0 & b_+ & 0 & 0 & 0 & 0 & 0 & b_- & 0 \\ 0 & 0 & a_+ & 0 & 0 & 0 & a_- & 0 & 0 \\ 0 & 0 & 0 & c_+ & 0 & c_- & 0 & 0 & 0 \\ 0 & 0 & 0 & 0 & 1 & 0 & 0 & 0 & 0 \\ 0 & 0 & 0 & c_- & 0 & c_+ & 0 & 0 & 0 \\ 0 & 0 & a_- & 0 & 0 & 0 & a+ & 0 & 0 \\ 0 & b_- & 0 & 0 & 0 & 0 & 0 & b_+ & 0 \\ a_- & 0 & 0 & 0 & 0 & 0 & 0 & 0 & a_+ \end{vmatrix} \tag{4.385}$$

where

$$a_\pm = \tfrac{1}{2}(e^{m_{11}^+\theta} \pm e^{m_{11}^-\theta}), \; b_\pm = \tfrac{1}{2}(e^{m_{21}^+\theta} \pm e^{m_{21}^-\theta}), \; c_\pm = \tfrac{1}{2}(e^{m_{22}^+\theta} \pm e^{m_{22}^-\theta}), \tag{4.386}$$

and $m_{ij}^\pm$ are parameters. The above braid matrix satisfies baxterized braid equation:

$$\widehat{R}_{12}(\theta)\widehat{R}_{23}(\theta + \theta')\widehat{R}_{12}(\theta') = \widehat{R}_{23}(\theta')\widehat{R}_{12}(\theta + \theta')\widehat{R}_{23}(\theta). \tag{4.387}$$

For the RTT relations of Faddeev–Reshetikhin–Takhtajan [272], we need the corresponding baxterized R-matrix, $R = P\hat{R}$ (P is the permutation matrix):

$$R(\theta) = \begin{pmatrix} a_+ & 0 & 0 & 0 & 0 & 0 & 0 & 0 & a_- \\ 0 & 0 & 0 & c_+ & 0 & c_- & 0 & 0 & 0 \\ 0 & 0 & a_- & 0 & 0 & 0 & a_+ & 0 & 0 \\ 0 & b_+ & 0 & 0 & 0 & 0 & 0 & b_- & 0 \\ 0 & 0 & 0 & 0 & 1 & 0 & 0 & 0 & 0 \\ 0 & b_- & 0 & 0 & 0 & 0 & 0 & b_+ & 0 \\ 0 & 0 & a_+ & 0 & 0 & 0 & a_- & 0 & 0 \\ 0 & 0 & 0 & c_- & 0 & c_+ & 0 & 0 & 0 \\ a_- & 0 & 0 & 0 & 0 & 0 & 0 & 0 & a_+ \end{pmatrix}, \tag{4.388}$$

which satisfies the *baxterized Yang–Baxter equation*:

$$R_{12}(\theta)R_{13}(\theta + \theta')R_{23}(\theta') = R_{23}(\theta')R_{13}(\theta + \theta')R_{12}(\theta). \tag{4.389}$$

In fact, we need the solutions of the constant YBE, which are as follows:

$$a_+ = b_+ = c_+ = 1/2, \quad a_+ = \pm a_-, \quad b_+ = \pm b_-, \quad c_+ = \pm c_-. \tag{4.390}$$

In view of (4.386) we see that for $a_+ = a_-$ the proper limit is obtained, for example, by taking the following limits: first $m_{11}^- = -\infty$, and then $\theta = 0$, while for $a_+ = -a_-$ the limit may be obtained for $m_{11}^+ = -\infty$ first, and then $\theta = 0$. Similarly are obtained the limits for $b_\pm$ and $c_\pm$.

So we have eight R matrices satisfying the constant YBE:

$$\begin{aligned}&(+,+,+),\ (-,+,+),\ (+,-,+),\ (+,+,-),\\ &(+,-,-),\ (-,+,-),\ (-,-,+),\ (-,-,-)\end{aligned} \tag{4.391}$$

where the ± signs denote, respectively, the signs of $a_+ = \pm a_-$, $b_+ = \pm b_-$ and $c_+ = \pm c_-$.

For the elements of the 3×3 T matrix we introduce the notation:

$$T = \begin{pmatrix} k & p & l \\ q & r & s \\ m & t & n \end{pmatrix} \tag{4.392}$$

4.8.9.2 Solutions of the RTT Equations and Exotic Bialgebras

We consider matrix bialgebras which are unital associative algebras generated by the nine elements from (4.392) $k, l, m, n, p, q, r, s, t$. The coproduct and counit relations are the classical ones:

$$\delta(T) = T \otimes T, \quad \varepsilon(T) = \mathbf{1}_3 \tag{4.393}$$

We expect the bialgebras under consideration not to be Hopf algebras, which, as in the $S03$ case [50], would be easier to check after we find the dual bialgebras.

In the next subsections we obtain the desired bialgebras by applying the RTT relations of [272]:

$$R\,T_1\,T_2 = T_2\,T_1\,R\,, \tag{4.394}$$

where $T_1 = T \otimes \mathbf{1}_2$, $T_2 = \mathbf{1}_2 \otimes T$, for $R = R(\theta)$ (4.388), and the parameters are the constants in (4.390) following the eight cases of (4.391).

4.8.9.3 Algebraic Relations

I) Relations which do not depend on the parameters $a_\pm$, $b_\pm, c_\pm$. We have the set of relations

$$\begin{aligned}(N) = \{k^2 = n^2,\ kn = nk,\ l^2 = m^2,\ lm = ml\\ km = nl,\ kl = nm,\ lk = mn,\ mk = ln\\ r(k-n) = (k-n)r = 0,\ r(l-m) = (l-m)r = 0\}\end{aligned} \tag{4.395}$$

The last two relations suggest to introduce the generators:

$$\begin{aligned}k = \tilde{k} + \tilde{n},\ n = \tilde{k} - \tilde{n};\ l = \tilde{l} + \tilde{m},\ m = \tilde{l} - \tilde{m},\\ p = \tilde{p} + \tilde{t},\ t = \tilde{p} - \tilde{t};\ q = \tilde{q} + \tilde{s},\ s = \tilde{q} - \tilde{s},\end{aligned} \tag{4.396}$$

In terms of these generators we have:

$$\begin{aligned}(N) = \{\tilde{k}\tilde{n} = \tilde{n}\tilde{k} = 0;\ \tilde{l}\tilde{m} = \tilde{m}\tilde{l} = 0;\\ \tilde{k}\tilde{m} = \tilde{n}\tilde{l} = 0;\ \tilde{l}\tilde{n} = \tilde{m}\tilde{k} = 0;\\ r\tilde{m} = r\tilde{n} = 0,\ \tilde{m}r = \tilde{n}r = 0\}.\end{aligned} \tag{4.397}$$

II) Relations that do not depend on the relative signs of (a_-, b_-), (a_-, c_-), and (b_-, c_-). In that case we have:
IIa) $a_+ = \pm a_-$:
If $a_+ = a_-$ we have the set of relations

$$A_+ = \{p^2 = t^2,\ pt = tp;\ q^2 = s^2,\ qs = sq\} \tag{4.398}$$

or in terms of the alternative generators we have:

$$A_+ = \{\tilde{p}\tilde{t} = \tilde{t}\tilde{p} = 0;\ \tilde{q}\tilde{s} = \tilde{s}\tilde{q} = 0\} \tag{4.399}$$

If $a+ = -a_-$ the set of relations is:

$$A_- = \{p^2 = -t^2,\ pt = -tp;\ q^2 = -s^2,\ qs = -sq\} \tag{4.400}$$

or alternatively:

$$A_- = \{\tilde{p}^2 = \tilde{t}^2 = 0;\ \tilde{q}^2 = \tilde{s}^2 = 0\}. \tag{4.401}$$

IIb) $b_+ = \pm b_-$:
If $b_+ = b_-$ we have the set of relations:

$$B_+ = \{rp = rt;\ rq = rs\} \tag{4.402}$$

Alternatively

$$B_+ = \{r\tilde{t} = 0;\ r\tilde{s} = 0\} \tag{4.403}$$

If $b+ = -b_-$

$$B_- = \{rp = -rt;\ \ rq = -rs\} \tag{4.404}$$

Alternatively

$$B_- = \{r\tilde{p} = 0;\ \ r\tilde{q} = 0\}. \tag{4.405}$$

IIc) $c_+ = c_\pm$:
If $c_+ = c_-$ we have

$$C_+ = \{pr = tr;\ \ qr = sr\} \tag{4.406}$$

Alternatively

$$C_+ = \{\tilde{t}r = 0;\ \ \tilde{s}r = 0\} \tag{4.407}$$

If $c_+ = -c_-$ we have

$$C_- = \{pr = -tr;\ \ qr = -sr\} \tag{4.408}$$

Alternatively

$$C_- = \{\tilde{p}r = 0;\ \ \tilde{q}r = 0\}. \tag{4.409}$$

III) Relations depending on the relative signs of (a_-, b_-), (a_-, b_-0), and (b_-, c_-).
IIIa) $a_- = \pm b_-$:
If $a_- = b_-$ we have the set of relations

$$\begin{aligned}(AB)_+ = \{&pk = tn,\ tk = pn;\ pl = tm,\ tl = pm;\\ &qk = sn,\ qn = sk;\ ql = sm,\ sl = qm\}\end{aligned} \tag{4.410}$$

Alternatively

$$(AB)_+ = \{\tilde{p}\tilde{m} = \tilde{p}\tilde{n} = \tilde{t}\tilde{k} = \tilde{t}\tilde{l} = 0;\ \ \tilde{q}\tilde{m} = \tilde{q}\tilde{n} = \tilde{s}\tilde{k} = \tilde{s}\tilde{l} = 0\} \tag{4.411}$$

If $a_- = -b_-$ we have

$$\begin{aligned}(AB)_- = \{&pk = -tn,\ tk = -pn;\ pl = -tm,\ tl = -pm;\\ &qk = -sn,\ qn = -sk;\ ql = -sm,\ sl = -qm\}\end{aligned} \tag{4.412}$$

Alternatively

$$(AB)_- = \{\tilde{p}\tilde{k} = \tilde{p}\tilde{l} = \tilde{t}\tilde{m} = \tilde{t}\tilde{n} = 0;\ \ \tilde{q}\tilde{k} = \tilde{q}\tilde{l} = \tilde{s}\tilde{m} = \tilde{s}\tilde{n} = 0\}. \tag{4.413}$$

IIIb) $a_- = \pm c_-$:
If $a_- = c_-$ we have

$$(AC)_+ = \{kp = nt,\ kt = np;\ lp = mt,\ lt = mp;\\ kq = ns,\ nq = ks;\ lq = ms,\ ls = mq\} \tag{4.414}$$

Alternatively

$$(AC)_+ = \{\tilde{k}\tilde{t} = \tilde{l}\tilde{t} = \tilde{m}\tilde{p} = \tilde{n}\tilde{p} = 0;\ \tilde{k}\tilde{s} = \tilde{l}\tilde{s} = \tilde{m}\tilde{q} = \tilde{n}\tilde{q} = 0\} \tag{4.415}$$

If $a_- = -c_-$ we have

$$(AC)_- = \{kp = -nt,\ kt = -np;\ lp = -mt,\ lt = -mp;\\ kq = -ns,\ nq = -ks;\ lq = -ms,\ ls = -mq\} \tag{4.416}$$

Alternatively

$$(AC)_- = \{\tilde{k}\tilde{p} = \tilde{l}\tilde{p} = \tilde{m}\tilde{t} = \tilde{n}\tilde{t} = 0;\ \tilde{k}\tilde{q} = \tilde{l}\tilde{q} = \tilde{m}\tilde{s} = \tilde{n}\tilde{s} = 0\}. \tag{4.417}$$

IIIc) $b_- = \pm c_-$:
If $b- = c_-$ we have

$$(BC)_+ = \{pq = ts,\ tq = ps;\ qp = st,\ qt = sp\} \tag{4.418}$$

Alternatively

$$(BC)_+ = \{\tilde{p}\tilde{s} = \tilde{t}\tilde{q} = 0;\ \tilde{s}\tilde{p} = \tilde{q}\tilde{t} = 0\} \tag{4.419}$$

If $b_- = -c_-$ we have

$$(BC)_- = \{pq = -ts,\ tq = -ps;\ qp = -st,\ qt = -sp\} \tag{4.420}$$

Alternatively

$$(BC)_- = \{\tilde{p}\tilde{q} = \tilde{t}\tilde{s} = 0;\ \tilde{q}\tilde{p} = \tilde{s}\tilde{t} = 0\} \tag{4.421}$$

4.8.9.4 Classification of Bialgebras

Thus we have the following solutions:

– For $a_+ = a_- = b_- = c_-$ we have the set of relations

$$(+,+,+) = \{N \cup A_+ \cup B_+ \cup C_+ \cup (AB)_+ \cup (AC)_+ \cup (BC)_+\} \tag{4.422}$$

Explicitly we have:

$$\begin{aligned}
\tilde{k}\tilde{m} = \tilde{m}\tilde{k} = 0;\ \ \tilde{k}\tilde{n} = \tilde{n}\tilde{k} = 0;\ \ \tilde{k}\tilde{t} = \tilde{t}\tilde{k} = 0;\ \ \tilde{k}\tilde{s} = \tilde{s}\tilde{k} = 0;\\
\tilde{l}\tilde{m} = \tilde{m}\tilde{l} = 0;\ \ \tilde{l}\tilde{n} = \tilde{n}\tilde{l} = 0;\ \ \tilde{l}\tilde{t} = \tilde{t}\tilde{l} = 0;\ \ \tilde{l}\tilde{s} = \tilde{s}\tilde{l} = 0;\\
\tilde{p}\tilde{m} = \tilde{m}\tilde{p} = 0;\ \ \tilde{p}\tilde{n} = \tilde{n}\tilde{p} = 0;\ \ \tilde{p}\tilde{t} = \tilde{t}\tilde{p} = 0;\ \ \tilde{p}\tilde{s} = \tilde{s}\tilde{p} = 0;\\
\tilde{q}\tilde{m} = \tilde{m}\tilde{q} = 0;\ \ \tilde{q}\tilde{n} = \tilde{n}\tilde{q} = 0;\ \ \tilde{q}\tilde{t} = \tilde{t}\tilde{q} = 0;\ \ \tilde{q}\tilde{s} = \tilde{s}\tilde{q} = 0;\\
r\tilde{m} = \tilde{m}r = 0;\ \ r\tilde{n} = \tilde{n}r = 0;\ \ r\tilde{t} = \tilde{t}r = 0;\ \ r\tilde{s} = \tilde{s}r = 0.
\end{aligned} \tag{4.423}$$

From (4.423) we see that the algebra $\mathcal{A}_{+++}$ is a direct sum of two subalgebras: $\mathcal{A}^1_{+++}$ with generators $\tilde{k}, \tilde{l}, \tilde{p}, \tilde{q}, r$, and $\mathcal{A}^2_{+++}$ with generators $\tilde{m}, \tilde{n}, \tilde{s}, \tilde{t}$. Both subalgebras are free, with no relations, and thus, no PBW bases.

– For $a_+ = a_- = -b_- = -c_-$ we have the set of relations

$$(+, -, -) = \{N \cup A_+ \cup B_- \cup C_- \cup (AB)_- \cup (AC)_- \cup (BC)_+\} \tag{4.424}$$

We omit the relations of the resulting algebra denoted $\mathcal{A}_{+--}$ since it is a conjugate of the previous algebra $\mathcal{A}_{+++}$, obtained by the exchange of the pairs of generators $(\tilde{p}, \tilde{q})$ and $(\tilde{s}, \tilde{t})$.

– For $a_+ = -a_- = b_- = c_-$ we have the set of relations

$$(-, +, +) = \{N \cup A_- \cup B_+ \cup C_+ \cup (AB)_- \cup (AC)_- \cup (BC)_+\} \tag{4.425}$$

Explicitly we have:

$$\begin{aligned}
\tilde{k}\tilde{m} = \tilde{m}\tilde{k} = 0;\ \ \tilde{k}\tilde{n} = \tilde{n}\tilde{k} = 0;\ \ \tilde{k}\tilde{p} = \tilde{p}\tilde{k} = 0;\ \ \tilde{k}\tilde{q} = \tilde{q}\tilde{k} = 0;\\
\tilde{l}\tilde{m} = \tilde{m}\tilde{l} = 0;\ \ \tilde{l}\tilde{n} = \tilde{n}\tilde{l} = 0;\ \ \tilde{l}\tilde{p} = \tilde{p}\tilde{l} = 0;\ \ \tilde{l}\tilde{q} = \tilde{q}\tilde{l} = 0;\\
r\tilde{m} = \tilde{m}r = 0;\ \ r\tilde{n} = \tilde{n}r = 0;\ \ r\tilde{t} = \tilde{t}r = 0;\ \ r\tilde{s} = \tilde{s}r = 0;\\
\tilde{t}\tilde{m} = \tilde{m}\tilde{t} = 0;\ \ \tilde{t}\tilde{n} = \tilde{n}\tilde{t} = 0;\ \ \tilde{t}\tilde{q} = \tilde{q}\tilde{t} = 0;\\
\tilde{s}\tilde{m} = \tilde{m}\tilde{s} = 0;\ \ \tilde{s}\tilde{n} = \tilde{n}\tilde{s} = 0;\ \ \tilde{s}\tilde{p} = \tilde{p}\tilde{s} = 0;\\
\tilde{p}^2 = \tilde{q}^2 = \tilde{s}^2 = \tilde{t}^2 = 0.
\end{aligned} \tag{4.426}$$

The structure of this algebra, denoted $\mathcal{A}_{-++}$, is more complicated. There are two quasi-free subalgebras: $\mathcal{A}^1_{-++}$ with generators $\tilde{k}, \tilde{l}, \tilde{t}, \tilde{s}$, and $\mathcal{A}^2_{-++}$ with generators $\tilde{m}, \tilde{n}, \tilde{p}, \tilde{q}$. They are quasi-free due to the last line of (4.426). They do not form a direct sum due to the existence of the following 12 two-letter building blocks of the basis $\mathcal{A}_{-++}$: $r\tilde{k}$, $r\tilde{l}$, $r\tilde{p}$, $r\tilde{q}$, $\tilde{p}\tilde{t}$, $\tilde{q}\tilde{s}$ plus the reverse order.

– For $a_+ = -a_- = -b_- = -c_-$ we have the set of relations

$$(-, -, -) = \{N \cup A_- \cup B_- \cup C_- \cup (AB)_+ \cup (AC)_+ \cup (BC)_+\} \tag{4.427}$$

We omit the relations of the resulting algebra denoted $\mathcal{A}_{---}$ since it is a conjugate of the previous algebra $\mathcal{A}_{-++}$, obtained by the exchanges of generators: $\tilde{p} \longrightarrow \tilde{s}$ and $\tilde{q} \longrightarrow \tilde{t}$.

– For $a_+ = a_- = -b_- = c_-$ we have

$$(+,-,+) = \{N \cup A_+ \cup B_- \cup C_+ \cup (AB)_- \cup (AC)_+ \cup (BC)_-\} \tag{4.428}$$

Explicitly we have:

$$\begin{aligned}
\tilde{k}\tilde{m} = \tilde{m}\tilde{k} = 0;\ \ \tilde{k}\tilde{n} = \tilde{n}\tilde{k} = 0;\ \ \tilde{k}\tilde{t} = \tilde{k}\tilde{s} = 0;\ \ \tilde{p}\tilde{k} = \tilde{q}\tilde{k} = 0;\\
\tilde{l}\tilde{m} = \tilde{m}\tilde{l} = 0;\ \ \tilde{l}\tilde{n} = \tilde{n}\tilde{l} = 0;\ \ \tilde{l}\tilde{t} = \tilde{l}\tilde{s} = 0;\ \ \tilde{p}\tilde{l} = \tilde{q}\tilde{l} = 0;\\
r\tilde{m} = \tilde{m}r = 0;\ \ r\tilde{n} = \tilde{n}r = 0;\ \ r\tilde{p} = r\tilde{q} = 0;\ \ \tilde{t}r = \tilde{s}r = 0;\\
\tilde{t}\tilde{p} = \tilde{p}\tilde{t} = 0;\ \ \tilde{t}\tilde{s} = \tilde{s}\tilde{t} = 0;\ \ \tilde{t}\tilde{m} = \tilde{t}\tilde{n} = 0;\\
\tilde{q}\tilde{p} = \tilde{p}\tilde{q} = 0;\ \ \tilde{q}\tilde{s} = \tilde{s}\tilde{q} = 0;\ \ \tilde{m}\tilde{q} = \tilde{n}\tilde{q} = 0;\\
\tilde{s}\tilde{m} = \tilde{s}\tilde{n} = 0;\ \ \tilde{m}\tilde{p} = \tilde{n}\tilde{p} = 0.
\end{aligned} \tag{4.429}$$

The structure of this algebra, denoted $\mathcal{A}_{+-+}$, is also complicated. There are four free subalgebras: $\mathcal{A}^1_{+-+}$ with generators $\tilde{k}, \tilde{l}, r$, $\mathcal{A}^2_{+-+}$ with generators $\tilde{m}, \tilde{n}$, $\mathcal{A}^3_{+-+}$ with generators $\tilde{p}, \tilde{s}$, $\mathcal{A}^4_{+-+}$ with generators $\tilde{q}, \tilde{t}$. Only the first two are in direct sum, otherwise all are related by the following 20 two-letter building blocks: $\tilde{k}\tilde{p}$, $\tilde{k}\tilde{q}$, $\tilde{s}\tilde{k}$, $\tilde{t}\tilde{k}$, $\tilde{l}\tilde{p}$, $\tilde{l}\tilde{q}$, $\tilde{s}\tilde{l}$, $\tilde{t}\tilde{l}$, $r\tilde{s}$, $r\tilde{t}$, $\tilde{p}r$, $\tilde{q}r$, $\tilde{m}\tilde{s}$, $\tilde{m}\tilde{t}$, $\tilde{p}\tilde{m}$, $\tilde{q}\tilde{m}$, $\tilde{n}\tilde{s}$, $\tilde{n}\tilde{t}$, $\tilde{p}\tilde{n}$, $\tilde{q}\tilde{n}$.

There is no overall ordering. There is some partial order if we consider the subalgebra formed by the generators of the latter three subalgebras: $\tilde{m}, \tilde{n}, \tilde{p}, \tilde{s}, \tilde{q}, \tilde{t}$, namely, we have:

$$\tilde{p}, \tilde{q} > \tilde{m}, \tilde{n} > \tilde{s}, \tilde{t} \tag{4.430}$$

But for the natural subalgebra formed by generators $\tilde{k}, \tilde{l}, r$, $\tilde{p}, \tilde{s}$, $\tilde{q}, \tilde{t}$, we have cyclic ordering:

$$\tilde{p}, \tilde{q} > r > \tilde{s}, \tilde{t} > \tilde{k}, \tilde{l} > \tilde{p}, \tilde{q}, \tag{4.431}$$

that is, no ordering. We have seen this phenomenon in the simpler exotic bialgebra $S03$ considered earlier.

– For $a_+ = a_- = b_- = -c_-$ we have the set of relations:

$$(+,+,-) = \{N \cup A_+ \cup B_+ \cup C_- \cup (AB)_+ \cup (AC)_- \cup (BC)_-\} \tag{4.432}$$

We omit the relations of the resulting algebra denoted $\mathcal{A}_{++-}$ since it is a conjugate of the previous algebra. It has the same four free algebras, and the only difference is that the subalgebras are related by 20 two-letter building blocks which are in reverse order w.r.t. the previous case: $\tilde{k}\tilde{s}$, $\tilde{k}\tilde{t}$, $\tilde{p}\tilde{k}$, $\tilde{q}\tilde{k}$, $\tilde{l}\tilde{s}$, $\tilde{l}\tilde{t}$, $\tilde{p}\tilde{l}$, $\tilde{q}\tilde{l}$, $r\tilde{p}$, $r\tilde{q}$, $\tilde{s}r$, $\tilde{s}r$, $\tilde{m}\tilde{p}$, $\tilde{m}\tilde{q}$, $\tilde{s}\tilde{m}$, $\tilde{t}\tilde{m}$, $\tilde{n}\tilde{p}$, $\tilde{n}\tilde{q}$, $\tilde{s}\tilde{n}$, $\tilde{t}\tilde{n}$.

– For $a_+ = -a_- = b_- = -c_-$ we have the set of relations

$$(-, +, -) = \{N \cup A_- \cup B_+ \cup C_- \cup (AB)_- \cup (AC)_+ \cup (BC)_-\} \tag{4.433}$$

Explicitly we have:

$$\begin{aligned}
\tilde{k}\tilde{m} = \tilde{m}\tilde{k} = 0;\ \ \tilde{k}\tilde{n} = \tilde{n}\tilde{k} = 0;\ \ \tilde{k}\tilde{t} = \tilde{k}\tilde{s} = 0;\ \ \tilde{p}\tilde{k} = \tilde{q}\tilde{k} = 0;\\
\tilde{l}\tilde{m} = \tilde{m}\tilde{l} = 0;\ \ \tilde{l}\tilde{n} = \tilde{n}\tilde{l} = 0;\ \ \tilde{l}\tilde{t} = \tilde{l}\tilde{s} = 0;\ \ \tilde{p}\tilde{l} = \tilde{q}\tilde{l} = 0;\\
r\tilde{m} = \tilde{m}r = 0;\ \ r\tilde{n} = \tilde{n}r = 0;\ \ r\tilde{t} = r\tilde{s} = 0;\ \ \tilde{p}r = \tilde{q}r = 0;\\
\tilde{s}\tilde{t} = \tilde{t}\tilde{s} = 0;\ \ \tilde{s}\tilde{m} = \tilde{s}\tilde{n} = 0;\ \ \tilde{t}\tilde{m} = \tilde{t}\tilde{n} = 0;\\
\tilde{p}\tilde{q} = \tilde{q}\tilde{p} = 0;\ \ \tilde{m}\tilde{p} = \tilde{n}\tilde{p} = 0;\ \ \tilde{m}\tilde{q} = \tilde{n}\tilde{q} = 0;\\
\tilde{p}^2 = \tilde{q}^2 = \tilde{s}^2 = \tilde{t}^2 = 0.
\end{aligned} \tag{4.434}$$

The structure of this algebra, denoted $\mathscr{A}_{-+-}$, is very complicated. There are two free subalgebras: $\mathscr{A}^1_{-+-}$ with generators $\tilde{k}, \tilde{l}r$, $\mathscr{A}^2_{-+-}$ with generators $\tilde{m}, \tilde{n}$, and four quasi-free subalgebras: $\mathscr{A}^3_{-+-}$ with generators $\tilde{p}, \tilde{s}$, $\mathscr{A}^4_{-+-}$ with generators $\tilde{q}, \tilde{t}$, $\mathscr{A}^5_{-+-}$ with generators $\tilde{p}, \tilde{t}$, $\mathscr{A}^6_{-+-}$ with generators $\tilde{q}, \tilde{s}$. The first four subalgebras have generators as in the $\mathscr{A}_{+-+}$ case (but taking into account the last line of (4.434)). Only the first two subalgebras are in direct sum, and there are intersections between the last four. Furthermore, all are related by the following 20 two-letter building blocks: $\tilde{k}\tilde{p}$, $\tilde{k}\tilde{q}$, $\tilde{s}\tilde{k}$, $\tilde{t}\tilde{k}$, $\tilde{l}\tilde{p}$, $\tilde{l}\tilde{q}$, $\tilde{s}\tilde{l}$, $\tilde{t}\tilde{l}$, $r\tilde{p}$, $r\tilde{q}$, $\tilde{s}r$, $\tilde{t}r$, $\tilde{m}\tilde{s}$, $\tilde{m}\tilde{t}$, $\tilde{p}\tilde{m}$, $\tilde{q}\tilde{m}$, $\tilde{n}\tilde{s}$, $\tilde{n}\tilde{t}$, $\tilde{p}\tilde{n}$, $\tilde{q}\tilde{n}$, which are the same as in the $\mathscr{A}_{+-+}$ case, except those involving r.

The last difference makes things better. Indeed, there is no overall ordering, more precisely we have:

$$\tilde{p}, \tilde{q} > \tilde{m}, \tilde{n} > \tilde{s}, \tilde{t} > \tilde{k}, \tilde{l}, r > \tilde{p}, \tilde{q} \tag{4.435}$$

that is, we have some cyclic order.

Thus, the bialgebra $\mathscr{A}_{-+-}$ may turn out to be the easiest to handle, as the exotic bialgebra $S03$ considered earlier.

– For $a_+ = -a_- = -b_- = c_-$ we have the set of relations

$$(-, -, +) = \{N \cup A_- \cup B_- \cup C_+ \cup (AB)_+ \cup (AC)_- \cup (BC)_-\} \tag{4.436}$$

We omit the relations of the resulting algebra denoted $\mathscr{A}_{--+}$ since it is a conjugate of the previous algebra obtained by the exchange of the pairs of generators $(\tilde{p}, \tilde{q})$ and $(\tilde{s}, \tilde{t})$.

Summary. Thus, taking into account conjugation, we have found *four* different bialgebras originating from the braid matrices (4.385):

$$\mathscr{A}_{+++} \cong \mathscr{A}_{+--}, \quad \mathscr{A}_{---} \cong \mathscr{A}_{-++}, \quad \mathscr{A}_{+-+} \cong \mathscr{A}_{++-}, \quad \mathscr{A}_{-+-} \cong \mathscr{A}_{--+} \tag{4.437}$$

The first two bialgebras have no ordering. The first one is simpler, since it is split in two subalgebras with five and four generators. The third bialgebra has partial ordering in one subalgebra. The last one, is the most promising since it has partial cyclic ordering.

The next task in our line of research is to find the dual bialgebras, analogously, as done above for the four-element exotic bialgebras. We do this in the next subsection for the most interesting of the above: $\mathscr{A}_{-+-} \equiv \mathscr{A}_{--+}$.

4.8.9.5 The Dual Bialgebra of A_{-+-}

To start with we begin with the coproducts of the elements of the T-matrix. Until now we used the changed basis of "tilde" generators. But here it would be better to make a further change in new "hat" generators:

$$\tilde{k} = \hat{k} + \hat{l},\ \tilde{l} = \hat{k} - \hat{l};\ \ \tilde{m} = \hat{m} - \hat{n},\ \tilde{n} = \hat{m} + \hat{n}. \tag{4.438}$$

Thus we have:

$$\begin{aligned}
&\delta(\hat{k}) = 2\hat{k} \otimes \hat{k} - 2\hat{m} \otimes \hat{n} + \tilde{p} \otimes \tilde{q}, \quad \delta(\hat{l}) = 2\hat{l} \otimes \hat{l} - 2\hat{n} \otimes \hat{m} + \tilde{t} \otimes \tilde{s},\\
&\delta(\hat{m}) = 2\hat{k} \otimes \hat{m} - 2\hat{m} \otimes \hat{l} - \tilde{p} \otimes \tilde{s}, \quad \delta(\hat{n}) = 2\hat{l} \otimes \hat{n} + 2\hat{n} \otimes \hat{k} + \tilde{t} \otimes \tilde{q},\\
&\delta(\hat{p}) = 2\hat{k} \otimes \hat{p} - 2\hat{m} \otimes \hat{t} + \tilde{p} \otimes r, \quad \delta(\hat{q}) = 2\hat{q} \otimes \hat{k} + 2\hat{s} \otimes \hat{n} + r \otimes \tilde{q},\\
&\delta(\hat{s}) = 2\hat{s} \otimes \hat{l} - 2\hat{q} \otimes \hat{m} + r \otimes \tilde{s}, \quad \delta(\hat{t}) = 2\hat{l} \otimes \hat{t} + 2\hat{n} \otimes \hat{p} + \tilde{t} \otimes r,\\
&\delta(r) = r \otimes r + 2\tilde{q} \otimes \tilde{p} + 2\tilde{s} \otimes \tilde{t},\\
&\epsilon(\hat{k}) = \epsilon(\hat{l}) = 1/2,\ \ \epsilon(r) = 1, \epsilon(z) = 0,\ \text{for } z = (\hat{m}, \hat{n}, \tilde{p}, \tilde{q}, \tilde{s}, \tilde{t}).
\end{aligned} \tag{4.439}$$

The bialgebra relations are as follows:

$$\begin{aligned}
&\hat{k}\hat{m} = \hat{m}\hat{k} = \hat{k}\hat{n} = \hat{n}\hat{k} = 0,\ \hat{k}\tilde{t} = \hat{k}\tilde{s} = \tilde{p}\hat{k} = \tilde{q}\hat{k} = 0,\\
&\hat{l}\hat{m} = \hat{m}\hat{l} = \hat{l}\hat{n} = \hat{n}\hat{l} = 0,\ \hat{l}\tilde{t} = \hat{l}\tilde{s} = \tilde{p}\hat{l} = \tilde{q}\hat{l} = 0,\\
&r\hat{m} = \hat{m}r = r\hat{n} = \hat{n}r = 0,\ \ r\tilde{t} = r\tilde{s} = \tilde{p}r = \tilde{q}r = 0,\\
&\tilde{s}\tilde{t} = \tilde{t}\tilde{s} = 0,\ \ \tilde{s}\hat{m} = \tilde{s}\hat{n} = \tilde{t}\hat{m} = \tilde{t}\hat{n} = 0,\\
&\tilde{p}\tilde{q} = \tilde{q}\tilde{p} = 0,\ \ \hat{m}\tilde{p} = \hat{n}\tilde{p} = \hat{m}\tilde{q} = \hat{n}\tilde{q} = 0, \tilde{p}^2 = \tilde{q}^2 = \tilde{s}^2 = \tilde{t}^2 = 0.
\end{aligned} \tag{4.440}$$

The dual elements are defined by our standard procedure:

$$\langle Z, f\rangle = \epsilon\left(\frac{\partial f}{\partial z}\right), \quad \text{where } z = (\hat{k}, \hat{l}, \hat{m}, \hat{n}, \tilde{p}, \tilde{q}, \tilde{s}, \tilde{t}, r). \tag{4.441}$$

The basis we are working is essentially the following

$$\begin{aligned}
\hat{k}^{\kappa}\hat{l}^{\ell}r^{\tau}, &\ \text{and all permutations of } (\hat{k}\hat{l}r),\\
\hat{k}^{\kappa}\hat{l}^{\ell}r^{\tau}\tilde{p}, &\ \text{and all permutations of } (\hat{k}\hat{l}r),\\
\hat{k}\hat{l}r)\tilde{q}, &\ \text{and all permutations of } (\hat{k}\hat{l}r),\\
\tilde{s}\hat{k}^{\kappa}\hat{l}^{\ell}r^{\tau}, &\ \text{and all permutations of } (\hat{k}\hat{l}r),\\
\tilde{t}\hat{k}^{\kappa}\hat{l}^{\ell}r^{\tau} &\ \text{and all permutations of } (\hat{k}\hat{l}r),\\
&\qquad\qquad\qquad\qquad \hat{m},\ \hat{n}.
\end{aligned} \tag{4.442}$$

Thus the following dual bialgebra is obtained:

$$\begin{aligned}
&\left[\hat{K}^{\kappa}, \hat{L}^{\ell}\right] = 0, \hat{K}^{\kappa}\hat{M} = 2^{\kappa}\hat{M},\ \ \hat{N}\hat{K}^{\kappa} = 2^{\kappa}\hat{N},\ \ \hat{M}\hat{K} = \hat{K}\hat{N} = 0,\\
&\hat{M}\hat{L}^{\ell} = 2^{\ell}\hat{M},\ \ \hat{L}^{\ell}\hat{N} = 2^{\ell}\hat{N},\ \ \hat{N}\hat{L} = \hat{L}\hat{M} = 0,\\
&\hat{M}^2 = \hat{N}^2 = 0,\ \ \hat{M}\hat{N} = -2\hat{K},\ \ \hat{N}\hat{M} = -2\hat{L},\\
&\left[\hat{K}, \tilde{P}\right] = 2\tilde{P},\ \ \left[\hat{L}, \tilde{P}\right] = 0,\ \ \left[\hat{R}, \tilde{P}\right] = -\tilde{P},\\
&\left[\hat{K}, \tilde{Q}\right] = -2\tilde{Q},\ \ \left[\hat{L}, \tilde{Q}\right] = 0,\ \ \left[\hat{R}, \tilde{Q}\right] = \tilde{Q},\\
&\left[\hat{K}, \tilde{S}\right] = 0,\ \ \left[\hat{L}, \tilde{S}\right] = -2\tilde{S},\ \ \left[\hat{R}, \tilde{S}\right] = \tilde{S},\\
&\left[\hat{K}, \tilde{T}\right] = 0,\ \ \left[\hat{L}, \tilde{T}\right] = 2\tilde{T},\ \ \left[\hat{R}, \tilde{T}\right] = -\tilde{T},\\
&\hat{M}\tilde{T} = -2\tilde{P},\ \ \tilde{T}\hat{M} = 0,\ \ \tilde{Q}\hat{M} = -2\tilde{S},\ \ \hat{M}\tilde{Q} = 0,\\
&\hat{N}\tilde{P} = 2\tilde{T},\ \ \tilde{P}\hat{N} = 0,\ \ \tilde{S}\hat{N} = 2\tilde{Q},\ \ \hat{N}\tilde{S} = 0,\\
&\hat{M}\tilde{P} = \tilde{P}\hat{M} = \hat{M}\tilde{S} = \tilde{S}\hat{M} = 0, \hat{N}\tilde{Q} = \tilde{Q}\hat{N} = \hat{N}\tilde{T} = \tilde{T}\hat{N} = 0,\\
&\left[\tilde{S}, \tilde{P}\right] = \hat{M},\ \ \left[\tilde{Q}, \tilde{T}\right] = \hat{N},\ \ \left[\tilde{Q}, \tilde{S}\right] = \left[\tilde{P}, \tilde{T}\right] = 0,\\
&\tilde{P}\tilde{Q} = \tilde{T}\tilde{S},\ \ \tilde{Q}\tilde{P} = \tilde{S}\tilde{T},\\
&\tilde{P}^2 = \tilde{Q}^2 = \tilde{S}^2 = \tilde{T}^2 = 0,\ \ \tilde{P}\tilde{T} = \tilde{T}\tilde{P} = \tilde{Q}\tilde{S} = \tilde{S}\tilde{Q} - 0.
\end{aligned} \tag{4.443}$$

Finally we write down the coproducts of the dual bialgebra:

$$\begin{aligned}
&\delta(\hat{K}) = \hat{K}\otimes 1_U + 1_U\otimes\hat{K},\quad \delta(\hat{L}) = \hat{L}\otimes 1_U + 1_U\otimes\hat{L},\\
&\delta(\hat{M}) = \hat{M}\otimes 1_U + 1_U\otimes\hat{M},\quad \delta(\hat{N}) = \hat{N}\otimes 1_U + 1_U\otimes\hat{N},\\
&\delta(\tilde{P}) = \tilde{P}\otimes 1_U + 1_U\otimes\tilde{P},\quad \delta(\tilde{Q}) = \tilde{Q}\otimes 1_U + 1_U\otimes\tilde{Q},\\
&\delta(\tilde{S}) = \tilde{S}\otimes 1_U + 1_U\otimes\tilde{S},\quad \delta(\tilde{T}) = \tilde{T}\otimes 1_U + 1_U\otimes\tilde{T},\\
&\delta(R) = R\otimes 1_U + 1_U\otimes R.
\end{aligned} \tag{4.444}$$

Conclusions and Outlook

In the present subsection we have found a multitude of exotic bialgebras and the dual of one of them. More duals should be constructed. More importantly, may continue the programme fulfilled successfully for the exotic bialgebra S03 (cf. above). In particular, it is important to find the FRT duals [272] which are different from the standard duals for the exotic bialgebras. Further one should find the baxterization of the dual algebras. Their finite-dimensional representations should be considered. Diagonalizations of the braid matrices would be used to handle the representations of the corresponding L-algebras (in the FRT formalism) and to formulate the fusion of finite-dimensional representations. More general algebras should be considered, for example, using a more general 9×9 R matrix with 16 parameters considered in [10]. Possible applications may be considered, in particular, exotic vertex models and integrable spin-chain models.

5 Invariant q-Difference Operators

Summary
We construct induced infinite-dimensional representations of the two-parameter quantum algebras $U_{p,q}(gl(2))$ and $U_{g,h}(gl(2))$ which are in duality with the deformations $GL_{p,q}(2)$ and $GL_{g,h}(2)$, respectively. The representations split into one-parameter representations of a one-generator central algebra and a three-generator quantum algebra, the latter in each case being a deformation of $U(sl(2))$. In both cases the representations can be mapped to representations in one complex variable, which are deformations of the standard one-parameter vector-field realization of $sl(2)$. The deformation in the case of $U_{g,h}(gl(2))$, actually of the Jordanian $U_{\tilde{g}}(sl(2))$, $\tilde{g} = g + h$, is a new deformation. We also obtain canonically finite-dimensional representations which can be restricted to the one-parameter three-generator subalgebras in both cases. The deformations of the invariant differntial operators are playing an important role. We do the same program for a Lorentz quantum algebra and for the generalized Lie algebra $sl(2)_q$. Finally, we discuss representations of $U_q(so(3))$ of integer spin only. This chapter is based mainly on [164, 218, 228, 241, 243].

5.1 The Case of $GL_{p,q}(2)$

5.1.1 Left and Right Action of $U_{p,q}(gl(2))$ on $GL_{p,q}(2)$

In this section following [228] we construct induced representations of the quantum algebra $\mathscr{U}_{p,q} = U_{p,q}(gl(2))$, which was found in [209] and reviewed in Section 4.4 as the dual of the two-parameter matrix quantum group $GL_{p,q}(2)$ introduced in [183] and reviewed in Section 4.3. We follow (with some modifications) the similar construction for $U_q(sl(2))$ in [210] (see also [211]). The representation spaces are built on formal power series in the generating elements a, b, c, d of $\mathscr{A}_{p,q} = GL_{p,q}(2)$, with commutations relations shown in (4.42).

The following notations will be useful:

$$t \equiv \sqrt{pq}, \quad s \equiv pq^{-1}, \quad \Longleftrightarrow \quad q = ts^{-1/2}, \quad p = ts^{1/2}. \tag{5.1}$$

There exists a multiplicative quantum determinant $\mathscr{D}$ (cf. (4.44)). For our purposes we shall suppose that $\mathscr{D}$ is invertible. Then from (4.44) we have:

$$a = (\mathscr{D} + p^{-1}bc)d^{-1}. \tag{5.2}$$

Thus, we shall take the following basis of $\mathscr{A}_{p,q}$:

$$f = f_{mnk\ell} = b^m\, c^n\, d^\ell\, \mathscr{D}^k\ , \quad k, m, n \in \mathbb{Z}_+\ , \ \ell \in \mathbb{Z}. \tag{5.3}$$

DOI 10.1515/9783110427707-005

For the dual algebra $\mathcal{U}_{p,q}$ here, we shall use the following generating elements: $r, r^{-1}, h, h^{-1}, X^+, X^-$ with relations:

$$\begin{aligned} &[r^\epsilon, h^\varepsilon] = 0, \quad [r^\epsilon, X^+] = 0, \quad [r^\epsilon, X^-] = 0, \qquad \epsilon, \varepsilon = \pm 1 \\ &rr^{-1} = 1_{\mathcal{U}}, \qquad hh^{-1} = 1_{\mathcal{U}} \\ &h^\varepsilon X^+ = t^\varepsilon X^+ h^\varepsilon, \quad h^\varepsilon X^- = t^{-\varepsilon} X^- h^\varepsilon, \quad [X^+, X^-] = \frac{h^2 - h^{-2}}{t - t^{-1}}. \end{aligned} \tag{5.4}$$

Note that in [209] instead of the parameters p and q $p^{1/2}$ and $q^{1/2}$ were used, and instead of the generators r, r^{-1} and h, h^{-1} H, K were used so that: $h^\varepsilon = t^{\varepsilon H/2}$, $r^\epsilon = s^{\epsilon K/2}$. Thus, strictly speaking, our algebra $\mathcal{U}_{p,q}$ is a subalgebra of the one in [209]. To obtain from (5.4) the $U(gl(2))$ commutation relations one has to use the expansions: $h^\varepsilon \approx 1 + \varepsilon(\log t) H/2$, $r^\epsilon \approx 1 + \epsilon(\log s) K/2$, and then to set: $\log t \to 0, \log s \to 0$. Thus, one gets:

$$[X^+, X^-] = H, \qquad [H, X^\pm] = \pm 2H, \tag{5.5}$$

$$[K, X^+] = 0, \qquad [K, Y^-] = 0, \qquad [K, H] = 0. \tag{5.6}$$

The coalgebra relations are [209]:

$$\begin{aligned} &\delta_{\mathcal{U}}(h^\varepsilon) = h^\varepsilon \otimes h^\varepsilon, \quad \delta_{\mathcal{U}}(r^\epsilon) = r^\epsilon \otimes r^\epsilon, \\ &\delta_{\mathcal{U}}(X^\pm) = X^\pm \otimes r^{\pm 1} h + h^{-1} \otimes X^\pm, \end{aligned} \tag{5.7}$$

$$\begin{aligned} &\epsilon_{\mathcal{U}}(h^\varepsilon) = 1, \quad \epsilon_{\mathcal{U}}(r^\epsilon) = 1, \quad \epsilon_{\mathcal{U}}(X^+) = 0, \quad \epsilon_{\mathcal{U}}(X^-) = 0, \\ &\gamma_{\mathcal{U}}(r^\epsilon) = r^{-\epsilon}, \quad \gamma_{\mathcal{U}}(h^\varepsilon) = h^{-\varepsilon}, \\ &\gamma_{\mathcal{U}}(X^\pm) = -t^{\pm 1} X^\pm r^{\mp 1}. \end{aligned} \tag{5.8}$$

Further we shall give the formulae only for the generators r, h, when the analogous formulae for r^{-1}, h^{-1} follow trivially from those for r, h.

Recall that the two sets of generators $r^\epsilon, h^\varepsilon, X^+$ and $r^\epsilon, h^\varepsilon, X^-$ generate conjugate Borel Hopf subalgebras of $\mathcal{U}_{p,q}$, and that it is not possible to decouple the r^ϵ generators [209].

The duality for the algebras $\mathcal{U}_{p,q}$ and $\mathcal{A}_{p,q}$ is given standardly by the pairings:

$$\langle r, \begin{pmatrix} a & b \\ c & d \end{pmatrix} \rangle = \begin{pmatrix} s^{1/2} & 0 \\ 0 & s^{1/2} \end{pmatrix}, \tag{5.9a}$$

$$\langle h, \begin{pmatrix} a & b \\ c & d \end{pmatrix} \rangle = \begin{pmatrix} t^{1/2} & 0 \\ 0 & t^{-1/2} \end{pmatrix}, \tag{5.9b}$$

$$\langle X^+, \begin{pmatrix} a & b \\ c & d \end{pmatrix} \rangle = \begin{pmatrix} 0 & 1 \\ 0 & 0 \end{pmatrix}, \tag{5.9c}$$

$$\langle X^-, \begin{pmatrix} a & b \\ c & d \end{pmatrix} \rangle = \begin{pmatrix} 0 & 0 \\ 1 & 0 \end{pmatrix}, \tag{5.9d}$$

$$\langle y, 1_{\mathcal{A}} \rangle = \varepsilon_{\mathcal{U}}(y). \tag{5.9e}$$

Now we introduce the left regular representation of $\mathcal{U}_{p,q}$ which in the classical case is the infinitesimal version of: $\pi_L(Y)\,M = Y^{-1}\,M, \quad Y, M \in GL(2)$. Namely, we set:

$$\pi_L(X)\,M = \langle \gamma_{\mathcal{U}}(X), M\rangle M, \qquad X \in \mathcal{U} \tag{5.10}$$

Explicitly we get from (5.10) for the generators of $\mathcal{U}_{p,q}$:

$$\pi_L(r)\begin{pmatrix} a\,b \\ c\,d \end{pmatrix} = \begin{pmatrix} s^{-\frac{1}{2}}a & s^{-\frac{1}{2}}b, \\ s^{-\frac{1}{2}}c & s^{-\frac{1}{2}}d \end{pmatrix}, \tag{5.11a}$$

$$\pi_L(h)\begin{pmatrix} a\,b \\ c\,d \end{pmatrix} = \begin{pmatrix} t^{-\frac{1}{2}}a & t^{-\frac{1}{2}}b, \\ t^{\frac{1}{2}}c & t^{\frac{1}{2}}d \end{pmatrix}, \tag{5.11b}$$

$$\pi_L(X^+)\begin{pmatrix} a\,b \\ c\,d \end{pmatrix} = \begin{pmatrix} -ts^{-\frac{1}{2}}c & -ts^{-\frac{1}{2}}d \\ 0 & 0 \end{pmatrix}, \tag{5.11c}$$

$$\pi_L(X^-)\begin{pmatrix} a\,b \\ c\,d \end{pmatrix} = \begin{pmatrix} 0 & 0 \\ -t^{-1}s^{\frac{1}{2}}a & -t^{-1}s^{\frac{1}{2}}b \end{pmatrix}. \tag{5.11d}$$

In order to derive the action of π on arbitrary elements of the basis we use the following *twisted derivation rule* consistent with the coproduct and the representation structure. Namely, we use [210, 211]:

$$\pi_L(y)\varphi\psi = \hat{m}\left(\pi_L(\delta'_{\mathcal{U}}(y))(\varphi \otimes \psi)\right) \tag{5.12}$$

where $\hat{m}$ is the multiplication map: $\hat{m} : \mathcal{A} \otimes \mathcal{A} \longrightarrow \mathcal{A}$, $\hat{m}(f \otimes f') = ff'$; $\delta'_{\mathcal{U}} = \sigma \circ \delta_{\mathcal{U}}$ is the opposite coproduct (σ is the permutation operator). Thus, in our concrete situation we have:

$$\begin{aligned}
&\pi_L(r)\,\varphi\psi = \pi_L(r)\varphi \cdot \pi_L(r)\psi, \\
&\pi_L(h)\,\varphi\psi = \pi_L(h)\varphi \cdot \pi_L(h)\psi, \\
&\pi_L(X^+)\,\varphi\psi = \pi_L(rh)\varphi \cdot \pi_L(X^+)\psi \;+\; \pi_L(X^+)\varphi \cdot \pi_L(h^{-1})\psi, \\
&\pi_L(X^-)\,\varphi\psi = \pi_L(r^{-1}h)\varphi \cdot \pi_L(X^-)\psi \;+\; \pi_L(X^-)\varphi \cdot \pi_L(h^{-1})\psi, \\
&\pi_L\begin{pmatrix} r & X^+ \\ X^- & h \end{pmatrix} 1_{\mathcal{A}} = \begin{pmatrix} 1_{\mathcal{A}} & 0 \\ 0 & 1_{\mathcal{A}} \end{pmatrix}
\end{aligned} \tag{5.13}$$

Thus we obtain first:

$$\pi_L\begin{pmatrix} r & X^+ \\ X^- & h \end{pmatrix} \mathcal{D} = \begin{pmatrix} s^{-1}\mathcal{D} & 0 \\ 0 & \mathcal{D} \end{pmatrix} \tag{5.14}$$

and then we get:

$$\pi_L(r)\begin{pmatrix}\mathscr{D}^n & b^n\\ c^n & d^n\end{pmatrix} = \begin{pmatrix}s^{-n}\mathscr{D}^n & s^{-\frac{n}{2}}b^n\\ s^{-\frac{n}{2}}c^n & s^{-\frac{n}{2}}d^n\end{pmatrix}, \tag{5.15a}$$

$$\pi_L(h)\begin{pmatrix}\mathscr{D}^n & b^n\\ c^n & d^n\end{pmatrix} = \begin{pmatrix}\mathscr{D}^n & t^{-\frac{n}{2}}b^n\\ t^{\frac{n}{2}}c^n & t^{\frac{n}{2}}d^n\end{pmatrix}, \tag{5.15b}$$

$$\pi_L(X^+)\begin{pmatrix}\mathscr{D}^n & b^n\\ c^n & d^n\end{pmatrix} = -t^{\frac{n+1}{2}}s^{-\frac{n}{2}}[n]_t\begin{pmatrix}0 & b^{n-1}d\\ 0 & 0\end{pmatrix}, \tag{5.15c}$$

$$\pi_L(X^-)\begin{pmatrix}\mathscr{D}^n & b^n\\ c^n & d^n\end{pmatrix} = -t^{\frac{n-3}{2}}s^{\frac{1}{2}}[n]_t\begin{pmatrix}0 & 0\\ ac^{n-1} & bd^{n-1}\end{pmatrix}, \tag{5.15d}$$

$$[n]_t = \frac{t^n - t^{-n}}{t - t^{-1}} = \frac{t^n - t^{-n}}{\lambda}, \quad \lambda = t - t^{-1}.$$

Next we introduce the right regular representation $\pi_R(X)$ [210, 211] (which is used also in [465], though not given in this form, being called left action and denoted π_l):

$$\pi_R(X)M = M\langle X, M\rangle, \qquad X \in \mathscr{U}_{p,q} \tag{5.16}$$

Of course, as in [210, 211] we shall use (5.16) as right action in order to reduce the left regular representation (and we could have also reversed the role of left and right).

Explicitly we have:

$$\pi_R(r)\begin{pmatrix}a & b\\ c & d\end{pmatrix} = \begin{pmatrix}s^{1/2}a & s^{1/2}b\\ s^{1/2}c & s^{1/2}d\end{pmatrix}, \tag{5.17a}$$

$$\pi_R(h)\begin{pmatrix}a & b\\ c & d\end{pmatrix} = \begin{pmatrix}t^{1/2}a & t^{-1/2}b\\ t^{1/2}c & t^{-1/2}d\end{pmatrix}, \tag{5.17b}$$

$$\pi_R(X^+)\begin{pmatrix}a & b\\ c & d\end{pmatrix} = \begin{pmatrix}0 & a\\ 0 & c\end{pmatrix}, \tag{5.17c}$$

$$\pi_R(X^-)\begin{pmatrix}a & b\\ c & d\end{pmatrix} = \begin{pmatrix}b & 0\\ d & 0\end{pmatrix}. \tag{5.17d}$$

The *twisted (right) derivation rule* ([210, 211]) is given by:

$$\pi_R(y)\varphi\psi = \hat{m}\left(\pi_R(\delta_{\mathscr{U}_g}(y))(\varphi \otimes \psi)\right) \tag{5.18}$$

that is, in our concrete situation:

$$\begin{aligned}\pi_R(r)\,\varphi\psi &= \pi_R(r)\varphi \cdot \pi_R(r)\psi,\\ \pi_R(h)\,\varphi\psi &= \pi_R(h)\varphi \cdot \pi_R(h)\psi,\\ \pi_R(X^+)\,\varphi\psi &= \pi_R(h^{-1})\varphi \cdot \pi_R(X^+)\psi \;+\; \pi_R(X^+)\varphi \cdot \pi_R(rh)\psi,\end{aligned} \tag{5.19}$$

$$\pi_R(X^-)\,\varphi\psi = \pi_R(h^{-1})\varphi\cdot\pi_R(X^-)\psi \;+\; \pi_L(X^-)\varphi\cdot\pi_L(r^{-1}h)\psi,$$
$$\pi_R\begin{pmatrix} r & X^+ \\ X^- & h\end{pmatrix} 1_{\mathscr{A}} = \begin{pmatrix} 1_{\mathscr{A}} & 0 \\ 0 & 1_{\mathscr{A}}\end{pmatrix}.$$

Thus we obtain first:

$$\pi_R\begin{pmatrix} r & X^+ \\ X^- & h\end{pmatrix} \mathscr{D} = \begin{pmatrix} s\mathscr{D} & 0 \\ 0 & \mathscr{D}\end{pmatrix} \tag{5.20}$$

and then we get:

$$\pi_R(r)\begin{pmatrix} \mathscr{D}^n\, b^n \\ c^n\;\; d^n\end{pmatrix} = \begin{pmatrix} s^n\mathscr{D}^n & s^{\frac{n}{2}}b^n \\ s^{\frac{n}{2}}c^n & s^{\frac{n}{2}}d^n\end{pmatrix}, \tag{5.21a}$$

$$\pi_R(h)\begin{pmatrix} \mathscr{D}^n\, b^n \\ c^n\;\; d^n\end{pmatrix} = \begin{pmatrix} \mathscr{D}^n & t^{-\frac{n}{2}}b^n \\ t^{\frac{n}{2}}c^n & t^{-\frac{n}{2}}d^n\end{pmatrix}, \tag{5.21b}$$

$$\pi_R(X^+)\begin{pmatrix} \mathscr{D}^n\, b^n \\ c^n\;\; d^n\end{pmatrix} = t^{\frac{n-1}{2}}s^{\frac{n-1}{2}}[n]_t\begin{pmatrix} 0 & ab^{n-1} \\ 0 & cd^{n-1}\end{pmatrix}, \tag{5.21c}$$

$$\pi_R(X^-)\begin{pmatrix} \mathscr{D}^n\, b^n \\ c^n\;\; d^n\end{pmatrix} = t^{\frac{n-1}{2}}[n]_t\begin{pmatrix} 0 & 0 \\ c^{n-1}d & 0\end{pmatrix}. \tag{5.21d}$$

5.1.2 Induced Representations of $\mathscr{U}_{p,q}$ and Intertwining Operators

Here we give the actual construction of the induced representations of $\mathscr{U}_{p,q}$. The induction in the deformed setting is performed by the imposition of the right covariance conditions (cf. [210, 211]). (For the equivalence of this method and the usual one in the classical setting $p = q = 1$ we refer to [197].) Thus, we start with functions which are formal power series:

$$\varphi = \sum_{\substack{k,m,n\in\mathbb{Z}_+\\ \ell\in\mathbb{Z}}} \mu_{k,\ell,m,n}\, b^m\, c^n\, d^\ell\, \mathscr{D}^k. \tag{5.22}$$

Then the *right covariance* conditions [197] with respect to X^-, h, r are:

$$\pi_R(X^-)\varphi = 0,\;\; \pi_R(h)\varphi = t^{-\nu/2}\varphi,\;\; \pi_R(r)\varphi = s^{\rho/2}\varphi. \tag{5.23}$$

Their implementation leads to the conditions:

$$n = 0,\quad \ell + m = \nu,\quad 2k + \ell + m = \rho, \tag{5.24}$$

from which follows that $\nu, \rho \in \mathbb{Z}$ and that $\mu_{k,\ell,m,n} \sim \delta_{n0}\delta_{\ell+m,\nu}\delta_{2k+\ell+m,\rho}$. Thus our reduced functions now are:

$$\varphi(b, d, \mathscr{D}) = \sum_{m\in\mathbb{Z}_+} \mu_m b^m d^{\nu-m} \mathscr{D}^{(\rho-\nu)/2}. \tag{5.25}$$

It is clear, also if we recall the following classical *Gauss decomposition* of $GL(2)$, which holds also here:

$$\begin{pmatrix} a & b \\ c & d \end{pmatrix} = \begin{pmatrix} 1 & bd^{-1} \\ 0 & 1 \end{pmatrix} \begin{pmatrix} d^{-1}\mathscr{D} & 0 \\ 0 & d \end{pmatrix} \begin{pmatrix} 1 & 0 \\ d^{-1}c & 1 \end{pmatrix} \tag{5.26}$$

that the relevant variables are bd^{-1}, d, $\mathscr{D}$, since the variable c was already eliminated.

Thus we shall also introduce the variable $\eta = bd^{-1}$. Then our functions become:

$$\tilde{\varphi}(\eta, d, \mathscr{D}) = \varphi(b, d, \mathscr{D}) = \sum_{m\in\mathbb{Z}_+} \hat{m}_m\, \eta^m\, d^\nu\, \mathscr{D}^{(\rho-\nu)/2} = \hat{\varphi}(\eta)\, d^\nu\, \mathscr{D}^{(\rho-\nu)/2}. \tag{5.27}$$

Now using formulae (5.15) we obtain:

$$\pi_L(r)\, \eta^m d^\nu \mathscr{D}^{(\rho-\nu)/2} = s^{-\rho/2}\, \eta^m\, d^\nu\, \mathscr{D}^{(\rho-\nu)/2}, \tag{5.28a}$$

$$\pi_L(h)\, \eta^m d^\nu \mathscr{D}^{(\rho-\nu)/2} = t^{-m+\nu/2}\, \eta^m\, d^\nu\, \mathscr{D}^{(\rho-\nu)/2}, \tag{5.28b}$$

$$\pi_L(X^+)\, \eta^m d^\nu \mathscr{D}^{(\rho-\nu)/2} = -t^{(3-\nu)/2} s^{-1/2} [m]_t\, \eta^{m-1}\, d^\nu\, \mathscr{D}^{(\rho-\nu)/2} \tag{5.28c}$$

$$\pi_L(X^-)\, \eta^m d^\nu \mathscr{D}^{(\rho-\nu)/2} = t^{(\nu-3)/2} s^{1/2} [m-\nu]_t\, \eta^{m+1}\, d^\nu\, \mathscr{D}^{(\rho-\nu)/2} \tag{5.28d}$$

Then in terms of the functions $\tilde{\varphi}(\eta, d, \mathscr{D})$ we have:

$$\begin{aligned} \pi_L(r)\tilde{\varphi}(\eta, d, \mathscr{D}) &= s^{-\rho/2}\, \tilde{\varphi}(\eta, d, \mathscr{D}), \\ \pi_L(h)\tilde{\varphi}(\eta, d, \mathscr{D}) &= t^{\nu/2}\, T^\eta_{t^{-1}}\, \tilde{\varphi}(\eta, d, \mathscr{D}), \\ \pi_L(X^+)\tilde{\varphi}(\eta, d, \mathscr{D}) &= -t^{(3-\nu)/2} s^{-1/2}\, D^\eta_t\, \tilde{\varphi}(\eta, d, \mathscr{D}), \\ \pi_L(X^-)\tilde{\varphi}(\eta, d, \mathscr{D}) &= t^{(\nu-3)/2} s^{1/2}\, \frac{\eta}{\lambda} \left(t^{-\nu} T^\eta_t - t^\nu T^\eta_{t^{-1}}\right) \tilde{\varphi}(\eta, d, \mathscr{D}), \end{aligned} \tag{5.29}$$

$$T^\eta_t f(\eta) = f(t\eta), \qquad D^\eta_t f(\eta) = \frac{\eta^{-1}}{\lambda} \left(T^\eta_t f(\eta) - T^\eta_{t^{-1}} f(\eta)\right). \tag{5.30}$$

It is immediate to check that $\pi_L(r)$, $\pi_L(h)$, $\pi_L(X^+)$, $\pi_L(X^-)$ satisfy (5.4). It is also clear that if we redefine them by setting:

$$\begin{aligned} &\pi_{\nu,\rho}(r) = \pi_L(r), \qquad \pi_{\nu,\rho}(h) = \pi_L(h), \\ &\pi_{\nu,\rho}(X^+) = t^{(\nu-3)/2} s^{1/2}\, \pi_L(X^+), \qquad \pi_{\nu,\rho}(X^-) = t^{(3-\nu)/2} s^{-1/2}\, \pi_L(X^-), \end{aligned} \tag{5.31}$$

then $\pi_{\nu,\rho}$ shall also satisfy (5.4).

Explicitly, the representation is given by:

$$\pi_{\nu,\rho}(r)\,\tilde{\varphi} = s^{-\rho/2}\,\tilde{\varphi}, \tag{5.32a}$$

$$\pi_{\nu,\rho}(h)\,\tilde{\varphi} = t^{\nu/2}\,T^{\eta}_{t^{-1}}\,\tilde{\varphi}, \tag{5.32b}$$

$$\pi_{\nu,\rho}(X^+)\,\tilde{\varphi} = -D^{\eta}_{t}\,\tilde{\varphi}, \tag{5.32c}$$

$$\pi_{\nu,\rho}(X^-)\,\tilde{\varphi} = \frac{\eta}{\lambda}\left(t^{-\nu}T^{\eta}_{t} - t^{\nu}T^{\eta}_{t^{-1}}\right)\tilde{\varphi}. \tag{5.32d}$$

We denote the representation space of functions $\tilde{\varphi}(\eta, \delta, \mathscr{D})$ with covariance properties (5.23) and transformation laws (5.32) by $\mathscr{C}_{\nu,\rho}$. For $p = q = 1$ our representations coincide with the holomorphic representations induced from the lower diagonal *Borel subgroup B* of $GL(2)$ and acting on the one-dimensional coset G/B. We notice that each representation decouples into a representation of the central subalgebra with generator r, and a representation of the Jimbo quantum algebra $U_t(sl(2))$ (cf. Section 1.2.3). Further we discuss only generic t, that is, which are not at nontrivial roots of unity. For generic t and $\nu \notin \mathbb{Z}_+$ the representations $\pi_{\nu,\rho}$ are irreducible. For generic t and $\nu \in \mathbb{Z}_+$ the representations $\pi_{\nu,\rho}$ are reducible. Indeed, it is easily seen from formulae (5.28) that the vectors $\eta^m d^\nu \mathscr{D}^{(\rho-\nu)/2}$ with $m = 0, 1, \ldots, \nu$ span an invariant subspace of $\mathscr{C}_{\nu,\rho}$ (cf. (5.28d)). Let us denote these finite-dimensional invariant subspaces by $\mathscr{E}_{\nu,\rho}$. Thus, for fixed ρ and $\nu \in \mathbb{Z}_+$ there are two irreducible representations realized in the spaces $\mathscr{E}_{\nu,\rho}$ and $\mathscr{C}_{\nu,\rho}/\mathscr{E}_{\nu,\rho}$.

Furthermore, for $\nu \in \mathbb{Z}_+$ the representations $\pi_{\nu,\rho}$ and $\pi_{-\nu-2,\rho}$ are partially equivalent. This partial equivalence should be realized by the operator:

$$\left(\pi_R(X^+)\right)^{\nu+1} : \mathscr{C}_{\nu,\rho} \longrightarrow \mathscr{C}_{-\nu-2,\rho}, \tag{5.33}$$

since the monomial $(X^+)^{\nu+1}$ is giving the singular vector $(X^+)^{\nu+1}\,v_0$ of the corresponding Verma module [198, 210].

As in [197] one should be careful since $\pi_R(X^+)$ is taking us out of the representation space, which is of course a prerequisite for (5.33), that is, that exactly the $(\nu+1)$-st power will have the required intertwining property. The latter we have to check independently in our setting, since the nonreduced representation spaces depend on all variables. Thus we calculate:

$$\begin{aligned}\left(\pi_R(X^+)\right)^s \eta^m d^\ell \mathscr{D}^k = \sum_{j=0}^{s} & t^{-(2(s+j)m+(s-2j)\ell-s(2s-2j+1))/2} \times \\ & \times s^{(2sk+2jm+s\ell-(s+2js-j))/2} \times \\ & \times \binom{s}{j}_t \frac{[m]_t!\,[\ell+j-s]_t!}{[m+j-s]_t!\,[\ell-s]_t!}\,\zeta^j\,\eta^{m-s}\,d^{\ell-2s}\,\mathscr{D}^{k+s-j}\end{aligned} \tag{5.34}$$

where, $\zeta = cb$. So indeed, there appears dependence on the c variable through ζ. Using the covariance condition once more (no ζ in the expansion) we see that that this is only possible if $\ell - s = -1$ In that case we have:

$$\left(\pi_R(X^+)\right)^{\ell+1} \eta^m\, d^\ell\, \mathscr{D}^k = t^{-\frac{1}{2}(\ell+1)(2m-\ell-3)}\, s^{\frac{1}{2}(\ell+1)(2k+\ell-1)} \times \\ \times [m]_t [m-1]_t \dots [m-\ell]_t\, \eta^{m-\ell-1}\, d^{-\ell-2}\, \mathscr{D}^{k+\ell+1} \tag{5.35}$$

Thus finally if we use the covariance with respect to r and h, namely, we should set $\ell = \nu$ and $k = (\rho - \nu)/2$, we get:

$$\left(\pi_R(X^+)\right)^{\nu+1} \eta^m\, d^\nu\, \mathscr{D}^{(\rho-\nu)/2} = t^{-\frac{1}{2}(\nu+1)(2m-\nu-3)}\, s^{\frac{1}{2}(\nu+1)(\rho-1)} \times \\ \times [m]_t [m-1]_t \dots [m-\nu]_t\, \eta^{m-\nu-1}\, d^{-\nu-2}\, \mathscr{D}^{(\rho+\nu+2)/2} \tag{5.36}$$

Thus, finally, in terms of the functions $\tilde{\varphi}$ we have:

$$(\pi_R(X^+))^{\nu+1} \tilde{\varphi} = \left(t^{(\nu+3)/2}\, s^{(\rho-1)/2} D_t^\eta (T_t^\eta)^{-1}\right)^{\nu+1} \tilde{\varphi} \tag{5.37}$$

Further as in [197] we introduce the restricted functions $\hat{\varphi}(\eta)$ by the formula which is prompted in (5.27):

$$\hat{\varphi}(\eta) = \left(\hat{A}_{\nu,\rho}\, \tilde{\varphi}\right)(\eta) \equiv \tilde{\varphi}(\eta, 1_{\mathscr{A}}, 1_{\mathscr{A}}). \tag{5.38}$$

We denote the representation space of $\hat{\varphi}(\eta)$ by $\widetilde{\mathscr{C}}_{\nu,\rho}$ and the representation acting in $\widetilde{\mathscr{C}}_{\nu,\rho}$ by $\hat{\pi}_{\nu,\rho}$. Thus the operator $\hat{A}_{\nu,\rho}$ acts from $\mathscr{C}_{\nu,\rho}$ to $\widetilde{\mathscr{C}}_{\nu,\rho}$. We shall use also the inverse operator $\hat{A}^{-1}_{\nu,\rho}$ which is defined by:

$$\tilde{\varphi}(\eta, d, \mathscr{D}) = \left(\hat{A}^{-1}_{\nu,\rho}\, \hat{\varphi}\right)(\eta, d, \mathscr{D}) \equiv \hat{\varphi}(\eta)\, d^\nu\, \mathscr{D}^{(\rho-\nu)/2} \tag{5.39}$$

The properties of $\widetilde{\mathscr{C}}_{\nu,\rho}$ follow from the intertwining requirements [197]:

$$\hat{\pi}_{\nu,\rho} \circ \hat{A}_{\nu,\rho} = \hat{A}_{\nu,\rho} \circ \pi_{\nu,\rho}, \qquad \pi_{\nu,\rho} \circ \hat{A}^{-1}_{\nu,\rho} = \hat{A}^{-1}_{\nu,\rho} \circ \hat{\pi}_{\nu,\rho} \tag{5.40}$$

In particular, the representation $\hat{\pi}_{\nu,\rho}$ is given by:

$$\hat{\pi}_{\nu,\rho}(r)\, \hat{\varphi}(\eta) = s^{-\rho/2}\, \hat{\varphi}(\eta), \tag{5.41a}$$

$$\hat{\pi}_{\nu,\rho}(h)\, \hat{\varphi}(\eta) = t^{\nu/2}\, T^\eta_{t^{-1}}\, \hat{\varphi}(\eta), \tag{5.41b}$$

$$\hat{\pi}_{\nu,\rho}(X^+)\, \hat{\varphi}(\eta) = -D_t^\eta\, \hat{\varphi}(\eta), \tag{5.41c}$$

$$\hat{\pi}_{\nu,\rho}(X^-)\, \hat{\varphi}(\eta) = \frac{\eta}{\lambda}\left(t^{-\nu} T_t^\eta - t^\nu T^\eta_{t^{-1}}\right) \hat{\varphi}(\eta), \tag{5.41d}$$

or, using the decomposition $\hat{\varphi}(\eta) = \sum_{m\in\mathbb{Z}_+} \tilde{\mu}_m \eta^m$ inherited from (5.27), we get the analogue of (5.28):

$$\hat{\pi}_{\nu,\rho}(r)\,\eta^m = s^{-\rho/2}\,\eta^m, \tag{5.42a}$$

$$\hat{\pi}_{\nu,\rho}(h)\,\eta^m = t^{-m+\nu/2}\,\eta^m, \tag{5.42b}$$

$$\hat{\pi}_{\nu,\rho}(X^+)\,\eta^m = -[m]_t\,\eta^{m-1}, \tag{5.42c}$$

$$\hat{\pi}_{\nu,\rho}(X^-)\,\eta^m = [m-\nu]_t\,\eta^{m+1}. \tag{5.42d}$$

At this moment we notice that we can consider (5.41),(5.42) for arbitrary complex ν, ρ. Actually the representation has decoupled into a representation of the central generator r (cf. (5.42a)) and the well-known representation of $U_t(sl(2))$ (cf. (5.42b,c,d)) [210, 211]. We know that for generic $t, \nu \in \mathbb{C}$ the representations $\hat{\pi}_{\nu,\rho}$ are irreducible. For generic $t \in \mathbb{C}$ and $\nu \in \mathbb{Z}_+$ the representations $\hat{\pi}_{\nu,\rho}$ are reducible. Moreover, for $\nu \in \mathbb{Z}_+$ the representations $\hat{\pi}_{\nu,\rho}$ and $\hat{\pi}_{-\nu-2,\rho}$ are partially equivalent. The intertwining operators between these pairs is naturally obtained from the ones relating the pairs $\pi_{\nu,\rho}$ and $\pi_{-\nu-2,\rho}$, namely:

$$\mathscr{I}_\nu : \widetilde{\mathscr{C}}_{\nu,\rho} \longrightarrow \widetilde{\mathscr{C}}_{-\nu-2,\rho}\,, \tag{5.43}$$

$$\mathscr{I}_\nu \equiv \left(t^{-(\nu+3)/2}\, s^{(1-\rho)/2}\right)^{\nu+1} \hat{A}^{-1}_{\nu,\rho} \circ (\pi_R(X^+))^{\nu+1} \circ \hat{A}_{-\nu-2,\rho}$$

$$\mathscr{I}_\nu \hat{\varphi} = \left(D^\eta_t (T^\eta_t)^{-1}\right)^{\nu+1} \hat{\varphi}$$

where we have also made use of the fact that the intertwining operators are defined up to multiplicative factors. Formulae (5.43) (for $s = 1$) were obtained first in [210] (see also [211]). The kernel $\mathscr{E}_\nu$ of the operator (5.43) is an invariant subspace of $\widetilde{\mathscr{C}}_{\nu,\rho}$. It consists of polynomials of degree $\le \nu$, $\dim \mathscr{E}_\nu = \nu+1$. The basis of $\mathscr{E}_\nu$ may be taken as $1_{\mathscr{A}}, \eta, \dots, \eta^\nu$, on which $\hat{\pi}_{\nu,\rho}$ acts as in (5.42). Thus $\mathscr{E}_\nu$, $\nu \in \mathbb{Z}_+$, is a finite-dimensional representation space with highest-weight vector $1_{\mathscr{A}}$ ($\hat{\pi}_{\nu,\rho}(X^+)1_{\mathscr{A}} = 0$) and lowest-weight vector η^ν ($\hat{\pi}_{\nu,\rho}(X^-)\eta^\nu = 0$).

Finally, we should note that since we have functions of one variable η we can treat it as complex variable z. In these terms we can also recover from (5.41) the classical vector-field representation of $gl(2)$ by setting (as noted above) $h^\varepsilon = t^{\varepsilon H/2}$, $r^\varepsilon = s^{\epsilon K/2}$, expanding $h^\varepsilon \approx 1 + \varepsilon(\log t)\,H/2$, $r^\varepsilon \approx 1 + \epsilon(\log s)\,K/2$, and taking the limit $\log t \to 0, \log s \to 0$. Thus, we get:

$$K = -\rho, \quad H = \nu - 2z\partial_z, \quad X^+ = -\partial_z, \quad X^- = z^2\partial_z - \nu z, \qquad \partial_z \equiv \frac{d}{dz} \tag{5.44}$$

which fulfills (5.5).

5.1.3 The Case $U_q(sl(2))$

Here in the context of the preceding subsections we consider the one-parameter case $p = q$. As we have seen in this case the quantum algebra $U_{q,q}(gl(2))$ completely splits in two subalgebras $U_q(sl(2))$ (generated by $h, X^{\pm}$) and $U(\mathscr{Z})$ (generated by K); cf. Theorem 4.1 in Section 4.1. We restrict to the subalgebra $U_q(sl(2))$ and use the results from the previous subsection. We set $s = p/q = 1$, and we ignore the representation parameter ρ which is not relevant in this situation. Thus, for the representation functions we have:

$$\tilde{\varphi}(\eta, d, \mathscr{D}) = \varphi(b, d, \mathscr{D}) = \sum_{m \in \mathbb{Z}_+} \hat{m}_m\, \eta^m\, d^\nu\, \mathscr{D}^{-\nu/2} = \hat{\varphi}(\eta)\, d^\nu\, \mathscr{D}^{-\nu/2}. \tag{5.45}$$

Then, the representation is given by:

$$\pi_\nu(h)\, \tilde{\varphi} = t^{\nu/2}\, T^{\eta}_{t^{-1}}\, \tilde{\varphi}, \tag{5.46a}$$

$$\pi_\nu(X^+)\, \tilde{\varphi} = -D^{\eta}_t\, \tilde{\varphi}, \tag{5.46b}$$

$$\pi_\nu(X^-)\, \tilde{\varphi} = \frac{\eta}{\lambda}\left(t^{-\nu} T^{\eta}_t - t^{\nu} T^{\eta}_{t^{-1}}\right) \tilde{\varphi}. \tag{5.46c}$$

where T^{η}_t, D^{η}_t are defined in (5.30). The above representation was obtained first in [309] by another method and then in [210, 211] by the method we follow here.

Further we introduce the restricted functions $\hat{\varphi}(\eta)$ as in (5.38). We denote the representation space of $\hat{\varphi}(\eta)$ by $\widetilde{\mathscr{C}}_\nu$ and the representation acting in $\widetilde{\mathscr{C}}\nu$ by $\hat{\pi}_\nu$. In particular, the representation $\hat{\pi}_\nu$ acting on $\hat{\varphi}$ looks exactly as the representation π_ν acting on $\tilde{\varphi}$. Next. using the decomposition $\hat{\varphi}(\eta) = \sum_{m \in \mathbb{Z}_+} \tilde{\mu}_m\, \eta^m$ inherited from (5.45), we get the analogue of (5.28):

$$\hat{\pi}_\nu(h)\, \eta^m = t^{-m+\nu/2}\, \eta^m, \tag{5.47a}$$

$$\hat{\pi}_\nu(X^+)\, \eta^m = -[m]_t\, \eta^{m-1}, \tag{5.47b}$$

$$\hat{\pi}_\nu(X^-)\, \eta^m = [m - \nu]_t\, \eta^{m+1}. \tag{5.47c}$$

Now we notice that we can consider (5.47) for arbitrary complex ν. Further we discuss only generic q, that is, which are not at nontrivial roots of unity. For generic q and $\nu \notin \mathbb{Z}_+$ the representations $\hat{\pi}_\nu$ are irreducible. For generic q and $\nu \in \mathbb{Z}_+$ the representations $\hat{\pi}_\nu$ are reducible. Moreover, for $\nu \in \mathbb{Z}_+$ the representations $\hat{\pi}_\nu$ and $\hat{\pi}_{-\nu-2}$ are partially equivalent. The intertwining operators between these pairs is naturally obtained from the ones relating the pairs π_ν and $\pi_{-\nu-2}$, namely:

$$\mathscr{I}_\nu : \widehat{C}_\nu \longrightarrow \widehat{C}_{-\nu-2}, \tag{5.48a}$$

$$\mathscr{I}_\nu \equiv \left(t^{-(\nu+3)/2}\right)^{\nu+1} \hat{A}_\nu^{-1} \circ (\pi_R(X^+))^{\nu+1} \circ \hat{A}_{-\nu-2}, \tag{5.48b}$$

$$\mathscr{I}_\nu \hat{\varphi} = \left(D^{\eta}_t (T^{\eta}_t)^{-1}\right)^{\nu+1} \hat{\varphi} \tag{5.48c}$$

where we have also made use of the fact that the intertwining operators are defined up to multiplicative factors. Formulae (5.48) were obtained first in [210, 211]. The kernel $\widehat{\mathcal{E}}_\nu$ of the operator (5.43) is an invariant subspace of $\widehat{C}_\nu$. It consists of polynomials of degree $\leq \nu$, dim $\widehat{\mathcal{E}}_\nu = \nu + 1$. The basis of $\widehat{\mathcal{E}}_\nu$ may be taken as $1_{\mathcal{A}}, \eta, \ldots, \eta^\nu$, on which $\hat{\pi}_{\nu,\rho}$ acts as in (5.47). Thus $\widehat{\mathcal{E}}_\nu$, $\nu \in \mathbb{Z}_+$, is a finite-dimensional representation space with highest-weight vector $1_{\mathcal{A}}$ ($\hat{\pi}_{\nu,\rho}(X^+)1_{\mathcal{A}} = 0$) and lowest-weight vector η^ν ($\hat{\pi}_{\nu,\rho}(X^-)\eta^\nu = 0$). Thus, for fixed $\nu \in \mathbb{Z}_+$ there are two irreducible representations realized in the spaces $\widehat{\mathcal{E}}_\nu$ and $\widehat{C}_\nu / \widehat{\mathcal{E}}_\nu$.

5.2 The Case of $GL_{g,h}(2)$

5.2.1 Left and Right Action of $U_{g,h}(gl(2))$ on $GL_{g,h}(2)$

In this section following [228] we construct induced representations of the quantum algebra $\mathcal{U}_{g,h} = U_{g,h}(gl(2))$, which was found in [39] and reviewed in Section 4.7.2 as the dual of the Jordanian two-parameter matrix quantum group $GL_{g,h}(2)$ introduced in [13] (denoted there $GL_{h,h'}$) and reviewed in Section 4.7.1. We follow (with some modifications) the similar construction of the previous section. The representation spaces are built on formal power series in the generating elements a, b, c, d, of $\mathcal{A}_{g,h} = GL_{g,h}(2)$, with commutations relations shown in (4.172).

We start with the following basis of $\mathcal{A}_{g,h}$:

$$f = f_{k,\ell,m,n} = b^m\, a^\ell\, c^n\, d^k, \qquad k, \ell, m, n \in \mathbb{Z}_+. \tag{5.49}$$

As in [39] we use also the following change of generating elements and parameters of $\mathcal{U}_{g,h}$:

$$\begin{aligned} \tilde{a} &= \tfrac{1}{2}(a + d), & \tilde{d} &= \tfrac{1}{2}(a - d), \\ \tilde{g} &= \tfrac{1}{2}(g + h), & \tilde{h} &= \tfrac{1}{2}(g - h) \end{aligned} \tag{5.50}$$

with PBW basis:

$$f' = f'_{k,\ell,m,n} = \tilde{a}^k\, \tilde{d}^\ell\, c^n\, b^m, \qquad k, \ell, m, n \in \mathbb{Z}_+ \tag{5.51}$$

The generating elements A, B, C, D, of $\mathcal{U}_{g,h}$ are given as in [39], see also (4.187). Further, in (4.198) was made a one-parameter change of basis from A, B, C, D to A, B, Y, H. Finally, was introduced a subalgebra $\widetilde{\mathcal{U}}_{g,h}$ of $\mathcal{U}_{g,h}$ with the basis: A, $K = K^+ = e^{\tilde{g}B}$, $K^{-1} = K^- = e^{-\tilde{g}B}$, Y, H, with relations in (4.201) and coalgebra structure in (4.206),(4.207),(4.208).

For further use we recall that to obtain from the above formulae the classical $U(gl(2))$, one has to reintroduce the generator B by setting $K^\pm = e^{\pm\tilde{g}B}$, then to expand $K^\pm \approx 1_{\mathscr{U}} \pm \tilde{g}B$ and to take the limit $\tilde{g} \to 0$. Thus, we obtain from (4.201):

$$[B, Y] = H, \quad [H, B] = 2B, \quad [H, Y] = -2Y, \tag{5.52a}$$

$$[A, B] = 0, \quad [A, Y] = 0, \quad [A, H] = 0, \tag{5.52b}$$

The duality for the algebras $\mathscr{U}_{g,h}$ and $\mathscr{A}_{g,h}$ is given by the pairings:

$$\begin{aligned}
&\langle A, \begin{pmatrix} a & b \\ c & d \end{pmatrix} \rangle = \begin{pmatrix} 1 & 0 \\ 0 & 1 \end{pmatrix}, \\
&\langle K^\pm, \begin{pmatrix} a & b \\ c & d \end{pmatrix} \rangle = \begin{pmatrix} 1 & \pm\tilde{g} \\ 0 & 1 \end{pmatrix}, \\
&\langle H, \begin{pmatrix} a & b \\ c & d \end{pmatrix} \rangle = \begin{pmatrix} 1 & \mu - \nu \\ 0 & -1 \end{pmatrix} = \begin{pmatrix} 1 & 2\mu - \tilde{g} \\ 0 & -1 \end{pmatrix}, \\
&\langle Y, \begin{pmatrix} a & b \\ c & d \end{pmatrix} \rangle = \begin{pmatrix} \frac{1}{2}\tilde{g} - \mu & \mu\nu \\ 1 & \frac{1}{2}\tilde{g} - \nu \end{pmatrix} = \begin{pmatrix} \frac{1}{2}\tilde{g} - \mu & \mu(\tilde{g} - \mu) \\ 1 & \mu - \frac{1}{2}\tilde{g} \end{pmatrix}.
\end{aligned} \tag{5.53}$$

For the left regular representation we need the following formulae [228]:

$$\langle \gamma_{\mathscr{U}}(A), \begin{pmatrix} a & b \\ c & d \end{pmatrix} \rangle = \begin{pmatrix} -1 & 0 \\ 0 & -1 \end{pmatrix} = -I_2\,, \tag{5.54a}$$

$$\langle \gamma_{\mathscr{U}}(K^\pm), \begin{pmatrix} a & b \\ c & d \end{pmatrix} \rangle = \begin{pmatrix} 1 & \mp\tilde{g} \\ 0 & 1 \end{pmatrix}, \tag{5.54b}$$

$$\langle \gamma_{\mathscr{U}}(H), \begin{pmatrix} a & b \\ c & d \end{pmatrix} \rangle = \begin{pmatrix} -1 & -2\tilde{\mu} \\ 0 & 1 \end{pmatrix}, \qquad \tilde{\mu} \equiv \mu + \tilde{h} + \tfrac{1}{2}\tilde{g}, \tag{5.54c}$$

$$\langle \gamma_{\mathscr{U}}(Y), \begin{pmatrix} a & b \\ c & d \end{pmatrix} \rangle = \begin{pmatrix} \tilde{\mu} & G \\ -1 & -\tilde{\mu} \end{pmatrix}, \qquad G \equiv \tilde{\mu}^2 - \tilde{g}^2/4 \tag{5.54d}$$

The left regular representation of $\mathscr{U}_{g,h}$ is again given by (5.10). Explicitly we get here:

$$\begin{aligned}
&\pi_L(A)\begin{pmatrix} a\ b \\ c\ d \end{pmatrix} = \begin{pmatrix} -1 & 0 \\ 0 & -1 \end{pmatrix}\begin{pmatrix} a\ b \\ c\ d \end{pmatrix} = \begin{pmatrix} -a & -b \\ -c & -d \end{pmatrix} \\
&\pi_L(K^\pm)\begin{pmatrix} a\ b \\ c\ d \end{pmatrix} = \begin{pmatrix} 1 & \mp\tilde{g} \\ 0 & 1 \end{pmatrix}\begin{pmatrix} a\ b \\ c\ d \end{pmatrix} = \begin{pmatrix} a \mp \tilde{g}c & b \mp \tilde{g}d \\ c & d \end{pmatrix}
\end{aligned} \tag{5.55}$$

$$\pi_L(H)\begin{pmatrix} a & b \\ c & d \end{pmatrix} = \begin{pmatrix} -1 & -2\tilde{\mu} \\ 0 & 1 \end{pmatrix}\begin{pmatrix} a & b \\ c & d \end{pmatrix} = \begin{pmatrix} -a-2\tilde{\mu}c & -b-2\tilde{\mu}d \\ c & d \end{pmatrix}$$

$$\pi_L(Y)\begin{pmatrix} a & b \\ c & d \end{pmatrix} = \begin{pmatrix} \tilde{\mu} & G \\ -1 & -\tilde{\mu} \end{pmatrix}\begin{pmatrix} a & b \\ c & d \end{pmatrix} = \begin{pmatrix} \tilde{\mu}a+Gc & \tilde{\mu}b+Gd \\ -a-\tilde{\mu}c & -b-\tilde{\mu}d \end{pmatrix}$$

In order to derive the action of π_L on arbitrary elements of the basis we use the twisted derivation rule (5.12). In the situation here this gives:

$$\pi_L(A)\,\varphi\psi = \pi_L(A)\varphi\cdot\psi + \varphi\cdot\pi_L(A)\psi\,, \tag{5.56a}$$

$$\pi_L(K^{\pm})\,\varphi\psi = \pi_L(K^{\pm})\varphi\cdot\pi_L(K^{\pm})\psi\,, \tag{5.56b}$$

$$\begin{aligned}\pi_L(H)\,\varphi\psi = {}& \pi_L(K^{-1})\varphi\cdot\pi_L(H)\psi + \pi_L(H)\varphi\cdot\pi_L(K)\psi + \\ & +\tfrac{\tilde{h}}{\tilde{g}}\,\pi_L(AK^{-1})\varphi\cdot\pi_L(K^{-1}-K)\psi\,,\end{aligned} \tag{5.56c}$$

$$\begin{aligned}\pi_L(Y)\,\varphi\psi = {}& \pi_L(K^{-1})\varphi\cdot\pi_L(Y)\psi + \pi_L(Y)\varphi\cdot\pi_L(K)\psi + \\ & +\tfrac{\tilde{h}^2}{2\tilde{g}}\,\pi_L(A^2K^{-1})\varphi\cdot\pi_L(K^{-1}-K)\psi + \\ & +\tilde{h}\,\pi_L(AK^{-1})\varphi\cdot\pi_L(H)\psi\,,\end{aligned} \tag{5.56d}$$

$$\pi_L\begin{pmatrix} A & H \\ K & Y \end{pmatrix} 1_{\mathscr{A}} = \begin{pmatrix} 0 & 0 \\ 1_{\mathscr{A}} & 0 \end{pmatrix}. \tag{5.56e}$$

Thus, we have also:

$$\pi_L\begin{pmatrix} A & H \\ K & Y \end{pmatrix}\mathscr{D} = \begin{pmatrix} -2\mathscr{D} & 0 \\ \mathscr{D} & 0 \end{pmatrix}. \tag{5.57}$$

Next we introduce the right regular representation $\pi_R(X)$ as in (5.16). Explicitly we get here:

$$\pi_R(A)\begin{pmatrix} a & b \\ c & d \end{pmatrix} = \begin{pmatrix} a & b \\ c & d \end{pmatrix}\begin{pmatrix} 1 & 0 \\ 0 & 1 \end{pmatrix} = \begin{pmatrix} a & b \\ c & d \end{pmatrix}, \tag{5.58a}$$

$$\pi_R(K^{\pm})\begin{pmatrix} a & b \\ c & d \end{pmatrix} = \begin{pmatrix} a & b \\ c & d \end{pmatrix}\begin{pmatrix} 1 & \pm\tilde{g} \\ 0 & 1 \end{pmatrix} = \begin{pmatrix} a & b\pm\tilde{g}a \\ c & d\pm\tilde{g}c \end{pmatrix}, \tag{5.58b}$$

$$\begin{aligned}\pi_R(H)\begin{pmatrix} a & b \\ c & d \end{pmatrix} &= \begin{pmatrix} a & b \\ c & d \end{pmatrix}\begin{pmatrix} 1 & 2\mu-\tilde{g} \\ 0 & -1 \end{pmatrix} = \\ &= \begin{pmatrix} a & (2\mu-\tilde{g})a-b \\ c & (2\mu-\tilde{g})c-d \end{pmatrix},\end{aligned} \tag{5.58c}$$

$$\begin{aligned}\pi_R(Y)\begin{pmatrix} a & b \\ c & d \end{pmatrix} &= \begin{pmatrix} a & b \\ c & d \end{pmatrix}\begin{pmatrix} \frac{1}{2}\tilde{g}-\mu & \mu(\tilde{g}-\mu) \\ 1 & \mu-\frac{1}{2}\tilde{g} \end{pmatrix} = \\ &= \begin{pmatrix} b+(\frac{1}{2}\tilde{g}-\mu)a & \mu(\tilde{g}-\mu)a+(\mu-\frac{1}{2}\tilde{g})b \\ d+(\frac{1}{2}\tilde{g}-\mu)c & \mu(\tilde{g}-\mu)c+(\mu-\frac{1}{2}\tilde{g})d \end{pmatrix}\end{aligned} \tag{5.58d}$$

The twisted derivation rule is given again by (5.18). Here, this gives:

$$\pi_R(A)\,\varphi\psi = \pi_R(A)\varphi\cdot\psi + \varphi\cdot\pi_R(A)\psi\,, \tag{5.59a}$$

$$\pi_R(K^{\pm})\,\varphi\psi = \pi_R(K^{\pm})\varphi\cdot\pi_R(K^{\pm})\psi\,, \tag{5.59b}$$

$$\begin{aligned}\pi_R(H)\,\varphi\psi &= \pi_R(K)\varphi\cdot\pi_R(H)\psi + \pi_R(H)\varphi\cdot\pi_R(K^{-1})\psi + \\ &+\tfrac{\tilde h}{\tilde g}\,\pi_R(K^{-1}-K)\varphi\cdot\pi_R(AK^{-1})\psi\,,\end{aligned} \tag{5.59c}$$

$$\begin{aligned}\pi_R(Y)\,\varphi\psi &= \pi_R(K)\varphi\cdot\pi_R(Y)\psi + \pi_R(Y)\varphi\cdot\pi_R(K^{-1})\psi + \\ &+\tfrac{\tilde h^2}{2\tilde g}\,\pi_R(K^{-1}-K)\varphi\cdot\pi_R(A^2K^{-1})\psi + \\ &+\tilde h\,\pi_R(H)\varphi\cdot\pi_R(AK^{-1})\psi\,,\end{aligned} \tag{5.59d}$$

$$\pi_R\begin{pmatrix}A & H\\ K & Y\end{pmatrix} 1_{\mathcal{A}} = \begin{pmatrix}0 & 0\\ 1_{\mathcal{A}} & 0\end{pmatrix}. \tag{5.59e}$$

Thus, we have also:

$$\pi_R\begin{pmatrix}A & H\\ K & Y\end{pmatrix}\mathcal{D} = \begin{pmatrix}2\mathcal{D} & 0\\ \mathcal{D} & 0\end{pmatrix}. \tag{5.60}$$

5.2.2 Induced Representations of $\mathcal{U}_{g,h}$ and Intertwining Operators

As in the p, q case the construction of the induced representations of $\mathcal{U}_{g,h}$ is performed in the deformed setting by the imposition of the right covariance conditions (cf. [210, 211]). We start with functions which are formal power series of the kind:

$$\varphi = \sum_{k,\ell,m,n\in\mathbb{Z}_+} \mu_{k,\ell,m,n} b^m\, a^\ell\, c^n\, d^k \tag{5.61}$$

Above we have defined left and right action of $\mathcal{U}_{g,h}$ on φ and as before we shall use the right action to reduce the left regular representation. Note that unlike the $\mathcal{U}_{p,q}$ case, where we could have used either of the two conjugate Borel subalgebras for the reduction, here we have only the "upper diagonal" Borel subalgebra generated by $A, K^{\pm}, H$. (Note that this feature is true also in the one-parameter Jordanian deformation as used for its representation theory [218] and next subsection.) Thus right covariance conditions will look as follows:

$$\pi_R(A)\,\varphi = \rho\,\varphi, \tag{5.62a}$$

$$\pi_R(K)\,\varphi = \varphi \quad\Leftrightarrow\quad \pi_R(B)\,\varphi = 0, \tag{5.62b}$$

$$\pi_R(H)\,\varphi = \nu\,\varphi, \tag{5.62c}$$

Let us apply first (5.62b). From (5.58b) it is clear that only those functions which do not depend on b and d will satisfy (5.62b). However, our functions may depend on b and

d through the determinant $\mathscr{D}$ since the latter is preserved by the action of $\pi_R(K)$ (cf. (5.60)).

Thus, instead of (5.61) we take for our functions:

$$\tilde{\varphi} = \sum_{k,\ell,n\in\mathbb{Z}_+} \mu_{k,\ell,n} a^\ell\, c^n\, \mathscr{D}^k, \tag{5.63}$$

with the same conditions (5.62) (with $\varphi \longrightarrow \tilde{\varphi}$). By construction (5.62b) is fulfilled for all $\tilde{\varphi}$. To apply (5.62a) we first obtain using (5.58a) and (5.59a):

$$\pi_R(A)\, a^\ell\, c^n\, \mathscr{D}^k = (2k + \ell + n)\, a^\ell\, c^n\, \mathscr{D}^k. \tag{5.64}$$

Combining (5.64) with (5.62a) we obtain that: $2k + \ell + n = \rho$. Analogously, we obtain using (5.58c) and (5.59c):

$$\pi_R(H)\, a^\ell\, c^n\, \mathscr{D}^k = (\ell + n)\, a^\ell\, c^n\, \mathscr{D}^k. \tag{5.65}$$

Combining (5.65) with (5.62c) we obtain that: $\ell+n = \nu$. Thus, we conclude that $\rho, \nu, \frac{1}{2}(\rho-\nu) \in \mathbb{Z}_+$, and that the functions in (5.63) reduce to:

$$\tilde{\varphi} = \sum_{n\in\mathbb{Z}_+} \mu_n a^n\, c^{\nu-n}\, \mathscr{D}^{(\rho-\nu)/2}. \tag{5.66}$$

Thus, our functions are given in the bases:

$$u_n = a^n\, c^{\nu-n}\, \mathscr{D}^{(\rho-\nu)/2}, \qquad n \in \mathbb{Z}_+ \tag{5.67}$$

Now if neither a or c has an inverse, these bases will be finite-dimensional, in contrast to the undeformed case. However, these finite-dimensional representations we shall obtain also if we suppose that either a or c has an inverse (see below). Thus, further we shall suppose that c has an inverse. (The choice of a having an inverse would make the further formulae much more complicated, as can be anticipated from the comparison of the action of the algebra on a or c.)

The transformation rules for the bases (5.67) are (with $\tilde{\mu} = 0$):

$$\begin{aligned}
\pi_L(A)\, u_n &= -\rho\, u_n\,, \\
\pi_L(K^\varepsilon)\, u_n &= (a - \varepsilon\tilde{g}c)^n\, c^{\nu-n}\, \mathscr{D}^{(\rho-\nu)/2} = \sum_{s=0}^{n} (-\varepsilon)^s \binom{n}{s} \Delta_s^\varepsilon\, u_{n-s}\,, \\
\Delta_s^\varepsilon &= \prod_{j=0}^{s-1} (\tilde{g} + \varepsilon j g)\,, \qquad (\Delta_0^\varepsilon = 1)\,, \qquad \varepsilon = \pm\,,
\end{aligned} \tag{5.68}$$

$$\begin{aligned}\pi_L(H)\,u_n &= (\nu-n)\,(a+\tilde{g}c)^n\,c^{\nu-n}\,\mathscr{D}^{(\rho-\nu)/2}\;+\\ &+\sum_{s=0}^{n}\tilde{\beta}_{ns}\,a^{n-s}\,c^{\nu-n+s}\,\mathscr{D}^{(\rho-\nu)/2}\;=\\ &=\sum_{s=0}^{n}\binom{n}{s}\left((\nu-n)\,\Delta_s^-+\frac{(-1)^s}{s+1}\,\Delta_s^+\,\beta_{ns}\right)u_{n-s}\,,\\ \beta_{n0} &= -n\,,\qquad \beta_{n1}=2(n-1)\,\tfrac{\tilde{h}}{\tilde{g}}\,,\\ \beta_{ns} &= 2(n-s)\,\tfrac{\tilde{h}}{\tilde{g}}\;+\\ &+\sum_{\ell=1}^{s-1}(-1)^\ell\,\frac{\Delta_\ell^-}{\Delta_{\ell+1}^+}\,\left\{(n-s)(\tilde{g}+2\tilde{h})\;-\;\ell\,(n+1)\,(\tilde{g}-\tilde{h})\right\}\end{aligned}$$

$$\begin{aligned}\pi_L(Y)\,u_n &= (n-\nu)\,u_{n+1}\;+\\ &+\left[\binom{n}{2}\tilde{h}+\binom{\nu-n}{2}\tilde{g}+(n-\nu)n(\tilde{g}+\tilde{h})\right]u_n\;+\\ &+\sum_{k=1}^{n}\left\{(-1)^k\gamma_{nk}+\binom{n}{k}\Delta_k^-\left[\binom{\nu-n}{2}\tilde{g}+(n-\nu)(n\tilde{h}+kg)\right]+\right.\\ &\left.+\binom{n}{k+1}(n-\nu)\,\Delta_{k+1}^-\right\}u_{n-k}\,,\\ \gamma_{nk} &= \Delta_k^+\left\{\binom{n-2}{k-2}\tfrac{1}{4}\tilde{g}+\tfrac{1}{\tilde{g}}\left(\frac{(n+k-1)(n-2)!}{4(n-k-1)!k!}\,\tilde{g}^2-\binom{n}{k+2}\tilde{h}^2-\right.\right.\\ &\left.-\binom{n+1}{k+2}\tilde{g}\tilde{h}\right)+\sum_{\ell=2}^{k}(-1)^\ell\frac{\Delta_\ell^-}{\Delta_\ell^+}\left[\ell\binom{n+1}{k+2}+\binom{n}{k+2}\right]\tilde{h}\;+\\ &+\sum_{\ell=1}^{k-1}(-1)^\ell\frac{\Delta_\ell^-}{\Delta_{\ell+1}^+}\left[\binom{n}{k}\left(\tfrac{1}{4}\tilde{g}^2+\ell^2\,\tilde{g}\,\tilde{h}\right)+2\binom{n}{k+1}\binom{\ell+1}{2}\tilde{g}\,\tilde{h}\,-\right.\\ &\left.\left.-\frac{n!(\ell+1)(\ell+n\ell+2n-k)}{(k+2)!(n-k-1)!}\,\tilde{h}^2\right]\right\}\end{aligned}$$

Thus, we have obtained infinite-dimensional representations of $\mathscr{U}_{g,h}$ parametrized by two integers. We shall denote by $\widetilde{\mathscr{C}}_{\nu,\rho}$ the representation spaces of the functions $\tilde{\varphi}$ in (5.63). For $g = h = 0$ our representations coincide with the holomorphic representations induced from the upper diagonal Borel subgroup B of $GL(2)$ and acting on the one-dimensional coset G/B. Let us comment that the two subalgebras $\mathscr{Z}$ and $\mathscr{U}'$ are represented separately with parameters ρ and ν, respectively. First we notice from (5.68) that if $\nu \in \mathbb{Z}_+$ the representation space $\widetilde{\mathscr{C}}_{\nu,\rho}$ is reducible since the vectors $u_n, n = 0, 1, \dots, \nu$ span a finite-dimensional invariant subspace (see (5.68d)). For further use we shall denote these representation spaces by $\widetilde{\mathscr{E}}_{\nu,\rho}$, $\nu \in \mathbb{Z}_+$. Thus, if $\nu \in \mathbb{Z}_+$ we have two irreducible representations with representation spaces isomorphic to $\widetilde{\mathscr{E}}_{\nu,\rho}$ and to $\widetilde{\mathscr{C}}_{\nu,\rho}/\widetilde{\mathscr{E}}_{\nu,\rho}$.

We would like further to reduce the representation spaces. From the above we are prompted to use the variable $\chi \equiv ac^{-1}$. This is also related to the following classical Gauss decomposition of $GL(2)$:

$$\begin{pmatrix} a & b \\ c & d \end{pmatrix} = \begin{pmatrix} 1 & 0 \\ ca^{-1} & 1 \end{pmatrix} \begin{pmatrix} a & 0 \\ 0 & a^{-1}\mathcal{D} \end{pmatrix} \begin{pmatrix} 1 & a^{-1}b \\ 0 & 1 \end{pmatrix} \tag{5.69}$$

though instead of $\xi \equiv ca^{-1}$ we shall use χ as defined above. We note also the following connection between the two variable $\chi = \xi^{-1} - g$.

The action on the new variable is:

$$\pi_L \begin{pmatrix} A & H \\ K^{\pm} & Y \end{pmatrix} \chi^n = \begin{pmatrix} 0 & -(2\chi + \tilde{g})\, \frac{(\chi+\tilde{g})^n - (\chi-\tilde{g})^n}{2\tilde{g}} \\ (\chi \mp \tilde{g})^n & (\chi^2 + \tilde{g}\chi - \frac{\tilde{g}^2}{4})\, \frac{(\chi+\tilde{g})^n - (\chi-\tilde{g})^n}{2\tilde{g}} \end{pmatrix} \tag{5.70}$$

Thus, our bases and functions shall be:

$$w_n = \chi^n\, c^\nu\, \mathcal{D}^{(\rho-\nu)/2}, \qquad n \in \mathbb{Z}_+ \tag{5.71}$$

$$\phi = \phi(\chi, c, \mathcal{D}) = \sum_{n\in\mathbb{Z}_+} \mu_n \chi^n\, c^\nu\, \mathcal{D}^{(\rho-\nu)/2} = \hat{\varphi}(\chi)\, c^\nu\, \mathcal{D}^{(\rho-\nu)/2} \tag{5.72}$$

We shall denote these representation spaces by $\mathcal{C}_{\nu,\rho}$, and the representation acting in these spaces by $\pi_{\nu,\rho}$. Explicitly the action is:

$$\pi_{\nu,\rho}(A)\, w_n = -\rho\, w_n \tag{5.73a}$$

$$\pi_{\nu,\rho}(K^{\pm})\, w_n = (\chi \mp \tilde{g})^n\, c^\nu \mathcal{D}^{\frac{\rho-\nu}{2}} = \sum_{k=0}^{n} \binom{n}{k} (\mp\tilde{g})^k\, w_{n-k} \tag{5.73b}$$

$$\begin{aligned} \pi_{\nu,\rho}(H)\, w_n &= \left(\nu - \tfrac{1}{\tilde{g}}\chi - \tfrac{1}{2}\right) (\chi + \tilde{g})^n\, c^\nu \mathcal{D}^{\frac{\rho-\nu}{2}} + \\ &\quad + \left(\tfrac{1}{\tilde{g}}\chi + \tfrac{1}{2}\right) (\chi - \tilde{g})^n\, c^\nu \mathcal{D}^{\frac{\rho-\nu}{2}} = \\ &= \left(\nu - \tfrac{1}{\tilde{g}}\chi - \tfrac{1}{2}\right) \pi_{\nu,\rho}(K^-)\, w_n + \\ &\quad + \left(\tfrac{1}{\tilde{g}}\chi + \tfrac{1}{2}\right) \pi_{\nu,\rho}(K^+)\, w_n = \qquad (5.73c) \\ &= (\nu - 2n)\, w_n + \sum_{k=0}^{n-1} \omega(n,k,\nu,\tilde{g})\, w_k \qquad (5.73c') \end{aligned}$$

$$\begin{aligned} \pi_{\nu,\rho}(Y)\, w_n &= \left\{ \tfrac{1}{2\tilde{g}}\chi^2 + (\tfrac{1}{2} - \nu)\chi - \right. \\ &\quad \left. - \tfrac{\tilde{g}}{2}(\nu^2 + 3\nu + \tfrac{1}{4}) \right\} (\chi + \tilde{g})^n\, c^\nu \mathcal{D}^{\frac{\rho-\nu}{2}} - \\ &\quad - \left(\tfrac{1}{2\tilde{g}}\chi^2 + \tfrac{1}{2}\chi - \tfrac{\tilde{g}}{8} \right) (\chi - \tilde{g})^n\, c^\nu \mathcal{D}^{\frac{\rho-\nu}{2}} = \end{aligned}$$

$$= \left\{\tfrac{1}{2\tilde{g}}\chi^2 + (\tfrac{1}{2} - \nu)\chi - \tfrac{\tilde{g}}{2}(\nu^2 + 3\nu + \tfrac{1}{4})\right\} \pi_{\nu,\rho}(K^-)\, w_n -$$

$$- \left(\tfrac{1}{2\tilde{g}}\chi^2 + \tfrac{1}{2}\chi - \tfrac{\tilde{g}}{8}\right) \pi_{\nu,\rho}(K^+)\, w_n = \tag{5.73d}$$

$$= (n - \nu)\, w_{n+1} + \sum_{k=0}^{n} \omega'(n, k, \nu, \tilde{g})\, w_k \tag{5.73d$'$}$$

Also in these bases for $\nu \in \mathbb{Z}_+$ there exists a $(\nu + 1)$-dimensional invariant subspace spanned here by the vectors w_k, $k \le \nu$ (cf. (5.73d')). Also, from the results of [218] follows that for $\nu \in \mathbb{Z}_+$ the representations $\pi_{\nu,\rho}$ and $\pi_{-\nu-2,\rho}$ are partially equivalent. This partial equivalence should be realized by the operator:

$$Q_\nu(\pi_R(Y)) : \mathscr{C}_{\nu,\rho} \longrightarrow \mathscr{C}_{-\nu-2,\rho} \tag{5.74}$$

where Q_ν is the polynomial in Y given in formula (37) of [218] (denoted there as $Q_p(p - 1)$, $p = \Lambda(H_0) + 1 = \nu + 1$), which polynomial gives the singular vector of the reducible Verma module V^Λ. The general expression for the singular vectors is given in the next subsection following [218].

Further we introduce the restricted functions $\hat{\varphi}(\chi)$ by the formula which is prompted in (5.72):

$$\hat{\varphi}(\chi) = \left(\hat{A}_{\nu,\rho}\, \tilde{\varphi}\right)(\chi) \equiv \phi(\chi, 1_{\mathscr{A}}, 1_{\mathscr{A}}). \tag{5.75}$$

We denote the representation space of $\hat{\varphi}(\chi)$ by $\widetilde{\mathscr{C}}_{\nu,\rho}$ and the representation acting in $\widetilde{\mathscr{C}}_{\nu,\rho}$ by $\hat{\pi}_{\nu,\rho}$. Thus the operator $\hat{A}_{\nu,\rho}$ acts from $\mathscr{C}_{\nu,\rho}$ to $\widetilde{\mathscr{C}}_{\nu,\rho}$. We shall use also the inverse operator $\hat{A}^{-1}_{\nu,\rho}$ which is defined by:

$$\tilde{\varphi}(\chi, c, \mathscr{D}) = \left(\hat{A}^{-1}_{\nu,\rho}\, \hat{\varphi}\right)(\chi, c, \mathscr{D}) \equiv \hat{\varphi}(\chi)\, c^\nu\, \mathscr{D}^{(\rho-\nu)/2} \tag{5.76}$$

The properties of $\widetilde{\mathscr{C}}_{\nu,\rho}$ follow from the intertwining requirements [197]:

$$\hat{\pi}_{\nu,\rho}\hat{A}_{\nu,\rho} = \hat{A}_{\nu,\rho}\pi_{\nu,\rho}, \qquad \pi_{\nu,\rho} \circ \hat{A}^{-1}_{\nu,\rho} = \hat{A}^{-1}_{\nu,\rho} \circ \hat{\pi}_{\nu,\rho} \tag{5.77}$$

In particular, the representation $\hat{\pi}_{\nu,\rho}$ is given by:

$$\hat{\pi}_{\nu,\rho}(A)\chi^n = -\rho\chi^n, \tag{5.78a}$$

$$\hat{\pi}_{\nu,\rho}(K^\pm)\chi^n = (\chi \mp \tilde{g})^n, \tag{5.78b}$$

$$\hat{\pi}_{\nu,\rho}(H)\chi^n = \left(\nu - \tfrac{1}{\tilde{g}}\chi - \tfrac{1}{2}\right)(\chi + \tilde{g})^n + \left(\tfrac{1}{\tilde{g}}\chi + \tfrac{1}{2}\right)(\chi - \tilde{g})^n =$$

$$= \left(\nu - \tfrac{1}{\tilde{g}}\chi - \tfrac{1}{2}\right)\pi_{\nu,\rho}(K^-)\chi^n +$$

$$+ \left(\tfrac{1}{\tilde{g}}\chi + \tfrac{1}{2}\right)\pi_{\nu,\rho}(K^+)\chi^n = \tag{5.78c}$$

$$= (\nu - 2n)\chi^n + \sum_{k=0}^{n-1} \omega(n, k, \nu, \tilde{g})\chi^k \tag{5.78c$'$}$$

$$\begin{aligned}\hat{\pi}_{\nu,\rho}(Y)\chi^n &= \left\{\tfrac{1}{2\tilde{g}}\chi^2 + (\tfrac{1}{2}-\nu)\chi - \tfrac{\tilde{g}}{2}(\nu^2+3\nu+\tfrac{1}{4})\right\}(\chi+\tilde{g})^n - \\ &- \left(\tfrac{1}{2\tilde{g}}\chi^2 + \tfrac{1}{2}\chi - \tfrac{\tilde{g}}{8}\right)(\chi-\tilde{g})^n = \\ &= \left\{\tfrac{1}{2\tilde{g}}\chi^2 + (\tfrac{1}{2}-\nu)\chi - \tfrac{\tilde{g}}{2}(\nu^2+3\nu+\tfrac{1}{4})\right\}\hat{\pi}_{\nu,\rho}(K^-)\chi^n - \\ &- \left(\tfrac{1}{2\tilde{g}}\chi^2 + \tfrac{1}{2}\chi - \tfrac{\tilde{g}}{8}\right)\hat{\pi}_{\nu,\rho}(K^+)\chi^n = \qquad (5.78d)\\ &= (n-\nu)\chi^{n+1} + \sum_{k=0}^{n}\omega'(n,k,\nu,\tilde{g})\chi^k \qquad (5.79)\end{aligned}$$

or in terms of the functions $\hat{\varphi}$:

$$\hat{\pi}_{\nu,\rho}(A)\,\hat{\varphi}(\chi) = -\rho\,\hat{\varphi}(\chi), \qquad (5.80a)$$

$$\hat{\pi}_{\nu,\rho}(K^{\pm})\,\hat{\varphi}(\chi) = \hat{\varphi}(\chi\mp\tilde{g}), \qquad (5.80b)$$

$$\begin{aligned}\hat{\pi}_{\nu,\rho}(H)\,\hat{\varphi}(\chi) &= \left(\nu-\tfrac{1}{\tilde{g}}\chi-\tfrac{1}{2}\right)\hat{\varphi}(\chi+\tilde{g}) + \left(\tfrac{1}{\tilde{g}}\chi+\tfrac{1}{2}\right)\hat{\varphi}(\chi-\tilde{g}) = \\ &= \left(\nu-\tfrac{1}{\tilde{g}}\chi-\tfrac{1}{2}\right)\hat{\pi}_{\nu,\rho}(K^-)\,\hat{\varphi}(\chi) + \\ &+ \left(\tfrac{1}{\tilde{g}}\chi+\tfrac{1}{2}\right)\hat{\pi}_{\nu,\rho}(K^+)\,\hat{\varphi}(\chi), \qquad (5.80c)\end{aligned}$$

$$\begin{aligned}\hat{\pi}_{\nu,\rho}(Y)\,\hat{\varphi}(\chi) &= \left\{\tfrac{1}{2\tilde{g}}\chi^2 + (\tfrac{1}{2}-\nu)\chi - \tfrac{\tilde{g}}{2}(\nu^2+3\nu+\tfrac{1}{4})\right\}\hat{\varphi}(\chi+\tilde{g}) - \\ &- \left(\tfrac{1}{2\tilde{g}}\chi^2 + \tfrac{1}{2}\chi - \tfrac{\tilde{g}}{8}\right)\hat{\varphi}(\chi-\tilde{g}) = \\ &= \left\{\tfrac{1}{2\tilde{g}}\chi^2 + (\tfrac{1}{2}-\nu)\chi - \right.\\ &\left. - \tfrac{\tilde{g}}{2}(\nu^2+3\nu+\tfrac{1}{4})\right\}\hat{\pi}_{\nu,\rho}(K^-)\,\hat{\varphi}(\chi) - \\ &- \left(\tfrac{1}{2\tilde{g}}\chi^2 + \tfrac{1}{2}\chi - \tfrac{\tilde{g}}{8}\right)\hat{\pi}_{\nu,\rho}(K^+)\,\hat{\varphi}(\chi). \qquad (5.80d)\end{aligned}$$

Now we notice that we can consider (5.78) and (5.80) for arbitrary complex ν, ρ. Actually, the representation has decoupled into a representation of the central algebra with generator A (cf. (5.78a) and (5.80a)), and a *new* representation of the Jordanian $U_{\tilde{g}}(sl(2))$ (cf. (5.78b,c,d) and (5.80b,c,d)). Analogously to before for generic $\nu \in \mathbb{C}$ the representations $\hat{\pi}_{\nu,\rho}$ are irreducible. For $\nu \in \mathbb{Z}_+$ the representations $\hat{\pi}_{\nu,\rho}$ are reducible, since there is a $(\nu+1)$-dimensional invariant subspace of the polynomials in χ of degree up to ν (cf. (5.79)). Also, from the results of [218], follows that for $\nu \in \mathbb{Z}_+$ the representations $\hat{\pi}_{\nu,\rho}$ and $\hat{\pi}_{-\nu-2,\rho}$ are partially equivalent. The intertwining operators between these pairs is naturally obtained from the ones relating the pairs $\pi_{\nu,\rho}$ and $\pi_{-\nu-2,\rho}$, namely:

$$\mathcal{I}_\nu : \widehat{C}_{\nu,\rho} \longrightarrow \widehat{C}_{-\nu-2,\rho} \qquad (5.81a)$$

$$\mathcal{I}_\nu \equiv \hat{A}^{-1}_{\nu,\rho} \circ Q_\nu(\pi_R(Y)) \circ \hat{A}_{-\nu-2,\rho} \qquad (5.81b)$$

Finally, we should note that since we have functions of one variable χ we can replace it with an ordinary complex variable z, and then the transformation properties (5.80) can be rewritten as follows:

$$\hat{\pi}_{\nu,\rho}(A)\,\hat{\varphi}(z) = -\rho\,\hat{\varphi}(z), \tag{5.82a}$$

$$\hat{\pi}_{\nu,\rho}(K^{\pm})\,\hat{\varphi}(z) = e^{\mp\tilde{g}\partial_z}\,\hat{\varphi}(z), \tag{5.82b}$$

$$\hat{\pi}_{\nu,\rho}(H)\,\hat{\varphi}(z) = \left(-\tfrac{1}{\tilde{g}}\,z - \tfrac{1}{2} + \nu\right) e^{\tilde{g}\partial_z}\,\hat{\varphi}(z) + $$
$$+ \left(\tfrac{1}{\tilde{g}}\,z + \tfrac{1}{2}\right) e^{-\tilde{g}\partial_z}\,\hat{\varphi}(z) \tag{5.82c}$$

$$= -\left(\tfrac{2}{\tilde{g}}\,z + 1\right) \sinh(\tilde{g}\partial_z)\,\hat{\varphi}(z) + \nu\, e^{\tilde{g}\partial_z}\,\hat{\varphi}(z) \tag{5.82c$'$}$$

$$\hat{\pi}_{\nu,\rho}(Y)\,\hat{\varphi}(z) = \left\{\tfrac{1}{2\tilde{g}}\,z^2 + (\tfrac{1}{2} - \nu)\,z - \tfrac{\tilde{g}}{2}\,(\nu^2 + 3\nu + \tfrac{1}{4})\right\} e^{\tilde{g}\partial_z}\,\hat{\varphi}(z) - $$
$$- \left(\tfrac{1}{2\tilde{g}}\,z^2 + \tfrac{1}{2}\,z - \tfrac{\tilde{g}}{8}\right) e^{-\tilde{g}\partial_z}\,\hat{\varphi}(z) = \tag{5.82d}$$
$$= \left(\tfrac{1}{\tilde{g}}\,z^2 + z - \tfrac{\tilde{g}}{4}\right) \sinh(\tilde{g}\partial_z)\,\hat{\varphi}(z) - $$
$$-\nu \left\{z + \tfrac{\tilde{g}}{2}\,(\nu + 3)\right\} e^{\tilde{g}\partial_z}\,\hat{\varphi}(z) \tag{5.82d$'$}$$

In these terms we can also recover from (5.82) the classical vector-field representation of $gl(2)$ by setting (as noted above) $K^{\pm} = e^{\pm\tilde{g}B}$, expanding $K^{\pm} \approx 1_{\mathcal{U}} \pm \tilde{g}B$ and taking the limit $\tilde{g} \to 0$. Thus, we obtain:

$$A = -\rho\,, \quad H = \nu - 2z\partial_z\,, \quad B = -\partial_z\,, \quad Y = z^2\partial_z - \nu z\,, \tag{5.83}$$

which fulfills (5.52).

Thus, the representation (5.82), more precisely, formulae (5.82b,c,d), give a *new* deformation of the classical vector-field realization of $sl(2)$.

5.2.3 Representations of the Jordanian Algebra $U_h(sl(2))$

Here we construct highest-weight modules (HWMs) of the Jordanian algebra $U_h(sl(2))$ following [218]. This algebra was obtained first in [499]. Here we recovered it in Section 4.7.5.1 from the one-parameter case of the Jordanian algebra $U_{g,h}(gl(2))$ for $g = h$.

This one-parameter subalgebra is a subalgebra of $U_{h,h}(gl(2))$ with generators H, Y, B and commutation relations (4.199a,b,c). For our purposes here we exchange the generator B by the two related generators:

$$C = \cosh(hB), \quad S = \sinh(hB) \tag{5.84}$$

and thus the commutation relations of $U_h(sl(2))$ become:

$$[H,C] = 2S^2, \qquad [H,S] = 2CS\,, \tag{5.85a}$$
$$[C,Y] = hHS - hCS, \quad [S,Y] = hHC - hS^2, \tag{5.85b}$$
$$[H,Y] = -2YC - hHS + hCS, \tag{5.85c}$$
$$[C,S] = 0, \qquad C^2 - S^2 = 1. \tag{5.85d}$$

In [499] instead of C, S were used the generators $K^{\pm} = C \pm S = e^{\pm hB}$.

The Casimir of $U_h(sl(2))$ (and of the extension) is given by:

$$\mathscr{C}_2 = \frac{1}{2}(H^2 + C^2) + \frac{1}{h}(YS + SY) \tag{5.86}$$

Let us introduce the following grading:

$$\deg Y = \deg h = 1, \qquad \deg H = \deg C = \deg S = \deg 1_{\mathscr{U}} = 0 \tag{5.87}$$

Then we can show that the algebra $U_h(sl(2))$ is a *graded Hopf algebra* (cf. [218]′). In particular, the algebra relations (5.85) are graded w.r.t. deg. The Casimir $\mathscr{C}_2$ is homogeneous w.r.t. deg with $\deg \mathscr{C}_2 = 0$.

5.2.4 Highest-Weight Modules over $U_h(sl(2))$

Note that the generators H, C, S generate a Hopf subalgebra $\mathscr{B}$ of $U_h(sl(2))$. This Hopf subalgebra is the analogue of the (universal envelope of the) Borel subalgebra, generated by H, X, of $sl(2)$. Note that there is no Borel-like conjugate of $\mathscr{B}$, which in the classical case would be generated by H, Y, since here H, Y do not generate a subalgebra of $\mathscr{U} \equiv U_h(sl(2))$.

Consider the one-dimensional representation of $\mathscr{B}$ generated by a basis vector v_0 so that the generators act on it as:

$$Hv_0 = \Lambda(H)v_0, \qquad Sv_0 = 0, \qquad Cv_0 = v_0 \tag{5.88}$$

where $\Lambda(H) \in \mathbb{C}$ is called the highest weight. Then the Verma module V^{Λ} over $\mathscr{U}$ is defined as the HWM induced from the module (5.88), and it is given by:

$$V^{\Lambda} \cong \text{c.l.s.}\{Y^n \otimes_{\mathscr{B}} v_0 | n \in \mathbb{Z}_+\} \tag{5.89}$$

Further we shall omit the sign $\otimes_{\mathscr{B}}$ since no confusion may arise. We note now some properties of the basis $Y^k v_0$ which are the same as in the $sl(2)$ case and which we shall use for a direct construction of the finite-dimensional HWM.

The value of the Casimir is the classical one (cf. (5.86)):

$$\mathscr{C}_2 v = \frac{1}{2}(\Lambda(H) + 1)^2 v, \quad \forall v \in V^\Lambda \tag{5.90}$$

Further we can show [218]′:

$$S^n Y^n v_0 = h^n n! \frac{\Gamma(\Lambda(H) + 1)}{\Gamma(\Lambda(H) + 1 - n)} v_0, \qquad n \in \mathbb{Z}_+, \tag{5.91}$$

and $S^k Y^n v_0 = 0, \quad \text{if} k > n$.

Let $\tilde{X} \equiv \frac{1}{h} S$. Then we have:

$$\tilde{X}^n Y^n v_0 = n! \frac{\Gamma(\Lambda(H) + 1)}{\Gamma(\Lambda(H) + 1 - n)} v_0, \qquad n \in \mathbb{Z}_+. \tag{5.92}$$

If we use the presentation $S = \sinh hX$, then for $q \to 1$ we get $\tilde{X} \to X$, and (5.92) becomes a $sl(2)$ result.

As in the undeformed case we expect finite-dimensional HWM whenever the highest weight is integral dominant; that is,

$$\Lambda(H) = p - 1; \qquad p \in \mathbb{N} \tag{5.93}$$

then the dimension of the representation is expected to be p. This is so for the trivial one-dimensional irrep given by (5.88) with $\Lambda(H) = 0$, and also $Y v_0 = 0$. The two-dimensional irrep (denoting the two vectors by u_0, u_1) is given by:

$$H(u_0, u_1) = (u_0, -u_1), \qquad Y(u_0, u_1) = (u_1, 0), \tag{5.94}$$

$$S(u_0, u_1) = h(0, u_0), \qquad C(u_0, u_1) = (u_0, u_1) \tag{5.95}$$

This irrep looks deformed, however, the h dependence may be absorbed if we replace S by $\tilde{X}$. Thus, the fundamental irrep is undeformed as in the case of the Drinfel'd-Jimbo deformation $U_q(sl(2))$.

For the consideration of the general case we introduce (as in the classical case) the following basis:

$$u_n^p = \sqrt{\frac{(p - 1 - n)!}{h^n n!(p - 1)!}} Y^n v_0, \qquad p \in \mathbb{N}, n \in \mathbb{Z}_+, n < p \tag{5.96}$$

On this basis the generators Y act as in the classical case (up to multiple of $\sqrt{h}$):

$$Y u_n^p = \sqrt{h(n + 1)(p - 1 - n)} u_{n+1}^p \tag{5.97}$$

Thus, we have [218]′:

Proposition 1. *For fixed $p \in \mathbb{N}$ the vectors u_n^p, $n = 0, \dots, p-1$ provide a p-dimensional irreducible representation of $U_h(sl(2))$.* ◊

Remark 5.1. Another realization of the above finite-dimensional representation was given in Section 5.2.2 by the vectors w_k, $k = 0, 1, \dots, \nu = p - 1$ (cf. (5.74b,c,d)). There $U_h(sl(2))$ is generated by $H, Y, K^{\pm} = C \pm S$. ◊

Note that these representations are deformed for $p > 2$. We give the example of $p = 3$. We can show [218]′ that the three-dimensional irreducible representation of $U_h(sl(2))$ is given by (5.97) for $p = 3$ and the following formulae with $u_k = u_k^{p=3}$:

$$H(u_0, u_1, u_2) = (2u_0, 0, -2u_2 - hu_0), \tag{5.98a}$$

$$S(u_0, u_1, u_2) = \sqrt{2h}(0, u_0, u_1), \tag{5.98b}$$

$$C(u_0, u_1, u_2) = (u_0, u_1, u_2 + hu_0). \tag{5.98c}$$

5.2.5 Singular Vectors of $U_h(sl(2))$ Verma Modules

In this section we try to use the Verma modules as in the classical case. We are first interested in their reducibility. In the classical case an important tool for this are the so called singular vectors.

Let us recall that for $sl(2)$ with generators X_0, Y_0, H_0 a singular vector v_s of a Verma module V^Λ is defined as follows: $v_s \in V^\Lambda$, $v_s \neq v_0$ and it satisfies the following properties:

$$X_0 v_s = 0, \tag{5.99a}$$

$$H_0 v_s = \Lambda'(H_0) v_s. \tag{5.99b}$$

Moreover, v_s exists iff $\Lambda(H_0) = p - 1 \in \mathbb{Z}_+$; furthermore $\Lambda' = \Lambda - p\alpha$, where α the positive root of the $sl(2)$ root system so that $\alpha(H_0) = 2$.

To implement (5.99) here we have first to construct a basis homogeneous with respect to H. We note that the vectors $Y^k v_0$ are not homogeneous with respect to H, except for $k = 0, 1$ (cf. also (5.98a)). The necessary basis is provided by the following [218]′:

Proposition 2. *A basis homogeneous with respect to the generator H is:*

$$v_n = \sum_{k=0}^{[n/2]} \alpha_{nk} h^{2k} Y^{n-2k} v_0 \tag{5.100}$$

$$H v_n = (\Lambda(H) - 2n) v_n, \tag{5.101}$$

and the coefficients α_{nk} may depend on $\Lambda(H)$ but not on h. In particular, $\alpha_{n0} = 1$,

$$\alpha_{n1} = \frac{1}{120} n(n-1)\Big(6n^3 - 3n^2(5\Lambda(H)+8) + n(10\Lambda(H)^2 + \\ +35\Lambda(H)+36) - 5\Lambda(H)^2 - 25\Lambda(H) - 24\Big) \qquad \diamond \tag{5.102}$$

Note that α_{n1} does not vanish except for special values of $\Lambda(H)$. Besides the trivial vanishing for $n = 0, 1$ we have: $\alpha_{21}(0) = \alpha_{21}(1) = \alpha_{31}(1) = 0$, where the argument of $\alpha_{n,k}$ denotes the value of $\Lambda(H)$. Other explicit expressions for α_{nk} are given in [218]′. Analogously, one may prove the following (cf. [218]′):

$$CY^n v_0 = \sum_{k=0}^{[n/2]} \beta_{nk} h^{2k} Y^{n-2k} v_0 \tag{5.103a}$$

$$SY^n v_0 = \sum_{k=0}^{[(n-1)/2]} \gamma_{nk} h^{2k+1} Y^{n-1-2k} v_0 \tag{5.103b}$$

and the coefficients β_{nk}, γ_{nk} may depend on $\Lambda(H)$ but not on h. In particular, $\beta_{n0} = 1$,

$$\beta_{n1} = \frac{1}{2} n(n-1)\big(\Lambda(H) - n + 2\big)\big(\Lambda(H) - n + 1\big) \tag{5.104a}$$

$$\gamma_{n0} = n\big(\Lambda(H) - n + 1\big) \tag{5.104b}$$

Let us denote by $Q_n(\Lambda)$ the polynomial in Y so that in (5.100):

$$v_n = Q_n(\Lambda(H))\, v_0 \tag{5.105}$$

In our situation it would be natural to define the singular vector for fixed $p \in \mathbb{N}$ as the H-homogeneous element from (5.100) for the corresponding value of $\Lambda(H)$:

$$v_s^p = Q_p(p-1) v_0 \tag{5.106}$$

Thus, property (5.99b) is achieved. The analogue of (5.99a) is played by the condition:

$$S v_s^p = 0 \tag{5.107}$$

The basis for the submodule I^Λ shall be played by:

$$\hat{v}_k^p = Q_k(-p-1)\, v_s^p = Q_k(-p-1) Q_p(p-1) v_0, \qquad k \in \mathbb{Z}_+ \tag{5.108}$$

For the implementation of the above one has to prove that each vector $Q_{p+k}(p-1)v_0$, $k \in \mathbb{Z}_+$, can be expressed in terms of $\hat{v}_j^p$, $j \le k$. Moreover, in the lower degree cases one needs to take only $j = k$:

$$Q_{p+n}(p-1) = Q_n(-p-1) Q_p(p-1) \tag{5.109}$$

5.3 q-Difference Intertwining Operators for a Lorentz Quantum Algebra

5.3.1 A Matrix Lorentz Quantum Group

In this section following [164] we present q-difference intertwining operators for the matrix Lorentz quantum group $\mathscr{L}$ introduced by Woronowicz–Zakrzewski [601]. Note that in Section 4.6 we studied the duality question for another Lorentz quantum group.

The matrix Lorentz quantum group $\mathscr{L}$ introduced in [601] is generated by the elements $\alpha, \beta, \gamma, \delta, \bar\alpha, \bar\beta, \bar\gamma, \bar\delta$ with the following commutation relations ($q \in \mathbb{R}$, $\lambda = q - q^{-1}$):

$$\begin{aligned}
&\alpha\beta = q\beta\alpha,\ \alpha\gamma = q\gamma\alpha,\ \beta\delta = q\delta\beta,\ \gamma\delta = q\delta\gamma,\\
&\beta\gamma = \gamma\beta,\ \alpha\delta - q\beta\gamma = 1_{\mathscr{L}},\ \delta\alpha - q^{-1}\beta\gamma = 1_{\mathscr{L}}, \qquad (5.110a)\\
&\bar\alpha\bar\beta = q^{-1}\bar\beta\bar\alpha,\ \bar\alpha\bar\gamma = q^{-1}\bar\gamma\bar\alpha,\ \bar\beta\bar\delta = q^{-1}\bar\delta\bar\beta,\ \bar\gamma\bar\delta = q^{-1}\bar\delta\bar\gamma,\\
&\bar\beta\bar\gamma = \bar\gamma\bar\beta,\ \bar\alpha\bar\delta - q^{-1}\bar\beta\bar\gamma = 1_{\mathscr{L}},\ \bar\delta\bar\alpha - q\bar\beta\bar\gamma = 1_{\mathscr{L}}, \qquad (5.110b)\\
&\alpha\bar\alpha = \bar\alpha\alpha,\ \beta\bar\beta = \bar\beta\beta,\ \gamma\bar\gamma = \bar\gamma\gamma,\ \delta\bar\delta = \bar\delta\delta, \qquad (5.110c)\\
&\alpha\bar\beta = q\bar\beta\alpha,\ \alpha\bar\gamma = q^{-1}\bar\gamma\alpha,\ \alpha\bar\delta = \bar\delta\alpha,\\
&\beta\bar\gamma = \bar\gamma\beta,\ \beta\bar\delta = q^{-1}\bar\delta\beta,\ \gamma\bar\delta = q\bar\delta\gamma, \qquad (5.110d)\\
&\beta\bar\alpha = q\bar\alpha\beta,\ \gamma\bar\alpha = q^{-1}\bar\alpha\gamma,\ \delta\bar\alpha = \bar\alpha\delta,\\
&\gamma\bar\beta = \bar\beta\gamma,\ \delta\bar\beta = q^{-1}\bar\beta\delta,\ \delta\bar\gamma = q\bar\gamma\delta. \qquad (5.110e)
\end{aligned}$$

Considered as a Hopf algebra, $\mathscr{L}$ has the following comultiplication $\Delta_{\mathscr{L}}$, counit $\varepsilon_{\mathscr{L}}$, and antipode $S_{\mathscr{L}}$ given on its generating elements by:

$$\Delta_{\mathscr{L}}\begin{pmatrix}\alpha & \beta\\ \gamma & \delta\end{pmatrix} = \begin{pmatrix}\alpha\otimes\alpha + \beta\otimes\gamma & \alpha\otimes\beta + \beta\otimes\delta\\ \gamma\otimes\alpha + \delta\otimes\gamma & \gamma\otimes\beta + \delta\otimes\delta\end{pmatrix}, \qquad (5.111a)$$

$$\Delta_{\mathscr{L}}\begin{pmatrix}\bar\alpha & \bar\beta\\ \bar\gamma & \bar\delta\end{pmatrix} = \begin{pmatrix}\bar\alpha\otimes\bar\alpha + \bar\beta\otimes\bar\gamma & \bar\alpha\otimes\bar\beta + \bar\beta\otimes\bar\delta\\ \bar\gamma\otimes\bar\alpha + \bar\delta\otimes\bar\gamma & \bar\gamma\otimes\bar\beta + \bar\delta\otimes\bar\delta\end{pmatrix},$$

$$\varepsilon_{\mathscr{L}}\begin{pmatrix}\alpha & \beta\\ \gamma & \delta\end{pmatrix} = \begin{pmatrix}1 & 0\\ 0 & 1\end{pmatrix}, \quad \varepsilon_{\mathscr{L}}\begin{pmatrix}\bar\alpha & \bar\beta\\ \bar\gamma & \bar\delta\end{pmatrix} = \begin{pmatrix}1 & 0\\ 0 & 1\end{pmatrix}, \qquad (5.111b)$$

$$S_{\mathscr{L}}\begin{pmatrix}\alpha & \beta\\ \gamma & \delta\end{pmatrix} = \begin{pmatrix}\delta & -q^{-1}\beta\\ -q\gamma & \alpha\end{pmatrix},$$

$$S_{\mathscr{L}}\begin{pmatrix}\bar\alpha & \bar\beta\\ \bar\gamma & \bar\delta\end{pmatrix} = \begin{pmatrix}\bar\delta & -q\bar\beta\\ -q^{-1}\bar\gamma & \bar\alpha\end{pmatrix}. \qquad (5.111c)$$

With the conjugation

$$\alpha \mapsto \bar\alpha,\ \beta \mapsto \bar\beta,\ \gamma \mapsto \bar\gamma,\ \delta \mapsto \bar\delta, \qquad (5.112)$$

and $q \mapsto q$, which acts as algebra anti-involution and coalgebra involution, $\mathscr{L}$ is a Hopf $*$-algebra. The Hopf algebra $\mathscr{L}$ contains two conjugate Hopf subalgebras, $SL_{q^{-1}}(2)$ and $SL_q(2)$, generated by $\alpha, \beta, \gamma, \delta$ and $\bar{\alpha}, \bar{\beta}, \bar{\gamma}, \bar{\delta}$, respectively. Note that using (5.112) relations (5.110b) and (5.110e) may be obtained from (5.110a) and (5.110d), respectively, while relations (5.110c) are self-conjugate.

Notice that we can consider $\mathscr{L}$ also for q complex, such that $|q| = 1$; in this case (5.112) with $q \mapsto q^{-1}$ acts as algebra and coalgebra involution. We shall come back to this case in the last subsection, where we consider the important case of q being a root of unity.

For our purposes, we assume that the elements δ and $\bar{\delta}$ are invertible. In this case, one can express α and $\bar{\alpha}$ in terms of the remaining generators:

$$\alpha = (1_{\mathscr{L}} + q\beta\gamma)\delta^{-1}, \quad \bar{\alpha} = (1_{\mathscr{L}} + q^{-1}\bar{\beta}\bar{\gamma})\bar{\delta}^{-1}. \tag{5.113}$$

Then, as a basis of $\mathscr{L}$ we take the following ordered monomials:

$$\beta^{\ell}\, \gamma^{m}\, \delta^{n}\, \bar{\beta}^{\bar{\ell}}\, \bar{\gamma}^{\bar{m}}\, \bar{\delta}^{\bar{n}}, \quad \ell, m, \bar{\ell}, \bar{m} \in \mathbb{Z}_+, \ n, \bar{n} \in \mathbb{Z}. \tag{5.114}$$

5.3.2 The Lorentz Quantum Algebra

The Lorentz quantum algebra $\mathscr{U}$ is the Hopf algebra which is in duality to $\mathscr{L}$. It is generated by six elements, which we denote by $k, e, f, \bar{k}, \bar{e}, \bar{f}$. The pairing of these generators with those of $\mathscr{L}$ is defined through the fundamental representation M of $\mathscr{U}$. The abstract matrix elements of M generate the matrix quantum group $\mathscr{L}$,

$$M = \begin{pmatrix} \begin{pmatrix} \alpha & \beta \\ \gamma & \delta \end{pmatrix} & 0_2 \\ 0_2 & \begin{pmatrix} \bar{\alpha} & \bar{\beta} \\ \bar{\gamma} & \bar{\delta} \end{pmatrix} \end{pmatrix} \tag{5.115}$$

and the duality relations are:

$$\langle k, M \rangle = \begin{pmatrix} \begin{pmatrix} q^{1/2} & 0 \\ 0 & q^{-1/2} \end{pmatrix} & 0_2 \\ 0_2 & 1_2 \end{pmatrix}, \tag{5.116a}$$

$$\langle e, M \rangle = \begin{pmatrix} \begin{pmatrix} 0 & 1 \\ 0 & 0 \end{pmatrix} & 0_2 \\ 0_2 & 0_2 \end{pmatrix}, \tag{5.116b}$$

$$\langle f, M\rangle = \begin{pmatrix} \begin{pmatrix} 0 & 0 \\ 1 & 0 \end{pmatrix} & 0_2 \\ 0_2 & 0_2 \end{pmatrix}, \tag{5.116c}$$

$$\langle \bar{k}, M\rangle = \begin{pmatrix} 1_2 & 0_2 \\ 0_2 & \begin{pmatrix} q^{1/2} & 0 \\ 0 & q^{-1/2} \end{pmatrix} \end{pmatrix}, \tag{5.116d}$$

$$\langle \bar{e}, M\rangle = \begin{pmatrix} 0_2 & 0_2 \\ 0_2 & \begin{pmatrix} 0 & 1 \\ 0 & 0 \end{pmatrix} \end{pmatrix}, \tag{5.116e}$$

$$\langle \bar{f}, M\rangle = \begin{pmatrix} 0_2 & 0_2 \\ 0_2 & \begin{pmatrix} 0 & 0 \\ 1 & 0 \end{pmatrix} \end{pmatrix}. \tag{5.116f}$$

Notice that since M is a representation, we have:

$$\langle XY, M\rangle = \langle X, M\rangle\langle Y, M\rangle, \quad X, Y \in \mathscr{U}, \tag{5.117}$$

where in the r.h.s. matrix multiplication is understood. Using these relations, from (5.111a) one derives the algebra relations obeyed by $k, e, f, \bar{k}, \bar{e}, \bar{f}$:

$$\begin{aligned} &kek^{-1} = qe, \quad kfk^{-1} = q^{-1}f, \quad [e,f] = \left(k^2 - k^{-2}\right)/\lambda, \\ &\bar{k}\bar{e}\bar{k}^{-1} = q\bar{e}, \quad \bar{k}\bar{f}\bar{k}^{-1} = q^{-1}\bar{f}, \quad [\bar{e},\bar{f}] = \left(\bar{k}^2 - \bar{k}^{-2}\right)/\lambda, \\ &[X, Y] = 0, \quad X = e, f, k, \quad Y = \bar{e}, \bar{f}, \bar{k}. \end{aligned} \tag{5.118}$$

Note that the first two lines of (5.118) are two copies of $\mathscr{U}_q(sl(2,\mathbb{C}))$.

The coalgebra structure of $\mathscr{U}$ is instead fixed by (5.110), (5.110b), and (5.110c). Explicitly, we have:

$$\begin{aligned} &\Delta_{\mathscr{U}}(k) = k \otimes k, \\ &\Delta_{\mathscr{U}}(e) = e \otimes k^{-1}\bar{k} + k\bar{k}^{-1} \otimes e, \\ &\Delta_{\mathscr{U}}(f) = f \otimes k^{-1}\bar{k}^{-1} + k\bar{k} \otimes f, \\ &\Delta_{\mathscr{U}}(\bar{k}) = \bar{k} \otimes \bar{k}, \\ &\Delta_{\mathscr{U}}(\bar{e}) = \bar{e} \otimes k^{-1}\bar{k} + k\bar{k}^{-1} \otimes \bar{e}, \\ &\Delta_{\mathscr{U}}(\bar{f}) = \bar{f} \otimes k\bar{k} + k^{-1}\bar{k}^{-1} \otimes \bar{f}, \\ &\varepsilon_{\mathscr{U}}(k) = \varepsilon_{\mathscr{U}}(\bar{k}) = 1, \quad \varepsilon_{\mathscr{U}}(e) = \varepsilon_{\mathscr{U}}(f) = \varepsilon_{\mathscr{U}}(\bar{e}) = \varepsilon_{\mathscr{U}}(\bar{f}) = 0, \end{aligned} \tag{5.119}$$

$$S_{\mathscr{U}}(k) = k^{-1}, \quad S_{\mathscr{U}}(e) = -q^{-1}e, \quad S_{\mathscr{U}}(f) = -qf,$$
$$S_{\mathscr{U}}(\bar{k}) = \bar{k}^{-1}, \quad S_{\mathscr{U}}(\bar{e}) = -q\bar{e}, \quad S_{\mathscr{U}}(\bar{f}) = -q^{-1}\bar{f}.$$

The conjugation of $\mathscr{U}$ is given by:

$$k \mapsto \bar{k}, \quad e \mapsto \bar{e}, \quad f \mapsto \bar{f}, \tag{5.120}$$

which acts as an algebra involution for real q and as an algebra anti-involution for q complex, such that $|q| = 1$, and as coalgebra anti-involution in both cases.

Below we shall supplement the pairing (5.116) with

$$\langle X, 1_{\mathscr{L}} \rangle = \varepsilon_{\mathscr{U}}(X). \tag{5.121}$$

5.3.3 Representations of the Lorentz Quantum Algebra

We shall now define two actions of the dual algebra $\mathscr{U}$ on the basis (5.114) of $\mathscr{L}$ (see also [465]). As above we introduce the left regular representation of $\mathscr{U}$:

$$\pi(Y)\, M = Y^{-1}\, M, \quad Y, M \in \mathscr{L}. \tag{5.122}$$

Explicitly we set for the generators of $\mathscr{U}$:

$$\begin{aligned}
\pi(k)M &= \langle k^{-1}, M \rangle M = \\
&= \begin{pmatrix} \begin{pmatrix} q^{-1/2} & 0 \\ 0 & q^{1/2} \end{pmatrix} & 0_2 \\ 0_2 & 1_2 \end{pmatrix} \begin{pmatrix} \begin{pmatrix} \alpha & \beta \\ \gamma & \delta \end{pmatrix} & 0_2 \\ 0_2 & \begin{pmatrix} \bar{\alpha} & \bar{\beta} \\ \bar{\gamma} & \bar{\delta} \end{pmatrix} \end{pmatrix} = \\
&= \begin{pmatrix} \begin{pmatrix} q^{-1/2}\alpha & q^{-1/2}\beta \\ q^{1/2}\gamma & q^{1/2}\delta \end{pmatrix} & 0_2 \\ 0_2 & \begin{pmatrix} \bar{\alpha} & \bar{\beta} \\ \bar{\gamma} & \bar{\delta} \end{pmatrix} \end{pmatrix}, \\
\pi(e)M &= \langle -e, M \rangle M = \\
&= \begin{pmatrix} \begin{pmatrix} 0 & -1 \\ 0 & 0 \end{pmatrix} & 0_2 \\ 0_2 & 0_2 \end{pmatrix} \begin{pmatrix} \begin{pmatrix} \alpha & \beta \\ \gamma & \delta \end{pmatrix} & 0_2 \\ 0_2 & \begin{pmatrix} \bar{\alpha} & \bar{\beta} \\ \bar{\gamma} & \bar{\delta} \end{pmatrix} \end{pmatrix} = \\
&= \begin{pmatrix} \begin{pmatrix} -\gamma & -\delta \\ 0 & 0 \end{pmatrix} & 0_2 \\ 0_2 & 0_2 \end{pmatrix},
\end{aligned}$$

$$
\begin{aligned}
\pi(f)M &= \langle -f, M\rangle M = \\
&= \begin{pmatrix} \begin{pmatrix} 0 & 0 \\ -1 & 0 \end{pmatrix} & 0_2 \\ 0_2 & 0_2 \end{pmatrix} \begin{pmatrix} \begin{pmatrix} \alpha & \beta \\ \gamma & \delta \end{pmatrix} & 0_2 \\ 0_2 & \begin{pmatrix} \bar{\alpha} & \bar{\beta} \\ \bar{\gamma} & \bar{\delta} \end{pmatrix} \end{pmatrix} = \\
&= \begin{pmatrix} \begin{pmatrix} 0 & 0 \\ -\alpha & -\beta \end{pmatrix} & 0_2 \\ 0_2 & 0_2 \end{pmatrix},
\end{aligned}
$$

$$
\begin{aligned}
\pi(\bar{k})M &= \langle \bar{k}^{-1}, M\rangle M = \\
&= \begin{pmatrix} 1_2 & 0_2 \\ 0_2 & \begin{pmatrix} q^{-1/2} & 0 \\ 0 & q^{1/2} \end{pmatrix} \end{pmatrix} \begin{pmatrix} \begin{pmatrix} \alpha & \beta \\ \gamma & \delta \end{pmatrix} & 0_2 \\ 0_2 & \begin{pmatrix} \bar{\alpha} & \bar{\beta} \\ \bar{\gamma} & \bar{\delta} \end{pmatrix} \end{pmatrix} = \\
&= \begin{pmatrix} \begin{pmatrix} \alpha & \beta \\ \gamma & \delta \end{pmatrix} & 0_2 \\ 0_2 & \begin{pmatrix} q^{-1/2}\bar{\alpha} & q^{-1/2}\bar{\beta} \\ q^{1/2}\bar{\gamma} & q^{1/2}\bar{\delta} \end{pmatrix} \end{pmatrix},
\end{aligned}
\tag{5.123}
$$

$$
\begin{aligned}
\pi(\bar{e})M &= \langle -\bar{e}, M\rangle M = \\
&= \begin{pmatrix} 0_2 & 0_2 \\ 0_2 & \begin{pmatrix} 0 & -1 \\ 0 & 0 \end{pmatrix} \end{pmatrix} \begin{pmatrix} \begin{pmatrix} \alpha & \beta \\ \gamma & \delta \end{pmatrix} & 0_2 \\ 0_2 & \begin{pmatrix} \bar{\alpha} & \bar{\beta} \\ \bar{\gamma} & \bar{\delta} \end{pmatrix} \end{pmatrix} = \\
&= \begin{pmatrix} 0_2 & 0_2 \\ 0_2 & \begin{pmatrix} -\bar{\gamma} & -\bar{\delta} \\ 0 & 0 \end{pmatrix} \end{pmatrix},
\end{aligned}
$$

$$
\begin{aligned}
\pi(\bar{f})M &= \langle -\bar{f}, M\rangle M = \\
&= \begin{pmatrix} 0_2 & 0_2 \\ 0_2 & \begin{pmatrix} 0 & 0 \\ -1 & 0 \end{pmatrix} \end{pmatrix} \begin{pmatrix} \begin{pmatrix} \alpha & \beta \\ \gamma & \delta \end{pmatrix} & 0_2 \\ 0_2 & \begin{pmatrix} \bar{\alpha} & \bar{\beta} \\ \bar{\gamma} & \bar{\delta} \end{pmatrix} \end{pmatrix} = \\
&= \begin{pmatrix} 0_2 & 0_2 \\ 0_2 & \begin{pmatrix} 0 & 0 \\ -\bar{\alpha} & -\bar{\beta} \end{pmatrix} \end{pmatrix}.
\end{aligned}
$$

In order to derive the action of π on arbitrary elements of the basis, as above we use the twisted derivation rule $\pi(X)\varphi\psi = \pi(\Delta'_{\mathscr{U}}(X))(\varphi \otimes \psi)$ (cf. (5.12)). Thus, we have:

$$
\begin{aligned}
\pi(k)\varphi\psi &= \pi(k)\varphi \cdot \pi(k)\psi\,, \\
\pi(e)\varphi\psi &= \pi(e)\varphi \cdot \pi(k\bar{k}^{-1})\psi + \pi(k^{-1}\bar{k})\varphi \cdot \pi(e)\psi\,, \\
\pi(f)\varphi\psi &= \pi(f)\varphi \cdot \pi(k\bar{k})\psi + \pi(k^{-1}\bar{k}^{-1})\varphi \cdot \pi(f)\psi\,, \\
\pi(\bar{k})\varphi\psi &= \pi(\bar{k})\varphi \cdot \pi(\bar{k})\psi\,, \\
\pi(\bar{e})\varphi\psi &= \pi(\bar{e})\varphi \cdot \pi(k\bar{k}^{-1})\psi + \pi(k^{-1}\bar{k})\varphi \cdot \pi(\bar{e})\psi\,, \\
\pi(\bar{f})\varphi\psi &= \pi(\bar{f})\varphi \cdot \pi(k^{-1}\bar{k}^{-1})\psi + \pi(k\bar{k})\varphi \cdot \pi(\bar{f})\psi.
\end{aligned}
\tag{5.124}
$$

(Note that, though the generators α, $\bar{\alpha}$ are redundant, we shall write them sometimes.) Applying these rules one obtains:

$$
\begin{aligned}
\pi(k)\begin{pmatrix} \begin{pmatrix} \alpha^n & \beta^n \\ \gamma^n & \delta^n \end{pmatrix} & 0_2 \\ 0_2 & \begin{pmatrix} \bar{\alpha}^n & \bar{\beta}^n \\ \bar{\gamma}^n & \bar{\delta}^n \end{pmatrix} \end{pmatrix} &= \begin{pmatrix} \begin{pmatrix} q^{-n/2}\alpha^n & q^{-n/2}\beta^n \\ q^{n/2}\gamma^n & q^{n/2}\delta^n \end{pmatrix} & 0_2 \\ 0_2 & \begin{pmatrix} \bar{\alpha}^n & \bar{\beta}^n \\ \bar{\gamma}^n & \bar{\delta}^n \end{pmatrix} \end{pmatrix} \\
\pi(e)\begin{pmatrix} \begin{pmatrix} \alpha^n & \beta^n \\ \gamma^n & \delta^n \end{pmatrix} & 0_2 \\ 0_2 & \begin{pmatrix} \bar{\alpha}^n & \bar{\beta}^n \\ \bar{\gamma}^n & \bar{\delta}^n \end{pmatrix} \end{pmatrix} &= -a_n \begin{pmatrix} \begin{pmatrix} \alpha^{n-1}\gamma & \beta^{n-1}\delta \\ 0 & 0 \end{pmatrix} & 0_2 \\ 0_2 & 0_2 \end{pmatrix} \\
\pi(f)\begin{pmatrix} \begin{pmatrix} \alpha^n & \beta^n \\ \gamma^n & \delta^n \end{pmatrix} & 0_2 \\ 0_2 & \begin{pmatrix} \bar{\alpha}^n & \bar{\beta}^n \\ \bar{\gamma}^n & \bar{\delta}^n \end{pmatrix} \end{pmatrix} &= -a_n \begin{pmatrix} \begin{pmatrix} 0 & 0 \\ \alpha\gamma^{n-1} & \beta\delta^{n-1} \end{pmatrix} & 0_2 \\ 0_2 & 0_2 \end{pmatrix}, \\
\pi(\bar{k})\begin{pmatrix} \begin{pmatrix} \alpha^n & \beta^n \\ \gamma^n & \delta^n \end{pmatrix} & 0_2 \\ 0_2 & \begin{pmatrix} \bar{\alpha}^n & \bar{\beta}^n \\ \bar{\gamma}^n & \bar{\delta}^n \end{pmatrix} \end{pmatrix} &= \begin{pmatrix} \begin{pmatrix} \alpha^n & \beta^n \\ \gamma^n & \delta^n \end{pmatrix} & 0_2 \\ 0_2 & \begin{pmatrix} q^{-n/2}\bar{\alpha}^n & q^{-n/2}\bar{\beta}^n \\ q^{n/2}\bar{\gamma}^n & q^{n/2}\bar{\delta}^n \end{pmatrix} \end{pmatrix} \\
\pi(\bar{e})\begin{pmatrix} \begin{pmatrix} \alpha^n & \beta^n \\ \gamma^n & \delta^n \end{pmatrix} & 0_2 \\ 0_2 & \begin{pmatrix} \bar{\alpha}^n & \bar{\beta}^n \\ \bar{\gamma}^n & \bar{\delta}^n \end{pmatrix} \end{pmatrix} &= -\bar{a}_n \begin{pmatrix} 0_2 & 0_2 \\ 0_2 & \begin{pmatrix} \bar{\alpha}^{n-1}\bar{\gamma} & \bar{\beta}^{n-1}\bar{\delta} \\ 0 & 0 \end{pmatrix} \end{pmatrix}, \\
\pi(\bar{f})\begin{pmatrix} \begin{pmatrix} \alpha^n & \beta^n \\ \gamma^n & \delta^n \end{pmatrix} & 0_2 \\ 0_2 & \begin{pmatrix} \bar{\alpha}^n & \bar{\beta}^n \\ \bar{\gamma}^n & \bar{\delta}^n \end{pmatrix} \end{pmatrix} &= -\bar{a}_n \begin{pmatrix} 0_2 & 0_2 \\ 0_2 & \begin{pmatrix} 0 & 0 \\ \bar{\alpha}\bar{\gamma}^{n-1} & \bar{\beta}\bar{\delta}^{n-1} \end{pmatrix} \end{pmatrix},
\end{aligned}
\tag{5.125}
$$

$$
a_n = q^{(1-n)/2}[n]_q, \quad \bar{a}_n = q^{(n-1)/2}[n]_q, \quad [n]_q = (q^n - q^{-n})/\lambda. \tag{5.126}
$$

Analogously, we introduce the right action $\pi_R(X)$ (cf. (5.16)). Explicitly we have:

$$\pi_R(k)M = M\langle k, M\rangle = \begin{pmatrix} \begin{pmatrix} q^{1/2}\alpha & q^{-1/2}\beta \\ q^{1/2}\gamma & q^{-1/2}\delta \end{pmatrix} & 0_2 \\ 0_2 & \begin{pmatrix} \bar\alpha & \bar\beta \\ \bar\gamma & \bar\delta \end{pmatrix} \end{pmatrix},$$

$$\pi_R(e)M = M\langle e, M\rangle = \begin{pmatrix} \begin{pmatrix} 0 & \alpha \\ 0 & \gamma \end{pmatrix} & 0_2 \\ 0_2 & 0_2 \end{pmatrix}, \tag{5.127}$$

$$\pi_R(f)M = M\langle f, M\rangle = \begin{pmatrix} \begin{pmatrix} \beta & 0 \\ \delta & 0 \end{pmatrix} & 0_2 \\ 0_2 & 0_2 \end{pmatrix},$$

$$\pi_R(\bar k)M = M\langle \bar k, M\rangle = \begin{pmatrix} \begin{pmatrix} \alpha & \beta \\ \gamma & \delta \end{pmatrix} & 0_2 \\ 0_2 & \begin{pmatrix} q^{1/2}\bar\alpha & q^{-1/2}\bar\beta \\ q^{1/2}\bar\gamma & q^{-1/2}\bar\delta \end{pmatrix} \end{pmatrix},$$

$$\pi_R(\bar e)M = M\langle \bar e, M\rangle = \begin{pmatrix} 0_2 & 0_2 \\ 0_2 & \begin{pmatrix} 0 & \bar\alpha \\ 0 & \bar\gamma \end{pmatrix} \end{pmatrix},$$

$$\pi_R(\bar f)M = M\langle \bar f, M\rangle = \begin{pmatrix} 0_2 & 0_2 \\ 0_2 & \begin{pmatrix} \bar\beta & 0 \\ \bar\delta & 0 \end{pmatrix} \end{pmatrix}, \tag{5.128}$$

The twisted derivation rule is $\pi_R(X)\varphi\psi = \pi_R(\Delta_{\mathscr{U}}(X))(\varphi\otimes\psi)$; that is,

$$\pi_R(k)\varphi\psi = \pi_R(k)\varphi\cdot\pi_R(k)\psi\,, \tag{5.129a}$$

$$\pi_R(e)\varphi\psi = \pi_R(e)\varphi\cdot\pi_R(k^{-1}\bar k)\psi + \pi_R(k\bar k^{-1})\varphi\cdot\pi_R(e)\psi$$

$$\pi_R(f)\varphi\psi = \pi_R(f)\varphi\cdot\pi_R(k^{-1}\bar k^{-1})\psi + \pi_R(k\bar k)\varphi\cdot\pi_R(f)\psi$$

$$\pi_R(\bar k)\varphi\psi = \pi_R(\bar k)\varphi\cdot\pi_R(\bar k)\psi\,, \tag{5.129b}$$

$$\pi_R(\bar e)\varphi\psi = \pi_R(\bar e)\varphi\cdot\pi_R(k^{-1}\bar k)\psi + \pi_R(k\bar k^{-1})\varphi\cdot\pi_R(\bar e)\psi$$

$$\pi_R(\bar f)\varphi\psi = \pi_R(\bar f)\varphi\cdot\pi_R(k\bar k)\psi + \pi_R(k^{-1}\bar k^{-1})\varphi\cdot\pi_R(\bar f)\psi$$

Using this, we find:

$$\pi_R(k)\begin{pmatrix} \begin{pmatrix} \alpha^n & \beta^n \\ \gamma^n & \delta^n \end{pmatrix} & 0_2 \\ 0_2 & \begin{pmatrix} \bar{\alpha}^n & \bar{\beta}^n \\ \bar{\gamma}^n & \bar{\delta}^n \end{pmatrix} \end{pmatrix} =$$

$$= \begin{pmatrix} \begin{pmatrix} q^{n/2}\alpha^n & q^{-n/2}\beta^n \\ q^{n/2}\gamma^n & q^{-n/2}\delta^n \end{pmatrix} & 0_2 \\ 0_2 & \begin{pmatrix} \bar{\alpha}^n & \bar{\beta}^n \\ \bar{\gamma}^n & \bar{\delta}^n \end{pmatrix} \end{pmatrix} \tag{5.130a}$$

$$\pi_R(e)\begin{pmatrix} \begin{pmatrix} \alpha^n & \beta^n \\ \gamma^n & \delta^n \end{pmatrix} & 0_2 \\ 0_2 & \begin{pmatrix} \bar{\alpha}^n & \bar{\beta}^n \\ \bar{\gamma}^n & \bar{\delta}^n \end{pmatrix} \end{pmatrix} =$$

$$= a_n \begin{pmatrix} \begin{pmatrix} 0 & \alpha\beta^{n-1} \\ 0 & \gamma\delta^{n-1} \end{pmatrix} & 0_2 \\ 0_2 & 0_2 \end{pmatrix}$$

$$\pi_R(f)\begin{pmatrix} \begin{pmatrix} \alpha^n & \beta^n \\ \gamma^n & \delta^n \end{pmatrix} & 0_2 \\ 0_2 & \begin{pmatrix} \bar{\alpha}^n & \bar{\beta}^n \\ \bar{\gamma}^n & \bar{\delta}^n \end{pmatrix} \end{pmatrix} =$$

$$= a_n \begin{pmatrix} \begin{pmatrix} \alpha^{n-1}\beta & 0 \\ \gamma^{n-1}\delta & 0 \end{pmatrix} & 0_2 \\ 0_2 & 0_2 \end{pmatrix}$$

$$\pi_R(\bar{k})\begin{pmatrix} \begin{pmatrix} \alpha^n & \beta^n \\ \gamma^n & \delta^n \end{pmatrix} & 0_2 \\ 0_2 & \begin{pmatrix} \bar{\alpha}^n & \bar{\beta}^n \\ \bar{\gamma}^n & \bar{\delta}^n \end{pmatrix} \end{pmatrix} =$$

$$= \begin{pmatrix} \begin{pmatrix} \alpha^n & \beta^n \\ \gamma^n & \delta^n \end{pmatrix} & 0_2 \\ 0_2 & \begin{pmatrix} q^{n/2}\bar{\alpha}^n & q^{-n/2}\bar{\beta}^n \\ q^{n/2}\bar{\gamma}^n & q^{-n/2}\bar{\delta}^n \end{pmatrix} \end{pmatrix} \tag{5.130b}$$

$$\pi_R(\bar{e})\begin{pmatrix} \begin{pmatrix} \alpha^n & \beta^n \\ \gamma^n & \delta^n \end{pmatrix} & 0_2 \\ 0_2 & \begin{pmatrix} \bar{\alpha}^n & \bar{\beta}^n \\ \bar{\gamma}^n & \bar{\delta}^n \end{pmatrix} \end{pmatrix} =$$

$$= \bar{a}_n \begin{pmatrix} 0_2 & 0_2 \\ 0_2 & \begin{pmatrix} 0 & \bar{\alpha}\bar{\beta}^{n-1} \\ 0 & \bar{\gamma}\bar{\delta}^{n-1} \end{pmatrix} \end{pmatrix}$$

$$\pi_R(\bar{f}) \begin{pmatrix} \begin{pmatrix} \alpha^n & \beta^n \\ \gamma^n & \delta^n \end{pmatrix} & 0_2 \\ 0_2 & \begin{pmatrix} \bar{\alpha}^n & \bar{\beta}^n \\ \bar{\gamma}^n & \bar{\delta}^n \end{pmatrix} \end{pmatrix} =$$

$$= \bar{a}_n \begin{pmatrix} 0_2 & 0_2 \\ 0_2 & \begin{pmatrix} \bar{\alpha}^{n-1}\bar{\beta} & 0 \\ \bar{\gamma}^{n-1}\bar{\delta} & 0 \end{pmatrix} \end{pmatrix}.$$

Now we introduce the elements φ as formal power series:

$$\varphi = \sum_{\substack{\ell,m,\bar{\ell},\bar{m}\in\mathbb{Z}_+ \\ n,\bar{n}\in\mathbb{Z}}} \mu_{\ell,m,n,\bar{\ell},\bar{m},\bar{n}} \beta^\ell \gamma^m \delta^n \bar{\beta}^{\bar{\ell}} \bar{\gamma}^{\bar{m}} \bar{\delta}^{\bar{n}} . \tag{5.131}$$

By (5.125) and (5.130) we have defined left and right action of $\mathcal{U}$ on φ. As above we use the right action to reduce the left regular representation. First we calculate:

$$\begin{aligned} \pi_R(f)\, \beta^\ell \gamma^m \delta^n \bar{\beta}^{\bar{\ell}} \bar{\gamma}^{\bar{m}} \bar{\delta}^{\bar{n}} &= q^{(n-\ell+\bar{\ell}-\bar{m}+\bar{n})/2} a_m\, \beta^\ell \gamma^{m-1} \delta^{n+1} \bar{\beta}^{\bar{\ell}} \bar{\gamma}^{\bar{m}} \bar{\delta}^{\bar{n}}, \\ \pi_R(\bar{f})\, \beta^\ell \gamma^m \delta^n \bar{\beta}^{\bar{\ell}} \bar{\gamma}^{\bar{m}} \bar{\delta}^{\bar{n}} &= q^{(n+\ell-m+\bar{\ell}-\bar{n})/2} \bar{a}_m\, \beta^\ell \gamma^m \delta^n \bar{\beta}^{\bar{\ell}} \bar{\gamma}^{\bar{m}-1} \bar{\delta}^{\bar{n}+1} \end{aligned}$$

from which imposing right covariance with respect to $f, \bar{f}$; that is:

$$\pi_R(f)\varphi = 0, \pi_R(\bar{f})\varphi = 0, \qquad \Rightarrow \qquad \mu_{\ell,m,n,\bar{\ell},\bar{m},\bar{n}} \sim \delta_{m0}\delta_{\bar{m}0} , \tag{5.132}$$

we obtain that there is no γ and $\bar{\gamma}$ dependence in φ. Next, we impose right covariance with respect to $k, \bar{k}$:

$$\pi_R(k)\varphi = q^{-r/2}\varphi, \quad \pi_R(\bar{k})\varphi = q^{-\bar{r}/2}\varphi, \tag{5.133}$$

where $r, \bar{r}$ are parameters to be specified below. On the other hand, using (5.129a,b) and (5.130a,b) one has:

$$\begin{aligned} \pi_R(k)\, \beta^\ell \delta^n \bar{\beta}^{\bar{\ell}} \bar{\delta}^{\bar{n}} &= q^{-(\ell+n)/2} \beta^\ell \delta^n\, \bar{\beta}^{\bar{\ell}} \bar{\delta}^{\bar{n}}, \\ \pi_R(\bar{k})\, \beta^\ell \delta^n \bar{\beta}^{\bar{\ell}} \bar{\delta}^{\bar{n}} &= q^{-(\bar{\ell}+\bar{n})/2} \beta^\ell \delta^n\, \bar{\beta}^{\bar{\ell}} \bar{\delta}^{\bar{n}}. \end{aligned} \tag{5.134}$$

Thus, we obtain: $n + \ell = r$, $\bar{n} + \bar{\ell} = \bar{r}$, and as a consequence: $r, \bar{r} \in \mathbb{Z}$ and $\mu_{\ell,0,n,\bar{\ell},0,\bar{n}} \sim \delta_{\ell+n,r}\delta_{\bar{\ell}+\bar{n},\bar{r}}$. Our reduced φ can now be written as:

$$\varphi(\beta, \delta, \bar{\beta}, \bar{\delta}) = \sum_{\ell,\bar{\ell}\in\mathbb{Z}_+} \mu_{\ell\bar{\ell}}\beta^\ell\delta^{r-\ell}\bar{\beta}^{\bar{\ell}}\bar{\delta}^{\bar{r}-\bar{\ell}}. \tag{5.135}$$

From (5.135) we are prompted to introduce the variables $\eta = \beta\delta^{-1}$, $\bar{\eta} = \bar{\beta}\bar{\delta}^{-1}$, which are noncommuting: $\eta\bar{\eta} = q^2\bar{\eta}\eta$. Then we can rewrite (5.135) as:

$$\begin{aligned}\tilde{\varphi}(\eta, \bar{\eta}, \delta, \bar{\delta}) &= \varphi(\beta, \delta, \bar{\beta}, \bar{\delta}) = \sum_{\ell,\bar{\ell}\in\mathbb{Z}_+} \tilde{\mu}_{\ell\bar{\ell}}\eta^\ell\bar{\eta}^{\bar{\ell}}\delta^r\bar{\delta}^{\bar{r}} = \\ &= \tilde{\varphi}(\eta, \bar{\eta}, 1_{\mathcal{L}}, 1_{\mathcal{L}})\delta^r\bar{\delta}^{\bar{r}}.\end{aligned} \tag{5.136}$$

Note that $\tilde{\varphi}$ obey the same covariance properties (5.132) and (5.133).

Now we can derive the $\mathcal{U}$-action π on φ. First, we find using (5.125) and (5.124):

$$\begin{aligned}\pi(k)\eta^\ell\bar{\eta}^{\bar{\ell}}\delta^r\bar{\delta}^{\bar{r}} &= q^{-\ell+r/2}\eta^\ell\bar{\eta}^{\bar{\ell}}\delta^r\bar{\delta}^{\bar{r}},\\ \pi(e)\eta^\ell\bar{\eta}^{\bar{\ell}}\delta^r\bar{\delta}^{\bar{r}} &= -q^{(2\bar{\ell}+r-\bar{r}-1)/2}[\ell]_q\eta^{\ell-1}\bar{\eta}^{\bar{\ell}}\delta^r\bar{\delta}^{\bar{r}},\\ \pi(f)\eta^\ell\bar{\eta}^{\bar{\ell}}\delta^r\bar{\delta}^{\bar{r}} &= q^{-(2\bar{\ell}+r-\bar{r}-1)/2}[\ell-r]_q\eta^{\ell+1}\bar{\eta}^{\bar{\ell}}\delta^r\bar{\delta}^{\bar{r}},\\ \pi(\bar{k})\eta^\ell\bar{\eta}^{\bar{\ell}}\delta^r\bar{\delta}^{\bar{r}} &= q^{-\bar{\ell}+\bar{r}/2}\eta^\ell\bar{\eta}^{\bar{\ell}}\delta^r\bar{\delta}^{\bar{r}},\\ \pi(\bar{e})\eta^\ell\bar{\eta}^{\bar{\ell}}\delta^r\bar{\delta}^{\bar{r}} &= -q^{(2\ell+r-\bar{r}+1)/2}[\bar{\ell}]_q\eta^\ell\bar{\eta}^{\bar{\ell}-1}\delta^r\bar{\delta}^{\bar{r}},\\ \pi(\bar{f})\eta^\ell\bar{\eta}^{\bar{\ell}}\delta^r\bar{\delta}^{\bar{r}} &= q^{-(2\ell+r-\bar{r}+1)/2}[\bar{\ell}-\bar{r}]_q\eta^\ell\bar{\eta}^{\bar{\ell}+1}\delta^r\bar{\delta}^{\bar{r}}.\end{aligned} \tag{5.137}$$

As a consequence, recalling (5.136), we find:

$$\begin{aligned}\pi(k)\tilde{\varphi}(\eta, \bar{\eta}, \delta, \bar{\delta}) &= q^{r/2}T_\eta^{-1}\tilde{\varphi}(\eta, \bar{\eta}, \delta, \bar{\delta}),\\ \pi(e)\tilde{\varphi}(\eta, \bar{\eta}, \delta, \bar{\delta}) &= -q^{(r-\bar{r}-1)/2}D_\eta T_{\bar{\eta}}\tilde{\varphi}(\eta, \bar{\eta}, \delta, \bar{\delta}),\\ \pi(f)\tilde{\varphi}(\eta, \bar{\eta}, \delta, \bar{\delta}) &= q^{-(r-\bar{r}-1)/2}\frac{\eta}{\lambda}\left(q^{-r}T_\eta - q^rT_\eta^{-1}\right)T_{\bar{\eta}}^{-1}\tilde{\varphi}(\eta, \bar{\eta}, \delta, \bar{\delta}),\\ \pi(\bar{k})\tilde{\varphi}(\eta, \bar{\eta}, \delta, \bar{\delta}) &= q^{\bar{r}/2}T_{\bar{\eta}}^{-1}\tilde{\varphi}(\eta, \bar{\eta}, \delta, \bar{\delta}),\\ \pi(\bar{e})\tilde{\varphi}(\eta, \bar{\eta}, \delta, \bar{\delta}) &= -q^{(r-\bar{r}+1)/2}D_{\bar{\eta}}T_\eta^{-1}\tilde{\varphi}(\eta, \bar{\eta}, \delta, \bar{\delta}),\\ \pi(\bar{f})\tilde{\varphi}(\eta, \bar{\eta}, \delta, \bar{\delta}) &= q^{-(r-\bar{r}+1)/2}\frac{\bar{\eta}}{\lambda}\left(q^{-\bar{r}}T_{\bar{\eta}} - q^{\bar{r}}T_{\bar{\eta}}^{-1}\right)T_\eta\tilde{\varphi}(\eta, \bar{\eta}, \delta, \bar{\delta}),\end{aligned} \tag{5.138}$$

where on any function g of η and $\bar{\eta}$

$$\begin{aligned}T_\eta g(\eta, \bar{\eta}) &= g(q\eta, \bar{\eta}), \quad D_\eta g(\eta, \bar{\eta}) = \frac{1}{\lambda\eta}(T_\eta - T_\eta^{-1})g(\eta, \bar{\eta}),\\ T_{\bar{\eta}} g(\eta, \bar{\eta}) &= g(\eta, q\bar{\eta}), \quad D_{\bar{\eta}} g(\eta, \bar{\eta}) = \frac{1}{\lambda\bar{\eta}}(T_{\bar{\eta}} - T_{\bar{\eta}}^{-1})g(\eta, \bar{\eta}).\end{aligned} \tag{5.139}$$

Notice that the operators T_η and $T_{\bar{\eta}}$ commute.

It is immediate to check that $\pi(k), \pi(e), \pi(f), \pi(\bar{k}), \pi(\bar{e}), \pi(\bar{f})$ satisfy (5.118). It is also clear that we can remove the inessential phases by setting:

$$\tilde{\pi}_{r,\bar{r}}(k) = \pi(k), \quad \tilde{\pi}_{r,\bar{r}}(e) = q^{(r-\bar{r}-1)/2}\pi(e), \quad \tilde{\pi}_{r,\bar{r}}(f) = q^{(\bar{r}-r+1)/2}\pi(f), \tag{5.140}$$

and the same settings for $k \mapsto \bar{k}$, $e \mapsto \bar{e}$, $f \mapsto \bar{f}$. Then $\tilde{\pi}_{r,\bar{r}}$ also satisfy (5.118).

We denote by $\mathscr{C}_{r,\bar{r}}$ the representation space of functions $\tilde{\varphi}(\eta, \bar{\eta}, \delta, \bar{\delta})$ with covariance properties (5.132) and (5.133) and transformation laws (5.138). Further, as in [198] we introduce the restricted functions $\hat{\varphi}(\eta, \bar{\eta})$ by the formula which is prompted in (5.136):

$$\hat{\varphi}(\eta, \bar{\eta}) \equiv \left(A\tilde{\varphi}\right)(\eta, \bar{\eta}) \doteq \tilde{\varphi}(\eta, \bar{\eta}, 1_{\mathscr{L}}, 1_{\mathscr{L}}). \tag{5.141}$$

We denote the representation space of $\hat{\varphi}(\eta, \bar{\eta})$ by $\hat{\mathscr{C}}_{r,\bar{r}}$ and the representation acting in $\hat{\mathscr{C}}_{r,\bar{r}}$ by $\hat{\pi}_{r,\bar{r}}$. Thus, the operator A acts from $\mathscr{C}_{r,\bar{r}}$ to $\hat{\mathscr{C}}_{r,\bar{r}}$. The properties of $\hat{\mathscr{C}}_{r,\bar{r}}$ follow from the intertwining requirement for A [198]:

$$\hat{\pi}_{r,\bar{r}}A = A\tilde{\pi}_{r,\bar{r}}. \tag{5.142}$$

In particular, the representation $\hat{\pi}_{r,\bar{r}}$ is given by:

$$\hat{\pi}_{r,\bar{r}}(k)\hat{\varphi}(\eta, \bar{\eta}) = q^{r/2}T_{\eta}^{-1}\hat{\varphi}(\eta, \bar{\eta}), \tag{5.143a}$$
$$\hat{\pi}_{r,\bar{r}}(e)\hat{\varphi}(\eta, \bar{\eta}) = -D_{\eta}T_{\bar{\eta}}\hat{\varphi}(\eta, \bar{\eta}), \tag{5.143b}$$
$$\hat{\pi}_{r,\bar{r}}(f)\hat{\varphi}(\eta, \bar{\eta}) = \frac{\eta}{\lambda}\left(q^{-r}T_{\eta} - q^{r}T_{\eta}^{-1}\right)T_{\bar{\eta}}^{-1}\hat{\varphi}(\eta, \bar{\eta}), \tag{5.143c}$$
$$\hat{\pi}_{r,\bar{r}}(\bar{k})\hat{\varphi}(\eta, \bar{\eta}) = q^{\bar{r}/2}T_{\bar{\eta}}^{-1}\hat{\varphi}(\eta, \bar{\eta}), \tag{5.143d}$$
$$\hat{\pi}_{r,\bar{r}}(\bar{e})\hat{\varphi}(\eta, \bar{\eta}) = -D_{\bar{\eta}}T_{\eta}^{-1}\hat{\varphi}(\eta, \bar{\eta}), \tag{5.143e}$$
$$\hat{\pi}_{r,\bar{r}}(\bar{f})\hat{\varphi}(\eta, \bar{\eta}) = \frac{\bar{\eta}}{\lambda}\left(q^{-\bar{r}}T_{\bar{\eta}} - q^{\bar{r}}T_{\bar{\eta}}^{-1}\right)T_{\eta}\hat{\varphi}(\eta, \bar{\eta}). \tag{5.143f}$$

or, using the decomposition:

$$\hat{\varphi}(\eta, \bar{\eta}) = \sum_{\ell,\bar{\ell}\in\mathbb{Z}_+} \tilde{\mu}_{\ell,\bar{\ell}}\,\eta^{\ell}\bar{\eta}^{\bar{\ell}}, \tag{5.144}$$

inherited from (5.136):

$$\hat{\pi}_{r,\bar{r}}(k)\eta^{\ell}\bar{\eta}^{\bar{\ell}} = q^{-\ell+r/2}\eta^{\ell}\bar{\eta}^{\bar{\ell}}, \tag{5.145a}$$
$$\hat{\pi}_{r,\bar{r}}(e)\eta^{\ell}\bar{\eta}^{\bar{\ell}} = -q^{\bar{\ell}}[\ell]_q\eta^{\ell-1}\bar{\eta}^{\bar{\ell}}, \tag{5.145b}$$
$$\hat{\pi}_{r,\bar{r}}(f)\eta^{\ell}\bar{\eta}^{\bar{\ell}} = q^{-\bar{\ell}}[\ell - r]_q\eta^{\ell+1}\bar{\eta}^{\bar{\ell}}, \tag{5.145c}$$
$$\hat{\pi}_{r,\bar{r}}(\bar{k})\eta^{\ell}\bar{\eta}^{\bar{\ell}} = q^{-\bar{\ell}+\bar{r}/2}\eta^{\ell}\bar{\eta}^{\bar{\ell}}, \tag{5.145d}$$
$$\hat{\pi}_{r,\bar{r}}(\bar{e})\eta^{\ell}\bar{\eta}^{\bar{\ell}} = -q^{\ell}[\bar{\ell}]_q\eta^{\ell}\bar{\eta}^{\bar{\ell}-1}, \tag{5.145e}$$
$$\hat{\pi}_{r,\bar{r}}(\bar{f})\eta^{\ell}\bar{\eta}^{\bar{\ell}} = q^{-\ell}[\bar{\ell} - \bar{r}]_q\eta^{\ell}\bar{\eta}^{\bar{\ell}+1}. \tag{5.145f}$$

Note that if we restrict to one $\mathscr{U}_q(sl(2,\mathbb{C}))$ subalgebra and to functions of one variable, for example, by setting $\bar{\eta} = 1_{\mathscr{L}}$, $T_{\bar{\eta}}$ =id, in formulae (5.143a,b,c), we obtain the q-difference realization of [309].

5.3.4 q-Difference Intertwining Operators

We have defined the representations $\hat{\pi}_{r,\bar{r}}$ for $r, \bar{r} \in \mathbb{Z}$. However, we notice that we can consider (5.143) and (5.145) for arbitrary complex $r, \bar{r}$. Now we make some statements which will be proved in the next section. For generic $r, \bar{r} \in \mathbb{C}$ the representations $\hat{\pi}_{r,\bar{r}}$ are irreducible. For $r \in \mathbb{Z}_+$ or $\bar{r} \in \mathbb{Z}_+$ the representations $\pi_{r,\bar{r}}$, $\hat{\pi}_{r,\bar{r}}$ are reducible. Moreover, for $r \in \mathbb{Z}_+$ the representations $\pi_{r,\bar{r}}$ and $\pi_{-r-2,\bar{r}}$ are partially equivalent, while for $\bar{r} \in \mathbb{Z}_+$ the representations $\pi_{r,\bar{r}}$ and $\pi_{r,-\bar{r}-2}$ are partially equivalent. The same statements hold for the restricted counterparts $\hat{\pi}_{r,\bar{r}}$. These partial equivalences are realized by operators:

$$\mathscr{I}_r : \mathscr{C}_{r,\bar{r}} \longrightarrow \mathscr{C}_{-r-2,\bar{r}}, \quad I_r : \hat{\mathscr{C}}_{r,\bar{r}} \longrightarrow \hat{\mathscr{C}}_{-r-2,\bar{r}}, \tag{5.146a}$$

$$\bar{\mathscr{I}}_{\bar{r}} : \mathscr{C}_{r,\bar{r}} \longrightarrow \mathscr{C}_{r,-\bar{r}-2}, \quad \bar{I}_{\bar{r}} : \hat{\mathscr{C}}_{r,\bar{r}} \longrightarrow \hat{\mathscr{C}}_{r,-\bar{r}-2}, \tag{5.146b}$$

where $r, \bar{r} \in \mathbb{Z}_+$, that is, one has:

$$\mathscr{I}_r \circ \pi_{r,\bar{r}} = \pi_{-r-2,\bar{r}} \circ \mathscr{I}_r, \quad I_r \circ \hat{\pi}_{r,\bar{r}} = \hat{\pi}_{-r-2,\bar{r}} \circ I_r, \tag{5.147a}$$

$$\bar{\mathscr{I}}_{\bar{r}} \circ \pi_{r,\bar{r}} = \pi_{r,-\bar{r}-2} \circ \bar{\mathscr{I}}_{\bar{r}}, \quad \bar{I}_{\bar{r}} \circ \hat{\pi}_{r,\bar{r}} = \hat{\pi}_{r,-\bar{r}-2} \circ \bar{I}_{\bar{r}}. \tag{5.147b}$$

We present now the explicit formulae for these intertwining operators. By the classical procedure of [198] one should take as intertwiners (up to nonzero multiplicative constants):

$$\mathscr{I}_r = \left(\pi_R(e)\right)^{r+1}, \quad I_r = \left(\hat{\pi}_R(e)\right)^{r+1}, \quad r \in \mathbb{Z}_+, \tag{5.148a}$$

$$\bar{\mathscr{I}}_{\bar{r}} = \left(\pi_R(\bar{e})\right)^{\bar{r}+1}, \quad \bar{I}_{\bar{r}} = \left(\hat{\pi}_R(\bar{e})\right)^{\bar{r}+1}, \quad \bar{r} \in \mathbb{Z}_+. \tag{5.148b}$$

The above is verified by straightforward calculation given in [164]. Furthermore, there it is found that in terms of the restricted functions $\hat{\varphi}$ holds:

$$I_r = \left(D_\eta T_\eta T_{\bar{\eta}}^2\right)^{r+1}, \quad r \in \mathbb{Z}_+, \tag{5.149a}$$

$$\bar{I}_{\bar{r}} = \left(D_{\bar{\eta}} T_{\bar{\eta}}^{-1} T_\eta^{-2}\right)^{\bar{r}+1}, \quad \bar{r} \in \mathbb{Z}_+. \tag{5.149b}$$

Finally we note that for $q = 1$ we recover the classical intertwining operators of Gelfand–Graev–Vilenkin [314] (see also [227] Appendix B):

$$I_r = \left(\frac{\partial}{\partial \eta}\right)^{r+1}, \quad r \in \mathbb{Z}_+, \qquad \bar{I}_{\bar{r}} = \left(\frac{\partial}{\partial \bar{\eta}}\right)^{\bar{r}+1}, \quad \bar{r} \in \mathbb{Z}_+. \tag{5.150}$$

5.3.5 Classification of Reducible Representations

Now we shall make complete the statements made at the beginning of last section about the representations $\pi_{r,\bar{r}}$, $\hat{\pi}_{r,\bar{r}}$. It is enough to work with the restricted representation $\hat{\pi}_{r,\bar{r}}$.

For $r \in \mathbb{Z}_+$ the operator I_r has a kernel $\mathcal{E}_{r,\bar{r}}$ which is a subspace of $\hat{\mathcal{C}}_{r,\bar{r}}$. This subspace consists of elements which are polynomials of degree $\le r$ with respect to η. The basis of $\mathcal{E}_{r,\bar{r}}$ may be taken as $\eta^{\ell}\bar{\eta}^{\bar{\ell}}$, $\ell \le r, \bar{\ell} \in \mathbb{Z}_+$. Note that from (5.145c) follows that $\hat{\pi}_{r,\bar{r}}(f)\eta^{r}\bar{\eta}^{\bar{\ell}} = 0$.

Analogously, for $\bar{r} \in \mathbb{Z}_+$ the operator $\bar{I}_{\bar{r}}$ has a kernel $\bar{\mathcal{E}}_{r,\bar{r}}$ which is a subspace of $\hat{\mathcal{C}}_{r,\bar{r}}$. This subspace consists of elements which are polynomials of degree $\le \bar{r}$ with respect to $\bar{\eta}$. The basis of $\bar{\mathcal{E}}_{r,\bar{r}}$ may be taken as $\eta^{\ell}\bar{\eta}^{\bar{\ell}}$, $\ell \in \mathbb{Z}_+, \bar{\ell} \le \bar{r}$. Also from (5.145f) follows that $\hat{\pi}_{r,\bar{r}}(\bar{f})\eta^{\ell}\bar{\eta}^{\bar{r}} = 0$.

Finally, for $r, \bar{r} \in \mathbb{Z}_+$ the intersection $\widetilde{\mathcal{E}}_{r,\bar{r}} = \mathcal{E}_{r,\bar{r}} \cap \bar{\mathcal{E}}_{r,\bar{r}}$ is a finite-dimensional subspace consisting of polynomials of degrees $\le r$ with respect to η and $\le \bar{r}$ with respect to $\bar{\eta}$. The basis of $\widetilde{\mathcal{E}}_{r,\bar{r}}$ may be taken as $\eta^{\ell}\bar{\eta}^{\bar{\ell}}$, $\ell \le r, \bar{\ell} \le \bar{r}$. Clearly, we have $\dim \widetilde{\mathcal{E}}_{r,\bar{r}} = (r+1)(\bar{r}+1)$.

Let us denote by $L_{r,\bar{r}}$ the irreducible subrepresentation of $\hat{\pi}_{r,\bar{r}}$. Clearly we have that $L_{r,\bar{r}} = \hat{\pi}_{r,\bar{r}}$ iff $r, \bar{r} \notin \mathbb{Z}_+$. Otherwise, $L_{r,\bar{r}}$ is a nontrivial subrepresentation of $\hat{\pi}_{r,\bar{r}}$ realized in $\mathcal{E}_{r,\bar{r}}$ when $r \in \mathbb{Z}_+, \bar{r} \notin \mathbb{Z}_+$, in $\bar{\mathcal{E}}_{r,\bar{r}}$ when $r \notin \mathbb{Z}_+, \bar{r} \in \mathbb{Z}_+$, in $\widetilde{\mathcal{E}}_{r,\bar{r}}$ when $r, \bar{r} \in \mathbb{Z}_+$. The last finite-dimensional irreducible representation has highest-weight vector $1_{\mathcal{L}}$ ($\hat{\pi}_{r,\bar{r}}(X)1_{\mathcal{L}} = 0, X = e, \bar{e}$) and lowest-weight vector $\eta^{r}\bar{\eta}^{\bar{r}}$ ($\hat{\pi}_{r,\bar{r}}(X)\eta^{r}\bar{\eta}^{\bar{r}} = 0, X = f, \bar{f}$). Note that all finite-dimensional irreducible representations of $\mathcal{U}$ are obtained in this way.

Finally, we may present all reducible representation spaces (together with some irreducible ones) in the following diagrams:

$$\begin{array}{ccc} \hat{\mathcal{C}}_{r,\bar{r}} & \longrightarrow & \hat{\mathcal{C}}_{-r-2,\bar{r}} \\ \downarrow & & \downarrow \\ \hat{\mathcal{C}}_{r,-\bar{r}-2} & \longrightarrow & \hat{\mathcal{C}}_{-r-2,-\bar{r}-2} \end{array} \qquad r, \bar{r} \in \mathbb{Z}_+, \tag{5.151}$$

$$\hat{\mathcal{C}}_{r,\bar{r}} \longrightarrow \hat{\mathcal{C}}_{-r-2,\bar{r}}, \qquad r \in \mathbb{Z}_+, \quad \bar{r} \notin \mathbb{Z}\backslash\{-1\}, \tag{5.152a}$$

$$\begin{array}{c} \hat{\mathcal{C}}_{r,\bar{r}} \\ \downarrow \\ \hat{\mathcal{C}}_{r,-\bar{r}-2} \end{array} \qquad r \notin \mathbb{Z}\backslash\{-1\}, \quad \bar{r} \in \mathbb{Z}_+, \tag{5.152b}$$

where the horizontal arrows represent the operators I_r, the vertical arrows represent the operators $\bar{I}_{\bar{r}}$. Note that (5.151) is a commutative diagram.

These diagrams also represent graphically a multiplet classification [193] of all representations which are either reducible or partially equivalent to reducible ones. Explicitly, this classification is as follows. All representation spaces $\hat{\mathcal{C}}_{r',\bar{r}'}$ when $r', \bar{r}' \in \mathbb{Z}\backslash\{-1\}$, are present in (5.151); all representation spaces $\hat{\mathcal{C}}_{r',\bar{r}}$ when $r' \in \mathbb{Z}\backslash\{-1\}$, $\bar{r} \notin \mathbb{Z}\backslash\{-1\}$ are present in (5.152a); all representation spaces $\hat{\mathcal{C}}_{r,\bar{r}'}$ when $r \notin \mathbb{Z}\backslash\{-1\}$, $\bar{r}' \in \mathbb{Z}\backslash\{-1\}$ are present in (5.152b).

Finally, we would make some comparison with the case $q = 1$, that is, when $\mathcal{U} = sl(2,\mathbb{C})$ [227, 314]. First, only representations with $r-\bar{r} \in \mathbb{Z}$ are integrable to representations of the group $SL(2,\mathbb{C})$ considered in [314] (or $SO_0(3,1)$ in [227]). Second, these representations are topological and though diagram (5.151) exists with the same operators, it has a little different content in [314]. In particular, the representation space $\hat{\mathcal{C}}_{-r-2,-\bar{r}-2}$ is also reducible due to the existence of an integral intertwining operator acting from $\hat{\mathcal{C}}_{-r-2,-\bar{r}-2}$ to $\hat{\mathcal{C}}_{r,\bar{r}}$ and having a nontrivial (infinite-dimensional) kernel.

5.3.6 The Roots of Unity Case

In the present section we treat the case when (5.112) is an involution and $q \in \mathbb{C}$, $|q| = 1$. Nothing is changed in all considerations for generic q. However, things change drastically when q is a root of unity, $q = e^{2\pi i/N}$, $N = 2, 3, \ldots$

First, all elements $\alpha^N, \beta^N, \gamma^N, \delta^N, \bar{\alpha}^N, \bar{\beta}^N, \bar{\gamma}^N, \bar{\delta}^N$, belong to the centre of $\mathcal{L}$. It is convenient to set:

$$\delta^N = \bar{\delta}^N = 1_{\mathcal{L}}, \beta^N = \gamma^N = \bar{\beta}^N = \bar{\gamma}^N = 0, \Rightarrow \alpha^N = \bar{\alpha}^N = 1_{\mathcal{L}}. \tag{5.153}$$

Then the basis of $\mathcal{L}$ instead of (5.114) is:

$$\beta^\ell \gamma^m \delta^n \bar{\beta}^{\bar{\ell}} \bar{\gamma}^{\bar{m}} \bar{\delta}^{\bar{n}}, \quad \ell, m, \bar{\ell}, \bar{m} \in \mathbb{Z}_+, \ell, m, \bar{\ell}, \bar{m} < N, \tag{5.154}$$
$$n, \bar{n} \in \mathbb{Z}, |n|, |\bar{n}| < N.$$

Note that (5.153) is consistent with the actions of $\mathcal{U}$ on $\mathcal{L}$, since, in particular, $a_N = \bar{a}_N = 0$ (cf. (5.126)).

Instead of (5.131) we have

$$\varphi = \sum_{\substack{\ell,m,\bar{\ell},\bar{m}<N \\ |n|,|\bar{n}|<N}} \mu_{\ell,m,n,\bar{\ell},\bar{m},\bar{n}} \beta^\ell \gamma^m \delta^n \bar{\beta}^{\bar{\ell}} \bar{\gamma}^{\bar{m}} \bar{\delta}^{\bar{n}}. \tag{5.155}$$

From the restrictions (5.133) of right covariance with respect to $k, \bar{k}$ we get that $|r|, |\bar{r}| < 2N$. Then, after the change of variables in (5.136) we have to restrict $|r|, |\bar{r}| < N$.

However, as at the beginning of Subsection 5.3.4, we shall consider (5.143) and (5.145) for arbitrary complex $r, \bar{r}$. Now all representations $\hat{\pi}_{r,\bar{r}}$ are finite-dimensional for any values of $r, \bar{r}$. This is clear from the analogue of (5.144):

$$\hat{\varphi}(\eta, \bar{\eta}) = \sum_{\ell,\bar{\ell}<N} \tilde{\mu}_{\ell,\bar{\ell}} \eta^{\ell} \bar{\eta}^{\bar{\ell}}. \tag{5.156}$$

Thus, the dimension of $\hat{\pi}_{r,\bar{r}}$ is at most N^2. This dimension is achieved when $r, \bar{r} \notin \mathbb{Z}$ or when $r+1, \bar{r}+1 \in N\mathbb{Z}$. Indeed, in these cases all elements in (5.156) are present in the representation space (cf. (5.145)). Further, for $x \in \mathbb{Z}$ let x_N be the smallest non-negative integer equal to x (mod N); thus, $0 \le x_N < N$. In the case $r+1 \in \mathbb{Z}\backslash N\mathbb{Z}$, from (5.145c) it follows that $\hat{\pi}_{r,\bar{r}}(f)\eta^{r_N}\bar{\eta}^{\bar{\ell}} = 0$. Thus, the basis of the representation space is given by the monomials $\eta^{\ell}\bar{\eta}^{\bar{\ell}}$ such that $\ell \le r_N$. Analogously, for $\bar{r}+1 \in \mathbb{Z}\backslash N\mathbb{Z}$, from (5.145f) it follows that $\hat{\pi}_{r,\bar{r}}(\bar{f})\eta^{\ell}\bar{\eta}^{\bar{r}_N} = 0$. Thus, the basis of the representation space is given by the monomials $\eta^{\ell}\bar{\eta}^{\bar{\ell}}$ such that $\bar{\ell} \le \bar{r}_N$.

Therefore, for the irreducible subrepresentation $L_{r,\bar{r}}$ of $\hat{\pi}_{r,\bar{r}}$ we have shown that:

$$\dim L_{r,\bar{r}} = \begin{cases} N^2, & \text{for} \quad r+1, \bar{r}+1 \notin \mathbb{Z}\backslash N\mathbb{Z}, \\ (r_N+1)N, & \text{for} \quad r+1 \in \mathbb{Z}\backslash N\mathbb{Z}, \bar{r}+1 \notin \mathbb{Z}\backslash N\mathbb{Z}, \\ (\bar{r}_N+1)N, & \text{for} \quad r+1 \notin \mathbb{Z}\backslash N\mathbb{Z}, \bar{r}+1 \in \mathbb{Z}\backslash N\mathbb{Z}, \\ (r_N+1)(\bar{r}_N+1), & \text{for} \quad r+1, \bar{r}+1 \in \mathbb{Z}\backslash N\mathbb{Z}. \end{cases} \tag{5.157}$$

From the point of view of the intertwiners, first one can check that if $r, \bar{r} \notin \mathbb{Z}$ there are no intertwining operators. Then we notice that the operator I_{N-1} acts from $\hat{\mathcal{C}}_{r,\bar{r}}$ to $\hat{\mathcal{C}}_{r-2N,\bar{r}}$ for any r, such that $r+1 \in N\mathbb{Z}$, while the operator $\bar{I}_{N-1}$ acts from $\hat{\mathcal{C}}_{r,\bar{r}}$ to $\hat{\mathcal{C}}_{r,\bar{r}-2N}$ for any $\bar{r}$, such that $\bar{r}+1 \in N\mathbb{Z}$. However, these operators are zero since their kernels are the whole spaces on which they act (cf. (5.64) and (5.37)). Thus when $r+1, \bar{r}+1 \notin \mathbb{Z}\backslash N\mathbb{Z}$ there are no nontrivial intertwining operators, the representations are irreducible with dimension N^2; that is, this is the first case in (5.157).

When $r+1 \in \mathbb{Z}\backslash N\mathbb{Z}$, each sequence $\hat{\mathcal{C}}_{r,\bar{r}} \longrightarrow \hat{\mathcal{C}}_{r-2N,\bar{r}}$ is replaced by $\hat{\mathcal{C}}_{r,\bar{r}} \longrightarrow \hat{\mathcal{C}}_{r-2r_N-2,\bar{r}} \longrightarrow \hat{\mathcal{C}}_{r-2N,\bar{r}}$, so that the operator acting in $\hat{\mathcal{C}}_{r,\bar{r}} \longrightarrow \hat{\mathcal{C}}_{r-2r_N-2,\bar{r}}$ is I_{r_N}, while the operators acting in $\hat{\mathcal{C}}_{r-2r_N-2,\bar{r}} \longrightarrow \hat{\mathcal{C}}_{r-2N,\bar{r}}$ is I_{N-r_N-2}. These operators have nontrivial kernels and all these representations spaces are reducible. Note, however, that these operators have zero composition: $I_{N-r_N-2} \circ I_{r_N} = I_{N-1} = 0$. Analogous statements hold when $\bar{r}+1 \in \mathbb{Z}\backslash N\mathbb{Z}$.

Therefore, we have *three* cases in which there are nontrivial intertwining operators and which correspond to the last three cases in (5.157). In the second and third case in (5.157), the representation spaces are grouped in one-dimensional lattices. (Such one-dimensional lattices were written for the $\mathcal{U}_q(sl(2,\mathbb{C}))$ case in [30, 209, 309, 506].) First we give explicitly the lattice corresponding to the second case in (5.157):

$$\cdots \longrightarrow \hat{\mathcal{C}}_{r,\bar{r}} \longrightarrow \hat{\mathcal{C}}_{r-2r_N-2,\bar{r}} \longrightarrow \hat{\mathcal{C}}_{r-2N,\bar{r}} \longrightarrow \cdots \tag{5.158}$$

The irreducible subrepresentations in (5.158) have the following dimensions:

$$\begin{aligned}\dim L_{r+2sN,\bar r} &= (r_N+1)N,\ s\in\mathbb{Z},\\ \dim L_{r-2r_N-2+2sN,\bar r} &= (N-r_N-1)N,\ s\in\mathbb{Z}.\end{aligned}\tag{5.159}$$

In exactly the same way one considers the third case in (5.157).

Finally, in the last case in (5.157) the representations are grouped in two-dimensional lattices as follows:

$$\begin{array}{ccccccc} & & \vdots & & \vdots & & \vdots & \\ & & \downarrow & & \downarrow & & \downarrow & \\ \cdots\longrightarrow & & \hat{\mathscr{C}}_{r,\bar r} & \longrightarrow & \hat{\mathscr{C}}_{r-2r_N-2,\bar r} & \longrightarrow & \hat{\mathscr{C}}_{r-2N,\bar r} & \longrightarrow\cdots \\ & & \downarrow & & \downarrow & & \downarrow & \\ \cdots\longrightarrow & & \hat{\mathscr{C}}_{r,\bar r-2\bar r_N-2} & \longrightarrow & \hat{\mathscr{C}}_{r-2r_N-2,\bar r-2\bar r_N-2} & \longrightarrow & \hat{\mathscr{C}}_{r-2N,\bar r-2\bar r_N-2} & \longrightarrow\cdots \\ & & \downarrow & & \downarrow & & \downarrow & \\ \cdots\longrightarrow & & \hat{\mathscr{C}}_{r,\bar r-2N} & \longrightarrow & \hat{\mathscr{C}}_{r-2r_N-2,\bar r-2N} & \longrightarrow & \hat{\mathscr{C}}_{r-2N,\bar r-2N} & \longrightarrow\cdots \\ & & \downarrow & & \downarrow & & \downarrow & \\ & & \vdots & & \vdots & & \vdots & \end{array}\tag{5.160}$$

The irreducible subrepresentations in (5.158) have the following dimensions:

$$\begin{aligned}\dim L_{r+2sN,\bar r+2\bar sN} &= (r_N+1)(\bar r_N+1),\\ \dim L_{r-2r_N-2+2sN,\bar r+2\bar sN} &= (N-r_N-1)(\bar r_N+1),\\ \dim L_{r+2sN,\bar r-2\bar r_N-2+2\bar sN} &= (r_N+1)(N-\bar r_N-1),\\ \dim L_{r-2r_N-2+2sN,\bar r-2\bar r_N-2+2\bar sN} &= (N-r_N-1)(N-\bar r_N-1),\end{aligned}\tag{5.161}$$

where in all four cases $s, \bar s \in \mathbb{Z}$.

5.4 Representations of the Generalized Lie Algebra $sl(2)_q$

5.4.1 Preliminaries

A number of authors [180, 182, 447, 455] have suggested definitions of "quantum Lie algebras", the aim being to obtain structures which bear the same relation to quantized enveloping algebras as Lie algebras do to their enveloping algebras. It is of interest to determine the representations of such quantum Lie algebras, in those cases where a notion of "representation" is defined, and compare them to the classical representation theory. For generic values of the deformation parameter q it is to be expected that the representations will resemble those of the classical Lie algebras which are deformed into the quantum versions, since the representation theory of a quantized enveloping algebra is essentially the same as that of the classical Lie algebra, but the details of this resemblance will help to illuminate the nature of a quantum Lie algebra. This relationship breaks down if q is a root of unity, which is of much interest in physics, and it is therefore particularly significant to determine the representations of a quantum Lie algebra in this case.

Here we construct finite-dimensional representations of the simplest example of the generalized Lie algebras introduced in [447]. A representation of this algebra, in the sense defined in [447], is nothing but a representation of an associative algebra, the enveloping algebra of the quantum Lie algebra. This is obtained from a larger algebra with a central element by imposing a relation giving the central element as a function of Casimir-like elements. We investigate the representations also of this larger algebra, which is possibly more natural in the context of generalized Lie algebras, and find that it has additional one-dimensional representations.

5.4.2 The Quantum Lie Algebra $\mathfrak{sl}(2)_q$

The generalized Lie algebra $\mathfrak{sl}(2)_q$ was introduced in [447](cf. also [566–568]). Its enveloping algebra $\mathscr{A} \equiv U(\mathfrak{sl}(2)_q)$ is defined by Eq. (3.5) of [447]. For the purposes of developing the representation theory it is enough to work with the algebras $\mathscr{B}$, $\mathscr{C}$ (cf. [447]). The algebra $\mathscr{B}$ is generated by four generators: $X_0, X_\pm, C$ with relations ($\lambda = q - q^{-1}$):

$$\begin{aligned} q^2 X_0 X_+ - X_+ X_0 &= qCX_+ \\ q^{-2} X_0 X_- - X_- X_0 &= -q^{-1} CX_- \\ X_+ X_- - X_- X_+ &= (q + q^{-1})\,(C - \lambda X_0) X_0 \\ CX_m &= X_m C, \qquad m = 0, \pm 1 \end{aligned} \tag{5.162}$$

The algebra $\mathscr{B}$ is related to the locally finite part $\mathscr{F}$ of the simply-connected quantized enveloping algebra $\bar{U}_q(sl(2))$. The algebra $\mathscr{F}$ was obtained in [447] from $\mathscr{B}$ by putting C equal to a function of the second-order Casimir:

$$C_2 = (q + q^{-1})\, X_0^2 + qX_- X_+ + q^{-1} X_+ X_- \tag{5.163}$$

namely,

$$C^2 = 1 + \frac{\lambda^2}{q + q^{-1}}\, C_2 \tag{5.164}$$

For shortness we shall call $\mathscr{C}$ the restricted algebra. The enveloping algebra $\mathscr{A}$, on the other hand, is obtained by putting $C = 1$ [447].

We shall need a triangular decomposition of $\mathscr{B}$:

$$\mathscr{B} = \mathscr{B}_+ \otimes \mathscr{B}_0 \otimes \mathscr{B}_- \tag{5.165}$$

where $\mathscr{B}_\pm$ is generated by $X_\pm$, while $\mathscr{B}_0$ is generated by X_0, C. We shall call the abelian Lie algebra $\mathscr{H}$ generated by X_0, C the Cartan subalgebra of $\mathscr{B}$. Note that $\mathscr{B}_0$ is the enveloping algebra of $\mathscr{H}$. The same decomposition is used for the algebra $\mathscr{C}$ with the relation (5.164) enforced.

Further we shall analyse the algebras $\mathscr{B}$ and $\mathscr{C}$ separately.

5.4.3 Highest-Weight Representations

HWMs of $\mathscr{B}$ are standardly determined by a highest-weight vector v_0 which is annihilated by the raising generator X_+ and on which the Cartan generators act by the corresponding value of the highest weight $\Lambda \in \mathscr{H}^*$:

$$\begin{aligned} X_+\, v_0 &= 0 \\ H\, v_0 &= \Lambda(H)\, v_0, \qquad H \in \mathscr{H} \\ \mu &\equiv \Lambda(X_0), \quad c \equiv \Lambda(C) \end{aligned} \tag{5.166}$$

In particular, we shall be interested in Verma modules over $\mathscr{C}$. As in the classical case a Verma module V^Λ is an HWM of weight Λ induced from one-dimensional representation of a Borel subalgebra $\tilde{\mathscr{B}}$, for example, $\tilde{\mathscr{B}} = \mathscr{B}_+ \otimes \mathscr{B}_0$, on the highest-weight vector, for example, v_0. As vector spaces we have:

$$V^\Lambda \cong \mathscr{B} \otimes_{\tilde{\mathscr{B}}} v_0 = \mathscr{B}_- \otimes v_0 = l.s.\{v_k \equiv X_-^k \otimes v_0 | k \in \mathbb{Z}_+\} \tag{5.167}$$

where we have identified $1_{\mathscr{B}} \otimes v_0$ with v_0.

The action of the generators of $\mathscr{B}$ on the basis of V^Λ is given as follows:

$$X_+\, v_k = q^{2k-2}\,(c - \lambda\mu)\,\big([2k]_q\mu - q[k]_q[k-1]_q c\big)\, v_{k-1} \tag{5.168a}$$

$$X_-\, v_k = v_{k+1} \tag{5.168b}$$

$$X_0\, v_k = (q^{2k}\mu - q^k[k]_q c)\, v_k \tag{5.168c}$$

$$C\, v_k = c\, v_k \tag{5.168d}$$

$$[k]_q \equiv (q^k - q^{-k})/\lambda$$

To obtain (5.168a,c) we have used the following calculations which follow from (5.162):

$$X_0 X_-^k = X_-^k \left(q^{2k} X_0 - q^k [k]_q C\right) \tag{5.169a}$$

$$[X_+, X_-^k] = X_-^{k-1}\, q^{2k-2}\, (C - \lambda X_0) \times \left([2k]_q X_0 - q[k]_q[k-1]_q C\right) \tag{5.169b}$$

As usually we start in the analysis of reducibility of Verma modules with the search for singular vectors $v_s \in V^\Lambda$ which are standardly defined as:

$$X_+ \, v_s = 0 \tag{5.170a}$$

$$H \, v_s = \Lambda'(H)\, v_s, \qquad H \in \mathscr{H}, \Lambda' \in \mathscr{H}^* \tag{5.170b}$$

First we note that since C is central its value is the same as on v_0: $c' \equiv \Lambda'(C) = c$. Further, we proceed to find the possible singular vectors using that they are eigenvectors of X_0. But the eigenvectors of X_0 are $X_-^n \otimes v_0$, all with different eigenvalues. Thus, a singular vector will be given by the classical expression (omitting the overall normalization): $v_s = X_-^n \otimes v_0$; for some fixed $n \in \mathbb{N}$, and we have:

$$X_0 \, v_s = \mu' \, v_s, \qquad \mu' \equiv \Lambda'(X_0) = q^{2n}\mu - q^n [n]_q c \tag{5.171}$$

Finally, we have to impose (5.170a) for which we calculate (using (5.169b)):

$$X_+ \, v_s = X_-^{n-1}\, q^{2n-2}\, (c - \lambda\mu)\, \left([2n]_q \mu - q[n]_q[n-1]_q c\right) \otimes v_0 \tag{5.172}$$

For the further analysis we suppose that the deformation parameter q is not a nontrivial root of unity. Then there are two possibilities for (5.172) to be zero, and, thus, we have two possibilities to fulfil (5.170a):

$$\mu = q[n]_q[n-1]_q \, c/[2n]_q \tag{5.173a}$$

$$\text{or} \quad c = \lambda\mu \tag{5.173b}$$

We shall analyse the two possibilities in (5.173) separately since they have very different implications; moreover, they are incompatible unless $c = \mu = 0$ when they coincide and which we shall treat as partial case of (5.173b).

5.4.3.1 Case $c \neq 0$

The first possibility (5.173a) (with $c \neq 0$) corresponds to the classical relation between n and the highest weight Λ (obtained for $q, c \to 1$): $\mu = (n-1)/2$. Thus, if we fix $n \in \mathbb{N}$ then $v_s = X_-^n \otimes v_0$ is a singular vector when μ has the value (5.173a). The shifted weight Λ' corresponds to another Verma module $V^{\Lambda'}$ which is the naximal invariant submodule of V^Λ. The corresponding eigenvalue of X_0 is (cf. (5.171)):

$$\mu' = -q[n]_q[n+1]_q c/[2n]_q \tag{5.174}$$

Note that the Verma module $V^{\Lambda'}$ does not have a singular vector.

The factor-module $L_{n,c} \cong V^\Lambda/V^{\Lambda'}$ is irreducible and finite-dimensional of dimension n. It has a highest-weight vector $|n, c\rangle$ such that:

$$\begin{aligned} X_+ \, |n, c\rangle &= 0, \qquad X_-^n \, |n, c\rangle = 0, \\ H \, |n, c\rangle &= \Lambda(H) \, |n, c\rangle, \qquad H \in \mathscr{H} \end{aligned} \tag{5.175}$$

Let us denote by $w_k \equiv X_-^k |n, c\rangle$, $k = 0, 1, \dots, n-1$, the states of $L_{n,c}$. The transformation rules for w_k are:

$$\begin{aligned} X_+ \, w_k &= q^{2k-n} \, [k]_q [n-k]_q \left(\frac{c \, [2]_q \, [n]_q}{[2n]_q} \right)^2 w_{k-1} \\ X_- \, w_k &= w_{k+1}, \qquad k < n-1 \\ X_- \, w_{n-1} &= 0 \\ X_0 \, w_k &= \frac{c \, q^k \, [n]_q}{[2n]_q} \left([n-k]_q - q^{1-n} [k+1]_q \right) w_k \\ C \, w_k &= c \, w_k \end{aligned} \tag{5.176}$$

Thus, the vector w_{n-1} is the lowest-weight vector of $L_{n,c}$.

Next we introduce a bilinear form in $L_{n,c}$ by the formula:

$$(w_j, w_k) \equiv \langle n, c | X_+^j X_-^k | n, c\rangle \tag{5.177}$$

where $\langle n, c|$ is such that $\langle n, c| \, |n, c\rangle = 1$ and:

$$\begin{aligned} \langle n, c| \, X_- &= 0 \\ \langle n, c| \, H &= \Lambda(H) \, \langle n, c|, \qquad H \in \mathscr{H} \\ \langle n, c| \, X_+^n &= 0 \end{aligned} \tag{5.178}$$

Then we obtain:

$$(w_j, w_k) = \delta_{jk} \, q^{k(k+1-n)} \, \frac{[k]_q! \, [n-1]_q!}{[n-1-k]_q!} \left(\frac{c \, [2]_q \, [n]_q}{[2n]_q} \right)^{2k} \tag{5.179}$$

where $[k]_q! \equiv [k]_q [k-1]_q \dots [1]_q$, $\quad [0]_q! \equiv 1$. Clearly (5.179) is real-valued for real q, c. Thus, for $q, c \in \mathbb{R}$ we can turn (5.177) into a scalar product and define the norm of the basis vectors:

$$|w_k| \equiv \sqrt{(w_k, w_k)} = q^{k(k+1-n)/2} \sqrt{\frac{[k]_q! \, [n-1]_q!}{[n-1-k]_q!}} \left(\frac{c \, [2]_q \, [n]_q}{[2n]_q} \right)^k \tag{5.180}$$

where have chosen the root that is positive for positive c, q. We can also introduce orthonormal basis:

$$u_k \equiv \frac{1}{|w_k|} w_k \,, \qquad (u_j, u_k) = \delta_{jk} \tag{5.181}$$

The transformation rules for the basis vectors u_k are:

$$\begin{aligned}
X_+ u_k &= q^{k-n/2} \sqrt{[k]_q [n-k]_q \frac{c\,[2]_q\,[n]_q}{[2n]_q}}\; u_{k-1} \\
X_- u_k &= q^{k+1-n/2} \sqrt{[n-1-k]_q\,[k+1]_q \frac{c\,[2]_q\,[n]_q}{[2n]_q}}\; u_{k+1} \\
X_0\, u_k &= \frac{c\, q^k\,[n]_q}{[2n]_q} \left([n-k]_q - q^{1-n}[k+1]_q\right) u_k \\
C\, u_k &= c\, u_k
\end{aligned} \tag{5.182}$$

The above scalar product is invariant under the real form $\mathscr{B}_r$ of $\mathscr{B}$ defined by the antilinear antiinvolution:

$$\omega(X^\pm) = X^\mp, \qquad \omega(X_0) = X_0, \qquad \omega(C) = C \tag{5.183}$$

Indeed, the algebraic relations (5.162) are preserved by ω for real q. The $\mathscr{B}_r$ invariance of the scalar product means that:

$$(w_j, X w_k) = (\omega(X) w_j, w_k), \qquad X \in \mathscr{B}, \tag{5.184}$$

which is automatically satisfied with the definition (5.177). (Note that (5.184) defines $(\,,)$ as the Shapovalov bilinear form [550].)

Thus, for every $n \in \mathbb{N}$ we have constructed n-dimensional irreducible representations (irreps) of $\mathscr{B}$ parametrized by $c \in \mathbb{C}$, $c \neq 0$, with basis w_k or u_k $(k = 0, \dots, n-1)$. For $q, c \in \mathbb{R}$ these are irreps of the real form $\mathscr{B}_r$, which are unitary when $q, c > 0$.

5.4.3.2 Case $c = \lambda\mu$

The second possibility (5.173b) has no classical analogue. It tells us that if c and μ are related as in (5.173b) then each vector of the basis of V^Λ is a singular vector. Moreover, all of them have the same weight since $\mu' = \mu$ (cf. (5.171)). This is clear also from the transformation rules (5.168) when $c = \lambda\mu$:

$$\begin{aligned}
X_+ v_k &= 0 \\
X_- v_k &= v_{k+1} \\
X_0\, v_k &= \mu\, v_k \\
C\, v_k &= \lambda\mu\, v_k
\end{aligned} \tag{5.185}$$

Clearly, we have an infinite sequence of embedded reducible Verma modules $V_n = l.s.\{v_k | k \in \mathbb{Z}_+, \quad k \geq n\}$ for $n \in \mathbb{Z}_+$ as follows: $V_n \supset V_{n+1}$, the latter being the maximal invariant submodule of the former. Note that V_n is isomorphic to a submodule of all V_m with $n > m$. Furthermore, because of the coincidence of the weights these modules are also all isomorphic to each other: $V_n \cong V_m$ for all m, n. It is also clear that for every μ there is only one irreducible module, namely, the one-dimensional $L_\mu \cong V_n/V_{n+1}$, for any n. Denoting by $|\mu\rangle$ the only state in L_μ we have for the action on it:

$$\begin{aligned} X_+ \, |\mu\rangle &= 0 \\ X_- \, |\mu\rangle &= 0 \\ X_0 \, |\mu\rangle &= \mu \, |\mu\rangle \\ C \, |\mu\rangle &= \lambda\mu \, |\mu\rangle \end{aligned} \tag{5.186}$$

Note that the above one-dimensional irrep is different from the one-dimensional $L_{1,c}$ from the previous subsection. Indeed, though the action of $X_\pm$ is the same, the ratio of eigenvalues of C to X_0 here is λ, while there it is $-[2]_q/q$.

5.4.4 Highest-Weight Representations of the Restricted Algebra

The highest-weight representations of the restricted algebra $\mathscr{C}$ are obtained from those of $\mathscr{B}$ imposing the relation (5.164). In particular, there is the following relation between the values of the Cartan generators:

$$c^2 = 1 + \lambda^2 \left(\frac{\mu^2}{q^2} + c\frac{\mu}{q} \right) \tag{5.187}$$

This relation has to be imposed on all formulae of the previous section. There are no essential consequences of this for the generic Verma modules. For the reducible Verma modules there are more interesting consequences. First we notice that the reducibility condition (5.173b) is incompatible with (5.187), and thus there would be no special one-dimensional irreps like L_μ, (cf. (5.186)). So it remains to consider the combination of the reducibility condition (5.173a) with (5.187) from which we obtain that:

$$c = \frac{\epsilon\,[2n]_q}{[2]_q\,[n]_q}, \qquad \mu = \frac{q[n]_q[n-1]_q\,c}{[2n]_q} = \frac{\epsilon\, q[n-1]_q}{[2]_q}, \qquad \epsilon = \pm 1 \tag{5.188}$$

In this case the analogue of (5.174) is:

$$\mu' = -\epsilon\, q[n+1]_q/[2]_q. \tag{5.189}$$

Let us denote the finite-dimensional representations of $\mathscr{C}$ by $\tilde{L}_{n,\epsilon}$ and the basis by $\tilde{w}_k$, $k = 0, \dots, n-1$. The transformation rules are:

$$\begin{aligned} X_+ \, \tilde{w}_k &= q^{2k-n}\, [k]_q [n-k]_q \, \tilde{w}_{k-1} \\ X_- \, \tilde{w}_k &= \tilde{w}_{k+1}, \qquad k < n-1 \end{aligned} \tag{5.190}$$

$$
\begin{aligned}
X_- \tilde{w}_{n-1} &= 0 \\
X_0 \tilde{w}_k &= \frac{\epsilon q^k}{[2]_q} \left([n-k]_q - q^{1-n}[k+1]_q\right) \tilde{w}_k \\
C \tilde{w}_k &= \frac{\epsilon [2n]_q}{[2]_q [n]_q} \tilde{w}_k
\end{aligned}
$$

Further, the analogues of (5.179) and (5.180) are:

$$
(\tilde{w}_j, \tilde{w}_k) = \delta_{jk}\, q^{k(k+1-n)} \frac{[k]_q!\,[n-1]_q!}{[n-1-k]_q!} \tag{5.191}
$$

$$
|\tilde{w}_k| \equiv \sqrt{(\tilde{w}_k, \tilde{w}_k)} = q^{k(k+1-n)/2}\, [k]_q! \sqrt{\frac{[k]_q!\,[n-1]_q!}{[n-1-k]_q!}} \tag{5.192}
$$

We can also introduce orthonormal basis:

$$
\tilde{u}_k \equiv \frac{1}{|\tilde{w}_k|} \tilde{w}_k, \qquad (\tilde{u}_j, \tilde{u}_k) = \delta_{jk} \tag{5.193}
$$

for which the transformation rules are:

$$
\begin{aligned}
X_+ \tilde{u}_k &= q^{k-n/2} \sqrt{[k]_q [n-k]_q}\, \tilde{u}_{k-1} \\
X_- \tilde{u}_k &= q^{k+1-n/2} \sqrt{[n-1-k]_q\, [k+1]_q}\, \tilde{u}_{k+1} \\
X_0 \tilde{u}_k &= \frac{\epsilon q^k}{[2]_q} \left([n-k]_q - q^{1-n}[k+1]_q\right) \tilde{u}_k \\
C \tilde{u}_k &= \frac{\epsilon [2n]_q}{[2]_q [n]_q} \tilde{u}_k
\end{aligned} \tag{5.194}
$$

Thus, for every $n \in \mathbb{N}$ we have constructed n-dimensional irreducible representations of $\mathscr{C}$ parametrized by $\epsilon = \pm 1$, with bases $\tilde{w}_k$ or $\tilde{u}_k$ $(k = 0, \ldots, n-1)$.

5.4.5 Highest-Weight Representations at Roots of Unity

Here we consider representations of the algebra $\mathscr{B}$ in the case when the deformation parameter is at roots of unity. More precisely, first we consider the cases when q^2 is a primitive N-th root of unity: $q = e^{\pi i/N}$, $N \in \mathbb{N} + 1$. Then we have:

$$
[x]_q = \frac{\sin \pi x/N}{\sin \pi/N} \tag{5.195}
$$

In such cases there are additional reducibility conditions coming from (5.172) besides (5.173a,b). For this we rewrite (5.173a) in a more general fashion:

$$\mu[2n]_q = q[n]_q[n-1]_q c \qquad (5.173a')$$

Then we note that from (5.195) follows that $[N]_q = [2N]_q = 0$, so (5.173a') is satisfied for $n \to N$. Thus, $v_s^N = X_-^N \otimes v_0$ is a singular vector independently of the highest weight Λ. Similarly to the analysis done in [198] and Section 2.7.2 for the quantized enveloping algebra $U_q(sl(2))$ all $v_s^{pN} = X_-^{pN} \otimes v_0$ for $p \in \mathbb{N}$ are singular vectors. The Verma modules they realize we denote by $\tilde{V}_p$, $p \in \mathbb{Z}_+$, $\tilde{V}_0 \equiv V^\Lambda$. These are embedded reducible Verma modules $\tilde{V}_p \supset \tilde{V}_{p+1}$ with the same highest weight Λ. Indeed, for any $\tilde{V}_p$ using (5.171) with $n \to pN$ we have: $\mu' = q^{2pN}\mu - q^{pN}[pN]_q c = \mu$.

The further analysis depends on whether there are additional singular vectors besides those just displayed. There are four cases.

5.4.5.1 Case when (5.173a,b) do not hold

We start with the case when μ, c do not satisfy either of (5.173a,b). We also suppose that $c \neq 0$ when N is *even*. Then there are no additional singular vectors and there is only one irreducible N-dimensional HWM $L_{\Lambda,N} \cong \tilde{V}_p/\tilde{V}_{p+1}$ (for any p), parametrized by all pairs μ, c not satisfying (5.173a,b). The action of the generators of $\mathscr{B}$ on the basis of $L_{\Lambda,N}$, which we denote by $\tilde{\varphi}_k$ $(k = 0, \ldots, N-1)$, is given as follows:

$$\begin{aligned}
X_+\, \tilde{\varphi}_k &= q^{2k-2}\,(c - \lambda\mu)\,\big([2k]_q\mu - q[k]_q[k-1]_q c\big)\,\tilde{\varphi}_{k-1} \\
X_-\, \tilde{\varphi}_k &= \tilde{\varphi}_{k+1}, \qquad k < N-1 \\
X_-\, \tilde{\varphi}_{N-1} &= 0 \\
X_0\, \tilde{\varphi}_k &= (q^{2k}\mu - q^k[k]_q c)\,\tilde{\varphi}_k \\
C\, \tilde{\varphi}_k &= c\,\tilde{\varphi}_k
\end{aligned} \qquad (5.196)$$

However, unlike the Drinfeld-Jimbo case, these finite-dimensional representations are not unitarizable, which is easily seen if one considers the analogue of the bilinear form (5.177).

5.4.5.2 Case when (5.173a) holds

Next we consider the case when μ, c satisfy (5.173a); for some $n \in \mathbb{N}$, $n < N$. We also suppose that $c \neq 0$ (for any N). First we note that $n < N$ is not a restriction, since then (5.173a) holds also for all $n + pN$, $p \in \mathbb{Z}$. Indeed, we have:

$$\begin{aligned}
&q[n+pN]_q[n+pN-1]_q c/[2(n+pN)]_q = \\
&= q[n]_q[n-1]_q \cos^2(\pi p)\, c/[2n]_q \cos(2\pi p) = \\
&= q[n]_q[n-1]_q c/[2n]_q = \mu
\end{aligned} \qquad (5.197)$$

Thus, we have another infinite series of singular vectors $v_s'^{pN} = X_-^{n+pN} \otimes v_0$ for $p \in \mathbb{Z}_+$. They realize reducible Verma modules which we denote by $\tilde{V}_p'$, $p \in \mathbb{Z}_+$; ($\tilde{V}_0'$ is the analogue of $V^{\Lambda'}$ introduced in the non-root-of-unity case, but here it is reducible). They all have the same highest weight Λ' determined by μ', c with μ' given by (5.171). Indeed, substituting n with $n + pN$ does not change the value of μ':

$$\begin{aligned} & q^{2(n+pN)}\mu - q^{n+pN}[n+pN]_q c = \\ & = q^{2n}\mu - q^{n+pN}\, e^{\pi i p}\, [n]_q\, \cos(\pi p)\, c = \\ & = q^{2n}\mu - q^n[n]_q c = \mu' \end{aligned} \tag{5.198}$$

Of course, after substituting μ with its value from (5.173a) we obtain the expression for μ' in (5.174). We have the following infinite embedding chain:

$$V^\Lambda \equiv \tilde{V}_0 \supset \tilde{V}_0' \supset \tilde{V}_1 \supset \tilde{V}_1' \supset \cdots \tag{5.199}$$

where all embeddings are noncomposite: the embeddings $\tilde{V}_p \supset \tilde{V}_p'$ are realized by singular vectors: $X_-^n \otimes v_p$, v_p being the highest-weight vector of $\tilde{V}_p$, while the embeddings $\tilde{V}_p' \supset \tilde{V}_{p+1}$ are realized by singular vectors: $X_-^{N-n} \otimes v_p'$, v_p' being the highest-weight vector of $\tilde{V}_p'$.

Now, factorizing each reducible Verma module by its maximal invariant submodule we obtain that for each $n \in \mathbb{N}$, $n < N$ there are two finite dimensional irreps parametrized by $c \in \mathbb{C}$, $c \neq 0$: $L_{n,N} \cong \tilde{V}_p/\tilde{V}_p'$ (for any p) which is n-dimensional, and $L_{n,N}' \cong \tilde{V}_p'/\tilde{V}_{p+1}$ (for any p) which is $(N-n)$-dimensional. However, it turns out that the irreps from one series are isomorphic to those of the other: $L_{n,N}' \cong L_{N-n,N}$. Indeed, note that the value of μ' for the Verma modules $\tilde{V}_p'$ given by (5.174) should be obtained (for consistency) also from the formula for μ with n substituted by $N-n$ (since this is the reducibility condition w.r.t. the noncomposite singular vector $X_-^{N-n} \otimes v_p'$) and indeed this is the case:

$$\begin{aligned} & q[N-n]_q[N-n-1]_q c/[2(N-n)]_q = \\ & = -q[n-N]_q[n+1-N]_q c/[2(n-N)]_q = \\ & = -q[n]_q[n+1]_q c\, \cos^2(\pi N)/[2n]_q\, \cos(2\pi N) = \\ & -q[n]_q[n+1]_q c/[2n]_q = \mu' \end{aligned}$$

Furthermore, the transformation rules for $L_{n,N}$ are the same as for $L_{n,c}$ (cf. (5.176)), while the transformation rules for $L_{n,N}'$ are obtained from (5.176) by substituting $n \to N-n$.

Thus, we are left with one series of finite-dimensional irreps $L_{n,N}$.

5.4.5.3 Case when (5.173b) holds

Next, we consider the case when μ, c satisfy (5.173b) for arbitrary c. Actually, nothing is changed from the non-root-of-unity case since the relevant formulae (5.185) and (5.186) are not changed.

5.4.5.4 Case N even and $c = 0$

Finally, we consider the case when N is even and $c = 0$. Let $\tilde{N} = N/2 \in \mathbb{N}$. In these cases there are additional reducibility conditions coming from (5.173a'). Indeed, from (5.195) follows that $[2\tilde{N}]_{\tilde{q}} = 0$ and $[\tilde{N}]_{\tilde{q}} \neq 0$. But if $c = 0$ then (5.173a') is again satisfied. Thus, the vector $\hat{\varphi}_s^{\tilde{N}} = X_-^{\tilde{N}} \otimes v_0$ is a singular vector independently of the value of μ. Similarly to the analysis of the first subsection also all $\hat{\varphi}_s^{p\tilde{N}} = X_-^{p\tilde{N}} \otimes v_0$ for $p \in \mathbb{N}$ are singular vectors. Note that for p even these are the singular vectors that we already have: $\hat{\varphi}_s^{p\tilde{N}} = v_s^{\tilde{p}N}$, $\tilde{p} = p/2$. The Verma modules they realize we denote by $\hat{V}_p$, $p \in \mathbb{Z}_+$, $\hat{V}_0 \equiv V^\Lambda$. These are embedded reducible Verma modules $\hat{V}_p \supset \hat{V}_{p+1}$ with the same value of μ up to sign. Indeed, for any $\hat{V}_p$ using (5.171) with $n \to p\tilde{N}$ we have: $\mu' = q^{2p\tilde{N}}\mu - q^{p\tilde{N}}[p\tilde{N}]_q c = (-1)^p \mu$. Certainly, for even p these are Verma modules from the first subsection: $\hat{V}_p = V_{\tilde{p}}$.

As above the further analysis depends on whether μ, c satisfy some of (5.173a,b). However, since $c = 0$ then the only additional possibility is that also $\mu = 0$, which is a partial case of (5.173b), which was considered in the previous subsection. Thus, further, we suppose that μ, c do not satisfy either of (5.173a,b) and that $\mu \neq 0$.

Then there are no additional singular vectors besides $\hat{\varphi}_s^{p\tilde{N}}$. Then for each $\mu \neq 0$ there is only one irreducible HWM $L_{\mu,\tilde{N}} \cong \hat{V}_p/\hat{V}_{p+1}$ (for any p) which is $\tilde{N}$-dimensional. The action of the generators of $\mathscr{B}$ on the basis of $L_{\mu,\tilde{N}}$, which we denote by $\hat{\varphi}_k$ ($k = 0, \ldots, \tilde{N} - 1$), is given as follows:

$$\begin{aligned} X_+ \hat{\varphi}_k &= -q^{2k-2}\lambda[2k]_q \mu^2 \hat{\varphi}_{k-1} \\ X_- \hat{\varphi}_k &= \hat{\varphi}_{k+1}, \qquad k < \tilde{N} - 1 \\ X_- \hat{\varphi}_{\tilde{N}-1} &= 0 \\ X_0 \hat{\varphi}_k &= q^{2k}\mu \hat{\varphi}_k \\ C \hat{\varphi}_k &= 0 \end{aligned} \tag{5.200}$$

Note that if $\tilde{N}$ is odd it seems that formulae (5.200) may be obtained from (5.196) for N odd and $c = 0$ by the substitution $N \to \tilde{N}$. However, this is not the same irrep since with the same replacement the parameter q there becomes $e^{\pi i/N} \to e^{\pi i/\tilde{N}}$ while the parameter q here is $e^{\pi i/2\tilde{N}}$.

5.4.6 Highest-Weight Representations at Roots of Unity of the Restricted Algebra

Here we consider representations of the restricted algebra $\mathscr{C}$ in the case when the deformation parameter is at roots of unity. We start with the case: $q = e^{\pi i/N}$, $N \in \mathbb{N} + 1$, and so (5.173a') holds. The analysis is as for the algebra $\mathscr{B}$ but imposing the relation (5.187), that is, combining the considerations of the previous two subsections.

5.4.6.1 Case when (5.173a) does not hold

We start with the case when μ, c do not satisfy (5.173a), that is, (5.188) does not hold. We also suppose that $c \neq 0$ when N is *even*. Then there is only one irreducible N-dimensional HWM parametrized by μ, c related by (5.187), which irrep we denote by $\tilde{L}_{\Lambda,N}$. For the transformation rules we can use formulae (5.196) with (5.187) imposed.

5.4.6.2 Case when (5.173a) holds and $c \neq 0$

Next we consider the case when μ, c satisfy (5.173a) and $c \neq 0$. Here we should be nore careful so we replace n by $n + pN$ with $n < N$. Combining the reducibility condition (5.173b) with (5.187) we first obtain that:

$$c^2 = \frac{[2(n+pN)]_q^2}{[2]_q^2\,[n+pN]_q^2} = \frac{[2n]_q^2}{[2]_q^2\,[n]_q^2} \tag{5.201}$$

Then we recover (5.188) and (5.189) for $n < N$ which means that we have the same situation as for the unrestricted algebra at roots of unity. Thus, for each $n \in \mathbb{N}$, $n < N$ and $\epsilon = \pm 1$ there is a finite-dimensional irrep: $\tilde{L}_{n,\epsilon,N}$ which is n-dimensional. The transformation rules for $\tilde{L}_{n,\epsilon,N}$ are the same as in the non-root-of-unity case (cf. (5.190)).

5.4.6.3 Case N even and $c = 0$

Finally, we consider the case when N is even and $c = 0$. Let $\tilde{N} = N/2 \in \mathbb{N}$. As for the unrestricted algebra there are additional reducibility conditions; that is, again the vector $v_s^{\tilde{N}} = X_-^{\tilde{N}} \otimes v_0$ is a singular vector. However, because of (5.187) the value of μ^2 is fixed:

$$\mu^2 = -\tilde{q}^2/\lambda^2, \qquad \mu = \epsilon\, i\tilde{q}/\lambda, \qquad \epsilon = \pm 1 \tag{5.202}$$

Otherwise, the analysis goes through and there is only two irreducible $\tilde{N}$-dimensional HWMs $\tilde{L}_{\epsilon,\tilde{N}}$ parametrized by ϵ. The action of the generators of $\mathscr{B}$ on the basis of $\tilde{L}_{\epsilon,\tilde{N}}$, which we denote by $\hat{\varphi}'_k$ ($k = 0, \ldots, \tilde{N}-1$), is given as follows:

$$\begin{aligned}
X_+\,\hat{\varphi}'_k &= \frac{\tilde{q}^{2k}[2k]_{\tilde{q}}}{\lambda}\,\hat{\varphi}'_{k-1} \\
X_-\,\hat{\varphi}'_k &= \hat{\varphi}'_{k+1}, \qquad k < \tilde{N}-1 \\
X_-\,\hat{\varphi}'_{\tilde{N}-1} &= 0 \\
X_0\,\hat{\varphi}'_k &= \frac{\epsilon\, i\tilde{q}^{2k+1}}{\lambda}\,\hat{\varphi}'_k \\
C\,\hat{\varphi}'_k &= 0
\end{aligned} \tag{5.203}$$

The crucial feature of these two irreps is that they do not have a classical limit for $\tilde{q} \to 1$ (obtained for $N \to \infty$).

5.5 Representations of $U_q(so(3))$ of Integer Spin Only

We construct induced representations of $\mathcal{U} = U_q(so(3)) \cong U_q(sl(2))$ on suitable q-cosets of the matrix quantum group $SO_q(3)$. From these we obtain canonically finite-dimensional representations of $\mathcal{U}$ only of odd dimension, that is, of integer spin. The matrix elements of these finite-dimensional representations are different from the standard $\mathcal{U}$ ones, which will be essential at least for the roots of unity case.

5.5.1 Preliminaries

Already from the papers of Drinfeld [251] and Jimbo [360] was clear that the quantum algebras $U_q(sl(2))$ and $U_q(so(3))$ are isomorphic since the above constructions use only the info about the root systems of $sl(2, \mathbb{C}) \cong so(3, \mathbb{C})$.

On the other hand, the corresponding matrix quantum groups $SL_q(2)$ and $SO_q(3)$ are not isomorphic. More precisely, as in the classical case, the matrix quantum group $SL_q(2)$ is a double cover of $SO_q(3)$ (cf. e. g., [258, 510]). Thus, one may expect that the induced holomorphic representations of $\mathcal{U} = U_q(sl(2))$ realized on suitable q-cosets of $SO_q(3)$ will have the feature of usual $SO(3, \mathbb{C})$ holomorphic irreps to be of integer spin only.

This is exactly what is shown in this section following [241]. For the applications it is also important that the matrix elements of these finite-dimensional representations are different from the standard $\mathcal{U}$ ones, which will be essential at least for the roots of unity case.

5.5.2 Matrix Quantum Group $SO_q(3)$ and the Dual $U_q(\mathcal{G})$

The matrix quantum group $\mathcal{A} = SO_q(n)$ is the q-deformed analog of the complex Lie group $SO(n, \mathbb{C})$ [272]. It is generated by n^2 elements which may be collected in a $n \times n$ matrix:

$$T = (t_{ij}) \tag{5.204}$$

and are subject to the following relations [272]:

$$R_q T_1 T_2 = T_2 T_1 R_q \tag{5.205}$$

$$TCT^t C^{-1} = CT^t C^{-1} T = I_n \tag{5.206}$$

where R_q is a certain $n^2{\times}n^2$ matrix, $T_1 = T \otimes I_n$, $T_2 = I_n \otimes T$, I_n is the identity $n{\times}n$ matrix, C is a certain $n{\times}n$ matrix. The coalgebra structure is given by [272] the following formulae for the coproduct $\delta_{\mathscr{A}}$, counit $\varepsilon_{\mathscr{A}}$, and antipode $\gamma_{\mathscr{A}}$,:

$$\delta_{\mathscr{A}}(t_{ik}) = \sum_{j=1}^{n} t_{ij} \otimes t_{jk}, \tag{5.207a}$$

$$\varepsilon_{\mathscr{A}}(t_{ik}) = \delta_{ik}, \tag{5.207b}$$

$$\gamma_{\mathscr{A}}(T) = C\,T^t\,C^{-1}, \tag{5.207c}$$

the antipode given in matrix form for compactness. Using these relations (5.206) are rewritten in the general form:

$$T\,\gamma_{\mathscr{A}}(T) = \gamma_{\mathscr{A}}(T)\,T = I_n \tag{5.208}$$

In the case $n = 3$ the R-matrix R_q has the form [272]:

$$R_q = \left[\begin{array}{ccc|ccc|ccc}
q & & & & & & & & \\
 & 1 & & & & & & & \\
 & & q^{-1} & & & & & & \\
\hline
 & \lambda & & 1 & & & & & \\
 & & \alpha & & 1 & & & & \\
 & & & & & 1 & & & \\
\hline
 & & \beta & & \alpha & & q^{-1} & & \\
 & & & & & \lambda & & 1 & \\
 & & & & & & & & q
\end{array}\right] \tag{5.209}$$

where

$$\lambda = q - q^{-1}, \quad \alpha = -q^{-1/2}\lambda, \quad \beta = (1 - q^{-1})\lambda$$

and the matrix C is:

$$C = \begin{pmatrix} 0 & 0 & q^{-1/2} \\ 0 & 1 & 0 \\ q^{1/2} & 0 & 0 \end{pmatrix}, \qquad C^2 = I_3 \tag{5.210}$$

With these choices from (5.205) and (5.206) we can derive the explicit relations which the the nine elements t_{ij} obey.

$$\begin{aligned}
t_{ik}t_{i\ell} &= q^{\ell-k}\,t_{i\ell}t_{ik}, && i = 1, 3, \quad k < \ell \\
t_{kj}t_{\ell j} &= q^{\ell-k}\,t_{\ell j}t_{kj}, && j = 1, 3, \quad k < \ell \\
t_{ij}t_{k\ell} &= q^{k+\ell-i-j}\,t_{k\ell}t_{ij}, && i < k, \quad j > \ell \\
t_{k\ell}t_{k+1,\ell+1} &= t_{k+1,\ell+1}t_{k\ell} + \lambda t_{k,\ell+1}t_{k+1,\ell}, && k, \ell = 1, 2
\end{aligned} \tag{5.211}$$

$$\begin{aligned}
t_{k,1}t_{k+1,3} &= qt_{k+1,3}t_{k,1} + \lambda t_{k+1,1}t_{k,3}, \qquad k = 1,2\\
t_{1,k}t_{3,k+1} &= qt_{3,k+1}t_{1,k} + \lambda t_{1,k+1}t_{3,k}, \qquad k = 1,2\\
t_{12}t_{32} &= q(t_{32}t_{12} + \mathscr{A}_{3\,13}t_{31})\\
t_{21}t_{23} &= q(t_{23}t_{21} + \mathscr{A}_{3\,13}t_{31})\\
t_{11}t_{33} &= q^2t_{33}t_{11} + q\lambda(t_{13}t_{31} - 1)\\
t_{12}t_{22} &= t_{22}t_{12} + \mathscr{A}_{3\,21}t_{13}, \quad t_{22}t_{23} = t_{23}t_{22} + \mathscr{A}_{3\,13}t_{32}\\
t_{21}t_{22} &= t_{22}t_{21} + \mathscr{A}_{3\,12}t_{31}, \quad t_{22}t_{32} = t_{32}t_{22} + \mathscr{A}_{3\,31}t_{23}\\
t_{12}^2 &= -q^{-1}[2]t_{11}t_{13}, \quad t_{23}^2 = -q^{-1}[2]t_{13}t_{33},\\
t_{21}^2 &= -q^{-1}[2]t_{11}t_{31}, \quad t_{32}^2 = -q^{-1}[2]t_{31}t_{33},\\
t_{12}t_{32} &= t_{21}t_{23}, \quad t_{32}t_{12} = t_{23}t_{21}
\end{aligned}$$

The quantum algebra in duality with $SO_q(n)$ is $U_q(so(n))$. For $n = 3$ one has $\mathscr{U} = U_q(so(3)) \cong U_q(sl(2))$ (cf. [272]). We use a rational basis of $\mathscr{U}$ extracted from the L-operators of [272]. It differs from the basis of [360] by an algebraic transformation. In terms of this basis of $\mathscr{U}$, which we denote by $X^\pm$, $k^\pm$, the algebraic relations are:

$$\begin{aligned}
&X^+X^- - X^-X^+ = (k^+ - k^-)/\lambda, \quad k^+k^- = k^-k^+ = 1_{\mathscr{U}},\\
&k^\pm X^+ = q^{\mp 1}X^+k^\pm, \quad k^\pm X^- = q^{\pm 1}X^-k^\pm,
\end{aligned} \tag{5.212}$$

the coalgebra relations are:

$$\begin{aligned}
&\delta_{\mathscr{U}}(k^\pm) = k^\pm \otimes k^\pm, \quad \varepsilon_{\mathscr{U}}(k^\pm) = 1, \quad \gamma_{\mathscr{U}}(k^\pm) = k^\mp,\\
&\delta_{\mathscr{U}}(X^+) = X^+ \otimes k^+ + 1_{\mathscr{U}} \otimes X^+,\\
&\delta_{\mathscr{U}}(X^-) = k^- \otimes X^- + X^- \otimes 1_{\mathscr{U}},\\
&\gamma_{\mathscr{U}}(X^+) = -X^+k^-, \quad \gamma_{\mathscr{U}}(X^-) = -k^+X^-,\\
&\varepsilon_{\mathscr{U}}(X^\pm) = 0.
\end{aligned} \tag{5.213}$$

The duality between the algebras $\mathscr{U}$ and $\mathscr{A}$ is given by the pairings between the generators which follows from [272] (formula (2.1) for $k = 1$, up to renormalization); explicitly, we have:

$$\langle X^+, T\rangle = \begin{pmatrix} 0 & -1 & 0\\ 0 & 0 & q^{1/2}\\ 0 & 0 & 0 \end{pmatrix}$$

$$\langle X^-, T\rangle = \begin{pmatrix} 0 & 0 & 0\\ 1 & 0 & 0\\ 0 & -q^{-1/2} & 0 \end{pmatrix}$$

$$\langle k^\pm, T\rangle = \begin{pmatrix} q^\mp & 0 & 0 \\ 0 & 1 & 0 \\ 0 & 0 & q^\pm \end{pmatrix} \tag{5.214}$$

These pairings are supplemented with the axiomatic pairing:

$$\langle X , 1_{\mathscr{A}} \rangle = \varepsilon_{\mathscr{U}}(X), \qquad \forall X \in \mathscr{U} \tag{5.215}$$

The pairing between arbitrary elements of $\mathscr{U}$ and $\mathscr{A}$ follows then from the properties of the duality pairing.

5.5.3 Representations of $U_q(so(3))$

Next we introduce the left regular representation of $\mathscr{U}$ which in the $q = 1$ case is the infinitesimal version of:

$$\pi(M')\, M = M'^{-1}\, M, \quad M', M \in SO(3, \mathbb{C}), \tag{5.216}$$

namely, we set:

$$\pi_L(X)\, t_{ij} = \sum_{k=1}^{3} \langle \gamma_{\mathscr{U}}(X), t_{ik}\rangle\, t_{kj}, \qquad X \in \mathscr{U} \tag{5.217}$$

Explicitly we get from (5.217) for the generators of $\mathscr{U}$:

$$\begin{aligned} \pi_L(k^\pm)\, t_{ij} &= q^{\pm(2-i)}\, t_{ij}\,, \\ \pi_L(X^+)\, T &= \begin{pmatrix} t_{21} & t_{22} & t_{23} \\ -q^{-1/2} t_{31} & -q^{-1/2} t_{32} & -q^{-1/2} t_{33} \\ 0 & 0 & 0 \end{pmatrix}, \\ \pi_L(X^-)\, T &= \begin{pmatrix} 0 & 0 & 0 \\ -t_{11} & -t_{12} & -t_{13} \\ q^{1/2} t_{21} & q^{1/2} t_{22} & q^{1/2} t_{23} \end{pmatrix} \end{aligned} \tag{5.218}$$

In order to derive the action of π_L on arbitrary elements of the basis we use the following twisted derivation rule consistent with the coproduct and the representation structure. Namely, we use [210, 211]:

$$\pi_L(y)\, ab = \hat{m}\left(\pi_L(\delta'_{\mathscr{U}}(y))(a \otimes b)\right) \tag{5.219}$$

where $\hat{m}$ is the multiplication map: $\hat{m} : \mathscr{A} \otimes \mathscr{A} \longrightarrow \mathscr{A}, \hat{m}(f \otimes f') = f \cdot f'$; $\delta'_{\mathscr{U}} = \sigma \circ \delta_{\mathscr{U}}$ is the opposite coproduct (σ is the permutation operator). Thus, in our concrete situation we have:

$$\begin{aligned}\pi_L(k^{\pm})\,ab &= \pi_L(k^{\pm})\,a\cdot\pi_L(k^{\pm})\,b\\ \pi_L(X^{+})\,ab &= \pi_L(k^{+})\,a\cdot\pi_L(X^{+})\,b+\pi_L(X^{+})a\cdot b\\ \pi_L(X^{-})\,ab &= \pi_L(X^{-})\,a\cdot\pi_L(k^{-})\,b+a\cdot\pi_L(X^{-})b\end{aligned}\tag{5.220}$$

Further we shall use that π_L is a representation; that is:

$$\begin{aligned}\pi_L(ZZ') &= \pi_L(Z)\cdot\pi_L(Z'),\\ \pi_L(\alpha Z+\beta Z') &= \alpha\pi_L(Z)\ +\ \beta\pi_L(Z'), \qquad \alpha,\beta\in\mathbb{C}.\end{aligned}\tag{5.221}$$

Next we introduce the right regular representation $\pi_R(X)$ [210, 211]:

$$\pi_R(X)\,t_{ij} = \sum_{k=1}^{3} t_{ik}\,\langle X, t_{kj}\rangle, \qquad X\in\mathcal{U}\tag{5.222}$$

Of course, as in all other cases we shall use (5.222) in order to reduce the left regular representation.

Explicitly we have:

$$\begin{aligned}\pi_R(k^{\pm})\,t_{ij} &= q^{\pm(j-2)}\,t_{ij}\\ \pi_R(X^{+})\,T &= \begin{pmatrix}0 & -t_{11} & q^{1/2}t_{12}\\ 0 & -t_{21} & q^{1/2}t_{22}\\ 0 & -t_{31} & q^{1/2}t_{32}\end{pmatrix}\\ \pi_R(X^{-})\,T &= \begin{pmatrix}t_{12} & -q^{-1/2}t_{13} & 0\\ t_{22} & -q^{-1/2}t_{23} & 0\\ t_{32} & -q^{-1/2}t_{33} & 0\end{pmatrix}\end{aligned}\tag{5.223}$$

The twisted derivation rule (cf. [211, 465]) is now given by:

$$\pi_R(y)\,ab = \hat{m}\,(\pi_R(\delta_{\mathcal{U}}(y))(a\otimes b))\tag{5.224}$$

that is, in our concrete situation:

$$\begin{aligned}\pi_R(k^{\pm})\,ab &= \pi_R(k^{\pm})a\cdot\pi_R(k^{\pm})b\\ \pi_R(X^{+})\,ab &= \pi_R(X^{+})a\cdot\pi_R(k^{+})b+a\cdot\pi_R(X^{+})b\\ \pi_R(X^{-})\,ab &= \pi_R(k^{-})a\cdot\pi_R(X^{-})b+\pi_R(X^{-})a\cdot b\end{aligned}\tag{5.225}$$

Further we note that since π_R is a representation we have, that is, (5.221) holds.

Further we need a PBW basis for $\mathcal{A}$. Due to the fact that there are many relations between the nine generators t_{ij} there are several ways to introduce such a basis. In particular, one may use the $2-to-1$ covering of $SO_q(3)$ by the matrix quantum group $SL_q(2)$ [258]. However, there is a more economic and simpler way to introduce such a basis via the use of a Gauss decomposition. Moreover, the approach of [198] would require the use of a Gauss decomposition anyway. To obtain this decomposition we suppose now that there exists an element t_{33}^{-1}. Explicitly, we have:

$$T = \begin{pmatrix} t_{11} & t_{12} & t_{13} \\ t_{21} & t_{22} & t_{23} \\ t_{31} & t_{32} & t_{33} \end{pmatrix} = \tag{5.226}$$

$$= \begin{pmatrix} 1 & -q^{1/2}\xi & -[2]^{-1}\xi^2 \\ 0 & 1 & \xi \\ 0 & 0 & 1 \end{pmatrix} \begin{pmatrix} t_{33}^{-1} & 0 & 0 \\ 0 & \eta & 0 \\ 0 & 0 & t_{33} \end{pmatrix} \begin{pmatrix} 1 & 0 & 0 \\ -q^{-1/2}\bar{z} & 1 & 0 \\ -[2]^{-1}\bar{z}^2 & \bar{z} & 1 \end{pmatrix} =$$

$$= \begin{pmatrix} t_{33}^{-1} + \xi\eta\bar{z} + q^{-2}[2]^{-2}\xi^2\bar{z}^2 t_{33} & -q^{1/2}\xi\eta - q^{-1}[2]^{-1}\xi^2\bar{z}t_{33} & -[2]^{-1}\xi^2 t_{33} \\ -q^{-1/2}\eta\bar{z} - q^{-2}[2]^{-1}\xi\bar{z}^2 t_{33} & \eta + q^{-1}\xi\bar{z}t_{33} & \xi t_{33} \\ -q^{-2}[2]^{-1}\bar{z}^2 t_{33} & q^{-1}\bar{z}t_{33} & t_{33} \end{pmatrix}$$

where

$$\begin{aligned} \xi &= t_{23}\, t_{33}^{-1}, \quad \zeta = t_{33}^{-1}\, t_{32}, \quad t = t_{33} \\ \eta &= t_{33}^{-1}\, d_{11}, \quad d_{11} = t_{22}t_{33} - qt_{23}t_{32} \\ [n] &= [n]_q = (q^{n/2} - q^{-n/2})/\lambda', \quad \lambda' = q^{1/2} - q^{-1/2} \end{aligned} \tag{5.227}$$

and the following formulae are used to check (5.226):

$$\begin{aligned} & t_{33}d_{11} = d_{11}t_{33}, \qquad t_{33}^2 = d_{11}^2, \qquad \Rightarrow \qquad \eta^2 = 1_{\mathscr{A}}, \\ & \xi^2 = -[2]\, t_{13}\, t_{33}^{-1}, \qquad t_{13}d_{11} = q^2\, d_{11}t_{13}, \qquad t_{23}d_{11} = q\, d_{11}t_{23} \\ & \bar{z}^2 = -[2]\, t_{33}^{-1}\, t_{31}, \qquad t_{31}d_{11} = q^2\, d_{11}t_{31}, \qquad t_{32}d_{11} = q\, d_{11}t_{32} \\ & t_{23}\, d_{11}\, t_{32} = q^{-3}\, \{t_{11}\, t_{33} - 1_{\mathscr{A}} - q^2\, t_{13}\, t_{31}\}\, t_{33}^2 \\ & t_{23}d_{11}t_{33}^{-1} = q^{1/2}t_{13}t_{32} - q^{-1/2}t_{12}t_{33} \\ & t_{33}^{-1}d_{11}t_{32} = q^{-1/2}t_{23}t_{31} - q^{1/2}t_{33}t_{21} \end{aligned} \tag{5.228}$$

The above relations in turn are verified by use of the explicit form of the algebraic relations of $SO_q(3)$ (5.211).

Thus, we see that the relevant variables are ξ, η, t, ζ and so a possible PBW basis is:

$$f = f_{m\epsilon p\ell} = \xi^m\, \eta^\epsilon\, t^p\, \zeta^\ell\,, \qquad m, \ell \in \mathbb{Z}_+\,, \quad \epsilon = 0, 1, \quad p \in \mathbb{Z}. \tag{5.229}$$

The commutation relations in this basis are:

$$\begin{aligned} & t\xi = q^{-1}\xi t\,, \qquad t\eta = \eta t\,, \qquad t\zeta = q^{-1}\zeta t\,, \\ & \eta\xi = \xi\eta\,, \qquad \zeta\xi = \xi\zeta\,, \qquad \zeta\eta = \eta\zeta\,. \end{aligned} \tag{5.230}$$

We see that this basis is very convenient since it is almost commutative.

Following our procedure the representation spaces will have elements which are formal power series in the basis (5.229) obeying right covariance conditions. By abuse of the notion we shall call these elements functions; explicitly, we write:

$$\tilde{\varphi} = \sum_{\substack{m,\ell\in\mathbb{Z}_+ \\ \epsilon=0,1,p\in\mathbb{Z}}} \mu_{m\epsilon p\ell}\, \xi^m\, \eta^\epsilon\, t^p\, \zeta^\ell \tag{5.231}$$

The right covariance conditions for the holomorphic representations are with respect to X^-, k^+:

$$\pi_R(X^-)\, \tilde{\varphi} = 0, \tag{5.232a}$$

$$\pi_R(k^+)\, \tilde{\varphi} = q^r\, \tilde{\varphi} \tag{5.232b}$$

where r is a parameter to be specified later. Note that from (5.232b) follows: $\pi_R(k^-)\,\tilde{\varphi} = q^{-r}\,\tilde{\varphi}$. First we calculate:

$$\pi_R(X^-)\begin{pmatrix}\xi & \eta \\ \zeta & t\end{pmatrix} = \begin{pmatrix}0 & 0 \\ -q^{1/2} & 0\end{pmatrix} \tag{5.233}$$

which means that in order to fulfil (5.232a) our functions should not depend on the variable ζ; that is, the functions become:

$$\tilde{\varphi} = \sum_{\substack{m\in\mathbb{Z}_+ \\ \epsilon=0,1,p\in\mathbb{Z}}} \mu_{m\epsilon p}\, \xi^m\, \eta^\epsilon\, t^p \tag{5.234}$$

Note that the algebra $\mathcal{Y}_q$ with PBW basis $\xi^m\eta^\epsilon t^p$ may be viewed as the q-deformation of (the local coordinates submanifold of) the q-*coset* $\mathcal{Y} = SO(3,\mathbb{C})/G^-$, where G^- is the subgroup of lower diagonal matrices with main diagonal entries equal to 1. Further note the decompostion $\mathcal{Y}_q = \mathcal{Y}_q^0 \oplus \mathcal{Y}_q^1$, where $\mathcal{Y}_q^0$, $\mathcal{Y}_q^1$ are isomorphic subalgebras with bases $\xi^m t^p$, $\xi^m \eta t^p$, respectively.

Next we obtain by direct calculation:

$$\pi_R(k^+)\, \xi^m\, \eta^\epsilon\, t^p = q^p\, \xi^m\, \eta^\epsilon\, t^p \tag{5.235}$$

From the latter and (5.232b) follows that in (5.234) there is no summation in p since $p = r$; consequently the parameter r should be integer and our functions become:

$$\tilde{\varphi} = \sum_{\substack{m\in\mathbb{Z}_+ \\ \epsilon=0,1}} \mu_{m\epsilon}\, \xi^m\, \eta^\epsilon\, t^r, \qquad r \in \mathbb{Z} \tag{5.236}$$

Further we suppose that q is not a root of unity. We calculate the transformation action:

$$\pi_L(k^{\pm})\, \xi^m\, \eta^\epsilon\, t^r = q^{\pm(m-r)}\, \xi^m\, \eta^\epsilon\, t^r, \tag{5.237a}$$

$$\pi_L(X^{+})\, \xi^m\, \eta^\epsilon\, t^r = -\, q^{m/2-1}\, [m]\, \xi^{m-1}\, \eta^\epsilon\, t^r, \tag{5.237b}$$

$$\pi_L(X^{-})\, \xi^m\, \eta^\epsilon\, t^r = q^{(1-m)/2}\, \frac{[2r-m]}{[2]}\, \xi^{m+1}\, \eta^\epsilon\, t^r. \tag{5.237c}$$

It is easy to check that $\pi_L(k^{\pm})$, $\pi_L(X^{\pm})$, satisfy (5.212). Note that these transformations are not changing the parameters r, ϵ; that is, we have obtained representations parametrized by $r \in \mathbb{Z}$, $\epsilon = 0, 1$. However, we see that the parameter ϵ is fictitious since the transformation rules do not depend on it. Furthermore, the variable η is passive also w.r.t. the right action: $\pi_R(X^{\pm})\, \eta = 0$, $\pi_R(k^{\pm})\, \eta = \eta$. Thus, for fixed ϵ the representation acts in the q-coset $\mathscr{Y}_q^{\epsilon}$; that is, our functions become:

$$\varphi = \varphi(\xi, \eta, t) = \sum_{m \in \mathbb{Z}_+} \mu_m\, \xi^m\, \eta^\epsilon\, t^r, \qquad r \in \mathbb{Z}, \quad \epsilon = 0, 1. \tag{5.238}$$

For simplicity, we shall further set $\epsilon = 0$ and denote our functions as $\varphi(\xi, t)$. Further, we denote the representation action by π_r which in terms of the functions $\varphi(\xi, t)$ may be written as:

$$\pi_r(k^{\pm})\, \varphi(\xi, t) = q^{\mp r}\, T^{\xi}_{q^{\pm}}\, \varphi(\xi, t), \tag{5.239}$$

$$\pi_r(X^{+})\, \varphi(\xi, t) = -\, q^{-1}\, T^{\xi}_{q^{1/2}}\, D^{\xi}_q\, \varphi(\xi, t),$$

$$\pi_r(X^{-})\, \varphi(\xi, t) = \frac{q^{1/2}\xi}{\lambda}\, T^{\xi}_{q^{-1/2}} \left(q^r T^{\xi}_{q^{-1/2}} - q^{-r} T^{\xi}_{q^{1/2}} \right) \varphi(\xi, t),$$

$$T^{\xi}_q\, f(\xi) = f(q\xi), \qquad D^{\xi}_q\, f(\xi) = \frac{\xi^{-1}}{\lambda'} \left(T^{\xi}_{q^{1/2}} - T^{\xi}_{q^{-1/2}} \right) f(\xi). \tag{5.240}$$

We denote with $\mathscr{C}_r$ the representation space of functions $\varphi(\xi, t)$ with covariance properties (5.232) and transformation laws (5.237) (with $\epsilon = 0$) and (5.239). For generic $q \in \mathbb{C}$ and $r \in \mathbb{Z}_+$ the representation π_r is reducible. Indeed, for $r \in \mathbb{Z}_+$ the representation space $\mathscr{C}_r$ has an invariant subspace $\mathscr{E}_r$ of dimension $2r + 1$ consisting of the vectors $\xi^m\, t^r$ for $m = 0, 1, \dots, 2r$ ($\xi^0\, t^0 \equiv 1_{\mathscr{A}}$). The latter statement is obvious, as from (5.237c) follows: $\pi_L(X^{-})\, \xi^{2r}\, t^r = 0$. Thus $\xi^{2r}\, t^r$ ($\xi^0\, t^0 = 1_{\mathscr{A}}$), is the lowest-weight vector, while t^r is the highest-weight vector: $\pi_L(X^{+})\, t^r = 0$.

Thus the set of finite-dimensional representations of the $\mathscr{U}$ obtained as subrepresentations of the elementary representations realized on the coset $\mathscr{Y}_q^0$ (or $\mathscr{Y}_q^1$) of $SO_q(3)$ is parametrized by the non-negative integers and for fixed $r \in \mathbb{Z}_+$ the corresponding finite-dimensional representation is of dimension $2r + 1$; that is, all dimensions are **odd**.

The latter result should be put in contrast with the fact that the set of finite-dimensional representations of the $\mathscr{U}$ obtained as subrepresentations of the elementary representations realized on cosets of $SL_q(2)$ is parametrized by the non-negative integers and for fixed $r \in \mathbb{Z}_+$ the corresponding finite-dimensional representation is of dimension $r + 1$; that is, all integer dimensions are possible.

Thus, we recover the classical result that the finite-dimensional irreps of $SO(3, \mathbb{C})$ are only of **integer** spin $j \in \mathbb{Z}_+$ $(j = r)$, and hence of *odd* dimension $2j + 1$, while the finite-dimensional irreps of $SL(2)$ (which is a double covering group of $SO(3, \mathbb{C})$) are of (half-)integer spin $j \in \mathbb{Z}_+/2$ $(j = r/2)$, and hence of any integer dimension $2j + 1$. (Of course, physicists consider finite-dimensional irreps of $SO(3, \mathbb{C})$ also of half-integer spin, calling them two-valued irreps; moreover, infinitesimally such considerations are also mathematically correct since $so(3) \cong sl(2)$.)

Otherwise, other results are in parallel with the $SL_q(2)$ case. In particular, the finite-dimensional invariant subspace $\mathscr{E}_r$ discussed above is the kernel of an operator $\mathscr{I}_r$ intertwining the representations π_r and $\pi_{r'}$; that is,

$$\mathscr{I}_r \, \pi_r(Y) = \pi_{r'}(Y) \, \mathscr{I}_r \,, \qquad Y \in \mathscr{U} \tag{5.241}$$

where r' is expected to be $-r - 1$. According to the general prescription [198] this operator should be given by $(\pi_R(X^+))^s$ where the parameter s is expected to be $2r + 1$ (= dim $\mathscr{E}_r$). This can be checked directly. Indeed, let $s \in \mathbb{N}$ and let us suppose that $\varphi' = (\pi_R(X^+))^s \, \varphi \in \mathscr{C}_{r'}$. The latter means first (by right covariance (5.232a)) that $\pi_R(X^-) \, \varphi' = 0$. We calculate:

$$\begin{aligned}
\pi_R(X^-) \, \varphi' &= \pi_R(X^-) \, (\pi_R(X^+))^s \, \varphi = \\
&= [\pi_R(X^-) \, , (\pi_R(X^+))^s] \, \varphi = \\
&= \pi_R \left([X^- \, , (X^+)^s]\right) \, \varphi = \\
&= \pi_R \left([X^- \, , (X^+)^s]\right) \, \varphi = \\
&= \pi_R \left([s] \, (X^+)^{s-1} \left(q^{(s-1)/2} \, k^- - q^{-(s-1)/2} \, k^+\right) / \lambda\right) \, \varphi = \\
&= \frac{[s]}{\lambda} \, \pi_R \left((X^+)^{s-1}\right) \, \pi_R \left(q^{(s-1)/2} \, k^- - q^{-(s-1)/2} \, k^+\right) \, \varphi = \\
&= \frac{[s]}{\lambda} \, \pi_R \left((X^+)^{s-1}\right) \left(q^{(s-1)/2-r} - q^{r-(s-1)/2}\right) \, \varphi = \\
&= \frac{[s] \, [s - 1 - 2r]}{[2]} \, \pi_R \left((X^+)^{s-1}\right) \, \varphi
\end{aligned} \tag{5.242}$$

For q not a root of unity the last quantity may be zero only for $s = 2r + 1$, as expected. Further we use the other condition of right covariance (5.232b): $\pi_R(k^+) \, \varphi' = q^{r'} \, \varphi'$; that is:

$$\begin{aligned}
\pi_R(k^+)\,\varphi' &= \pi_R(k^+)\ (\pi_R(X^+))^s\ \varphi\ = \\
&= \pi_R\left(k^+\,(X^+)^s\right)\ \varphi\ = \\
&= \pi_R\left(q^{-s}\,(X^+)^s\,k^+\right)\ \varphi\ = \\
&= q^{-s}\ \pi_R\left((X^+)^s\right)\ \pi_R\left(k^+\right)\ \varphi\ = \\
&= q^{r-s}\ \pi_R\left((X^+)^s\right)\ \varphi = q^{r-s}\,\varphi', \\
&\Longrightarrow\quad r' = r - s = -r - 1.
\end{aligned} \tag{5.243}$$

Thus, indeed the intertwining operator $\mathscr{I}_r$ is (up to multiplicative nonzero constant):

$$\mathscr{I}_r = \pi_R(X^+)^{2r+1}. \tag{5.244}$$

Finally, as in [198] we introduce the restricted functions $\hat{\varphi}(\xi)$ by the formula:

$$\hat{\varphi}(\xi) = \left(A\varphi\right)(\xi) \equiv \varphi(\xi, 1_{\mathscr{A}}) = \sum_{m\in\mathbb{Z}_+} \mu_m\,\xi^m. \tag{5.245}$$

Note that the basis ξ^m may be viewed as (the local coordinates submanifold of) the coset $\mathscr{Z} = SO(3,\mathbb{C})/B^-$, where $B^- = HG^-$ is the subgroup of lower diagonal matrices, H being the subgroup of diagonal matrices.

We denote the representation space of $\hat{\varphi}(\xi)$ by $\hat{\mathscr{C}}_r$ and the representation acting in $\hat{\mathscr{C}}_r$ by $\hat{\pi}_r$. Thus the operator A acts from $\mathscr{C}_r$ to $\hat{\mathscr{C}}_r$. The properties of $\hat{\mathscr{C}}_r$ follow from the intertwining requirement for A [198]:

$$\hat{\pi}_r A = A\pi_r. \tag{5.246}$$

In particular, the representation action $\hat{\pi}_r$ on the basis ξ^m is given by:

$$\begin{aligned}
\hat{\pi}_r(k^\pm)\,\xi^m &= q^{\pm(m-r)}\,\xi^m, \\
\hat{\pi}_r(X^+)\,\xi^m &= -\,q^{m/2-1}\,[m]\,\xi^{m-1}, \\
\hat{\pi}_r(X^-)\,\xi^m &= q^{(1-m)/2}\,\frac{[2r-m]}{[2]}\,\xi^{m+1}.
\end{aligned} \tag{5.247}$$

In terms of the functions $\hat{\varphi}$ the representation $\hat{\pi}_r$ acts as:

$$\begin{aligned}
\hat{\pi}_r(k^\pm)\,\hat{\varphi}(\xi) &= q^{\mp r}\,T^\xi_{q^\pm}\,\hat{\varphi}(\xi), \\
\hat{\pi}_r(X^+)\,\hat{\varphi}(\xi) &= -\,q^{-1}\,T^\xi_{q^{1/2}}\,D^\xi_q\,\hat{\varphi}(\xi), \\
\hat{\pi}_r(X^-)\,\hat{\varphi}(\xi) &= \frac{q^{1/2}\xi}{\lambda}\,T^\xi_{q^{-1/2}}\left(q^r T^\xi_{q^{-1/2}} - q^{-r} T^\xi_{q^{1/2}}\right)\hat{\varphi}(\xi).
\end{aligned} \tag{5.248}$$

These functions have the property that we can extend (5.247) and (5.248) for arbitrary complex r. For generic $q, r \in \mathbb{C}$ the representations $\hat{\pi}_r$ are irreducible. For generic

$q \in \mathbb{C}$ and $r \in \mathbb{Z}_+/2$ the representations $\hat{\pi}_r$ are reducible. In the latter case all properties parallel the infinitesimal version of the classical case; that is, on the coset $\mathscr{X}$ the restricted representations of the algebra $\mathscr{U}$ may have subrepresentations also of half-integer spin. Otherwise, the description is as for $\mathscr{C}_r$: the representation space $\hat{\mathscr{C}}_r$ has an invariant subspace $\tilde{e}_r$ of dimension $2r+1$ consisting of the vectors ξ^m for $m = 0, 1, \ldots, 2r$ ($\xi^0 \equiv 1_{\mathscr{A}}$), ξ^{2r} being the lowest-weight vector, $1_{\mathscr{A}}$ being the highest-weight vector.

6 Invariant q-Difference Operators Related to $GL_q(n)$

Summary

This chapter is devoted to the detailed consideration of the q-difference operators related to $GL_q(n)$. We consider in detail several special cases, in particular, the case of $U_q(sl(3))$ and the polynomial solutions of q-difference equations. The relation of these solutions with the Gelfand–(Weyl)–Zetlin basis is studied in detail, also in the case of roots of unity, where new features are discovered. The case of $U_q(sl(4))$ is developed also in preparation for the subsequent chapter. This chapter is based mainly on [211, 220, 230, 244–246].

6.1 Representations Related to $GL_q(n)$

In this section we follow mainly [211, 220]. We consider again the matrix quantum group $\mathcal{A}_g = GL_q(n)$, $q \in \mathbb{C}$, introduced in Section 4.1 though replacing $q^{1/2}$ by q. Thus, we set instead of (4.4) ($\lambda = q - q^{-1}$):

$$M_{i\ell}M_{ij} = qM_{ij}M_{i\ell}\,, \quad \text{for} \quad \ell > j\,, \tag{6.1a}$$

$$M_{kj}M_{ij} = qM_{ij}M_{kj}\,, \quad \text{for} \quad k > i\,, \tag{6.1b}$$

$$M_{kj}M_{i\ell} = M_{i\ell}M_{kj}\,, \quad \text{for} \quad k > i\,, \ell > j\,, \tag{6.1c}$$

$$M_{ij}M_{k\ell} = M_{k\ell}M_{ij} - \lambda M_{i\ell}M_{kj}\,, \quad \text{for} \quad k > i\,, \ell > j. \tag{6.1d}$$

This algebra has determinant $\mathcal{D}$ given by (4.6) but with

$$\epsilon(w) = \prod_{\substack{j<k\\ w(j)>w(k)}} (-q^{-1}) = (-q^{-1})^{\ell(w)}\,. \tag{6.2}$$

Next one defines the left and right quantum cofactor matrix A_{ij} [462]:

$$A_{ij} = \textstyle\sum_{w(i)=j} \frac{\epsilon(w\circ\sigma_i)}{\epsilon(\sigma_i)} M_{1,w(1)} \dots \widehat{M}_{ij} \dots M_{n,w(n)} =$$

$$= \textstyle\sum_{w(j)=i} \frac{\epsilon(w\circ\sigma'_j)}{\epsilon(\sigma'_j)} M_{w(1),1} \dots \widehat{M}_{ij} \dots M_{w(n),n}\,, \tag{6.3}$$

where σ_i and σ'_j denote the cyclic permutations:

$$\sigma_i = \{i, \dots, 1\}\,, \sigma'_j = \{j, \dots, n\}\,, \tag{6.4}$$

DOI 10.1515/9783110427707-006

and the notation $\hat{x}$ indicates that x is to be omitted. Now one can show that [462]:

$$\sum_j M_{ij}A_{\ell j} = \sum_j A_{ji}M_{j\ell} = \delta_{i\ell}\mathscr{D}\,, \tag{6.5}$$

and obtain the left and right inverse [462]:

$$M^{-1} = \mathscr{D}^{-1}A = A\mathscr{D}^{-1}. \tag{6.6}$$

Thus, the antipode in $GL_q(n)$ is [462] (cf. also (4.10)):

$$\gamma_{\mathscr{A}}(M_{ij}) = \mathscr{D}^{-1}A_{ji} = A_{ji}\mathscr{D}^{-1}. \tag{6.7}$$

Next we introduce a basis of $GL_q(n)$ which consists of monomials

$$\begin{aligned} f &= (M_{21})^{p_{21}} \dots (M_{n,n-1})^{p_{n,n-1}}(M_{11})^{\ell_1} \dots (M_{nn})^{\ell_n} \times \\ &\quad \times (M_{n-1,n})^{n_{n-1,n}} \dots (M_{12})^{n_{12}} = \\ &= f_{\bar{\ell},\bar{p},\bar{n}}\,, \end{aligned} \tag{6.8}$$

where $\bar{\ell}, \bar{p}, \bar{n}$ denote the sets $\{\ell_i\}$, $\{p_{ij}\}$, $\{n_{ij}\}$, respectively, $\ell_i, p_{ij}, n_{ij} \in \mathbb{Z}_+$ and we have used the so-called normal ordering of all elements M_{ij} $(1 \le i, j \le n)$. Namely, we first put the elements M_{ij} with $i > j$ in lexicographic order; that is, if $i < k$ then M_{ij} $(i > j)$ is before $M_{k\ell}$ $(k > \ell)$ and M_{ti} $(t > i)$ is before M_{tk} $(t > k)$; then we put the elements M_{ii}; finally we put the elements M_{ij} with $i < j$ in antilexicographic order; that is, if $i > k$ then M_{ij} $(i < j)$ is before $M_{k\ell}$ $(k < \ell)$ and M_{ti} $(t < i)$ is before M_{tk} $(t < k)$. Note that the basis (6.8) includes also the unit element $1_{\mathscr{A}_g}$ of $\mathscr{A}_g$ when all $\{\ell_i\}$, $\{p_{ij}\}$, $\{n_{ij}\}$ are equal to zero; that is:

$$f_{\bar{0},\bar{0},\bar{0}} = 1_{\mathscr{A}_g}. \tag{6.9}$$

We need the algebra in duality with $GL_q(n)$. This is the algebra $\mathscr{U}_g = U_q(sl(n)) \otimes U_q(\mathscr{Z})$, where $U_q(\mathscr{Z})$ is central in $\mathscr{U}_g$ [209, 233]. Let us denote the Chevalley generators of $sl(n)$ by H_i, $X_i^{\pm}$, $i = 1, \dots, n-1$. Then we take (as in (1.52)) for the rational "Chevalley" generators of $\mathscr{U} = U_q(sl(n)) : k_i = q^{H_i/2}$, $k_i^{-1} = q^{-H_i/2}$, $X_i^{\pm}$, $i = 1, \dots, n-1$, with the following *algebra* relations:

$$k_i k_j = k_j k_i\,, \quad k_i k_i^{-1} = k_i^{-1} k_i = 1_{\mathscr{U}_g}\,, \quad k_i X_j^{\pm} = q^{\pm a_{ij}} X_j^{\pm} k_i \tag{6.10a}$$

$$[X_i^+, X_j^-] = \delta_{ij}\left(k_i^2 - k_i^{-2}\right)/\lambda\,, \tag{6.10b}$$

$$\left(X_i^{\pm}\right)^2 X_j^{\pm} - [2]_q X_i^{\pm} X_j^{\pm} X_i^{\pm} + X_j^{\pm}\left(X_i^{\pm}\right)^2 = 0\,, |i - j| = 1 \tag{6.10c}$$

$$[X_i^{\pm}, X_j^{\pm}] = 0\,, |i - j| \neq 1 \tag{6.10d}$$

where a_{ij} is the Cartan matrix of $sl(n)$, and *coalgebra* relations:

$$\delta_{\mathcal{U}}(k_i^{\pm}) = k_i^{\pm} \otimes k_i^{\pm}, \tag{6.11a}$$
$$\delta_{\mathcal{U}}(X_i^{\pm}) = X_i^{\pm} \otimes k_i + k_i^{-1} \otimes X_i^{\pm}, \tag{6.11b}$$
$$\varepsilon_{\mathcal{U}}(k_i^{\pm}) = 1, \quad \varepsilon_{\mathcal{U}}(X_i^{\pm}) = 0, \tag{6.11c}$$
$$\gamma_{\mathcal{U}}(k_i) = k_i^{-1}, \quad \gamma_{\mathcal{U}}(X_i^{\pm}) = -q^{\pm 1} X_i^{\pm}, \tag{6.11d}$$

where $k_i^+ = k_i$, $k_i^- = k_i^{-1}$. Further, we denote the generator of $\mathcal{Z}$ by H and the generators of $U_q(\mathcal{Z})$ by $k = q^{H/2}$, $k^{-1} = q^{-H/2}$, $kk^{-1} = k^{-1}k = 1_{\mathcal{U}_g}$. The generators k, k^{-1} commute with the generators of $\mathcal{U}$, and their coalgebra relations are as those of any k_i. From now on we shall give most formulae only for the generators $k_i, X_i^{\pm}, k$, since the analogous formulae for k_i^{-1}, k^{-1} follow trivially from those for k_i, k, respectively.

The bilinear form giving the duality pairing between $\mathcal{U}_g$ and $\mathcal{A}_g$ is given by [233]:

$$\langle k_i, M_{j\ell}\rangle = \delta_{j\ell} q^{(\delta_{ij} - \delta_{i+1,j})/2}, \tag{6.12a}$$
$$\langle X_i^+, M_{j\ell}\rangle = \delta_{j+1,\ell}\delta_{ij}, \tag{6.12b}$$
$$\langle X_i^-, M_{j\ell}\rangle = \delta_{j-1,\ell}\delta_{i\ell}, \tag{6.12c}$$
$$\langle k, M_{j\ell}\rangle = \delta_{j\ell} q^{1/2}. \tag{6.12d}$$

The pairing between arbitrary elements of $\mathcal{U}_g$ and f follows then from the properties of the duality pairing. The pairing (6.12) is standardly supplemented with

$$\langle y, 1_{\mathcal{A}_g}\rangle = \varepsilon_{\mathcal{U}_g}(y). \tag{6.13}$$

It is well known that the pairing provides the fundamental representation of $\mathcal{U}_g$:

$$F(y)_{j\ell} = \langle y, M_{j\ell}\rangle, \quad y = k_i, X_i^{\pm}, k. \tag{6.14}$$

Of course, $F(k) = q^{1/2} I_n$, where I_n is the unit $n \times n$ matrix.

6.1.1 Actions of $U_q(gl(n))$ and $U_q(sl(n))$

We begin by defining two actions of the quantum algebra in duality $\mathcal{U}_g$ on the basis (6.8) of $\mathcal{A}_g$.

First we introduce the *left regular representation* of $\mathcal{U}_g$ which in the $q = 1$ case is the infinitesimal version of:

$$\pi(Y) M = Y^{-1} M, \quad Y, M \in GL(n). \tag{6.15}$$

Explicitly, we define the action of $\mathscr{U}_g$ as follows:

$$\pi(y)M_{i\ell} \doteq \left(F\left(\gamma^0_{\mathscr{U}_g}(y)\right)M\right)_{i\ell} = \sum_j F\left(\gamma^0_{\mathscr{U}_g}(y)\right)_{ij} M_{j\ell} =$$
$$= \sum_j \langle \gamma^0_{\mathscr{U}_g}(y), M_{ij}\rangle M_{j\ell} \tag{6.16}$$

where y denotes the generators of $\mathscr{U}_g$ and $\gamma^0_{\mathscr{U}_g}$ is the antipode $\gamma_{\mathscr{U}_g}$ for $q = 1$, the possible pairs being given explicitly by:

$$(y, \gamma^0_{\mathscr{U}_g}(y)) = (k_i, k_i^{-1}), (X_i^{\pm}, -X_i^{\pm}), (k, k^{-1}). \tag{6.17}$$

From (6.16) we find the explicit action of the generators of $\mathscr{U}_g$:

$$\pi(k_i)M_{j\ell} = q^{(\delta_{i+1,j}-\delta_{ij})/2} M_{j\ell}\,, \tag{6.18a}$$
$$\pi(X_i^+)M_{j\ell} = -\delta_{ij} M_{j+1\ell}\,, \tag{6.18b}$$
$$\pi(X_i^-)M_{j\ell} = -\delta_{i+1,j} M_{j-1\ell}\,, \tag{6.18c}$$
$$\pi(k)M_{j\ell} = q^{-1/2} M_{j\ell}. \tag{6.18d}$$

The above is supplemented with the following action on the unit element of $\mathscr{A}_g$:

$$\pi(k_i)1_{\mathscr{A}_g} = 1_{\mathscr{A}_g}\,, \quad \pi(X_i^{\pm})1_{\mathscr{A}_g} = 0\,, \quad \pi(k)1_{\mathscr{A}_g} = 1_{\mathscr{A}_g}. \tag{6.19}$$

In order to derive the action of $\pi(y)$ on arbitrary elements of the basis (6.8), we use the twisted derivation rule consistent with the coproduct and the representation structure, namely, we take: $\pi(y)\varphi\psi = \pi(\delta'_{\mathscr{U}_g}(y))(\varphi \otimes \psi)$, where $\delta'_{\mathscr{U}_g} = \sigma \circ \delta_{\mathscr{U}_g}$ is the opposite coproduct (σ is the permutation operator). Thus, we have:

$$\pi(k_i)\varphi\psi = \pi(k_i)\varphi \cdot \pi(k_i)\psi\,, \tag{6.20a}$$
$$\pi(X_i^{\pm})\varphi\psi = \pi(X_i^{\pm})\varphi \cdot \pi(k_i^{-1})\psi + \pi(k_i)\varphi \cdot \pi(X_i^{\pm})\psi\,, \tag{6.20b}$$
$$\pi(k)\varphi\psi = \pi(k)\varphi \cdot \pi(k)\psi\,. \tag{6.20c}$$

From now on we suppose that q is not a nontrivial root of unity.

Applying the above rules one obtains:

$$\pi(k_i)(M_{j\ell})^n = q^{n(\delta_{i+1,j}-\delta_{ij})/2}(M_{j\ell})^n\,, \tag{6.21a}$$
$$\pi(X_i^+)(M_{j\ell})^n = -\delta_{ij} c_n (M_{j\ell})^{n-1} M_{j+1\ell}\,, \tag{6.21b}$$
$$\pi(X_i^-)(M_{j\ell})^n = -\delta_{i+1,j} c_n M_{j-1\ell} (M_{j\ell})^{n-1}\,, \tag{6.21c}$$
$$\pi(k)(M_{j\ell})^n = q^{-n/2}(M_{j\ell})^n\,, \tag{6.21d}$$

where

$$c_n = q^{(n-1)/2}[n]_q, \quad [n]_q = (q^n - q^{-n})/\lambda\,. \tag{6.22}$$

Note that (6.19) and (6.18) are partial cases of (6.21) for $n = 0$ and $n = 1$ respectively (cf. (6.9)).

Analogously, we introduce the right action (see also [465]) which in the classical case is the infinitesimal counterpart of:

$$\pi_R(Y)M = MY\,, \quad Y, M \in GL(n). \tag{6.23}$$

Thus, we define the right action of $\mathcal{U}_g$ as follows:

$$\pi_R(y)M_{i\ell} = (MF(y))_{i\ell} = \sum_j M_{ij}F(y)_{j\ell} = \sum_j M_{ij}\langle y, M_{j\ell}\rangle\,, \tag{6.24}$$

where y denotes the generators of $\mathcal{U}_g$.

From (6.24) we find the explicit right action of the generators of $\mathcal{U}_g$:

$$\pi_R(k_i)M_{j\ell} = q^{(\delta_{i\ell}-\delta_{i+1,\ell})/2}M_{j\ell}\,, \tag{6.25a}$$
$$\pi_R(X_i^+)M_{j\ell} = \delta_{i+1,\ell}M_{j,\ell-1}\,, \tag{6.25b}$$
$$\pi_R(X_i^-)M_{j\ell} = \delta_{i\ell}M_{j,\ell+1}\,, \tag{6.25c}$$
$$\pi_R(k)M_{j\ell} = q^{1/2}M_{j\ell}\,, \tag{6.25d}$$

supplemented by the right action on the unit element:

$$\pi_R(k_i)1_{\mathcal{A}_g} = 1_{\mathcal{A}_g}\,, \quad \pi_R(X_i^\pm)1_{\mathcal{A}_g} = 0\,, \quad \pi_R(k)1_{\mathcal{A}_g} = 1_{\mathcal{A}_g}. \tag{6.26}$$

The twisted derivation rule is now given by $\pi_R(y)\varphi\psi = \pi_R(\delta_{\mathcal{U}_g}(y))(\varphi \otimes \psi)$; that is,

$$\pi_R(k_i)\varphi\psi = \pi_R(k_i)\varphi \cdot \pi_R(k_i)\psi\,, \tag{6.27a}$$
$$\pi_R(X_i^\pm)\varphi\psi = \pi_R(X_i^\pm)\varphi \cdot \pi_R(k_i)\psi + \pi_R(k_i^{-1})\varphi \cdot \pi_R(X_i^\pm)\psi\,, \tag{6.27b}$$
$$\pi_R(k)\varphi\psi = \pi_R(k)\varphi \cdot \pi_R(k)\psi\,. \tag{6.27c}$$

Using this, we find:

$$\pi_R(k_i)(M_{j\ell})^n = q^{n(\delta_{i\ell}-\delta_{i+1,\ell})/2}(M_{j\ell})^n\,, \tag{6.28a}$$
$$\pi_R(X_i^+)(M_{j\ell})^n = \delta_{i+1,\ell}c_n M_{j,\ell-1}(M_{j\ell})^{n-1}\,, \tag{6.28b}$$
$$\pi_R(X_i^-)(M_{j\ell})^n = \delta_{i\ell}c_n(M_{j\ell})^{n-1}M_{j,\ell+1}\,, \tag{6.28c}$$
$$\pi_R(k)(M_{j\ell})^n = q^{n/2}(M_{j\ell})^n. \tag{6.28d}$$

6.1.2 Representation Spaces

Let us now introduce the elements φ as formal power series of the basis (6.8):

$$\varphi = \sum_{\bar{\ell},\bar{m},\bar{n}\in\mathbb{Z}_+} \mu_{\bar{\ell},\bar{m},\bar{n}} (M_{21})^{m_{21}} \dots (M_{n,n-1})^{m_{n,n-1}} (M_{11})^{\ell_1} \dots (M_{nn})^{\ell_n} \times \\ \times (M_{n-1,n})^{n_{n-1,n}} \dots (M_{12})^{n_{12}}. \tag{6.29}$$

By (6.21) and (6.28) we have defined left and right action of $\mathcal{U}_g$ on φ. As in the classical case the left and right actions commute, and as in [197] we shall use the right covariance to reduce the left regular representation. In particular, we would like the right action to mimic some properties of a highest-weight module, that is, annihilation by the raising generators X_i^+ and scalar action by the (exponents of the) Cartan operators k_i, k. However, first we have to make a change of basis using the q-analogue of the classical Gauss decomposition. For this we have to suppose that the principal minor determinants of M:

$$\begin{aligned} \mathcal{D}_m &= \sum_{w\in S_m} \epsilon(w) M_{1,w(1)} \dots M_{m,w(m)} = \\ &= \sum_{w\in S_m} \epsilon(w) M_{w(1),1} \dots M_{w(m),m} , \quad m \le n , \end{aligned} \tag{6.30}$$

are invertible; note that $\mathcal{D}_n = \mathcal{D}$, $\mathcal{D}_{n-1} = A_{nn}$.

Further, for the ordered sets $I = \{i_1 < \dots < i_r\}$ and $J = \{j_1 < \dots < j_r\}$, let ξ^I_J be the r-minor determinant with respect to rows I and columns J such that

$$\xi^I_J = \sum_{w\in S_r} \epsilon(w) M_{i_{w(1)}j_1} \cdots M_{i_{w(r)}j_r} . \tag{6.31}$$

Note that $\xi^{1\dots i}_{1\dots i} = \mathcal{D}_i$. Then one has [63] ($i, j, \ell = 1, \dots, n$):

$$M_{i\ell} = \sum_j B_{ij} Z_{j\ell} , \quad B_{i\ell} = \xi^{1\dots\ell-1 i}_{1\dots\ell} \mathcal{D}^{-1}_{\ell-1} , \quad Z_{i\ell} = \mathcal{D}^{-1}_i \xi^{1\dots i}_{1\dots i-1\ell} , \tag{6.32}$$

$B_{i\ell} = 0$ for $i < \ell$, $Z_{i\ell} = 0$ for $i > \ell$, (which follows from the obvious extension of (6.31) to the case when I, resp. J, is not ordered). Then Z_{ij}, $i < j$, may be regarded as a q-analogue of local coordinates of the coset $B\backslash GL(n)$.

For our purposes we need a refinement of this decomposition:

$$B_{i\ell} = \tilde{Y}_{i\ell} \mathcal{D}_{\ell\ell} , \quad \tilde{Y}_{i\ell} = \xi^{1\dots\ell-1 i}_{1\dots\ell} \mathcal{D}^{-1}_{\ell} , \quad \mathcal{D}_{\ell\ell} = \mathcal{D}_\ell \mathcal{D}^{-1}_{\ell-1} , \quad (\mathcal{D}_0 \equiv 1_{\mathcal{A}_g}) , \tag{6.33}$$

where $\tilde{Y}_{j\ell}$, $j > \ell$, may be regarded as a q-analogue of local coordinates of the coset $GL(n)/DZ$.

Clearly, we can replace the basis (6.8) of $\mathscr{A}_g$ with a basis in terms of $\tilde{Y}_{i\ell}$, $i > \ell$, $\mathscr{D}_\ell$, $Z_{i\ell}$, $i < \ell$. (Note that we set $\tilde{Y}_{ii} = Z_{ii} = 1_{\mathscr{A}_g}$.) Thus, we consider formal power series:

$$\varphi = \sum_{\substack{\bar{m},\bar{n}\in\mathbb{Z}_+\\ \bar{\ell}\in\mathbb{Z}}} \mu'_{\bar{\ell},\bar{m},\bar{n}} (\tilde{Y}_{21})^{m_{21}} \dots (\tilde{Y}_{n,n-1})^{m_{n,n-1}} (\mathscr{D}_1)^{\ell_1} \dots (\mathscr{D}_n)^{\ell_n} \times \\ \times (Z_{n-1,n})^{n_{n-1,n}} \dots (Z_{12})^{n_{12}}. \tag{6.34}$$

Now, let us impose right covariance (cf. [197]) with respect to X_i^+ , that is, we require:

$$\pi_R(X_i^+)\varphi = 0. \tag{6.35}$$

First we notice by a direct calculation that:

$$\pi_R(X_i^+)\xi_J^I = 0\,, \quad \text{for} \quad J = \{1, \dots, j\}\,, \forall I\,, \tag{6.36}$$

from which follow:

$$\pi_R(X_i^+)\mathscr{D}_j = 0\,, \quad \pi_R(X_i^+)\tilde{Y}_{j\ell} = 0. \tag{6.37}$$

On the other hand $\pi_R(X_i^+)$ acts nontrivially on $Z_{j\ell}$:

$$\pi_R(X_i^+)Z_{j\ell} = \delta_{i+1,\ell}\, q^{\delta_{ij}/2} Z_{j,\ell-1}. \tag{6.38}$$

Thus, (6.35) simply means that our functions φ do not depend on $Z_{j\ell}$. Thus, the functions obeying (6.35) are:

$$\varphi = \sum_{\bar{\ell}\in\mathbb{Z}\,,\bar{m}\in\mathbb{Z}_+} \mu_{\bar{\ell},\bar{m}} (\tilde{Y}_{21})^{m_{21}} \dots (\tilde{Y}_{n,n-1})^{m_{n,n-1}} (\mathscr{D}_1)^{\ell_1} \dots (\mathscr{D}_n)^{\ell_n}. \tag{6.39}$$

Next, we impose right covariance with respect to k_i, k:

$$\pi_R(k_i)\varphi = q^{r_i/2}\varphi\,, \tag{6.40a}$$
$$\pi_R(k)\varphi = q^{\hat{r}/2}\varphi\,, \tag{6.40b}$$

where $r_i, \hat{r}$ are parameters to be specified below. On the other hand using (6.27a,c), ((6.28)a,c) we have:

$$\pi_R(k_i)\xi_J^I = q^{\delta_{ij}/2}\xi_J^I\,, \quad \pi_R(k)\xi_J^I = q^{j/2}\xi_J^I\,, \quad \text{for} \quad J = \{1, \dots, j\}\,, \forall I\,, \tag{6.41}$$

from which follows:

$$\pi_R(k_i)\mathscr{D}_j = q^{\delta_{ij}/2}\mathscr{D}_j\,, \quad \pi_R(k)\mathscr{D}_j = q^{j/2}\mathscr{D}_j\,, \tag{6.42a}$$
$$\pi_R(k_i)\tilde{Y}_{j\ell} = \tilde{Y}_{j\ell}\,, \quad \pi_R(k)\tilde{Y}_{j\ell} = \tilde{Y}_{j\ell}\,, \tag{6.42b}$$

and thus we have:

$$\pi_R(k_i)\varphi = q^{\ell_i/2}\varphi\,, \tag{6.43a}$$

$$\pi_R(k)\varphi = q^{\sum_{j=1}^n j\ell_j/2}\varphi. \tag{6.43b}$$

Remark 6.1. For $q = 1$ the elementary representations (in particular, the right covariance conditions) for a complex semisimple Lie group G_c are given by (cf. Volume 1):

$$\mathscr{C}_{\Lambda,\Lambda'} = \{\mathscr{F} \in C^\infty(G_c) | \mathscr{F}(gxn) = e^{\Lambda(X)+\Lambda'(\bar{X})} \cdot \mathscr{F}(g)\}\,, \quad \Lambda(X) - \Lambda'(X) \in \mathbb{Z}\,, \tag{6.44}$$

where $x = \exp(X)$, $X \in \mathscr{H}_c$, $n \in G_c^+ = \exp(\mathscr{G}_c^+)$, using the Gauss decomposition $\mathscr{G}_c = \mathscr{G}_c^+ \oplus \mathscr{H}_c \oplus \mathscr{G}_c^-$ of the Lie algebra $\mathscr{G}_c$ of G_c, and the last condition in (6.44) is necessary to ensure uniqueness on the Cartan subgroup $H_c = \exp(\mathscr{H}_c)$ of G_c. In the quantum group setting above, for simplicity, we are using infinitesimal holomorphic representations for which $\Lambda' = 0$. For $U_q(sl(2))$ with $\Lambda' \neq 0$ we refer to Section 5.3 where this construction was carried out for a q-deformed Lorentz algebra. ◊

Comparing right covariance conditions (6.40) with the direct calculations (6.43) we obtain $\ell_i = r_i$, for $i < n$, $\sum_{j=1}^n j\ell_j = \hat{r}$. This means that $r_i, \hat{r} \in \mathbb{Z}$ and that there is no summation in ℓ_i, also $\ell_n = (\hat{r} - \sum_{i=1}^{n-1} ir_i)/n$.

Thus, the reduced functions obeying (6.35) and (6.40) are:

$$\varphi = \sum_{\bar{m}\in\mathbb{Z}_+} \mu_{\bar{m}} (\tilde{Y}_{21})^{m_{21}} \dots (\tilde{Y}_{n,n-1})^{m_{n,n-1}} (\mathscr{D}_1)^{r_1} \dots (\mathscr{D}_{n-1})^{r_{n-1}} (\mathscr{D}_n)^{\hat{\ell}}\,, \tag{6.45}$$

where $\hat{\ell} = (\hat{r} - \sum_{i=1}^{n-1} ir_i)/n$.

Next we would like to derive the $\mathscr{U}_g$ – action π on φ. First, we notice that $\mathscr{U}$ acts trivially on $\mathscr{D}_n = \mathscr{D}$:

$$\pi(X_i^\pm)\mathscr{D} = 0\,, \quad \pi(k_i)\mathscr{D} = \mathscr{D}. \tag{6.46}$$

Then we note:

$$\pi(k)\mathscr{D}_j = q^{-j/2}\mathscr{D}_j\,, \quad \pi(k)\tilde{Y}_{j\ell} = \tilde{Y}_{j\ell}\,, \tag{6.47}$$

from which follows:

$$\pi(k)\varphi = q^{-\hat{r}/2}\varphi. \tag{6.48}$$

Thus, the action of $\mathscr{U}$ involves only the parameters r_i, $i < n$, while the action of $U_q(\mathscr{Z})$ involves only the parameter $\hat{r}$. Thus we can consistently also from the representation theory point of view restrict to the matrix quantum group $SL_q(n)$; that is, we set:

$$\mathscr{D} = \mathscr{D}^{-1} = 1_{\mathscr{A}_g}. \tag{6.49}$$

Then the quantum algebra in duality is $\mathscr{U} = U_q(sl(n))$. This is justified as in the $q = 1$ case [197] since for our considerations only the semisimple part of the algebra is important.

Thus, the reduced functions for the $\mathscr{U}$ action are:

$$\begin{aligned}\tilde{\varphi}(\bar{Y}, \bar{\mathscr{D}}) &= \sum_{\bar{m}\in\mathbb{Z}_+} \mu_{\bar{m}} (\tilde{Y}_{21})^{m_{21}} \dots (\tilde{Y}_{n,n-1})^{m_{n,n-1}} \times \\ &\quad \times (\mathscr{D}_1)^{r_1} \dots (\mathscr{D}_{n-1})^{r_{n-1}} = &(6.50a)\\ &= \hat{\varphi}(\bar{Y})(\mathscr{D}_1)^{r_1} \dots (\mathscr{D}_{n-1})^{r_{n-1}} &(6.50b)\end{aligned}$$

where $\bar{Y}, \bar{\mathscr{D}}$ denote the variables $\tilde{Y}_{il}$, $i > \ell$, $\mathscr{D}_i$, $i < n$.

Further we note the commutation relations of the $\tilde{Y}_{ij}$ and $\mathscr{D}_i$ variables:

$$\tilde{Y}_{i\ell}\tilde{Y}_{ij} = q\tilde{Y}_{ij}\tilde{Y}_{i\ell}\,, i > \ell > j\,, \tag{6.51a}$$

$$\tilde{Y}_{kj}\tilde{Y}_{ij} = q\tilde{Y}_{ij}\tilde{Y}_{kj}\,, k > i > j\,, \tag{6.51b}$$

$$\tilde{Y}_{kj}\tilde{Y}_{i\ell} = \tilde{Y}_{i\ell}\tilde{Y}_{kj}\,, k > i > \ell > j\,, \tag{6.51c}$$

$$\tilde{Y}_{ki}\tilde{Y}_{\ell j} = \tilde{Y}_{\ell j}\tilde{Y}_{ki}\,, k > i > \ell > j\,, \tag{6.51d}$$

$$\tilde{Y}_{k\ell}\tilde{Y}_{ij} = \tilde{Y}_{ij}\tilde{Y}_{k\ell} + \lambda\tilde{Y}_{i\ell}\tilde{Y}_{kj}\,, k > i > \ell > j\,, \tag{6.51e}$$

$$\tilde{Y}_{ki}\tilde{Y}_{ij} = q^{-1}\tilde{Y}_{ij}\tilde{Y}_{ki} + q^{-1}\lambda\tilde{Y}_{kj}\,, k > i > j\,, \tag{6.51f}$$

$$Y_{j\ell}\mathscr{D}_i = \mathscr{D}_i Y_{j\ell}\,, j > \ell > i\,, \tag{6.51g}$$

$$Y_{j\ell}\mathscr{D}_i = q\mathscr{D}_i Y_{j\ell}\,, j > i \geq \ell\,, \tag{6.51h}$$

$$Y_{j\ell}\mathscr{D}_i = \mathscr{D}_i Y_{j\ell}\,, i \geq j > \ell\,, \tag{6.51i}$$

where in (6.51d) we use $\tilde{Y}_{i\ell} = 0$ when $i < \ell$. Note that (6.51a-d) may be obtained by replacing $M_{i\ell}$ with $\tilde{Y}_{i\ell}$ in (6.1a-d). Note that the structure of the q-coset for general n is exhibited already for $n = 4$, while for $n = 3$ relations (6.51c,d) are not present. The commutation relations between the Z and $\mathscr{D}$ variables are obtained from (6.51) by just replacing Y_{st} by Z_{ts} in all formulae.

Note that for real q the q-coset is invariant under the antilinear anti-involution $\tilde{\omega}$ acting as:

$$\tilde{\omega}(\tilde{Y}_{j\ell}) = \tilde{Y}_{n+1-\ell,n+1-j}. \tag{6.52}$$

Thus it can be considered as a q-coset of the quantum group $SU_q([(n+1)/2]_{\text{int}}, [n/2]_{\text{int}})$, where $[x/2]_{\text{int}}$ is the biggest integer number not greater than x. The same invariance holds for the Z coordinate q-coset.

Next we calculate:

$$\begin{aligned}
\pi(k_i)\mathscr{D}_j &= q^{-\delta_{ij}/2}\mathscr{D}_j\,, &(6.53)\\
\pi(X_i^+)\mathscr{D}_j &= -\delta_{ij}\tilde{Y}_{j+1,j}\mathscr{D}_j\,,\\
\pi(X_i^-)\mathscr{D}_j &= 0\,,\\
\pi(k_i)\tilde{Y}_{j\ell} &= q^{\frac{1}{2}(\delta_{i+1,j}-\delta_{ij}-\delta_{i+1,\ell}+\delta_{i\ell})}\tilde{Y}_{j\ell} &(6.54)\\
\pi(X_i^+)\tilde{Y}_{j\ell} &= -\delta_{ij}\tilde{Y}_{j+1,\ell} + \delta_{i\ell}q^{1-\delta_{j,\ell+1}/2}\tilde{Y}_{\ell+1,\ell}\tilde{Y}_{j\ell} +\\
&\quad +\delta_{i+1,\ell}\left(q^{-1}\tilde{Y}_{j,\ell-1} - \tilde{Y}_{\ell,\ell-1}\tilde{Y}_{j\ell}\right),\\
\pi(X_i^-)\tilde{Y}_{j\ell} &= -\delta_{i+1,j}q^{-\delta_{i\ell}/2}\tilde{Y}_{j-1,\ell}.
\end{aligned}$$

These results have the important consequence that the degrees of the variables $\mathscr{D}_j$ are not changed by the action of $\mathscr{U}$; that is, the parameters r_i characterize this action. Furthermore it is easy to check that $\pi(y)$ satisfy (6.10). Thus, we have obtained representations of $\mathscr{U}$. These are analogues of the elementary representations of the classical case $q = 1$.

To obtain these representations more explicitly one just applies (6.53), (6.54) to the basis in (6.50) using (6.20). In particular, we have:

$$\begin{aligned}
\pi(k_i)(\mathscr{D}_j)^n &= q^{-n\delta_{ij}/2}(\mathscr{D}_j)^n\,, \quad n\in\mathbb{Z}\,, &(6.55)\\
\pi(X_i^+)(\mathscr{D}_j)^n &= -\delta_{ij}\bar{c}_n\tilde{Y}_{j+1,j}(\mathscr{D}_j)^n\,, \quad n\in\mathbb{Z}\,,\\
\pi(X_i^-)(\mathscr{D}_j)^n &= 0\,, \quad n\in\mathbb{Z}\,,\\
\pi(k_i)(\tilde{Y}_{j\ell})^n &= q^{\frac{n}{2}(\delta_{i+1,j}-\delta_{ij}-\delta_{i+1,\ell}+\delta_{i\ell})}(\tilde{Y}_{j\ell})^n\,, \quad n\in\mathbb{Z}_+\,, &(6.56)\\
\pi(X_i^+)(\tilde{Y}_{j\ell})^n &= -\delta_{ij}c_n(\tilde{Y}_{j\ell})^{n-1}\tilde{Y}_{j+1,\ell} +\\
&\quad +\delta_{i+1,\ell}\bar{c}_n\left(q^{-1}\tilde{Y}_{j,\ell-1}(\tilde{Y}_{j\ell})^{n-1} - \tilde{Y}_{\ell,\ell-1}(\tilde{Y}_{j\ell})^n\right) +\\
&\quad +\delta_{i\ell}q^{1-n\delta_{j,\ell+1}/2}c_n\tilde{Y}_{\ell+1,\ell}(\tilde{Y}_{j\ell})^n\,, \quad n\in\mathbb{Z}_+\,,\\
\pi(X_i^-)(\tilde{Y}_{j\ell})^n &= -\delta_{i+1,j}q^{-\delta_{j,\ell+1}n/2}c_n\tilde{Y}_{j-1,\ell}(\tilde{Y}_{j\ell})^{n-1}\,, \quad n\in\mathbb{Z}_+\,,\\
\bar{c}_n &= q^{(1-n)/2}[n]_q. &(6.57)
\end{aligned}$$

We shall denote by $\mathscr{C}_{\bar{r}}$ the representation space of functions in (6.50) which have covariance properties (6.35), (6.40a). The representation acting in $\mathscr{C}_{\bar{r}}$ we denote by $\tilde{\pi}_{\bar{r}}$ doing also a renormalization to simplify things later, namely, we set:

$$\tilde{\pi}_{\bar{r}}(k_i) = \pi(k_i)\,, \qquad \tilde{\pi}_{\bar{r}}(X_i^\pm) = q^{\pm(r_i-1)/2}\pi(X_i^\pm). \tag{6.58}$$

Then $\tilde{\pi}_{\bar{r}}$ also satisfy (6.10).

Further, since the action of $\mathscr{U}$ is not affecting the degrees of $\mathscr{D}_i$, we introduce (as in [197]) the restricted functions $\hat{\varphi}(\bar{Y})$ by the formula which is prompted in (6.50b):

$$\hat{\varphi}(\bar{Y}) \equiv \left(\hat{\mathscr{A}}\tilde{\varphi}\right)(\bar{Y}) \doteq \tilde{\varphi}(\bar{Y}, \mathscr{D}_1 = \dots = \mathscr{D}_{n-1} = 1_{\mathscr{A}_g}), \tag{6.59a}$$

$$\hat{\varphi}(\bar{Y}) = \textstyle\sum_{\bar{m}\in\mathbb{Z}_+} \mu_{\bar{m}} (\tilde{Y}_{21})^{m_{21}} \dots (\tilde{Y}_{n,n-1})^{m_{n,n-1}}. \tag{6.59b}$$

We denote the representation space of $\hat{\varphi}(\bar{Y})$ by $\hat{\mathscr{C}}_{\bar{r}}$ and the representation acting in $\hat{\mathscr{C}}_{\bar{r}}$ by $\hat{\pi}_{\bar{r}}$. Thus, the operator $\hat{\mathscr{A}}$ acts from $\mathscr{C}_{\bar{r}}$ to $\hat{\mathscr{C}}_{\bar{r}}$. The properties of $\hat{\mathscr{C}}_{\bar{r}}$ follow from the intertwining requirement for $\hat{\mathscr{A}}$ [197]:

$$\hat{\pi}_{\bar{r}} \circ \hat{\mathscr{A}} = \hat{\mathscr{A}} \circ \tilde{\pi}_{\bar{r}}. \tag{6.60}$$

For the more compact exposition of the representation formulae we shall need below also the following operators (corresponding to each of the variables $\tilde{Y}_{j\ell}$):

$$\hat{M}_{j\ell}\hat{\varphi}(\bar{Y}) = \sum_{\bar{m}\in\mathbb{Z}_+} \mu_{\bar{m}} \hat{M}_{j\ell} (\tilde{Y}_{21})^{m_{21}} \dots (\tilde{Y}_{n,n-1})^{m_{n,n-1}}, \tag{6.61}$$

$$\begin{aligned}
\hat{M}_{j\ell}(\tilde{Y}_{21})^{m_{21}} \dots (\tilde{Y}_{n,n-1})^{m_{n,n-1}} &= (\tilde{Y}_{21})^{m_{21}} \dots (\tilde{Y}_{j\ell})^{m_{j\ell}+1} \dots \times \\
&\quad \times \dots (\tilde{Y}_{n,n-1})^{m_{n,n-1}}, \\
T_{j\ell}\hat{\varphi}(\bar{Y}) &= \sum_{\bar{m}\in\mathbb{Z}_+} \mu_{\bar{m}} T_{j\ell} (\tilde{Y}_{21})^{m_{21}} \dots (\tilde{Y}_{n,n-1})^{m_{n,n-1}}, \\
T_{j\ell}(\tilde{Y}_{21})^{m_{21}} \dots (\tilde{Y}_{n,n-1})^{m_{n,n-1}} &= q^{m_{j\ell}} (\tilde{Y}_{21})^{m_{21}} \dots (\tilde{Y}_{n,n-1})^{m_{n,n-1}}
\end{aligned} \tag{6.62}$$

Next we introduce also the homogeneity (number) operators $N_{j\ell}$ for the variable $\tilde{Y}_{j\ell}$:

$$\begin{aligned}
N_{j\ell}\hat{\varphi}(\bar{Y}) &= \sum_{\bar{m}\in\mathbb{Z}_+} \mu_{\bar{m}} N_{j\ell} (\tilde{Y}_{21})^{m_{21}} \dots (\tilde{Y}_{n,n-1})^{m_{n,n-1}} \\
N_{j\ell}(\tilde{Y}_{21})^{m_{21}} \dots (\tilde{Y}_{n,n-1})^{m_{n,n-1}} &= m_{j\ell} (\tilde{Y}_{21})^{m_{21}} \dots (\tilde{Y}_{n,n-1})^{m_{n,n-1}}
\end{aligned} \tag{6.63}$$

Clearly, we have the relation:

$$T_{j\ell} = q^{N_{j\ell}}. \tag{6.64}$$

Using the above we define the q-difference operators which admit a general definition on a larger domain than polynomials, but on polynomials are well defined as follows:

$$\begin{aligned}
\hat{\mathscr{D}}_{j\ell}\hat{\varphi}(\bar{Y}) &= \frac{1}{\lambda} \hat{M}_{j\ell}^{-1} \left(T_{j\ell} - T_{j\ell}^{-1}\right) \hat{\varphi}(\bar{Y}) = \\
&= \hat{M}_{j\ell}^{-1} [N_{j\ell}] \hat{\varphi}(\bar{Y})
\end{aligned} \tag{6.65}$$

from which follows:

$$\hat{\mathscr{D}}_{j\ell}(\tilde{Y}_{21})^{m_{21}}\dots(\tilde{Y}_{n,n-1})^{m_{n,n-1}} = [m_{j\ell}]_q\,(\tilde{Y}_{21})^{m_{21}}\dots(\tilde{Y}_{j\ell})^{m_{j\ell}-1}\dots \\ \dots(\tilde{Y}_{n,n-1})^{m_{n,n-1}} \tag{6.66}$$

Note that although $\hat{M}_{j\ell}^{-1}$ is not defined on $(\tilde{Y}_{21})^{m_{21}}\dots(\tilde{Y}_{n,n-1})^{m_{n,n-1}}$ for $m_{j\ell} = 0$, the operator $\hat{\mathscr{D}}_{j\ell}$ is well defined on such terms, and the result is zero (given by the action of $(T_{j\ell} - T_{j\ell}^{-1})$). Of course, for $q \to 1$ we have $\hat{\mathscr{D}}_{j\ell} \to \partial_{Y_{j\ell}} \equiv \partial/\partial Y_{j\ell}$. Note that the above operators for different variables commute; that is, with these we have actually passed to commuting variables.

For the intertwining operators between partially equivalent representations we need the action of $\pi_R(X_i^-)$ on $\tilde{Y}_{j\ell}$ and $\mathscr{D}_\ell$. Using (6.28) and (6.27) we obtain:

$$\pi_R(X_i^-)(\mathscr{D}_\ell)^n = \delta_{i\ell}c_n(\mathscr{D}_\ell)^n Z_{\ell,\ell+1}\,, \tag{6.67a}$$

$$\pi_R(X_i^-)(\tilde{Y}_{j\ell})^n = \delta_{i\ell}q^{n-3/2}[n]_q(\tilde{Y}_{j\ell})^{n-1}\tilde{Y}_{j,\ell+1}\mathscr{D}_{\ell+1}\mathscr{D}_\ell^{-2}\mathscr{D}_{\ell-1} \tag{6.67b}$$

where, as usual, we use $\tilde{Y}_{jj} = 1_{\mathscr{A}} = \mathscr{D}_0$. We shall use also the repeated action of $\pi_R(X_i^-)$ so in addition we need:

$$\pi_R(X_i^-)Z_{j\ell} = \delta_{i\ell}Z_{j,\ell+1} - \delta_{ij}q^{-\delta_{j+1,\ell}/2}Z_{j,j+1}Z_{j\ell} + \delta_{i,j-1}\mathscr{D}_j^{-1}\xi^{1\dots j}_{1\dots j-2,j,\ell}\,, \tag{6.68}$$

$$\pi_R(k_i)Z_{j\ell} = q^{(\delta_{i+1,j}-\delta_{ij}+\delta_{i\ell}-\delta_{i+1,\ell})/2}Z_{j\ell}. \tag{6.69}$$

6.1.3 Reducibility and Partial Equivalence

We have defined the representations $\hat{\pi}_{\bar{r}}$ for $r_i \in \mathbb{Z}$. However, notice that we can consider the restricted functions $\hat{\varphi}(\tilde{Y})$ for arbitrary complex r_i. We shall make this extension from now on, since this gives the same set of (holomorphic) representations for $U_q(sl(n))$ as in the case $q = 1$.

Now we make some statements which are true in the classical case [197], and will be illustrated below. For any i, j, such that $1 \le i \le j \le n-1$, define:

$$m_{ij} \equiv r_i + \dots + r_j + j - i + 1\,, \tag{6.70}$$

note $m_i = m_{ii} = r_i + 1$, $m_{ij} = m_i + \dots + m_j$. Note that the possible choices of i, j are in one-to-one correspondence with the positive roots $\alpha = \alpha_{ij} = \alpha_i + \dots + \alpha_j$ of the root system of $sl(n)$, the cases $i = j = 1\dots, n-1$ enumerating the simple roots $\alpha_i = \alpha_{ii}$. In general, $m_{ij} \in \mathbb{C}$ for the representations $\hat{\pi}_{\bar{r}}$, while $m_{ij} \in \mathbb{Z}$ for the representations $\pi_{\bar{r}}$. If $m_{ij} \notin \mathbb{N}$, for all possible i, j the representations $\hat{\pi}_{\bar{r}}$, $\pi_{\bar{r}}$ are irreducible. If $m_{ij} \in \mathbb{N}$, for some i, j the representations $\hat{\pi}_{\bar{r}}$, $\pi_{\bar{r}}$ are reducible. The corresponding irreducible subrepresentations are still infinite-dimensional unless $m_i \in \mathbb{N}$ for all $i = 1, \dots, n-1$.

The representation spaces of the irreducible subrepresentations are invariant irreducible subspaces of our representation spaces. These invariant subspaces are spanned by functions depending on all variables $Y_{j\ell}$, except when for some $s \in \mathbb{N}, 1 \le s \le n-1$, we have $m_s = m_{s+1} = \cdots = m_{n-1} = 1$. In the latter case these functions depend only on the $(s-1)(2n-s)/2$ variables $Y_{j\ell}$ with $\ell < s$, (the unrestricted subrepresentation functions depend still on $\mathscr{D}_\ell$ with $\ell < s$). In particular, for $s = 2$ the restricted subrepresentation functions depend only on the $n-1$ variables Y_{j1}. The latter situation is relatively simple also in the q case since these variables are q-commuting: $Y_{j1}Y_{k1} = qY_{k1}Y_{j1}$, $j > k$. (For $s = 1$ the irreducible subrepresentation is one-dimensional, hence no dependence on any variables.)

Furthermore, for $m_{ij} \in \mathbb{N}$ the representation $\hat{\pi}_{\bar{r}}$, $\pi_{\bar{r}}$ respectively is partially equivalent to the representation $\hat{\pi}_{\bar{r}'}$, $\pi_{\bar{r}'}$, respectively with $m'_\ell = r'_\ell + 1$ being explicitly given as follows [197]:

$$m'_\ell = \begin{cases} m_\ell \,, & \text{for } \ell \neq i-1, i, j, j+1 \,, \\ m_{\ell j} \,, & \text{for } \ell = i-1 \,, \\ -m_{\ell+1,j} \,, & \text{for } \ell = i < j \,, \\ -m_{i,\ell-1} \,, & \text{for } \ell = j > i \,, \\ -m_\ell \,, & \text{for } \ell = i = j \,, \\ m_{i\ell} \,, & \text{for } \ell = j+1 \,. \end{cases} \tag{6.71}$$

These partial equivalences are realized by intertwining operators:

$$\mathscr{I}_{ij} : \mathscr{C}_{\bar{r}} \longrightarrow \mathscr{C}_{\bar{r}'} \,, \quad m_{ij} \in \mathbb{N} \,, \tag{6.72a}$$

$$I_{ij} : \hat{\mathscr{C}}_{\bar{r}} \longrightarrow \hat{\mathscr{C}}_{\bar{r}'} \,, \quad m_{ij} \in \mathbb{N} \,, \tag{6.72b}$$

that is, one has:

$$\mathscr{I}_{ij} \circ \pi_{\bar{r}} = \pi_{\bar{r}'} \circ \mathscr{I}_{ij} \,, \quad m_{ij} \in \mathbb{N} \,, \tag{6.73a}$$

$$I_{ij} \circ \hat{\pi}_{\bar{r}} = \hat{\pi}_{\bar{r}'} \circ I_{ij} \,, \quad m_{ij} \in \mathbb{N}. \tag{6.73b}$$

The invariant irreducible subspace of $\hat{\pi}_{\bar{r}}$ (respectively, $\pi_{\bar{r}}$) discussed above is the intersection of the kernels of all intertwining operators acting from $\hat{\pi}_{\bar{r}}$ (respectively, $\pi_{\bar{r}}$). When all $m_i \in \mathbb{N}$ the invariant subspace is finite-dimensional with dimension $\prod_{1\le i\le j\le n-1} m_{ij} / \prod_{t=1}^{n-1} t!$, and all finite-dimensional (holomorphic) irreps of $U_q(sl(n))$ can be obtained in this way.

We restate now the canonical procedure for the derivation of these intertwining operators (cf. [197, 211]) in the current setting. By the procedure one should take as intertwiners (up to nonzero multiplicative constants):

$$\mathscr{I}^m_{ij} = \mathscr{P}^m_{ij}\left(\pi_R(X^-_i), \ldots, \pi_R(X^-_j)\right), m = m_{ij} \in \mathbb{N} \,, \tag{6.74a}$$

$$I^m_{ij} = \mathscr{P}^m_{ij}\left(\hat{\pi}_R(X^-_i), \ldots, \hat{\pi}_R(X^-_j)\right), m = m_{ij} \in \mathbb{N} \,, \tag{6.74b}$$

where $\mathscr{P}^m_{ij}$ is a homogeneous polynomial in each of its $(j - i + 1)$ variables of degree m, while the operators I^m_{ij} are defined through $\mathscr{I}^m_{ij}$ and the operator $\hat{\mathscr{A}}$. The polynomial $\mathscr{P}^m_{ij}$ gives a singular vector v_{ij} in the Verma module $V^{\Lambda(\bar{r})}$ with highest weight $\Lambda(\bar{r})$ determined by $\bar{r}$, ($\bar{r}$ plays the role of χ), that is,

$$v_{ij} = \mathscr{P}^m_{ij}\left(X^-_i, \ldots, X^-_j\right) v_0 , \tag{6.75}$$

where v_0 is the highest-weight vector of $V^{\Lambda(\bar{r})}$. The explicit expression for v_{ij} with $j = i + p - 1$ is given in (2.95). In particular, in the case of the simple roots, that is, when $m_i = m_{ii} = r_i + 1 \in \mathbb{N}$, we have:

$$\mathscr{I}^{m_i}_i = (\pi_R(X^-_i))^{m_i} , m_i \in \mathbb{N}. \tag{6.76}$$

Implementing the above one should be careful since $\hat{\pi}_R(X^-_i)$ is not preserving the reduced spaces $\mathscr{C}_{\bar{r}}, \hat{\mathscr{C}}_{\bar{r}}$, which is of course a prerequisite for (6.73), (6.74), (6.76).

6.2 The Case of $U_q(sl(3))$

In this section we also follow [211]. In this section we consider in more detail the case $n = 3$. (The case $n = 2$ was discussed in Section 5.1.3. It can also be obtained by restricting the construction for the complexification of the Lorentz quantum algebra to one of its $U_q(sl(2))$ subalgebras, see Section 5.3.)

Let us now for $n = 3$ denote the coordinates on the q-flag manifold by: $\xi = Y_{21}$, $\eta = Y_{32}$, $\zeta = Y_{31}$. We note for future use the commutation relations between these coordinates:

$$\xi\eta = q\eta\xi - \lambda\zeta, \quad \eta\zeta = q\zeta\eta, \quad \zeta\xi = q\xi\zeta. \tag{6.77}$$

The reduced functions for the $U_q(sl(3))$ action are (cf. (6.50)):

$$\tilde{\varphi}(\bar{Y}, \bar{\mathscr{D}}) = \sum_{j,n,\ell\in\mathbb{Z}_+} \mu_{j,n,\ell}\xi^j\zeta^n\eta^\ell(\mathscr{D}_1)^{r_1}(\mathscr{D}_2)^{r_2} = \tag{6.78a}$$

$$= \sum_{j,n,\ell\in\mathbb{Z}_+} \mu_{j,n,\ell}\tilde{\varphi}_{jn\ell}, \tag{6.78b}$$

$$\tilde{\varphi}_{jn\ell} = \xi^j\zeta^n\eta^\ell(\mathscr{D}_1)^{r_1}(\mathscr{D}_2)^{r_2}. \tag{6.78c}$$

Now the action of $U_q(sl(3))$ on (6.78) is given explicitly by:

$$\pi(k_1)\tilde{\varphi}_{jn\ell} = q^{j+(n-\ell-r_1)/2}\tilde{\varphi}_{jn\ell}, \tag{6.79a}$$

$$\pi(k_2)\tilde{\varphi}_{jn\ell} = q^{\ell+(n-j-r_2)/2}\tilde{\varphi}_{jn\ell}, \tag{6.79b}$$

$$\pi(X_1^+)\tilde{\varphi}_{jn\ell} = q^{(1+n-\ell-r_1)/2}[n+j-\ell-r_1]_q\tilde{\varphi}_{j+1,n\ell} + +q^{j+(n-\ell-3r_1-1)/2}[\ell]_q\tilde{\varphi}_{j,n+1,\ell-1}, \tag{6.79c}$$

$$\pi(X_2^+)\tilde{\varphi}_{jn\ell} = q^{(1+n-j-r_2)/2}[\ell-r_2]_q\tilde{\varphi}_{jn,\ell+1} - -q^{-\ell+(j-n+r_2-1)/2}[j]_q\tilde{\varphi}_{j-1,n+1,\ell}, \tag{6.79d}$$

$$\pi(X_1^-)\tilde{\varphi}_{jn\ell} = q^{(\ell-n+r_1-1)/2}[j]_q\tilde{\varphi}_{j-1,n\ell}, \tag{6.79e}$$

$$\pi(X_2^-)\tilde{\varphi}_{jn\ell} = -q^{(n-j+r_2-1)/2}[\ell]_q\tilde{\varphi}_{jn,\ell-1} - -q^{-\ell+(n-j+r_2-1)/2}[n]_q\tilde{\varphi}_{j+1,n-1,\ell}. \tag{6.79f}$$

It is easy to check that $\pi(k_i)$, $\pi(X_i^\pm)$ satisfy (6.10). It is also clear that we can remove the inessential phases by setting:

$$\tilde{\pi}_{r_1,r_2}(k_i) = \pi(k_i), \quad \tilde{\pi}_{r_1,r_2}(X_i^\pm) = q^{\pm(r_i-1)/2}\pi(X_i^\pm). \tag{6.80}$$

Then $\tilde{\pi}_{r_1,r_2}$ also satisfy (6.10).

Then we consider the restricted functions (cf. (6.59)):

$$\hat{\varphi}(\bar{Y}) = \sum_{j,n,\ell\in\mathbb{Z}_+} \mu_{j,n,\ell}\xi^j\zeta^n\eta^\ell = \tag{6.81a}$$

$$= \sum_{j,n,\ell\in\mathbb{Z}_+} \mu_{j,n,\ell}\hat{\varphi}_{jn\ell}, \tag{6.81b}$$

$$\hat{\varphi}_{jn\ell} = \xi^j\zeta^n\eta^\ell. \tag{6.81c}$$

As a consequence of the intertwining property (5.40) we obtain that $\hat{\varphi}_{jn\ell}$ obey the same transformation rules (6.79) as $\tilde{\varphi}_{jn\ell}$, that is, we have:

$$\hat{\pi}_{r_1,r_2}(k_1)\hat{\varphi}_{jn\ell} = q^{j+(n-\ell-r_1)/2}\hat{\varphi}_{jn\ell}, \tag{6.82a}$$

$$\hat{\pi}_{r_1,r_2}(k_2)\hat{\varphi}_{jn\ell} = q^{\ell+(n-j-r_2)/2}\hat{\varphi}_{jn\ell}, \tag{6.82b}$$

$$\hat{\pi}_{r_1,r_2}(X_1^+)\hat{\varphi}_{jn\ell} = q^{(n-\ell)/2}[n+j-\ell-r_1]_q\hat{\varphi}_{j+1,n\ell} + +q^{j-r_1-1+(n-\ell)/2}[\ell]_q\hat{\varphi}_{j,n+1,\ell-1}, \tag{6.82c}$$

$$\hat{\pi}_{r_1,r_2}(X_2^+)\hat{\varphi}_{jn\ell} = q^{(n-j)/2}[\ell-r_2]_q\hat{\varphi}_{jn,\ell+1} - -q^{r_2-1-\ell+(j-n)/2}[j]_q\hat{\varphi}_{j-1,n+1,\ell}, \tag{6.82d}$$

$$\hat{\pi}_{r_1,r_2}(X_1^-)\hat{\varphi}_{jn\ell} = q^{(\ell-n)/2}[j]_q\hat{\varphi}_{j-1,n\ell}, \tag{6.82e}$$

$$\hat{\pi}_{r_1,r_2}(X_2^-)\hat{\varphi}_{jn\ell} = -q^{(n-j)/2}[\ell]_q\hat{\varphi}_{jn,\ell-1} - -q^{-\ell+(n-j)/2}[n]_q\hat{\varphi}_{j+1,n-1,\ell}. \tag{6.82f}$$

Let us introduce the following operators acting on our functions:

$$\hat{M}^{\pm}_{\kappa}\hat{\varphi}(\bar{Y}) = \sum_{j,n,\ell\in\mathbb{Z}_+} \mu_{j,n,\ell}\hat{M}^{\pm}_{\kappa}\hat{\varphi}_{jn\ell}, \tag{6.83a}$$

$$T_{\kappa}\hat{\varphi}(\bar{Y}) = \sum_{j,n,\ell\in\mathbb{Z}_+} \mu_{j,n,\ell}T_{\kappa}\hat{\varphi}_{jn\ell}, \tag{6.83b}$$

where $\kappa = \xi, \eta, \zeta$, and the explicit action on $\hat{\varphi}_{jn\ell}$ is defined by:

$$\hat{M}^{\pm}_{\xi}\hat{\varphi}_{jn\ell} = \hat{\varphi}_{j\pm1,n\ell}, \tag{6.84a}$$

$$\hat{M}^{\pm}_{\eta}\hat{\varphi}_{jn\ell} = \hat{\varphi}_{jn,\ell\pm1}, \tag{6.84b}$$

$$\hat{M}^{\pm}_{\zeta}\hat{\varphi}_{jn\ell} = \hat{\varphi}_{j,n\pm1,\ell}, \tag{6.84c}$$

$$T_{\xi}\hat{\varphi}_{jn\ell} = q^{j}\hat{\varphi}_{jn\ell}, \tag{6.84d}$$

$$T_{\eta}\hat{\varphi}_{jn\ell} = q^{\ell}\hat{\varphi}_{jn\ell}, \tag{6.84e}$$

$$T_{\zeta}\hat{\varphi}_{jn\ell} = q^{n}\hat{\varphi}_{jn\ell}. \tag{6.84f}$$

Now we define the q-difference operators by:

$$\hat{\mathscr{D}}_{\kappa}\hat{\varphi}(\bar{Y}) = \frac{1}{\lambda}\hat{M}^{-}_{\kappa}\left(T_{\kappa} - T_{\kappa}^{-1}\right)\hat{\varphi}(\bar{Y}), \quad \kappa = \xi, \eta, \zeta. \tag{6.85}$$

Thus, we have:

$$\hat{\mathscr{D}}_{\xi}\hat{\varphi}_{jn\ell} = [j]\hat{\varphi}_{j-1,n\ell}, \tag{6.86a}$$

$$\hat{\mathscr{D}}_{\eta}\hat{\varphi}_{jn\ell} = [\ell]\hat{\varphi}_{jn,\ell-1}, \tag{6.86b}$$

$$\hat{\mathscr{D}}_{\zeta}\hat{\varphi}_{jn\ell} = [n]\hat{\varphi}_{j,n-1,\ell}. \tag{6.86c}$$

Of course, for $q \to 1$ we have $\mathscr{D}_{\kappa} \to \partial_{\kappa} \equiv \partial/\partial\kappa$.

In terms of the above operators the transformation rules (6.82) are written as follows:

$$\hat{\pi}_{r_1,r_2}(k_1)\hat{\varphi}(\bar{Y}) = q^{-r_1/2}T_{\xi}T_{\zeta}^{1/2}T_{\eta}^{-1/2}\hat{\varphi}(\bar{Y}), \tag{6.87a}$$

$$\hat{\pi}_{r_1,r_2}(k_2)\hat{\varphi}(\bar{Y}) = q^{-r_2/2}T_{\eta}T_{\zeta}^{1/2}T_{\xi}^{-1/2}\hat{\varphi}(\bar{Y}), \tag{6.87b}$$

$$\begin{aligned}\hat{\pi}_{r_1,r_2}(X_1^{+})\hat{\varphi}(\bar{Y}) &= (1/\lambda)\hat{M}_{\xi}T_{\zeta}^{1/2}T_{\eta}^{-1/2} \times \\ &\quad\times \left(q^{-r_1}T_{\xi}T_{\zeta}T_{\eta}^{-1} - q^{r_1}T_{\xi}^{-1}T_{\zeta}^{-1}T_{\eta}\right)\hat{\varphi}(\bar{Y}) + \\ &\quad+ q^{-r_1-1}\hat{M}_{\zeta}\mathscr{D}_{\eta}T_{\xi}T_{\zeta}^{1/2}T_{\eta}^{-1/2}\hat{\varphi}(\bar{Y}),\end{aligned} \tag{6.87c}$$

$$\begin{aligned}\hat{\pi}_{r_1,r_2}(X_2^{+})\hat{\varphi}(\bar{Y}) &= (1/\lambda)\hat{M}_{\eta}T_{\zeta}^{1/2}T_{\xi}^{-1/2}\left(q^{-r_2}T_{\eta} - q^{r_2}T_{\eta}^{-1}\right)\hat{\varphi}(\bar{Y}) - \\ &\quad- q^{r_2-1}\hat{M}_{\zeta}\mathscr{D}_{\xi}T_{\xi}^{1/2}T_{\zeta}^{-1/2}T_{\eta}^{-1}\hat{\varphi}(\bar{Y}),\end{aligned} \tag{6.87d}$$

$$\hat{\pi}_{r_1,r_2}(X_1^-)\hat{\varphi}(\bar{Y}) = \mathscr{D}_\xi T_\zeta^{-1/2} T_\eta^{1/2} \hat{\varphi}(\bar{Y}), \tag{6.87e}$$

$$\begin{aligned}\hat{\pi}_{r_1,r_2}(X_2^-)\hat{\varphi}(\bar{Y}) &= -\mathscr{D}_\eta T_\zeta^{1/2} T_\xi^{-1/2} \hat{\varphi}(\bar{Y}) - \\ &\quad -\hat{M}_\xi \mathscr{D}_\zeta T_\xi^{-1/2} T_\zeta^{1/2} T_\eta^{-1} \hat{\varphi}(\bar{Y}),\end{aligned} \tag{6.87f}$$

where $\hat{M}_\kappa = \overset{+}{\hat{M}_\kappa}$.

Notice that it is possible to obtain a realization of the representation $\hat{\pi}_{r_1,r_2}$ on monomials in three commuting variables x, y, z. Indeed, one can relate the noncommuting algebra $\mathbb{C}[\xi, \eta, \zeta]$ with the commuting one $\mathbb{C}[x, y, z]$ by fixing an ordering prescription. However, such realization in commuting variables may be obtained much more directly as is done by other methods (cf. Section 6.3.1 below). Here we are interested in the noncommutative case and we continue to work with the noncommuting variables ξ, η, ζ.

Now we can illustrate some of the general statements of the previous section. Let $m_2 = r_2 + 1 \in \mathbb{N}$. Then it is clear that functions $\hat{\varphi}$ from (6.81) with $\mu_{j,n,\ell} = 0$ if $\ell \geq m_2$ form an invariant subspace since:

$$\hat{\pi}_{r_1,r_2}(X_2^+)\hat{\varphi}_{jnr_2} = -q^{-1+(j-n)/2}[j]_q \hat{\varphi}_{j-1,n+1,r_2}, \tag{6.88}$$

and all other operators in (6.82) either preserve or lower the index ℓ. The same is true for the functions $\tilde{\varphi}$. In particular, for $m_2 = 1$ the functions in the invariant subspace do not depend on the variable η. In this case we have functions of two q-commuting variables $\zeta\xi = q\xi\zeta$ which are much easier to handle that the general noncommutative case (6.77).

The intertwining operator (6.76) for $m_2 \in \mathbb{N}$ is given as follows. First we calculate:

$$\begin{aligned}(\pi_R(X_2^-))^s \tilde{\varphi}_{jn\ell} &= (\pi_R(X_2^-))^s \xi^j \zeta^n \eta^\ell \mathscr{D}_1^{r_1} \mathscr{D}_2^{r_2} = \\ &= \xi^j \zeta^n \sum_{t=0}^{s} a_{st} \eta^{\ell-t} \mathscr{D}_1^{r_1+t} \mathscr{D}_2^{r_2-s-t} (\xi_{13}^{12})^{s-t}, \\ a_{st} &= q^{t\ell+r_2 s/2-(s+t)(s+t+1)/4} \binom{s}{t}_q \frac{[r_2-t]_q![\ell]_q!}{[r_2-s]_q![\ell-t]_q!}\end{aligned} \tag{6.89}$$

where $\binom{n}{k}_q \equiv [n]_q!/[k]_q![n-k]_q!$, $[m]_q! \equiv [m]_q[m-1]_q \ldots [1]_q$. Thus, indeed $\pi_R(X_2^-)$ is not preserving the reduced space $\mathscr{C}_{r_1,r_2}$, and furthermore there is the additional variable ξ_{13}^{12}. Since we would like $\pi_R(X_2^-)$ to some power to map to another reduced space this is only possible if the coefficients a_{st} vanish for $s \neq t$. This happens iff $s = r_2 + 1 = m_2$. Thus we have (in terms of the representation parameters $m_i = r_i + 1$):

$$\begin{aligned}&(\pi_R(X_2^-))^{m_2} \xi^j \zeta^n \eta^\ell \mathscr{D}_1^{m_1-1} \mathscr{D}_2^{m_2-1} = \\ &= q^{m_2(\ell-1-m_2/2)} \frac{[\ell]_q!}{[\ell-m_2]_q!} \xi^j \zeta^n \eta^{\ell-m_2} \mathscr{D}_1^{m_{12}-1} \mathscr{D}_2^{-m_2-1}.\end{aligned} \tag{6.90}$$

Comparing the powers of $\mathscr{D}_i$ we recover at once (2.77) for our situation, namely, $m'_1 = m_{12}$, $m'_2 = -m_2$. Thus, we have shown (6.72a) and (6.73a). Then (6.72b) and (6.73b) follow using (5.40). This intertwining operator has a kernel which is just the invariant subspace discussed above – from the factor $1/[\ell - m_2]_q!$ in (6.90) it is obvious that all monomials with $\ell < m_2$ are mapped to zero.

For the restricted functions we have:

$$\begin{aligned}(\pi_R(X_2^-))^{m_2}\,\hat{\varphi}_{jn\ell} &= q^{m_2(\ell-1-m_2/2)}\frac{[\ell]_q!}{[\ell-m_2]_q!}\hat{\varphi}_{jn,\ell-m_2} = \\ &= q^{-3m_2/2}\left(\mathscr{D}_\eta T_\eta\right)^{m_2}\hat{\varphi}_{jn\ell}.\end{aligned} \tag{6.91}$$

Thus, renormalizing (6.76b) by $q^{-3m_2/2}$ we finally have:

$$I_2^{m_2} = \left(\mathscr{D}_\eta T_\eta\right)^{m_2}. \tag{6.92}$$

For $q = 1$ this operator reduces to the known result: $I_2 = (\partial_\eta)^{m_2}$ [202].

Let now $m_1 \in \mathbb{N}$. In a similar way, though the calculations are more complicated, we find:

$$\begin{aligned}&(\pi_R(X_1^-))^{m_1}\,\xi^j\zeta^n\eta^\ell\mathscr{D}_1^{m_1-1}\mathscr{D}_2^{m_2-1} = \\ &= q^{m_1(j+n-\ell-1-m_1/2)}\sum_{t=0}^{m_1} q^{-t(t+3+2j)/2}\times \\ &\times\binom{m_1}{t}_q\frac{[j]_q![n]_q!}{[j-m_1+t]_q![n-t]_q!}\xi^{j+t-m_1}\zeta^{n-t}\eta^{\ell+t}\mathscr{D}_1^{-m_1-1}\mathscr{D}_2^{m_{12}-1}.\end{aligned} \tag{6.93}$$

Comparing the powers of $\mathscr{D}_i$ we recover (2.77) for our situation, namely, $m'_1 = -m_1$, $m'_2 = m_{12}$. Thus, we have shown (6.72) and (6.73).

For the restricted functions we have:

$$\begin{aligned}(\pi_R(X_1^-))^{m_1}\,\hat{\varphi}_{jn\ell} &= q^{m_1(j+n-\ell-1-m_1/2)}\sum_{t=0}^{m_1} q^{t(t+3+2j)/2}\times \\ &\times\binom{m_1}{t}_q\frac{[j]_q![n]_q!}{[j-m_1+t]_q![n-t]_q!}\hat{\varphi}_{j+t-m_1,n-t,\ell+t} = \\ &= q^{-m_1(3/2+m_1)}T_\zeta^{m_1}\sum_{t=0}^{m_1}\hat{M}_\eta^t\mathscr{D}_\zeta^t(q\mathscr{D}_\xi T_x)^{m_1-t}T_\eta^{-m_1}\hat{\varphi}_{jn\ell}.\end{aligned} \tag{6.94}$$

Then, renormalizing (5.43b) we finally have:

$$I_1^{m_1} = T_\zeta^{m_1}\sum_{t=0}^{m_1}\hat{M}_\eta^t\mathscr{D}_\zeta^t(q\mathscr{D}_\xi T_x)^{m_1-t}T_\eta^{-m_1}. \tag{6.95}$$

For $q = 1$ this operator reduces to the known result: $I_1^{m_1} = (\partial_\xi + \eta\partial_\zeta)^{m_1}$ [202].

Finally, let us consider the case $m = m_{12} = m_1 + m_2 \in \mathbb{N}$, first with $m_1, m_2 \notin \mathbb{Z}_+$. In this case the intertwining operator is given by (6.74), (6.75) using singular vector from Section 2.4 and [198]:

$$\mathscr{P}^m_{12}(X_1^-, X_2^-) = \sum_{s=0}^{m} a_s (X_1^-)^{m-s}(X_2^-)^m (X_1^-)^s, \tag{6.96}$$
$$a_s = (-1)^s a \frac{[m_1]_q}{[m_1 - s]_q} \binom{m}{s}_q, \; s = 0, \ldots, m, a \neq 0.$$

Let us illustrate the resulting intertwining operator in the case $m = 1$. Then, we have, setting in (6.96) $a = [1 - m_1]_q$:

$$\mathscr{I}^1_{12} = [1 - m_1]_q \pi_R(X_1^-)\pi_R(X_2^-) + [m_1]_q \pi_R(X_2^-)\pi_R(X_1^-). \tag{6.97}$$

Then we can see at once the intertwining properties of $\mathscr{I}^1_{12}$ by calculating:

$$\begin{aligned} &\mathscr{I}^1_{12}\xi^j\zeta^n\eta^\ell \mathscr{D}_1^{m_1-1}\mathscr{D}_2^{m_2-1} = \\ &= q^{j+n-2-m_1}[j]_q[\ell]_q \xi^{j-1}\zeta^n\eta^{\ell-1}\mathscr{D}_1^{m_1-2}\mathscr{D}_2^{m_2-2} + \\ &+ q^{n-2}[n]_q[\ell + m_1]_q \xi^j \zeta^{n-1}\eta^\ell \mathscr{D}_1^{m_1-2}\mathscr{D}_2^{m_2-2}. \end{aligned} \tag{6.98}$$

Comparing the powers of $\mathscr{D}_i$ we recover (2.77) for our situation, namely, $m_1' = -m_2 = m_1 - 1$, $m_2' = -m_1 = m_2 - 1$.

For the restricted functions we have:

$$\begin{aligned} &\left([1 - m_1]_q \, \pi_R(X_1^-)\, \pi_R(X_2^-) \; + \; [m_1]_q \, \pi_R(X_2^-)\, \pi_R(X_1^-)\right) \hat{\varphi}_{jn\ell} \; = \\ &= \; q^{n-2+j-m_1} \, [j]_q \, [\ell]_q \, \hat{\varphi}_{j-1,n,\ell-1} \; + \; q^{n-2} \, [n]_q \, [\ell + m_1]_q \, \hat{\varphi}_{j,n-1,\ell} \; = \\ &= \; q^{-2} \left(q^{-m_1} \, \mathscr{D}_\xi \, T_\xi \, \mathscr{D}_\eta \; + \; (1/\lambda) \, \mathscr{D}_\zeta \, (q^{m_1} T_\eta - q^{-m_1} T_\eta^{-1})\right) T_\zeta \, \hat{\varphi}_{jn\ell} \end{aligned} \tag{6.99}$$

Rescaling (6.74b) we finally have:

$$I^1_{12} = \left(q^{-m_1}\mathscr{D}_\xi T_\xi \mathscr{D}_\eta + (1/\lambda)\mathscr{D}_\zeta (q^{m_1} T_\eta - q^{-m_1} T_\eta^{-1})\right) T_\zeta. \tag{6.100}$$

For $q = 1$ this operator is: $I_{12} = \partial_\xi \partial_\eta + (m_1 + \eta\partial_\eta)\partial_\zeta$ [202].

Above we have supposed that $m_1, m_2 \notin \mathbb{Z}_+$. However, after the proper choice of a in (6.96), (e. g., as made above in (6.97) we can consider the singular vector (6.96) and the resulting intertwining operator also when $m_1, m_2 \in \mathbb{Z}_+$. In these cases the singular vector is reduced in four different ways (cf. (2.89)). Accordingly, the intertwining operator becomes composite, that is, it can be expressed as the composition of the intertwiners introduced so far as follows:

$$I_{12}^{m} = c_1 I_1^{m_2} I_2^{m} I_1^{m_1} = \tag{6.101a}$$
$$= c_2 I_2^{m_1} I_1^{m} I_2^{m_2} =$$
$$= c_3 I_2^{m_1} I_{12}^{m_2} I_1^{m_1} = \tag{6.101b}$$
$$= c_4 I_1^{m_2} I_{12}^{m_1} I_2^{m_2}.$$

The four expressions were used to prove commutativity of the hexagon diagram of $U_q(sl(3,\mathbb{C})))$ [198]. This diagram involves six representations which are denoted by V_{00}, V_{00}^1, V_{00}^2, V_{00}^{12}, V_{00}^{21}, V_{00}^3, in (29) of [198] and which in our notation are connected by the intertwiners in (6.101) as follows:

$$\hat{\mathcal{C}}_{m_1,m_2} \overset{I_1^{m_1}}{\longrightarrow} \hat{\mathcal{C}}_{-m_1,m} \overset{I_2^{m}}{\longrightarrow} \hat{\mathcal{C}}_{m_2,-m} \overset{I_1^{m_2}}{\longrightarrow} \hat{\mathcal{C}}_{-m_2,-m_1} \tag{6.102a}$$

$$\hat{\mathcal{C}}_{m_1,m_2} \overset{I_2^{m_2}}{\longrightarrow} \hat{\mathcal{C}}_{m,-m_2} \overset{I_1^{m}}{\longrightarrow} \hat{\mathcal{C}}_{-m,m_1} \overset{I_2^{m_1}}{\longrightarrow} \hat{\mathcal{C}}_{-m_2,-m_1} \tag{6.102b}$$

$$\hat{\mathcal{C}}_{m_1,m_2} \overset{I_1^{m_1}}{\longrightarrow} \hat{\mathcal{C}}_{-m_1,m} \overset{I_{12}^{m_2}}{\longrightarrow} \hat{\mathcal{C}}_{-m,m_1} \overset{I_2^{m_1}}{\longrightarrow} \hat{\mathcal{C}}_{-m_2,-m_1} \tag{6.102c}$$

$$\hat{\mathcal{C}}_{m_1,m_2} \overset{I_2^{m_2}}{\longrightarrow} \hat{\mathcal{C}}_{m,-m_2} \overset{I_{12}^{m_1}}{\longrightarrow} \hat{\mathcal{C}}_{m_2,-m} \overset{I_1^{m_2}}{\longrightarrow} \hat{\mathcal{C}}_{-m_2,-m_1} \tag{6.102d}$$

Of these six representations only $\hat{\mathcal{C}}_{m_1,m_2}$ has a finite-dimensional irreducible subspace iff $m_1 m_2 > 0$, the dimension being $m_1 m_2 m/2$ [198]. If $m_1 = 0$ the intertwining operators with superscript m_1 become the identity (since in these cases the intertwined spaces coincide) and the compositions in (6.101) and (6.102) are shortened to two arrows in cases (a,b,d) and one arrow in case (c) (respectively, for $m_2 = 0$, two arrows in cases (a,b,c), one arrow in (d)). (Such considerations are part of the multiplet classification given in [198].)

6.3 Polynomial Solutions of q-Difference Equations in Commuting Variables

In this section we follow mainly [246]. A new approach to the theory of polynomial solutions of q-difference equations is proposed. The approach is based on the representation theory of simple Lie algebras $\mathcal{G}$ and their q-deformations and is presented here for $U_q(sl(n))$. First a q-difference realization of $U_q(sl(n))$ in terms of $n(n-1)/2$ commuting variables and depending on $n-1$ complex representation parameters r_i, is constructed. From this realization lowest-weight modules (LWMs) are obtained which are studied in detail for the case $n = 3$ (the well-known $n = 2$ case is also recovered). All reducible LWM are found and the polynomial bases of their invariant irreducible subrepresentations are explicitly given. This also gives a classification of the quasi-exactly solvable operators in the present setting. The invariant subspaces are obtained as solutions of certain invariant q-difference equations, that is, these are kernels of invariant q-difference operators, which are also explicitly given. Such operators were not used until now in the theory of polynomial solutions. Finally the states in all subrepresentations are depicted graphically via the so called Newton diagrams.

6.3.1 Procedure for the Construction of the Representations

The procedure is iterative. In fact, we have to use also $U_q(gl(n))$. Let us introduce first some notation. The basic q-number notation $[a] = \frac{q^{a/2}-q^{-a/2}}{q^{1/2}-q^{-1/2}}$ will be used also for diagonal operators H replacing a. Following Biedenharn and Lohe [101] our representations will be given in terms of $n(n-1)/2$ variables. For our purposes we denote these variables by z_i^k, $2 \le k \le n$, $1 \le i \le k-1$. Next we introduce the number operator N_i^k for the coordinate z_i^k, that is, $N_i^k z_j^m = \delta_{mk}\delta_{ij} z_j^m$ and the q-difference operators D_i^k, which admit a general definition on a larger domain than polynomials, but on polynomials are well defined as follows:

$$D_i^k = \frac{1}{z_i^k}[N_i^k]. \tag{6.103}$$

Further we note that the representations of $U_q(sl(n))$ will be characterized by $n-1$ complex parameters $r_k \in \mathbb{C}$, $1 \le k \le n-1$.

We rewrite formulae (5.3), (6.10), and (6.22) from [101] in the following way :

$$\Gamma_n(E_{ij}) = \Gamma_{n-1}(E_{ij}) q^{\frac{1}{4}(N_i^n - N_j^n)} + q^{\frac{1}{4}\Gamma_{n-1}(E_{jj}-E_{ii})} z_i^n D_j^n \qquad i < j < n \tag{6.104a}$$

$$\Gamma_n(E_{ij}) = \Gamma_{n-1}(E_{ij}) q^{\frac{1}{4}(N_j^n - N_i^n)} + q^{\frac{1}{4}\Gamma_{n-1}(E_{ii}-E_{jj})} z_i^n D_j^n \qquad n > i > j, \tag{6.104b}$$

$$\Gamma_n(E_{ii}) = \Gamma_{n-1}(E_{ii}) + N_i^n, \; i < n, \tag{6.104c}$$

$$\Gamma_n(E_{nn}) = \Gamma_{n-1}(E_{nn}) - \sum_{i=1}^{n-1} N_i^n, \tag{6.104d}$$

$$\Gamma_n(E_{ni}) = q^{\frac{1}{4}\left(\sum_{j=1}^{i-1}\Gamma_{n-1}(E_{jj}) - \sum_{j=i+1}^{n-1}\Gamma_{n-1}(E_{jj})\right)} D_i^n, \; i < n \tag{6.104e}$$

$$\Gamma_n(E_{in}) = q^{\alpha_{ii}^n} z_i^n \Big[\Gamma_{n-1}(E_{nn}) - \Gamma_{n-1}(E_{ii}) - \sum_{k=1}^{n-1} N_k^n\Big] - \\ - \sum_{\substack{j=1 \\ j \ne i}}^{n-1} q^{\alpha_{ij}^n} z_j^n \Gamma_{n-1}(E_{ij}), \; i < n, \tag{6.104f}$$

where

$$
\begin{aligned}
q^{\alpha_{ij}^n} &\equiv q^{-\frac{1}{4}\left(\sum_{k=1}^{j-1}\Gamma_{n-1}(E_{kk})-\sum_{k=j+1}^{n-1}\Gamma_{n-1}(E_{kk})\right)} q^{\left(\frac{+}{0}\right)\left(\frac{1}{2}(\Gamma_{n-1}(E_{nn})-\sum_{k=1}^{n-1}N_k^n)+\frac{3}{4}\right)} \\
&\times q^{\left(\frac{+}{0}\right)\left(\frac{1}{4}(N_i^n+N_j^n)\right)} q^{\left(\frac{+}{0}\right)\frac{1}{2}\left(\sum_{\substack{i<k<j \text{if} i<j \\ j<k<i \text{if} j<i}} N_k^n\right)}, \\
\left(\frac{+}{0}\right) &= \begin{cases} + & \text{for i < j,} \\ - & \text{for i > j,} \\ 0 & \text{for i = j,} \end{cases}
\end{aligned}
\tag{6.105}
$$

$\Gamma_{n-1}(E_{ij})$ are defined at the previous step, except $\Gamma_{n-1}(E_{nn})$ which adds the representation parameter r_{n-1} and is given by:

$$
\Gamma_{n-1}(E_{nn}) = \sum_{k=0}^{n-1} r_k = r^{n-1} + r_0. \tag{6.106}
$$

The parameter r_0 represents the center of $U_q(gl(n))$ and is decoupled later. The additional input with respect to [101] is : 1) notation – we put indices on Γ corresponding to the case we consider – thus, this is made an iterative procedure; 2) we give values to the Cartan generators which values are consistent with previous knowledge from representation theory; 3) we introduce q-difference operators D_i^n to replace $\bar{z}_i^n$; 4) we have done also something artificial by using $\Gamma_{n-1}(E_{nn})$ – this is given by the "total number" r^{n-1} (which for finite-dimensional representations is the number of boxes of the Young tableaux minus $n-1$) plus the number r_0 representing the center of $U_q(gl(n))$.

Denoting $\mathcal{Z}^n = \sum_{i=1}^n E_{ii}$ we have:

$$
\begin{aligned}
\Gamma_n(\mathcal{Z}^n) &= \sum_{i=1}^{n} \Gamma_n(E_{ii}) = r^{n-1} + r_0 + \Gamma_{n-1}(\mathcal{Z}^{n-1}) = \\
&= \sum_{i=1}^{n-1} r^i + n r_0 = \sum_{i=0}^{n-1} (n-i) r_i.
\end{aligned}
\tag{6.107}
$$

Thus, as expected, $\Gamma_n(\mathcal{Z}^n)$ is central. Then the generators $H_i^n = \Gamma_n(E_{ii} - E_{i+1,i+1})$, $1 \le i < n$, $\Gamma_n(E_{ij})$, $i \ne j$, form a q-difference operator realization of $U_q(sl(n))$.

It is straightforward to obtain the explicit expressions for $\Gamma_n(E_{ij})$. In particular, we have

$$
\Gamma_n(E_{ii}) = \sum_{j=0}^{n-i-1} N_i^{n-j} - \sum_{j=1}^{i-1} N_j^i + \sum_{j=0}^{i-1} r_j, \tag{6.108}
$$

with the usual convention that a sum is zero if the upper limit is smaller than the lower limit. From this we obtain the expressions for the Cartan generators H_i^n (as defined above):

$$H_i^n = 2N_i^{i+1} - r_i + \sum_{j=0}^{n-i-2}(N_i^{n-j} - N_{i+1}^{n-j}) + \sum_{j=1}^{i-1}(N_j^{i+1} - N_j^i) \quad i < n. \tag{6.109}$$

Let us illustrate things for $n = 2, 3$.

For $n = 2$ we have (we use only (6.104e,f) and (6.109)):

$$\begin{aligned}
\Gamma_2(E_{12}) &= z_1^2[r_1 - N_1^2] = x[r - N_x],\\
\Gamma_2(E_{21}) &= D_1^2 = D_x,\\
H_1^2 &= 2N_x - r_1,
\end{aligned} \tag{6.110}$$

where we have denoted $z_1^2 = x$, $N_1^2 = N_x$. This reproduces the known realization [309] of the $U_q(sl(2))$ representations with $X^+ = \Gamma_2(E_{12})$, $X^- = \Gamma_2(E_{21})$, $H = H_1^2$, depending on the representation parameter r_1 (r_0 being cancelled as expected). For $q = 1$ this coincides with the classical $sl(2)$ vector-filed realization.

Next we take $n = 3$ setting $z_1^3 = z$, $z_2^3 = y$, $N_1^3 = N_z$, $N_2^3 = N_y$, $D_1^3 = D_z$, $D_2^3 = D_y$, $r = r^2 = r_1 + r_2$. Due to our recursive procedure we inherit form the case $n = 2$ the variable $z_1^2 = x$, and the operators N_x, D_x. Besides this we renormalize the generators $\Gamma_3(E_{13})$ and $\Gamma_3(E_{31})$ so that they obey the standard $U_q(sl(3))$ relations (these are different in [101], cf. (6.17) and (6.20)):

$$\begin{aligned}
\Gamma_3(E_{13}) &= \Gamma_3(E_{12})\Gamma_3(E_{23}) - q^{1/2}\Gamma_3(E_{23})\Gamma_3(E_{12}),\\
\Gamma_3(E_{31}) &= \Gamma_3(E_{32})\Gamma_3(E_{21}) - q^{-1/2}\Gamma_3(E_{21})\Gamma_3(E_{32}).
\end{aligned} \tag{6.111}$$

Thus, we have:

$$\begin{aligned}
H_1^3 &= 2N_1^2 - r_1 + N_1^3 - N_2^3 = 2N_x - r_1 + N_z - N_y,\\
H_2^3 &= 2N_2^3 - r_2 + N_1^3 - N_1^2 = 2N_y - r_2 + N_z - N_x,\\
\Gamma_3(E_{12}) &= \Gamma_2(E_{12})q^{\frac{1}{4}(N_1^3-N_2^3)} + q^{\frac{1}{4}\Gamma_2(E_{22}-E_{11})}z_1^3D_2^3 =\\
&= x[r_1 - N_x]q^{\frac{1}{4}(N_z-N_y)} + zD_yq^{\frac{1}{4}(r_1-2N_x)},\\
\Gamma_3(E_{21}) &= \Gamma_2(E_{21})q^{\frac{1}{4}(N_1^3-N_2^3)} + q^{\frac{1}{4}\Gamma_2(E_{22}-E_{11})}z_2^3D_1^3 =\\
&= D_xq^{\frac{1}{4}(N_z-N_y)} + yD_zq^{\frac{1}{4}(r_1-2N_x)},\\
\Gamma_3(E_{23}) &= q^{-\frac{1}{4}\Gamma_2(E_{11})}z_2^3\Big[\Gamma_2(E_{33}) - \Gamma_2(E_{22}) - \sum_{k=1}^{2}N_k^3\Big] -\\
&\quad - q^{\alpha_{21}}z_1^3\Gamma_2(E_{21}) =\\
&= y[r_2 + N_x - N_z - N_y]q^{-\frac{1}{4}(N_x+r_0)} -\\
&\quad - zD_xq^{-\frac{1}{4}(2r-r_1+r_0+N_x-N_z-N_y+1)},\\
\Gamma_3(E_{32}) &= q^{\frac{1}{4}\Gamma_2(E_{11})}D_2^3 = D_yq^{\frac{1}{4}(r_0+N_x)},
\end{aligned} \tag{6.112}$$

$$\Gamma_3(E_{13}) = q^{\frac{1}{4}(\Gamma_2(E_{22})-2\Gamma_3(E_{22}))} z_1^3\Big[\Gamma_2(E_{33}) - \Gamma_2(E_{11}) - \sum_{k=1}^{2} N_k^3\Big] -$$
$$-q^{-\frac{1}{2}\Gamma_3(E_{22})} q^{\alpha_{12}} z_2^3 \Gamma_2(E_{12}) =$$
$$= z[r - N_x - N_z - N_y] q^{\frac{1}{4}(N_x - r_1 - 2N_y - r_0)} -$$
$$-xy[r_1 - N_x] q^{\frac{1}{4}(2r_2 - r_0 + N_x - N_z - 3N_y + 1)}$$
$$\Gamma_3(E_{31}) = q^{\frac{1}{4}(2\Gamma_3(E_{22}) - \Gamma_2(E_{22}))} D_1^3 = D_z q^{\frac{1}{4}(r_1 + r_0 - N_x + 2N_y)}.$$

We now rescale the generators E_{3i}, E_{i3} $(i = 1, 2)$ so as to absorb the parameter r_0. (Such a rescaling should be done also for the general $U_q(gl(n))$ case.) Thus the realization of $U_q(sl(3))$ depends only on the parameters r_1, r_2, as in the classical case which may be obtained from (6.112) by setting $q = 1$.

6.3.2 Reducibility of the Representations and Invariant Subspaces

6.3.2.1 Lowest-Weight Representations

Let us apply the realization (6.104) to the function 1. Using the fact that $N_i^n 1 = 0 = D_i^n 1$ we have:

$$\Gamma_n(E_{ii})\,1 = \sum_{j=0}^{i-1} r_j = r^{i-1}, \quad H_i^n\,1 = -r_i, \quad i \le n, \tag{6.113a}$$
$$\Gamma_n(E_{ni})\,1 = 0, \quad i < n, \tag{6.113b}$$
$$\Gamma_n(E_{ij})\,1 = \Gamma_{n-1}(E_{ij})\,1 = \cdots = \Gamma_i(E_{ij})\,1 = 0, \quad j < i < n \tag{6.113c}$$
$$\Gamma_n(E_{ij})\,1 = \Gamma_{n-1}(E_{ij})\,1 = \cdots = \Gamma_j(E_{ij})\,1, \quad i < j < n \tag{6.113d}$$
$$\begin{aligned}\Gamma_n(E_{in})\,1 &= q^{\frac{1}{4}\left(\sum_{k=i+1}^{n-1} r^{k-1} - \sum_{k=1}^{i-1} r^{k-1}\right)} z_i^n \left[r_i + \cdots + r_{n-1}\right] - \\ &- \sum_{s=i+1}^{n-1} q^{\alpha_{is}^n} z_s^n\, \Gamma_s(E_{is})\,1,\; i < n,\end{aligned} \tag{6.113e}$$

where

$$\begin{aligned} q^{\alpha_{is}^n} &= q^{\frac{1}{4}\left(\sum_{k=s+1}^{n-1} r^{k-1} - \sum_{k=1}^{s-1} r^{k-1}\right)} q^{\frac{1}{2}(r^{n-1} - \sum_{k=1}^{n-1} N_k^n) + \frac{3}{4}} \times \\ &\times q^{\frac{1}{4}(N_i^n + N_s^n)} q^{\frac{1}{2}\sum_{k=i+1}^{k=s-1} N_k^n}, \quad i < s. \end{aligned} \tag{6.114}$$

It is straightforward to obtain the explicit expressions for $\Gamma_n(E_{ij})1$, $i < j \le n$, applying recursively (6.113d,e). In particular, we have:

$$\Gamma_n(E_{i,i+1})1 = \Gamma_{i+1}(E_{i,i+1})1 = q^{-\frac{1}{4}\sum_{k=1}^{i-1} r^{k-1}} [r_i] z_i^{i+1}. \tag{6.115}$$

Thus, we have obtained an LWM with lowest-weight vector 1 (it is annihilated by the lowering generators $\Gamma_n(E_{ij})$, $j < i \le n$), and lowest-weight Λ such that $\Lambda(H_i) = -r_i$ (cf.

(6.113a)). Generically this LWM is irreducible and then it is isomorphic to the Verma module with this lowest weight. The states in it correspond to the monomials of the Poincaré–Birkhoff–Witt basis of $U_q(\mathscr{G}^+)$, where $\mathscr{G}^+$ is the subalgebra of the raising generators. This is isomorphic to the monomials in the variables z_i^k. When the representation parameters r_i or certain combinations thereof are non-negative integers our representations are reducible. Below we consider in detail the cases $n = 2$ and $n = 3$.

6.3.2.2 Case $U_q(sl(2))$

We start with $n = 2$ (though this example is well known). Using (6.110) we apply H, X^+, X^- to the function 1. We use the fact that $N_x 1 = 0 = D_x 1$. Thus:

$$H1 = -r, X^+1 = x[r], X^-1 = 0. \tag{6.116}$$

Thus, we obtain an LWM with lowest-weight vector 1 and lowest-weight Λ such that $\Lambda(H) = -r$. All states are given by powers of x, that is, the basis is x^k with $k \in \mathbb{Z}_+$ and the representation is infinite dimensional. The action of $U_q(sl(2))$ is given by:

$$X^+x^k = [r-k]x^{k+1}, \quad X^-x^k = [k]x^{k-1}, \quad Hx^k = (2k-r)x^k. \tag{6.117}$$

Clearly, if $r \notin \mathbb{Z}_+$ this representation is irreducible. Furthermore all states may be obtained by the application of X^+ to the LWV; that is :

$$(X^+)^k 1 = x^k[r][r-1]\dots[r-k+1], k \in \mathbb{Z}_+. \tag{6.118}$$

Let $r \in \mathbb{Z}_+$, then $(X^+)^{r+1}1 = X^+x^r[r]! = 0$. Thus, the states x^k with $k = 0, 1, \dots r$ form a finite-dimensional subrepresentation with dim $= r + 1$. Note that the complement of this subrepresentation; that is, the states x^k with $k > r$, is not an invariant subspace.

Clearly, any polynomial in $H, X^\pm$, will preserve this invariant subspace and thus would be a quasi-exactly solvable operator.

The invariant subspace may be obtained as the solution of either one of the following equations:

$$(X^+)^{r+1}f(x) = 0, \tag{6.119a}$$
$$(X^-)^{r+1}f(x) = 0, \tag{6.119b}$$

in the space of formal power series $f(x) = \sum_{k\in\mathbb{Z}_+} \mu_k x^k$. Note, however, that only (6.119b) (which is enough) was expected – this is an artefact of $n = 2$ simplifications. Indeed, only the operator in (6.119b) has the intertwining property (as in the classical case [202]):

$$(X^-)^{r+1}\Gamma_2(X)_r = \Gamma_2(X)_{r'}(X^-)^{r+1}, \qquad r' = -r-2, \tag{6.120}$$

where $X = H, X^\pm$, and $\Gamma_2(X)_r$ is from (6.110) with explicit notation for the representation parameter of the two representations which are intertwined.

6.3.2.3 Case $U_q(sl(3))$

Let us apply (6.112) to the function 1:

$$\begin{aligned} H_1 1 &= -r_1, \quad H_2 1 = -r_2, \\ \Gamma_3(E_{12})1 &= x[r_1], \quad \Gamma_3(E_{21})1 = 0, \\ \Gamma_3(E_{23})1 &= y[r_2], \quad \Gamma_3(E_{32})1 = 0, \\ \Gamma_3(E_{13})1 &= q^{-\frac{1}{4}r_1} z[r] - q^{\frac{1}{4}(2r_2+1)} yx[r_1], \qquad \Gamma_3(E_{31})1 = 0. \end{aligned} \tag{6.121}$$

Thus, we obtain a lowest-weight module with lowest-weight vector 1 and lowest-weight Λ such that $\Lambda(H_k) = -r_k$. All states are given by powers of x, y, z; that is, the basis is generated by $x^j z^k y^\ell$ with $j, k, \ell \in \mathbb{Z}_+$. The action of $U_q(sl(3))$ is given by:

$$H_1 x^j z^k y^\ell = (-r_1 + 2j - \ell + k) x^j z^k y^\ell, \tag{6.122a}$$

$$H_2 x^j z^k y^\ell = (-r_2 - j + 2\ell + k) x^j z^k y^\ell, \tag{6.122b}$$

$$\begin{aligned} \Gamma_3(E_{12}) x^j z^k y^\ell &= [r_1 - j] q^{\frac{1}{4}(k-\ell)} x^{j+1} z^k y^\ell + \\ &+ [\ell] q^{\frac{1}{4}(r_1 - 2j)} x^j z^{k+1} y^{\ell-1}, \end{aligned} \tag{6.122c}$$

$$\begin{aligned} \Gamma_3(E_{21}) x^j z^k y^\ell &= [j] q^{\frac{1}{4}(k-\ell)} x^{j-1} z^k y^\ell + \\ &+ [k] q^{\frac{1}{4}(r_1 - 2j)} x^j z^{k-1} y^{\ell+1}, \end{aligned} \tag{6.122d}$$

$$\begin{aligned} \Gamma_3(E_{23}) x^j z^k y^\ell &= q^{-\frac{1}{4}j} [r_2 + j - k - \ell] x^j z^k y^{\ell+1} - \\ &- [j] q^{-\frac{1}{4}(r_1 + 2r_2 + j - k - \ell + 1)} x^{j-1} z^{k+1} y^\ell, \end{aligned} \tag{6.122e}$$

$$\Gamma_3(E_{32}) x^j z^k y^\ell = [\ell] q^{\frac{1}{4}j} x^j z^k y^{\ell-1}, \tag{6.122f}$$

$$\begin{aligned} \Gamma_3(E_{13}) x^j z^k y^\ell &= q^{\frac{1}{4}(j - r_1 - 2\ell)} [r - j - k - \ell] x^j z^{k+1} y^\ell - \\ &- q^{\frac{1}{4}(2r_2 + j - k - 3\ell + 1)} [r_1 - j] x^{j+1} z^k y^{\ell+1}, \end{aligned} \tag{6.122g}$$

$$\Gamma_3(E_{31}) x^j z^k y^\ell = [k] q^{\frac{1}{4}(r_1 - j + 2\ell)} x^j z^{k-1} y^\ell \tag{6.122h}$$

Further, in this section we show the following results which parallel the classical situation (cf. [202]):

1. If r_1, or r_2, or $r + 1 \in \mathbb{Z}_+$ this representation is reducible. It contains an irreducible subrepresentation which is infinite-dimensional, except when both $r_1, r_2 \in \mathbb{Z}_+$;
2. If $r_1, r_2, r + 1 \notin \mathbb{Z}_+$ this representation is irreducible and infinite-dimensional.

Clearly, if $r_1 \in \mathbb{Z}_+$ the representation (6.122) becomes reducible since the monomials $x^j z^k y^\ell$ with $j \le r_1$ form an invariant subspace since from (6.122c,g) we have:

$$\begin{aligned} \Gamma_3(E_{12}) x^{r_1} z^k y^\ell &= [\ell] q^{-\frac{1}{4}r_1} x^{r_1} z^{k+1} y^{\ell-1}, \\ \Gamma_3(E_{13}) x^{r_1} z^k y^\ell &= [r_2 - k - \ell] q^{-\frac{1}{2}\ell} x^{r_1} z^{k+1} y^\ell, \end{aligned} \tag{6.123}$$

and all other operators are either lowering or preserving the powers of x. This invariant subspace may be described as the solution of the following q-difference equation:

$$(D_x)^{r_1+1} f(x,y,z) = 0. \tag{6.124}$$

Note that the operator in (6.124) has the intertwining property (as in the classical case [202]):

$$(D_x)^{r_1+1} \Gamma_3(X)_{r_1,r_2} = \Gamma_3(X)_{r_1',r_2'} (D_x)^{r_1+1}, \qquad r_1' = -r_1 - 2, r_2' = r + 1, \tag{6.125}$$

where $X = E_{ii} - E_{i+1,i+1}, E_{ij}, i \neq j, \Gamma_3(X)_{r_1,r_2}$ is taken from (6.112) with explicit dependence of the representation parameters of the two representations which are intertwined.

The subrepresentation obtained is infinite-dimensional if $r_2 \notin \mathbb{Z}_+$ since the powers of y, z are still unrestricted by (6.122e,g).

If $r_2 \in \mathbb{Z}_+$ the representation in (6.122) becomes reducible. In the classical case ($q = 1$) the equation which singles out the invariant subspace is [202]:

$$\left(x\partial_z + \partial_y\right)^{r_2+1} f(x,y,z) = 0, q = 1. \tag{6.126}$$

For the quantum case we have the following expression:

$$\begin{aligned} {}_q\mathscr{D}_2^{r_2+1} f(x,y,z) &= 0, \\ {}_q\mathscr{D}_2^{k} &= \sum_{s=0}^{k} \binom{k}{s}_q x^{k-s} D_z^{k-s} D_y^s q^{\frac{1}{4}s(N_z-r_1)+\frac{1}{4}(s-k)N_y-\frac{1}{4}kN_x}, \end{aligned} \tag{6.127}$$

which coincides with (6.126) for $q = 1$. The invariant subspace is infinite-dimensional if $r_1 \notin \mathbb{Z}_+$.

As in the classical case [202] the explicit form (6.127) of this operator may be checked by the intertwining property:

$${}_q\mathscr{D}_2^{r_2+1} \Gamma_3(X)_{r_1,r_2} = \Gamma_3(X)_{r_1',r_2'} {}_q\mathscr{D}_2^{r_2+1}, \qquad r_1' = r + 1, r_2' = -r_2 - 2, \tag{6.128}$$

where $X = E_{ij}, i \neq j, E_{ii} - E_{i+1,i+1}, \Gamma_3(X)_{r_1,r_2}$ is from (6.112) as in (6.125).

Further in this subsection we consider the case when both $r_k \in \mathbb{Z}_+$. Then there is a finite-dimensional irreducible subspace of dimension:

$$d_{r_1,r_2} = \frac{1}{2}(r_1+1)(r_2+1)(r+2). \tag{6.129}$$

Thus, we recover the complete list of the finite-dimensional irreps of $U_q(sl(3))$ and $SL(3)$, nd by default, also the complete list of the finite-dimensional unitary irreps of $U_q(su(3))$ and $SU(3)$ (we have assumed that q is not a nontrivial root of 1).

Next we use the following general formula valid for arbitrary r_k:

$$v_{\ell kj} \equiv \Gamma_3(E_{23})^\ell \Gamma_3(E_{13})^k \Gamma_3(E_{12})^j 1 =$$
$$= \sum_{s=0}^{k} \sum_{n=0}^{\ell} (-1)^{s-n} \binom{k}{s}_q \binom{\ell}{n}_q q^{\frac{1}{4}\{(j-r_1)k-\ell j+(s-n)(r_1+2r_2-k-\ell+2)\}} \times$$
$$\times \frac{\Gamma_q(r_1+1)\Gamma_q(r-j-s+1)}{\Gamma_q(r_1-j-s+1)\Gamma_q(r-j-k+1)} \times$$
$$\times \frac{\Gamma_q(r_2+j+s-k-n+1)}{\Gamma_q(r_2+j+s-k-\ell+1)} \frac{[j+s]!}{[j+s-n]!} \times$$
$$\times y^{\ell+s-n} z^{k-s+n} x^{j+s-n},$$
$$\ell+k+j \le r, \quad 0 \le j \le r_1, \quad 0 \le \ell \le r_2 . \tag{6.130}$$

One of the main results of [246] is that the basis of the finite-dimensional irrep with dimension d_{r_1,r_2} for $r_1, r_2 \in \mathbb{Z}+$ (cf. (6.129)) is given by $v_{\ell kj}$ iff $\ell + k + j \le r$, $0 \le j \le r_1$, $0 \le \ell \le r_2$. In the next section we relate this basis to the standard Gel'fand–Zetlin basis.

For later reference we note the special polynomial v_{0r0} which corresponds to the highest-weight vector (as we shall see later):

$$v_{0r0} \equiv \Gamma_3(E_{13})^r 1 = q^{-\frac{1}{4}rr_1} [r]_q! \sum_{s=0}^{r_1} (-1)^s \binom{r_1}{s}_q q^{\frac{1}{4}s(r-r_1+2)} x^s z^{r-s} y^s \tag{6.131}$$

Note also that when $r_1 = 0$ there is no dependence on x in (6.130), all states being the monomials $v_{\ell k0} \sim y^\ell z^k$.

Also for later reference we note the explicit value of $v_{\ell kj}$ for $z = 0$ (given by the term $s = k$ and $n = 0$):

$$v_{\ell kj}\big|_{z=0} = (-1)^k q^{\frac{1}{4}\{(j-r_1)k-\ell j+k(r_1+2r_2-k-\ell+2)\}} \times \tag{6.132}$$
$$\times \frac{\Gamma_q(r_1+1)}{\Gamma_q(r_1-j-k+1)} \frac{\Gamma_q(r_2+j+1)}{\Gamma_q(r_2+j-\ell+1)} x^{j+k} y^{\ell+k}$$

Note that the RHS of (6.132) is equal to zero when $j + k \ge r_1 + 1$ (because of the $\Gamma_q(r_1 - j - k + 1)$ in the denominator). In this case one applies $(D_z)^{j+k-r_1}$ to both sides of (6.130) and then sets $z = 0$.

Next, we discuss the case when $r + 1 \in \mathbb{Z}_+$, but $r_k \notin \mathbb{Z}_+$. Following our procedure for invariant differential operators, we use the $U_q(sl(3))$ singular vector in Chevalley basis (cf. f-la (27) from [198] or Section 2.4):

$$v_3^{r+2} = \sum_{s=0}^{r+2} \frac{(-1)^s}{[r_1+1-s]_q} \binom{r+2}{s}_q (\hat{X}_1^+)^{r+2-s} (\hat{X}_2^+)^{r+2} (\hat{X}_1^+)^s \tag{6.133}$$

In the latter we substitute the corresponding action of the simple root generators $\hat{X}_1^+, \hat{X}_2^+$ by the operators ${}_q\mathscr{D}_1 = D_x$ and by ${}_q\mathscr{D}_2$ from (6.127) to obtain the invariant q-difference operator:

$${}_q\mathscr{D}_3^{r+2} = \sum_{s=0}^{r+2} \frac{(-1)^s}{[r_1+1-s]_q} \binom{r+2}{s}_q D_x^{r+2-s} ({}_q\mathscr{D}_2)^{r+2} D_x^s \tag{6.134}$$

Thus, our subspace is singled out by the following explicit equation [246]:

$${}_q\mathscr{D}_3^{r+2} f(x,y,z) = 0, \tag{6.135}$$

$$\begin{aligned} {}_q\mathscr{D}_3^{r+2} &= \sum_{s=0} \sum_{t=0} \frac{q^{\frac{1}{4}(r+2-t-2s)r_1 + \frac{1}{2}(r+2)t + \frac{1}{2}(r+1)(s-4)} \Gamma_q(-1-r_1)}{[s]![t]![r+2-s-t]!\Gamma_q(r-r_1+2-s)} \times \\ &\times D_z^{r+2-t} D_y^t D_x^t \prod_{u=1}^{r+2-s-t} [N_x - t + 1 - u] q^{\frac{(2s-r-2)}{4} N_x + \frac{t}{4} N_y + \frac{(t+r+2)}{4} N_z} \end{aligned}$$

As in the classical case [202] the explicit form of this operator may be checked by the intertwining property:

$${}_q\mathscr{D}_3^{r+2} \Gamma_3(X)_{r_1,r_2} = \Gamma_3(X)_{r_1',r_2'} {}_q\mathscr{D}_3^{r+2}, \quad r_1' = -r_2 - 2, \quad r_2' = -r_1 - 2, \tag{6.136}$$

where $X = E_{ii} - E_{i+1,i+1}, E_{ij}, i \neq j, \Gamma_3(X)_{r_1,r_2}$ is from (6.112) as in (6.125) and (6.128).

The states in the subrepresentation are given by $v_{\ell kj}$, with $\ell, k, j \in \mathbb{Z}_+$, $k \leq r+1$.

Let us illustrate the above by the simplest limiting case of $r = -1$. Then the states are:

$$\begin{aligned} v_{\ell 0 j}|_{r=-1} &= \sum_{n=0}^{\ell} (-1)^\ell \binom{\ell}{n}_q q^{\frac{1}{4}\{-\ell j + n(r_1+\ell)\}} \times \\ &\times \frac{\Gamma_q(r_1+1)}{\Gamma_q(r_1-j+1)} \frac{\Gamma_q(1+r_1-j+\ell)}{\Gamma_q(1+r_1-j+n)} \frac{[j]!}{[j-n]!} x^{j-n} z^n y^{\ell-n} = \\ &= q^{-\frac{1}{4}\ell j} \frac{\Gamma_q(r_1+1)}{\Gamma_q(r_1-j+1)} \frac{\Gamma_q(j-r_1)}{\Gamma_q(j-r_1-\ell)} \times \\ &\times x^j y_2^\ell F_1^q(-j, -\ell; r_1-j+1; q^{\frac{1}{4}(r_1+\ell)} \frac{z}{xy}). \end{aligned} \tag{6.137}$$

Alternatively one may check that this is the general solution of (6.135) for $r = -1$.

6.3.3 Newton Diagrams

It this section we give a visualization of the representation spaces. Each state is represented by a point on an integer lattice in $n(n-1)/2$ dimensions, that is, on $\mathbb{Z}_+^{n(n-1)/2}$. For a finite-dimensional subrepresentation the number of these points is finite and the hull of these points is a convex polyhedron in $\mathbb{R}_+^{n(n-1)/2}$. Such a polyhedron (not necessarily convex) was called a Newton diagram [43]. In the present context this notion was introduced in [580], where also some examples in the case of functions in one and two variables were given (for $q = 1$), when the figures are planar (polygons). Below, we give explicitly the Newton diagrams for $n = 3$. Moreover, we introduce also infinite Newton diagrams to depict the infinite-dimensional nontrivial subrepresentations.

6.3.3.1 Finite Newton Diagrams for n=3

Fix $r_k \in \mathbb{Z}_+$. Then the Newton diagram is given by the points with integer coordinates j, ℓ, k in $\mathbb{Z}_+^3$ such that:

$$0 \leq j + k + \ell \leq r, \tag{6.138a}$$

$$0 \leq j \leq r_1, \tag{6.138b}$$

$$0 \leq \ell \leq r_2, \tag{6.138c}$$

(cf. below **Figure** 6.1 taken from [246]). The polyhedron formed by these points is planar only for $r_1 = 0$ or $r_2 = 0$ in which case it is a triangle (only (6.138a) is relevant since $r = r_2$ or $r = r_1$). (The case $r_1 = 0$ was given in [580].)

Fix a point j, ℓ, k. This is represented by the state $v_{\ell kj}$. Then, the number of states is:

$$\sum_{j=0}^{r_1}\sum_{k=0}^{r-j}\sum_{\ell=0}^{\min(r-k-j,r_2)} 1 = \sum_{j=0}^{r_1}\sum_{k=0}^{r_1-j}\sum_{\ell=0}^{r_2} 1 + \sum_{j=0}^{r_1}\sum_{k=r_1-j+1}^{r-j}\sum_{\ell=0}^{r-k-j} 1 =$$

$$= \frac{(r_1+1)(r_1+2)(r_2+1)}{2} + \frac{(r_1+1)r_2(r_2+1)}{2} = d_{r_1,r_2}, \tag{6.139}$$

as expected (cf. (6.129)).

Note that such diagrams have an advantage over the usual weight diagrams for $sl(3)$ and $su(3)$ which are degenerate. For instance, consider the adjoint representation obtained for $r_1 = r_2 = 1$. The weight diagram consists of two orbits of the Weyl group, one with six points with multiplicity one, and the other with one point with multiplicity two. To the latter point in our diagram correspond the two linearly independent states:

$$v_{101} = q^{-\frac{1}{4}}([2]_q xy - q^{-1}z), \tag{6.140a}$$

$$v_{010} = q^{-\frac{1}{4}}(z - qxy), \tag{6.140b}$$

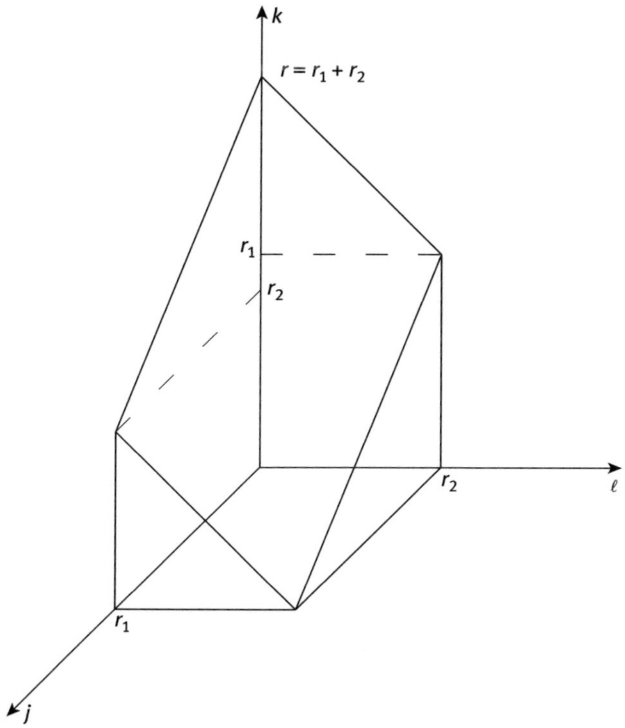

Figure 6.1: Newton diagram for the finite-dimensional representations of $U_q(\mathrm{sl}(3))$

6.3.3.2 Infinite Newton diagrams for n=3

Here either $r_1 \notin \mathbb{Z}_+$ or $r_2 \notin \mathbb{Z}_+$ and the considerations run in parallel with considerations of the polynomial basis. Below $j, \ell, k \in \mathbb{Z}_+$.

1. For $r_1 \in \mathbb{Z}_+$ and $r_2, r+1 \notin \mathbb{Z}_+$ the Newton diagram is given by the points with coordinates:

$$0 \le k, \qquad 0 \le j \le r_1, \qquad 0 \le \ell. \tag{6.141}$$

2. For $r_2 \in \mathbb{Z}_+$ and $r_1, r+1 \notin \mathbb{Z}_+$ the Newton diagram is given by the points with coordinates:

$$0 \le k, \qquad 0 \le j, \qquad 0 \le \ell \le r_2. \tag{6.142}$$

3. For $r+1 \in \mathbb{Z}_+$ and $r_1, r_2 \notin \mathbb{Z}_+$ the Newton diagram is given by the points with coordinates:

$$0 \le k \le r+1, \qquad 0 \le j, \qquad 0 \le \ell. \tag{6.143}$$

4. For $r_1, r+1 \in \mathbb{Z}_+$ and $r_2+1 \in -\mathbb{N}$ the Newton diagram is given by *two* sets of points with coordinates:

$$0 \le k, \qquad -r_2 - 1 \le j \le r_1, \qquad 0 \le \ell. \tag{6.144}$$

$$0 \le k \le r+1, \qquad 0 \le j \le -r_2 - 2, \qquad 0 \le \ell. \tag{6.145}$$

5. For $r_1 = r+1 \in \mathbb{Z}_+$ and $r_2 = -1$ the Newton diagram is given by (6.141). It can be obtained formally from the previous case by setting $r_2 = -1$, then (6.144) coincides with (6.141), while (6.145) is empty.
6. For $r_2, r+1 \in \mathbb{Z}_+$ and $r_1+1 \in -\mathbb{N}$ the Newton diagram is given by *two* sets of points with coordinates:

$$0 \le k, \qquad 0 \le j \qquad -r_1 - 1 \le j \le r_2, . \tag{6.146}$$

$$0 \le k \le r+1, \qquad 0 \le \ell, \qquad 0 \le j \le -r_2 - 2. \tag{6.147}$$

7. For $r_2 = r+1 \in \mathbb{Z}_+$ and $r_1 = -1$ the Newton diagram is given by (6.142). It can be obtained formally from the previous case by setting $r_1 = -1$, then (6.146) coincides with (6.142), while (6.147) is empty.

6.4 Application of the Gelfand–(Weyl)–Zetlin Basis

6.4.1 Correspondence with the GWZ Basis

In this section we follow [230, 244, 245]. We would like to establish the correspondence between our basis for the finite-dimensional irreducible representations given by the states $v_{\ell kj}$ (cf. (6.130)) and the $SU(3)$ Gel'fand–Weyl–Zetlin basis:

$$(\mathbf{m}) = \begin{pmatrix} m_{13} & & m_{23} & & m_{33} \\ & m_{12} & & m_{22} & \\ & & m_{11} & & \end{pmatrix} \tag{6.148}$$

(Note that in the literature this basis is most often called Gel'fand–Zetlin basis, here we keep the usage from [230, 244, 245].) In fact, the above is for $U(3)$, and we shall set $m_{33} = 0$ to restrict to $SU(3)$. Further we need the operators corresponding to isospin $\hat{I}^2$, third component of isospin $\hat{I}_z$, and hypercharge $\hat{Y}$:

$$\begin{aligned} \hat{I}_z &= \tfrac{1}{2}H_1\,, \quad \hat{Y} = \tfrac{1}{3}(H_1 + 2H_2) \\ \hat{I} &= E_{21}E_{12} + [\tfrac{1}{2}H_1]_q[\tfrac{1}{2}H_1 + 1]_q \end{aligned} \tag{6.149}$$

Note that $\hat{I}$ is the Casimir of the $U_q(sl(2))$ quantum subgroup generated by E_{21}, E_{12}, H_1. It is easy to see that, like the GWZ states, also the $v_{\ell kj}$ states are eigenvectors of $\hat{I}_z$ and $\hat{Y}$, but they are not eigenvectors of $\hat{I}$. In fact we have:

$$\begin{aligned} \Gamma_3(\tfrac{1}{2}H_1)\, v_{\ell kj} &= \big(j + k - \tfrac{1}{2}(r_1 + \ell + k)\big) v_{\ell kj} \\ \Gamma_3(\frac{1}{3}(H_1 + 2H_2))\, v_{\ell kj} &= \big(r_1 + k + \ell - \frac{2}{3}(r + r_1)\big)\, v_{\ell kj} \\ \Gamma_3(E_{21})\Gamma_3(E_{12}) v_{\ell kj} &= \big((j+1)(r_1 - j) + \ell(k+1)\big)\, v_{\ell kj} + \\ &\quad + k v_{\ell+1,k-1,j+1} + (r_1 - j + 1)\ell j v_{\ell-1,k+1,j-1} \end{aligned} \tag{6.150}$$

The last formula is given for $q = 1$ since it is only to illustrate our point. We shall diagonalize $\Gamma_3(E_{21})\Gamma_3(E_{12})$ in the next subsection and find explicit polynomial eigenvectors. Here we find alternatively an explicit correspondence between $(\mathbf{m})$ and the appropriate linear combination of $v_{\ell kj}$'s. But first we place the labels ℓ, k, j in a GWZ pattern.

First, we fix the correspondence between the two representations, namely, between the labels $\{m_{13}, m_{23}\}$ and $\{r, r_1\}$, by considering the lowest-weight vector. This is the GWZ vector [65]:

$$\begin{pmatrix} m_{13} & & m_{23} & & 0 \\ & m_{23} & & 0 & \\ & & 0 & & \end{pmatrix} \tag{6.151}$$

which has $I = -I_z = m_{23}/2$ and $Y = -\frac{1}{3}(2m_{13} - m_{23})$. In our realization the lowest-weight vector is $v_{000} = 1$ and thus from (6.150) we get that $I_z = -r_1/2$ and $Y = -\frac{1}{3}(2r - r_1)$. Therefore we find $m_{13} = r$ and $m_{23} = r_1$.

For further use we record explicitly the patterns corresponding to the highest-weight state (h.w.s.) and to the lowest-weight state (l.w.s.):

$$(\text{h.w.s.}) = \begin{pmatrix} r & & r_1 & & 0 \\ & r & & r_1 & \\ & & r & & \end{pmatrix} \tag{6.152a}$$

$$(\text{l.w.s.}) = \begin{pmatrix} r & & r_1 & & 0 \\ & r_1 & & 0 & \\ & & 0 & & \end{pmatrix} \tag{6.152b}$$

Remark 6.2. Notice that the well-known conjugation of representation ($[m_{13}, m_{23}, 0] \to [m_{13}, m_{13} - m_{23}, 0]$) [65] corresponds to the exchange of r_1 with r_2 and that the dimension of the representation $[m_{13}, m_{23}, 0]$, namely, $\frac{1}{2}(m_{13}+2)(m_{23}+1)(m_{13}-m_{23}+1)$, matches (6.129). ◊

To place the $v_{\ell,k,j}$ states in a GWZ pattern we split them as in [246] (cf. (66)) in two subsets depending whether $j+k \le r_1$ or $j+k > r_1$. In the first case the correspondence is given by:

$$v_{\ell,k,j} = \begin{pmatrix} r & & r_1 & & 0 \\ & r_1+\ell & & k & \\ & & j+k & & \end{pmatrix}, \qquad j+k \le r_1. \tag{6.153}$$

In the second case the correspondence is given by:

$$v_{\ell,k,j} = \begin{pmatrix} r & & r_1 & & 0 \\ & j+k+\ell & & r_1-j & \\ & & j+k & & \end{pmatrix}, \qquad j+k > r_1, \tag{6.154}$$

which is valid also for the boundary case $j+k = r_1$, when it coincides with (6.153). The betweenness constraint

$$m_{ij} \ge m_{i,j-1} \ge m_{i+1,j}, \tag{6.155}$$

typical of the GWZ pattern then gives the constraints $0 \le j \le r_1$, $0 \le \ell \le r_2$ and $0 \le j+k+\ell \le r$ found above for the finite-dimensional representations.

The actual correspondence is proved using well-known techniques of raising and lowering operators developed for classical groups and adapted to quantum groups (cf. [65, 100, 101, 516, 582]). Identifying the lowest-weight states one can find explicitly a polynomial $p_{(\mathbf{m})}$ in $U_q(\mathcal{G}^+)$ which corresponds to $(\mathbf{m})$.

Let us denote by $\hat{1}$ the lowest-weight state of any realization of the $U_q(sl(3))$ finite-dimensional representation with parameters r_1, r_2 . Then we have (up to multiplicative normalization constant) [516, 582]:

$$p_{(\mathbf{m})}\,\hat{1} = (E_{21})^{m_{12}-m_{11}}\,\tilde{C}^{r-m_{12}}\,(E_{32})^{r_1-m_{22}}\,(E_{13})^{r}\,\hat{1} = \tag{6.156a}$$

$$\begin{aligned} &= [m_{12}+m_{22}-r_1]_q! \sum_{t=0}^{r-m_{12}} \sum_{u\in\mathbb{Z}_+} (-1)^t \binom{m_{12}-m_{11}+t}{u}_q \times \\ &\times \frac{q^{\frac{1}{2}(m_{22}-t-r_1)(m_{12}+m_{22}-r_1)+\frac{u}{2}(u-2m_{22}-m_{12}+m_{11}+r_1+t)}}{[t]_q!\,[m_{22}-t]_q!\,[m_{12}-m_{22}+1+t]_q!} \times \\ &\times \frac{[m_{12}-m_{22}+1]_q!\,[t+r_1-m_{22}]_q!}{[m_{11}-m_{12}-m_{22}+r_1+u]_q!} \times \\ &\times \frac{[m_{12}+m_{22}-m_{11}-u]_q!}{[m_{12}+m_{22}-r_1-u]_q!} \times \\ &\times (E_{23})^u\,(E_{13})^{m_{12}+m_{22}-r_1-u}\,(E_{12})^{m_{11}-m_{12}-m_{22}+r_1+u}\,\hat{1}\,, \end{aligned} \tag{6.156b}$$

$$\tilde{C} \equiv E_{31}\,[H_1+1] + E_{21}\,E_{32}\,q^{-\frac{H_1+1}{2}} = \\ = E_{32}\,E_{21}\,[H_1+1] - E_{21}\,E_{32}\,[H_1] \tag{6.156c}$$

Remark 6.3. We would like to stress the peculiarity of (6.156a). One gets (in (6.156b)) a correspondence of the GWZ states with polynomials in $U(\mathcal{G}^+)$ but formula (6.156a) first gives us a one-to-one correspondence of the GWZ states with *monomials* in the q-deformed enveloping algebra $U_q(\mathcal{G}^-)$ of the *lowering* generators. Note that the latter monomials are not in the standard Poincaré–Birkhoff–Witt basis of $U_q(\mathcal{G}^-)$, namely, instead of the generator E_{31} one has the generator of the same weight $\hat{\mathcal{C}}$ (cf. formula (6.3) of [516]). These monomials produce the polynomials of $U(\mathcal{G}^+)$ since they act on $(E^{13})^r$ which is in $U(\mathcal{G}^+)$ and $(E^{13})^r\ \hat{1}$ is the highest-weight vector. Finally, we note that there exists a similar description of this correspondence only in terms of raising generators, in particular, involving an analogue of $\hat{\mathcal{C}}$ in $U(\mathcal{G}^+)$. However, the present description is simpler for our purposes here, while the other is used in Section 6.4.7, where it is more useful. ◊

Finally, we get the correspondence we need using (6.156):

Theorem 6.1. *A realization of the GWZ basis as polynomials in three variables (real or complex) is given by the formula:*

$$\begin{aligned}\phi_{(\mathbf{m})} &= \Gamma_3(p_{(\mathbf{m})})1 = [m_{12}+m_{22}-r_1]_q! \times \\ &\times \sum_{t=0}^{r-m_{12}} \sum_{u\in\mathbb{Z}_+} \binom{m_{12}-m_{11}+t}{u}_q \times \\ &\times (-1)^t\, \frac{q^{\frac{1}{2}(m_{22}-t-r_1)(m_{12}+m_{22}-r_1)+\frac{u}{2}(u-2m_{22}-m_{12}+m_{11}+r_1+t)}}{[t]_q!\,[m_{22}-t]_q!\,[m_{12}-m_{22}+1+t]_q!} \times \\ &\times \frac{[m_{12}-m_{22}+1]_q!\,[t+r_1-m_{22}]_q!\,[m_{12}+m_{22}-m_{11}-u]_q!}{[m_{11}-m_{12}-m_{22}+r_1+u]_q!\,[m_{12}+m_{22}-r_1-u]_q!} \times \\ &\times v_{u,m_{12}+m_{22}-r_1-u,m_{11}-m_{12}-m_{22}+r_1+u}\end{aligned} \tag{6.157}$$

Proof. Straightforward using (6.156) and our formula for $v_{\ell kj}$ (6.130). ◊

For later reference we note the explicit value of $\phi_{(\mathbf{m})}$ for $z=0$ (using (6.132)):

$$\phi_{(\mathbf{m})}|_{z=0} = \frac{\mathcal{N}^+_{(\mathbf{m})}}{\Gamma_q(r_1-m_{11}+1)} x^{m_{11}} y^{m_{12}+m_{22}-r_1}\,, \tag{6.158a}$$

$$\mathcal{N}^+_{(\mathbf{m})} = [r_1]_q!\,[m_{12}+m_{22}-r_1]_q! \sum_{t=0}^{r-m_{12}} \sum_{u\in\mathbb{Z}_+} (-1)^{t+u+m_{12}+m_{22}+r_1} \times$$
$$\times \binom{m_{12}-m_{11}+t}{u}_q (r+m_{11}-m_{12}-m_{22}+1)^q_u \tag{6.158b}$$
$$\times \frac{q^{\frac{1}{2}\{(u-m_{12}-m_{22}+r_1)(m_{12}-r-1+t)+(m_{12}+m_{22}-r_1)(m_{11}/2-r_1)\}}}{[t]_q!\,[m_{22}-t]_q!\,[m_{12}-m_{22}+1+t]_q!} \times$$
$$\times \frac{[m_{12}-m_{22}+1]_q!\,[t+r_1-m_{22}]_q!\,[m_{12}+m_{22}-m_{11}-u]_q!}{[m_{11}-m_{12}-m_{22}+r_1+u]_q!\,[m_{12}+m_{22}-r_1-u]_q!}$$

which is useful for $r_1 - m_{11} + 1 > 0$. Otherwise it is zero (due to the singled out factor $\Gamma_q(r_1-m_{11}+1)$), and to obtain a nonzero value one first has to differentiate $m_{11}-r$ times (6.157) w.r.t. z.

We note also the expression for the lowest-weight state obtained from (6.156) and (6.157) for $m_{12} = r_1$, $m_{11} = m_{22} = 0$:

$$p_{\text{lws}}\,\hat{1} = (E_{21})^{r_1}\,\tilde{C}^{r-r_1}\,(E_{32})^{r_1}\,(E_{13})^{r}\,\hat{1} =$$
$$= \mathcal{N}^+_{\text{lws}}\,\hat{1} = ([r_1]_q!)^3\,\hat{1} \tag{6.159a}$$
$$\phi_{\text{lws}} = \Gamma_3(p_{\text{lws}})1 = ([r_1]_q!)^3 \tag{6.159b}$$

which of course differ from $\hat{1}$, 1, respectively, by a constant – the corresponding value of $\mathcal{N}^+_{(\mathbf{m})}$.

6.4.2 q-Hypergeometric Realization of the GWZ Basis

In the previous section we exhibited the relation of the GWZ basis and the polynomial basis $v_{\ell kj}$. By formula (6.157) this provides also a polynomial realization of the GWZ basis in the same variables x, y, z. However, (6.157) is not very explicit, since it contains a quadruple sum (a double sum in (6.157) and a double sum in (6.130)). Instead of partially summing (6.157), in this section we shall find a polynomial realization directly (not relying on the correspondence with $v_{\ell kj}$) using the fact that the GWZ states are eigenvectors of the operators $\hat{I}_z, \hat{Y}, \hat{I}$.

We shall proceed as follows. Let us denote (as in (6.157)) the unknown polynomial function corresponding to $(\mathbf{m})$ by:

$$\psi = \psi_{(\mathbf{m})}(x, y, z) \tag{6.160}$$

Naturally, $\psi_{(\mathbf{m})}$ can differ from $\phi_{(\mathbf{m})}$ in (6.157) only by a multiplicative constant which we shall fix later.

In order to use effectively the fact that ψ is an eigenfunction of $\hat{I}_z, \hat{Y}, \hat{I}$ we use their explicit q-difference realization (6.112). We write:

$$\tilde{I}_z \equiv \tfrac{1}{2}\Gamma_3(H_1) = \tfrac{1}{2}\left(2N_x - r_1 + N_z - N_y\right) \tag{6.161a}$$

$$\tilde{Y} \equiv \tfrac{1}{3}\Gamma_3\left(H_1 + 2H_2\right) = N_y + N_z - \tfrac{1}{3}(r_1 + 2r_2) \tag{6.161b}$$

$$\begin{aligned}\tilde{I}^2 \equiv\ & \Gamma_3(E_{21})\Gamma_3(E_{12}) + [\tilde{I}_z]_q[\tilde{I}_z + 1]_q = \\ =\ & [N_x + 1]_q[r_1 - N_x]_q q^{\frac{1}{2}(N_z - N_y)} + [N_z + 1]_q[N_y]_q q^{\frac{1}{2}(r_1 - 2N_x)} + \\ & + \frac{z}{xy}[N_x]_q[N_y]_q q^{\frac{1}{4}(r_1 - 2N_x + N_z - N_y + 2)} + \\ & + \frac{xy}{z}[N_z]_q[r_1 - N_x]_q q^{\frac{1}{4}(r_1 - 2N_x + N_z - N_y - 2)} + [\tilde{I}_z]_q[\tilde{I}_z + 1]_q\end{aligned} \tag{6.161c}$$

The eigenfunction conditions satisfied by ψ are:

$$\tilde{I}_z\,\psi = I_z\,\psi = \left(m_{11} - \tfrac{1}{2}(m_{12} + m_{22})\right)\psi \tag{6.162a}$$

$$\tilde{Y}\,\psi = Y\,\psi = \left(m_{12} + m_{22} - \tfrac{2}{3}(r + r_1)\right)\psi \tag{6.162b}$$

$$\begin{aligned}\tilde{I}^2\,\psi &= [I]_q\,[I + 1]_q\,\psi = \\ &= \left[\frac{m_{12} - m_{22}}{2}\right]_q \left[\frac{m_{12} - m_{22}}{2} + 1\right]_q \psi\end{aligned} \tag{6.162c}$$

Next we consider the operators $\tilde{I}_z + \frac{1}{2}\tilde{Y}$, $\tilde{Y}$, from which we obtain the following homogeneity conditions:

$$(N_x + N_z)\,\psi = \left(\tilde{I}_z + \tfrac{1}{2}\tilde{Y} + \tfrac{1}{3}(r + r_1)\right)\psi = m_{11}\,\psi \tag{6.163a}$$

$$\begin{aligned}\left(N_y + N_z\right)\psi &= \left(\tilde{Y} + \tfrac{1}{3}(r - r_1)\right) = \kappa\,\psi\,, \\ \kappa &\equiv m_{12} + m_{22} - r_1\end{aligned} \tag{6.163b}$$

From these homogeneity conditions and the explicit form of (6.162c) we are prompted to make the following change of variables:

$$x' = x, \qquad y' = y, \qquad \zeta = \frac{z}{xy} \tag{6.164}$$

from which follows:

$$N_x = N_{x'} - N_\zeta, \qquad N_y = N_{y'} - N_\zeta, \qquad N_z = N_\zeta \tag{6.165}$$

Thus, the homogeneity conditions (6.163) simplify to:

$$N_{x'}\psi = m_{11}\psi\,, \quad N_{y'}\psi = \kappa\psi\,, \tag{6.166}$$

that is, our polynomials actually have the form:

$$\psi = \psi_{(\mathbf{m})} = x'^{m_{11}} y'^{\kappa} \tilde{\psi}(\zeta) \tag{6.167}$$

Actually from this expression we can deduce that $\tilde{\psi}$ is a polynomial in ζ of degree at most $n_0 \equiv \min(m_{11}, \kappa)$. Indeed, if $\tilde{\psi}$ is a polynomial in ζ of higher degree, then ψ would not be a polynomial in x or y or both, contradicting our starting assumption.

Substituting now (6.167) in (6.162c) and taking into account (6.166) we obtain the following equation for $\tilde{\psi}$:

$$\begin{aligned}\Bigg(& [m_{11} + 1 - N_\zeta]_q [r_1 - m_{11} + N_\zeta]_q q^{N_\zeta - \frac{1}{2}\kappa} + \\ & + [1 + N_\zeta]_q [\kappa - N_\zeta]_q q^{\frac{1}{2} r_1 - m_{11} + N_\zeta} + \\ & + \zeta\, [m_{11} - N_\zeta]_q [\kappa - N_\zeta]_q q^{\frac{1}{4}(r_1 - \kappa) + \frac{1}{2}(1 - m_{11}) + N_\zeta} + \\ & + \zeta^{-1}\, [N_\zeta]_q [r_1 - m_{11} + N_\zeta]_q q^{\frac{1}{4}(r_1 - \kappa) - \frac{1}{2}(1 + m_{11}) + N_\zeta} + \\ & + [m_{11} - m_{12}]_q [m_{11} - m_{22} + 1]_q \Bigg) \tilde{\psi}(\zeta) = 0\end{aligned} \tag{6.168}$$

The unique (up to nonzero multiple) polynomial solution of (6.168) is given by q-Jacobi or, equivalently, by q-hypergeometric polynomials. In particular, if $\beta = r_1 - m_{11} + 1 \notin \mathbb{Z}_-$ then such a solution is:

$$\begin{aligned}\tilde{\psi}_1(\zeta) = {}_1F_0^q(-m_{22};\ q^{\frac{1}{4}(m_{22} - m_{12} - 2)} \zeta\,) \times \\ \times {}_2F_1^q(m_{22} - m_{11},\ r_1 - m_{12};\ r_1 - m_{11} + 1;\ q^{\frac{1}{4}(r_1 + \kappa)} \zeta\,)\end{aligned} \tag{6.169}$$

where ${}_2F_1^q$ is a q-hypergeometric polynomial:

$${}_2F_1^q(a, b; c; \zeta) = \sum_{s \in \mathbb{Z}_+} \frac{(a)_s^q\, (b)_s^q}{[s]_q!\, (c)_s^q} \zeta^s, \qquad c \notin \mathbb{Z}_- \tag{6.170}$$

${}_1F_0^q$ is a degenerate q-hypergeometric polynomial:

$${}_1F_0^q(a; \zeta) = \sum_{s \in \mathbb{Z}_+} \frac{(a)_s^q}{[s]_q!} \zeta^s = {}_2F_1^q(a, b; b; \zeta) \tag{6.171}$$

and $(v)^q_s$ is the *q-Pochhammer symbol*:

$$(v)^q_s \doteq [v+s-1]_q[v+s-2]_q \dots [v]_q = \frac{\Gamma_q(v+s)}{\Gamma_q(v)} \tag{6.172}$$

Note that (6.172) ensures that (6.170) and (6.171) are polynomials of degree min $(-a,-b)$, $-a$, respectively, when $a, b \in \mathbb{Z}_-$, as is in our case. Note that for $q = 1$ (6.170) goes into the standard hypergeometric polynomial, while (6.171) becomes just the binomial $(1-\zeta)^{m_{22}}$.

If $\beta = r_1 - m_{11} + 1 \in \mathbb{Z}_-$ then the polynomial solution of (6.168) is given by:

$$\begin{aligned}\tilde{\psi}_2(\zeta) = {} & \zeta_1^{m_{11}-r_1} F^q_0(-m_{22};\ q^{\frac{1}{4}(m_{22}-m_{12}-2)}\zeta\,) \times \\ & \times {}_2F^q_1(m_{22}-r_1,\ m_{11}-m_{12};\ m_{11}-r_1+1;\ q^{\frac{1}{4}(r_1+\kappa)}\zeta\end{aligned} \tag{6.173}$$

In order to relate (6.169) and (6.173) it is enough to replace in (6.169)

$$\begin{aligned}& {}_2F^q_1(m_{22}-m_{11},\ r_1-m_{12};\ r_1-m_{11}+1;\ q^{\frac{1}{4}(r_1+\kappa)}\zeta\,) \mapsto \\ & \mapsto \frac{1}{\Gamma_q(r_1-m_{11}+1)}{}_2F^q_1(m_{22}-m_{11},\ r_1-m_{12};\ r_1-m_{11}+1;\ q^{\frac{1}{4}(r_1+\kappa)}\zeta\,)\end{aligned} \tag{6.174}$$

Then this expression is valid for arbitrary $r_1 - m_{11} + 1$, and up to some multiplicative constant is equal to (6.173) when $r_1 - m_{11} + 1 \in \mathbb{Z}_-$. Thus, finally we shall write the polynomial solution of (6.168) as:

$$\begin{aligned}\tilde{\psi}(\zeta) = {} & \frac{1}{\Gamma_q(r_1-m_{11}+1)}{}_1F^q_0(-m_{22};\ q^{\frac{1}{4}(m_{22}-m_{12}-2)}\zeta\,) \times \\ & \times {}_2F^q_1(m_{22}-m_{11},\ r_1-m_{12};\ r_1-m_{11}+1;\ q^{\frac{1}{4}(r_1+\kappa)}\zeta\,)\end{aligned} \tag{6.175}$$

For the lowest-weight state ($m_{12} = r_1$, $m_{11} = m_{22} = 0$) we get:

$$\tilde{\psi}_{\text{lws}} = \frac{1}{[r_1]_q!} \tag{6.176}$$

For the highest-weight state ($m_{12} = m_{11} = r$, $m_{22} = r_1$) we get:

$$\begin{aligned}\tilde{\psi}_{\text{hws}}(\zeta) &= q^{\frac{1}{4}(r^2-r_1^2)}\,[r-r_1]_q!\,\zeta_1^{r-r_1} F^q_0(-r_1;\ q^{\frac{1}{4}(r_1-r-2)}\zeta\,) \\ \psi_{\text{hws}}(x,y,z) &= q^{\frac{1}{4}(r^2-r_1^2)}\,[r-r_1]_q!\,x^{r_1}\,y^{r_1}\,z^{r-r_1} \times \\ & \quad \times {}_1F^q_0(-r_1;\ q^{\frac{1}{4}(r_1-r-2)}\frac{z}{xy}\,)\end{aligned} \tag{6.177}$$

since from ${}_2F^q_1$ survives only the term $\zeta^{r-r_1} = \zeta^{r_2}$.

We shall write down the relation between the expressions (6.157) and (6.167) (with (6.175)) as:

$$\phi_{(\mathbf{m})} = \mathcal{N}_{(\mathbf{m})}\, \psi_{(\mathbf{m})} \tag{6.178}$$

For $r_1 - m_{11} + 1 \notin \mathbb{Z}_-$ we have $\mathcal{N}_{(\mathbf{m})} = \mathcal{N}^+_{(\mathbf{m})}$ which we find by comparing

$$\psi_{(\mathbf{m})}|_{z=0} = \frac{1}{\Gamma_q(r_1 - m_{11} + 1)} x^{m_{11}} y^{m_{12}+m_{22}-r_1}$$

with (6.158). Note that (6.159) is a partial case of (6.178). When $r_1 - m_{11} + 1 \in \mathbb{Z}_-$ (i. e., $m_{11} - r_1 \in \mathbb{N}$), one has first to differentiate $m_{11} - r_1$ times w.r.t. z both $\phi_{(\mathbf{m})}$ and $\psi_{(\mathbf{m})}$ and then to set $z = 0$. In particular, for the highest-weight state we compare (6.131) and (6.177), since then $\phi_{\text{hws}} = v_{0r0} = \Gamma_3(E_{13})^r 1$. Rewriting (6.131) as:

$$\phi_{\text{hws}} = (-1)^{r_1}\, q^{\frac{1}{4}r_1(2-r_1)}\, [r]_q! x^{r_1}\, y^{r_1}\, z_1^{r-r_1} F_0^q\left(-r_1;\, q^{\frac{1}{4}(r_1-r-2)}\frac{z}{xy}\right)$$

we get:

$$\phi_{\text{hws}} = (-1)^{r_1}\, q^{\frac{1}{4}(2r_1-r^2)}\, \frac{[r_1]_q!}{[r-r_1]_q!}\psi_{\text{hws}} \tag{6.179}$$

6.4.3 Explicit Orthogonality of the GWZ Basis

For the orthogonality of the GWZ basis we shall use an adaptation of the so called Shapovalov form [550]. This is a bilinear $\mathbb{C}$-valued form on Verma modules. The Verma module V^Λ of lowest-weight $\Lambda \in \mathcal{H}^*$ is the lowest-weight module such that $V^\Lambda = U_q(\mathcal{G}^+) \otimes v_0$, where $\mathcal{G}^+$ is the subalgebra of the raising generators E_{jk}, $j < k$, and v_0 is the lowest vector such that:

$$E_{jk}\, v_0 = 0, \quad j > k, \qquad H_k\, v_0 = \Lambda(H_k)\, v_0 \tag{6.180}$$

The states in a Verma module correspond to the monomials of the Poincaré–Birkhoff–Witt basis of $U_q(\mathcal{G}^+)$, namely:

$$\begin{aligned} u_{\ell kj} &\equiv p_{\ell kj} \otimes v_0, \\ p_{\ell kj} &\equiv (E_{23})^\ell (E_{13})^k (E_{12})^j, \qquad \ell, k, j \in \mathbb{Z}_+ \end{aligned} \tag{6.181}$$

that is, this basis is one-to-one with the basis $v_{\ell kj}$ for general r_k. Further, for simplicity se shall omit the sign $\otimes$; that is, we shall write: $u_{\ell kj} = p_{\ell kj}\, v_0$ or $u = p\, v_0$ for short. We need the involutive antiautomorphism of $U_q(\mathcal{G})$ such that:

$$\omega(H_k) = H_k, \qquad \omega(E_{jk}) = E_{kj}, \qquad \omega(q) = q^{-1} \tag{6.182}$$

Using the above conjugation the Shapovalov form can be defined as follows:

$$\begin{aligned}&(u\,,\,u')=(p\,v_0\,,\,p'\,v_0)\equiv(v_0\,,\,\omega(p)\,p'\,v_0)=\\&=(\omega(p')\,p\,v_0\,,\,v_0)\,,\\&u=p\,v_0\,,u'=p'\,v_0\,,\qquad p,p'\in U_q(\mathcal{G}^+),u,u'\in V^\Lambda\end{aligned}\tag{6.183}$$

supplemented by the normalization condition $(v_0, v_0) = 1$. More explicitly from (6.183) we have:

$$\begin{aligned}&\left(u_{\ell kj}\,,\,u_{\ell'k'j'}\right)=\left(p_{\ell kj}\,v_0\,,\,p_{\ell'k'j'}\,v_0\right)=\\&=\left(v_0\,,\,\omega(p_{\ell kj})\,p_{\ell'k'j'}\,v_0\right)=\left(\omega(p_{\ell'k'j'})\,p_{\ell kj}v_0\,,\,v_0\right)=\\&=\left(v_0\,,\,(E_{21})^j\,(E_{31})^k\,(E_{32})^\ell\,(E_{23})^{\ell'}\,(E_{13})^{k'}\,(E_{12})^{j'}\,v_0\right)\\&=\left((E_{21})^{j'}\,(E_{31})^{k'}\,(E_{32})^{\ell'}\,(E_{23})^{\ell}\,(E_{13})^{k}\,(E_{12})^{j}\,v_0\,,\,v_0\right)\end{aligned}\tag{6.184}$$

Note that subspaces with different weights are orthogonal w.r.t. to this form:

$$\left(u_{\ell kj},u_{\ell'k'j'}\right)\sim\;\delta_{k+\ell,k'+\ell'}\,\delta_{k+j,k'+j'}\tag{6.185}$$

To show (6.185) one uses (6.184b) when $k+\ell > k'+\ell'$ and/or $k+j > k'+j'$, while (6.184c) is used when $k+\ell < k'+\ell'$ and/or $k+j < k'+j'$.

We shall give a realization of the Shapovalov form in our setting in the following way. Using the one-to-one correspondence we replace $u_{\ell kj}$ by $v_{\ell kj}$ and the lowest-weight vector v_0 by the lowest-weight vector $\hat{1}$ of the abstract finite-dimensional irrep and by the function 1 in the polynomial realization. Namely, we shall use instead of (6.183) the following bilinear form:

$$(u\,,\,u')_f=(p\,\hat{1}\,,\,p'\,\hat{1})_f\equiv\left(\Gamma_3(\omega(p))\,\Gamma_3(p')\,1\right)|_{x=y=z=0}\tag{6.186}$$

More explicitly, we have:

$$\begin{aligned}&\left(u_{\ell kj}\,,\,u_{\ell'k'j'}\right)_f=\left(p_{\ell kj}\,\hat{1}\,,\,p_{\ell'k'j'}\,\hat{1}\right)_f=\\&=\left(\Gamma_3(\omega(p_{\ell kj}))\,\Gamma_3(p_{\ell'k'j'})\,1\right)|_{x=y=z=0}=\left(\hat{p}_{\ell kj}\,v_{\ell'k'j'}\right)|_{x=y=z=0}\,,\\&\hat{p}_{\ell kj}\equiv\Gamma_3(\omega(p_{\ell kj}))=(\Gamma_3(E_{21}))^j\,(\Gamma_3(E_{31}))^k\,(\Gamma_3(E_{32}))^\ell\end{aligned}\tag{6.187}$$

Clearly, when $k+\ell > k'+\ell'$ and/or $k+j > k'+j'$ we have $\hat{p}_{\ell kj}\,v_{\ell'k'j'} = 0$. When $k+\ell < k'+\ell'$ and $k+j < k'+j'$ the expression $\hat{p}_{\ell kj}\,v_{\ell'k'j'}$ is not zero but a homogeneous polynomial of x, y, z which vanishes after the substitution $x = y = z = 0$. Finally, when $k+\ell = k'+\ell'$ and $k+j = k'+j'$ the expression $\hat{p}_{\ell kj}\,v_{\ell'k'j'}$ is a numerical one coinciding with $\left(u_{\ell kj}\,,\,u_{\ell'k'j'}\right)$ because of the automorphism.

We can further simplify (6.187) if we set $x = y = z = 0$ in $\hat{p}_{\ell kj}$ from the very beginning, namely, we replace $\hat{p}_{\ell kj}$ by:

$$\begin{aligned}
\tilde{p}_{\ell kj} &\equiv (\tilde{\Gamma}_3(E_{21}))^j\,(\tilde{\Gamma}_3(E_{31}))^k\,(\tilde{\Gamma}_3(E_{32}))^\ell \\
\tilde{\Gamma}_3(E_{21}) &\equiv D_x q^{\frac{1}{4}(N_z-N_y)} \\
\tilde{\Gamma}_3(E_{32}) &\equiv D_y q^{\frac{1}{4}N_x} = \Gamma_3(E_{32}) \\
\tilde{\Gamma}_3(E_{31}) &\equiv D_z q^{\frac{1}{4}(r_1-N_x+2N_y)} = \Gamma_3(E_{31})
\end{aligned} \tag{6.188}$$

Note that this operation affects only $\Gamma_3(E_{21})$ and it is easy to check that

$$\left(u_{\ell kj}\,,\,u_{\ell' k' j'}\right)_f \equiv \left(\tilde{p}_{\ell kj}\,v_{\ell' k' j'}\right)|_{x=y=z=0} \tag{6.189}$$

Further we note that:

$$\tilde{p}_{\ell kj} = q^{\frac{1}{4}\left((\ell-k)N_x+(2k-j)N_y+jN_z+j\ell+k(r_1-j)\right)}\,(D_x)^j\,(D_z)^k\,(D_y)^\ell \tag{6.190}$$

We shall use the above to prove the main result in this section:

Theorem 6.2. *Let* $(\mathbf{m})$ *and* $(\mathbf{m}')$ *be two GWZ patterns. Then we have:*

$$\begin{aligned}
(\phi_{(\mathbf{m})},\phi_{(\mathbf{m}'))})_p &= (p_{(\mathbf{m})}\,\hat{1}, p_{((\mathbf{m}'))}\,\hat{1})_f = \\
&= \delta_{m_{11},m'_{11}}\,\delta_{m_{12},m'_{12}}\,\delta_{m_{22},m'_{22}}\,(-1)^{r_1}\,q^{\frac{1}{4}(m_{11}\kappa-2r_1)} \times \\
&\times \frac{[r]_q!\,[r-m_{12}]_q!\,[r-m_{22}+1]_q!}{[m_{12}-m_{22}+1]_q}\,\mathcal{N}_{(\bar{m})}
\end{aligned} \tag{6.191}$$

The proof is given in [244]. The appearance of the constant $\mathcal{N}_{(\bar{m})}$ in (6.191) is due to the fact that in the derivation $\phi_{(\bar{m}')}$ was substituted with $\mathcal{N}_{(\bar{m})}\,\psi_{(\bar{m}')}$. ◊

We can use the form (6.184) and (6.187) to define a scalar product if we consider our conjugation ω as antilinear. Then we actually restrict to the real form $U_q(su(3))$ and q is restricted to be a phase $|q| = 1$ (cf. (6.182)). Then we define the scalar product of the functions $\phi_{(\mathbf{m})} = \Gamma_3(p_{(\mathbf{m})})\,1$ or $\psi_{(\mathbf{m})}$

$$\left(\phi_{(\mathbf{m})},\phi_{(\mathbf{m}')}\right)_p \equiv \left(p_{(\mathbf{m})}\,\hat{1}, p_{(\mathbf{m}')}\,\hat{1}\right)_f \tag{6.192a}$$

$$\left(\psi_{(\mathbf{m})},\psi_{(\mathbf{m}')}\right) \equiv \frac{1}{|\mathcal{N}_{(\mathbf{m})}|^2}\,\left(p_{(\mathbf{m})}\,\hat{1}, p_{(\mathbf{m}')}\,\hat{1}\right)_f \tag{6.192b}$$

We note two partial cases:

$$(\phi_{\rm lws}, \phi_{\rm lws})_p = ([r_1]_q!)^4$$

$$(\psi_{\rm lws}, \psi_{\rm lws}) = \frac{1}{([r_1]_q!)^2} \tag{6.193a}$$

$$(\phi_{\rm hws}, \phi_{\rm hws})_p = [r]_q!\,[r_1]_q!$$

$$(\psi_{\rm hws}, \psi_{\rm hws}) = \frac{[r]_q!\,([r-r_1]_q!)^2}{[r_1]_q!} \tag{6.193b}$$

Using this scalar product we can introduce orthonormal GWZ polynomials by:

$$\hat{\varphi}_{(\mathbf{m})} \equiv \frac{\psi_{(\mathbf{m})}}{|\,(\psi_{(\mathbf{m})}, \psi_{(\mathbf{m})})\,|^{1/2}} \tag{6.194}$$

so that

$$\left(\hat{\varphi}_{(\mathbf{m})}, \hat{\varphi}_{(\mathbf{m}')}\right) = \delta_{(\mathbf{m}),(\mathbf{m}')} \tag{6.195}$$

In particular, we have:

$$\hat{\varphi}_{\rm lws}(x,y,z) = [r_1]_q!\,, \tag{6.196a}$$

$$\begin{aligned}\hat{\varphi}_{\rm hws}(x,y,z) &= \frac{1}{[r-r_1]_q!}\left(\frac{[r_1]_q!}{[r]_q!}\right)^{\frac{1}{2}} \psi_{\rm hws}(x,y,z) = \\ &= \left(\frac{[r]_q!}{[r_1]_q!}\right)^{\frac{1}{2}} q^{\frac{1}{4}(r^2-r_1^2)} x^{r_1}\, y^{r_1}\, z^{r-r_1} \times \\ &\quad \times {}_1F_0^q\left(-r_1;\, q^{\frac{1}{4}(r_1-r-2)}\frac{z}{xy}\right)\end{aligned} \tag{6.196b}$$

6.4.4 Normalized GWZ basis

6.4.4.1 Action on the Unnormalized GWZ Bases and Relations between Them

Our aim now is to obtain normalized GWZ basis. To achieve this first we consider the action of the Chevalley generators $X_j^{\pm}$, $j = 1, 2$, of $U_q(sl(3))$ on the two realizations of the unnormalized GWZ basis introduced in above . After deriving the action we shall use it in order to find the proportionality constant between the two realizations.

First we consider the "operatorial" GWZ basis. We recall formulae (6.156). We rewrite (6.156b) differing by by an overall multiplicative constant:

$$\begin{aligned}\tilde{\phi}_{(\mathbf{m})} &= (-1)^{r_1-m_{22}}\, q^{\frac{1}{2}(m_{22}-r_1)(m_{22}-1)}\, [r]_q![r-m_{22}+1]_q![r-m_{12}]_q! \times \\ &\times \sum_{t=0}^{r-m_{12}} \sum_{u\in\mathbb{Z}_+} (-1)^t \binom{m_{12}-m_{11}+t}{u}_q \times \\ &\times \frac{q^{\frac{t}{2}(r_1-m_{12}-m_{22})+\frac{u}{2}(u-2m_{22}-m_{12}+m_{11}+r_1+t)}}{[t]_q!\,[m_{22}-t]_q!\,[m_{12}-m_{22}+1+t]_q!} \times \\ &\times \frac{[t+r_1-m_{22}]_q!\,[m_{12}+m_{22}-m_{11}-u]_q!}{[m_{11}-m_{12}-m_{22}+r_1+u]_q!\,[m_{12}+m_{22}-r_1-u]_q!} \times \\ &\times (E_{23})^u\,(E_{13})^{m_{12}+m_{22}-r_1-u}\,(E_{12})^{m_{11}-m_{12}-m_{22}+r_1+u}\,\hat{1}\end{aligned} \tag{6.197}$$

The first result here is the explicit calculation of the action of the Chevalley generators on $\tilde{\phi}_{(\mathbf{m})}$ which we denote also by the variable numbers of the GWZ pattern:

$$\tilde{\phi}_{(m_{12},m_{11},m_{22})} \equiv \tilde{\phi}_{(\mathbf{m})}\,. \tag{6.198}$$

We have:

$$X_1^+\tilde{\phi}_{(\mathbf{m})} = [m_{12}-m_{11}]_q[m_{11}-m_{22}+1]_q\,\tilde{\phi}_{(m_{12},m_{11}+1,m_{22})} \tag{6.199a}$$

$$X_1^-\tilde{\phi}_{(\mathbf{m})} = \tilde{\phi}_{(m_{12},m_{11}-1,m_{22})} \tag{6.199b}$$

$$\begin{aligned}X_2^+\tilde{\phi}_{(\mathbf{m})} &= \frac{[r-m_{12}]_q[m_{12}-r_1+1]_q[m_{12}+2]_q}{[m_{12}-m_{22}+1]_q}\,\tilde{\phi}_{(m_{12}+1,m_{11},m_{22})} + \\ &+ \frac{[r-m_{22}+1]_q[r_1-m_{22}]_q[m_{22}+1]_q}{[m_{12}-m_{22}+1]_q}\,\tilde{\phi}_{(m_{12},m_{11},m_{22}+1)}\end{aligned} \tag{6.199c}$$

$$\begin{aligned}X_2^-\tilde{\phi}_{(\mathbf{m})} &= \frac{[m_{12}-m_{11}]_q}{[m_{12}-m_{22}+1]_q}\,\tilde{\phi}_{(m_{12}-1,m_{11},m_{22})} + \\ &+ \frac{[m_{11}-m_{22}+1]_q}{[m_{12}-m_{22}+1]_q}\,\tilde{\phi}_{(m_{12},m_{11},m_{22}-1)}\end{aligned} \tag{6.199d}$$

In these calculations we use only (6.156a) and abstract algebra: the commutation relations between the Chevalley generators H_j, $X_j^{\pm}$, $j = 1, 2$, the definitions of $\tilde{C}$ and E_{13}, and the fact that $\hat{1}$ is the lowest-weight vector.

Further, we shall use also the realization of $U_q(sl(3))$ given in (6.112) to obtain a polynomial in the variables x, y, z corresponding to the GWZ pattern $(\mathbf{m})$. For this we define:

$$\phi_{(\tilde{m})}(x,y,z) \doteq (\Gamma_3(E_{21}))^{m_{12}-m_{11}}\,(\Gamma_3(\tilde{C}))^{r-m_{12}}\,(\Gamma_3(E_{32}))^{r_1-m_{22}}\,(\Gamma_3(E_{13}))^{r}\,1 \tag{6.200}$$

For this quantity will hold the same formulae (6.199) we derived above – this follows just because Γ_3 is a representation. On the other hand we should stress that this quantity is a polynomial in the variables x, y, z. We give two examples which we shall use below – of h.w.s. (using (6.175)) and l.w.s.:

$$\begin{aligned}
\phi_{(h.w.s.)} &= (\Gamma_3(E_{13}))^r 1 = (-1)^{r_1} q^{\frac{1}{4}r_1(2-r_1)} [r]_q! x^{r_1} y^{r_1} z^{r-r_1} \times \\
&\quad \times {}_1F_0{}^q (-r_1; q^{\frac{1}{4}(r_1-r-2)} \frac{z}{xy}) \qquad (6.201a) \\
\phi_{(l.w.s.)} &= (\Gamma_3(E_{21}))^{r_1} (\Gamma_3(\hat{\mathscr{C}}))^{r-r_1} (\Gamma_3(E_{32}))^{r_1} (\Gamma_3(E_{13}))^r 1 = \\
&= (-1)^{r_1} q^{\frac{r_1}{2}} \frac{[r]_q![r+1]_q![r-r_1]_q![r_1]_q!}{[r_1+1]_q} \qquad (6.201b)
\end{aligned}$$

Next we find the action on the realization of the unnormalized GWZ states via hypergeometric functions (cf. (6.167) and (6.175)):

$$\psi_{(\mathbf{m})}(x, y, z) = \psi_{(m_{12},m_{11},m_{22})} = x^{m_{11}} y^{\kappa} \tilde{\psi}(\zeta) , \qquad (6.202)$$
$$\kappa = m_{12} + m_{22} - r_1 , \qquad \zeta = \frac{z}{xy}$$

The second result here is the following action of the generators:

$$\begin{aligned}
\Gamma_3(X_1^+)\, \psi_{(\mathbf{m})} &= q^{-\frac{1}{4}\kappa}\, \psi_{(m_{12},m_{11}+1,m_{22})} \qquad (6.203) \\
\Gamma_3(X_1^-)\, \psi_{(\mathbf{m})} &= q^{\frac{1}{4}\kappa}\, [m_{12} - m_{11} + 1]_q\, [m_{11} - m_{22}]_q\, \psi_{(m_{12},m_{11}-1,m_{22})} \\
\Gamma_3(X_2^+)\, \psi_{(\mathbf{m})} &= b_1^+\, \psi_{(m_{12}+1,m_{11},m_{22})} + b_2^+\, \psi_{(m_{12},m_{11},m_{22}+1)} \\
\Gamma_3(X_2^-)\, \psi_{(\mathbf{m})} &= b_1^-\, \psi_{(m_{12}-1,m_{11},m_{22})} + b_2^-\, \psi_{(m_{12},m_{11},m_{22}-1)}
\end{aligned}$$

where:

$$\begin{aligned}
b_1^+ &= q^{-\frac{1}{4}m_{11}} \frac{[r - m_{12}]_q\, [m_{12} - m_{11} + 1]_q}{[m_{12} - m_{22} + 1]_q} \qquad (6.204) \\
b_2^+ &= q^{-\frac{1}{4}m_{11}} \frac{[r - m_{22} + 1]_q\, [m_{11} - m_{22}]_q}{[m_{12} - m_{22} + 1]_q} \\
b_1^- &= q^{\frac{1}{4}m_{11}} \frac{[m_{12} - r_1]_q\, [m_{12} + 1]_q}{[m_{12} - m_{22} + 1]_q} \\
b_2^- &= q^{\frac{1}{4}m_{11}} \frac{[r_1 - m_{22} + 1]_q\, [m_{22}]_q}{[m_{12} - m_{22} + 1]_q}
\end{aligned}$$

We derive this action using only the explicit realization of $\Gamma_3(\cdot)$ given in (6.112) and using some relations between q-hypergeometric functions which are given in Appendix B of [230].

Now we use the action on the two unnormalized polynomial realizations of the GWZ states $\phi_{(\bar{m})}$ and $\psi_{(\mathbf{m})}$ in order to derive the proportionality constant between them. We set:

$$\phi_{(\bar{m})} = \mathcal{N}_{(\bar{m})}\, \psi_{(\mathbf{m})} \tag{6.205}$$

Before proceeding we note the two cases in which we already know this constant:

$$\mathcal{N}_{(h.w.s)} = (-1)^{r_1}\, q^{\frac{1}{4}(2r_1 - r^2)}\, \frac{[r]_q!}{[r-r_1]_q!} \tag{6.206a}$$

$$\mathcal{N}_{(l.w.s)} = (-1)^{r_1}\, q^{\frac{r_1}{2}}\, \frac{[r]_q!\,[r+1]_q!\,[r-r_1]_q!\,([r_1]_q!)^2}{[r_1+1]_q} \tag{6.206b}$$

where (6.206a) is obtained by using (6.177), while (6.206b) is obtained by using (6.176).

The idea is as follows (on the example of X_1^-): On one hand we have:

$$\Gamma_3(X_1^-)\phi_{(\bar{m})} = \phi_{(m_{12},m_{11}-1,m_{22})} = \mathcal{N}_{(m_{12},m_{11}-1,m_{22})}\, \psi_{(m_{12},m_{11}-1,m_{22})} \tag{6.207}$$

On the other hand (using (6.203b)):

$$\begin{aligned} \Gamma_3(X_1^-)\, \phi_{(\bar{m})} &= \mathcal{N}_{(\bar{m})}\, X_1^-\psi_{(\mathbf{m})} = \\ &= \mathcal{N}_{(\bar{m})}\, q^{\frac{1}{4}\kappa}\, [m_{12} - m_{11} + 1]_q\, [m_{11} - m_{22}]_q\, \psi_{(m_{12},m_{11}-1,m_{22})} \end{aligned} \tag{6.208}$$

If $m_{11} > m_{22}$ comparing (6.207) and (6.208) we get a relation expressing $\mathcal{N}_{(\bar{m})}$ in terms of $\mathcal{N}_{(m_{12},m_{11}-1,m_{22})}$. Besides the above there are five more relations, expressing $\mathcal{N}_{(\bar{m})}$ through $\mathcal{N}_{(\bar{m}')}$ with $m_{11} \to m_{11}+1$, $m_{12} \to m_{12} \pm 1$, $m_{22} \to m_{22} \pm 1$. It is enough to use the three relations which decrease m_{ij}, each relation affecting only a single m_{ij}:

$$\mathcal{N}_{(\bar{m})} = q^{-\frac{1}{4}\kappa}\, \frac{1}{[m_{12}-m_{11}+1]_q[m_{11}-m_{22}]_q}\, \mathcal{N}_{(m_{12},m_{11}-1,m_{22})} \tag{6.209a}$$

$$\mathcal{N}_{(\bar{m})} = q^{-\frac{1}{4}m_{11}}\, \frac{[m_{12}-m_{11}]_q}{[m_{12}-r_1]_q[m_{12}+1]_q}\, \mathcal{N}_{(m_{12}-1,m_{11},m_{22})} \tag{6.209b}$$

$$\mathcal{N}_{(\bar{m})} = q^{-\frac{1}{4}m_{11}}\, \frac{[m_{11}-m_{22}+1]_q}{[r_1-m_{22}+1]_q[m_{22}]_q}\, \mathcal{N}_{(m_{12},m_{11},m_{22}-1)} \tag{6.209c}$$

(For the three relations which increase m_{ij} we refer to [230].)

Now, we use relation (6.209c) until on the RHS we get $\mathcal{N}_{(m_{12},m_{11},0)}$, then we use (6.209a) until on the RHS we get $\mathcal{N}_{(m_{12},0,0)}$, finally we use (6.209b) until on the RHS we get $\mathcal{N}_{(r_1,0,0)} = \mathcal{N}_{(l.w.s.)}$; that is, we get:

$$\mathcal{N}_{(\bar{m})} = q^{-\frac{1}{4}m_{11}\kappa}\, \frac{[r_1+1]_q[r_1-m_{22}]_q![m_{12}-m_{11}]_q!}{[r_1]_q![m_{12}-r_1]_q![m_{11}-m_{22}]_q![m_{12}+1]_q![m_{22}]_q!}\, \mathcal{N}_{(l.w.s.)} \tag{6.210}$$

and using (6.206b) we finally obtain:

$$\mathcal{N}_{(\bar{m})} = (-1)^{r_1} q^{\frac{1}{4}(2r_1 - m_{11}(m_{12}+m_{22}-r_1))} [r]_q! [r+1]_q! [r-r_1]_q! [r_1]_q! \times \\ \times \frac{[r_1 - m_{22}]_q! [m_{12} - m_{11}]_q!}{[m_{12} - r_1]_q! [m_{11} - m_{22}]_q! [m_{12}+1]_q! [m_{22}]_q!} \tag{6.211}$$

From (6.211) follows also:

$$\mathcal{N}_{(\text{h.w.s.})} = q^{-\frac{1}{4}r^2} \frac{[r_1+1]_q}{\left([r_1]_q![r-r_1]_q!\right)^2 [r+1]_q!} \mathcal{N}_{(\text{l.w.s.})} \tag{6.212}$$

which is consistent with (6.206).

6.4.5 Scalar Product and Normalized GWZ States

By (6.192) we have defined a scalar product in terms of the constant $\mathcal{N}_{(\bar{m})}$. Now that we know this constant we can fix the scalar product completely, that is, we have:

$$\begin{aligned}\left(\phi_{(\mathbf{m})}, \phi_{(\mathbf{m}')}\right)_p = {} & \delta_{m_{11},m'_{11}} \delta_{m_{12},m'_{12}} \delta_{m_{22},m'_{22}} ([r]_q!)^2 [r+1]_q! [r-r_1]_q! [r_1]_q! \\ & \times \frac{[r-m_{12}]_q! [r-m_{22}+1]_q!}{[m_{12}-r_1]_q! [m_{11}-m_{22}]_q! [m_{12}+1]_q!} \times \\ & \times \frac{[m_{12}-m_{11}]_q! [r_1-m_{22}]_q!}{[m_{22}]_q! [m_{12}-m_{22}+1]_q}\end{aligned} \tag{6.213}$$

or in terms of $\psi_{(\mathbf{m})}$:

$$\begin{aligned}\left(\psi_{(\mathbf{m})}, \psi_{(\mathbf{m}')}\right) = {} & \frac{1}{|\mathcal{N}_{(\mathbf{m})}|^2} \left(\phi_{(\mathbf{m})}, \phi_{(\mathbf{m}')}\right)_p = \\ = {} & \delta_{m_{11},m'_{11}} \delta_{m_{12},m'_{12}} \delta_{m_{22},m'_{22}} \times \\ & \times \frac{[r-m_{12}]_q! [r-m_{22}+1]_q!}{[r+1]_q! [r-r_1]_q! [r_1]_q! [m_{12}-m_{22}+1]_q} \times \\ & \times \frac{[m_{12}-r_1]_q! [m_{11}-m_{22}]_q! [m_{12}+1]_q! [m_{22}]_q!}{[r_1-m_{22}]_q! [m_{12}-m_{11}]_q!}\end{aligned} \tag{6.214}$$

Further, we complete our program of finding explicit polynomial realizations of the normalized GWZ states. We set:

$$\begin{aligned}\hat{\phi}_{(\mathbf{m})} \doteq {} & \frac{\phi_{(\mathbf{m})}}{\sqrt{\left(\phi_{(\mathbf{m})}, \phi_{(\mathbf{m})}\right)_p}} = \\ = {} & \frac{1}{N_\phi(\bar{m})} (\Gamma_3(E_{21}))^{m_{12}-m_{11}} \times \\ & \times (\Gamma_3(\tilde{C}))^{r-m_{12}} (\Gamma_3(E_{32}))^{r_1-m_{22}} (\Gamma_3(E_{13}))^r \, 1\end{aligned} \tag{6.215}$$

$$N_\phi(\bar{m}) \doteq \sqrt{(\phi_{(\mathbf{m})}, \phi_{(\mathbf{m})})_p} = \tag{6.216}$$

$$= [r]_q! \sqrt{\frac{[r+1]_q!\,[r-r_1]_q!\,[r_1]_q!\,[r-m_{12}]_q!}{[m_{12}-r_1]_q!\,[m_{11}-m_{22}]_q!\,[m_{12}+1]_q!}} \times$$

$$\times \sqrt{\frac{[r-m_{22}+1]_q!\,[m_{12}-m_{11}]_q!\,[r_1-m_{22}]_q!}{[m_{22}]_q!\,[m_{12}-m_{22}+1]_q}}$$

Analogously, we set:

$$\hat{\psi}_{(\mathbf{m})} \doteq \frac{\psi_{(\mathbf{m})}}{\sqrt{(\psi_{(\mathbf{m})}, \psi_{(\mathbf{m})})}} = \tag{6.217}$$

$$= \frac{1}{N_\psi(\bar{m})} \frac{x^{m_{11}}\, y^{m_{12}+m_{22}-r_1}}{\Gamma_q(r_1-m_{11}+1)}\, {}_1F_0{}^q\,(-m_{22};\, q^{\frac{1}{4}(m_{22}-m_{12}-2)}\, \frac{z}{xy}) \times$$

$$\times\, {}_2F_1{}^q\,(m_{22}-m_{11},\, r_1-m_{12};\, r_1-m_{11}+1;\, q^{\frac{1}{4}(m_{12}+m_{22})}\, \frac{z}{xy})$$

$$N_\psi(\bar{m}) \doteq \sqrt{(\psi_{(\mathbf{m})}, \psi_{(\mathbf{m})})} = N_\phi(\bar{m})/|\mathcal{N}_{(\mathbf{m})}| = \tag{6.218}$$

$$= \sqrt{\frac{[r-m_{12}]_q!\,[r-m_{22}+1]_q!\,[m_{12}-r_1]_q!}{[r+1]_q!\,[r-r_1]_q!\,[r_1]_q!}} \times$$

$$\times \sqrt{\frac{[m_{11}-m_{22}]_q!\,[m_{12}+1]_q!\,[m_{22}]_q!}{[m_{12}-m_{22}+1]_q\,[r_1-m_{22}]_q!\,[m_{12}-m_{11}]_q!}}$$

Finally, we calculate the action of the Chevalley generators on our normalized GWZ states. We get:

$$\begin{aligned} X_1^+\,\hat{\phi}_{(\mathbf{m})} &= \sqrt{[m_{12}-m_{11}]_q\,[m_{11}-m_{22}+1]_q}\,\hat{\phi}_{(m_{12},m_{11}+1,m_{22})} \\ X_1^-\,\hat{\phi}_{(\mathbf{m})} &= \sqrt{[m_{12}-m_{11}+1]_q\,[m_{11}-m_{22}]_q}\,\hat{\phi}_{(m_{12},m_{11}-1,m_{22})} \\ X_2^+\,\hat{\phi}_{(\mathbf{m})} &= a_1^+\,\hat{\phi}_{(m_{12}+1,m_{11},m_{22})} + a_2^+\,\hat{\phi}_{(m_{12},m_{11},m_{22}+1)} \\ X_2^-\,\hat{\phi}_{(\mathbf{m})} &= a_1^-\,\hat{\phi}_{(m_{12}-1,m_{11},m_{22})} + a_2^-\,\hat{\phi}_{(m_{12},m_{11},m_{22}-1)} \end{aligned} \tag{6.219}$$

$$\begin{aligned} H_1\,\hat{\phi}_{(\mathbf{m})} &= (2m_{11}-m_{12}-m_{22})\,\hat{\phi}_{(\mathbf{m})} \\ H_2\,\hat{\phi}_{(\mathbf{m})} &= (2(m_{12}+m_{22})-m_{11}-r-r_1)\,\hat{\phi}_{(\mathbf{m})} \end{aligned} \tag{6.220}$$

$$a_1^+ = \sqrt{\frac{[r-m_{12}]_q\,[m_{12}-r_1+1]_q\,[m_{12}+2]_q\,[m_{12}-m_{11}+1]_q}{[m_{12}-m_{22}+1]_q\,[m_{12}-m_{22}+2]_q}}$$

$$a_2^+ = \sqrt{\frac{[r-m_{22}+1]_q\,[r_1-m_{22}]_q\,[m_{22}+1]_q\,[m_{11}-m_{22}]_q}{[m_{12}-m_{22}]_q\,[m_{12}-m_{22}+1]_q}} \tag{6.221}$$

$$a_1^- = \sqrt{\frac{[r-m_{12}+1]_q\,[m_{12}-r_1]_q\,[m_{12}+1]_q\,[m_{12}-m_{11}]_q}{[m_{12}-m_{22}]_q\,[m_{12}-m_{22}+1]_q}}$$

$$a_2^- = \sqrt{\frac{[r-m_{22}+2]_q\,[r_1-m_{22}+1]_q\,[m_{22}]_q\,[m_{11}-m_{22}+1]_q}{[m_{12}-m_{22}+1]_q\,[m_{12}-m_{22}+2]_q}}$$

Of course, the action of the Cartan generators (6.220) is the same for normalized and for unnormalized GWZ states. We note now that in (6.219) we have recovered the standard transformation rules which until now were written without derivation – for $q = 1$ in [65, 315], and for $q \neq 1$ in [143]. In fact, since the only restriction on the transformation rules were the commutation relations of $U_q(sl(3))$ later it was shown [47], that this restriction was very weak and one can generalize the above formulae by replacing the square roots, that is, the powers 1/2, in the matrix elements in (6.219) by the powers 0 and 1. There is no such freedom in our case. The only freedom we have is in phase factors, like the one relating $\hat{\phi}_{(\mathbf{m})}$ and $\hat{\psi}_{(\mathbf{m})}$. Indeed, the transformation rules for $\hat{\psi}_{(\mathbf{m})}$ are the same as (6.219) except for the $q^{\cdots}$ factors which are the same as in (6.203) and (6.204).

6.4.6 Summation Formulae

In this section we derive summation formulae using formula (6.211) for the constant $\mathcal{N}_{(\bar{m})}$ and another independent expression for $\mathcal{N}_{(\bar{m})}$. To find the latter we use formulae (6.130),(6.132), then we recall the polynomial $\phi_{(\bar{m})}$ at $z = 0$ using (6.158) Next we note the value of $\psi_{(\mathbf{m})}$ at $z = 0$ (using (6.170)):

$$(\psi_{(\mathbf{m})})_{|z=0} = \frac{x^{m_{11}} y^{m_{12}+m_{22}-r_1}}{\Gamma_q(r_1-m_{11}+1)} \tag{6.222}$$

Now we compare (6.205), (6.158a), and (6.222), and conclude that:

$$\mathcal{N}_{(\mathbf{m})} = \mathcal{N}^+_{(\mathbf{m})}, \qquad r_1 \geq m_{11} \tag{6.223}$$

From the latter using (6.158b) and (6.211) we get the following summation formula:

$$\sum_{t=0}^{r-m_{12}} \sum_{u\in\mathbb{Z}_+} (-1)^{t+u} \binom{m_{12}-m_{11}+t}{u}_q \times$$

$$\times \frac{q^{\frac{t}{2}(r_1-m_{12}-m_{22})+\frac{u}{2}(t+m_{12}-r-1)}\,[r_1-m_{22}+t]_q!}{[t]_q!\,[m_{22}-t]_q!\,[m_{12}-m_{22}+1+t]_q!} \times$$

$$\times \frac{[r+m_{11}-m_{12}-m_{22}+u]_q!\,[m_{12}+m_{22}-m_{11}-u]_q!}{[r_1+m_{11}-m_{12}-m_{22}+u]_q!\,[m_{12}+m_{22}-r_1-u]_q!} =$$

$$= \frac{[r+m_{11}-m_{12}-m_{22}]_q!\,[r_1-m_{22}]_q!\,[m_{12}+m_{22}-m_{11}]_q!}{[m_{12}+1-m_{22}]_q!\,[m_{12}+m_{22}-r_1]_q!\,[m_{22}]_q!} \times$$

$$\times \sum_{t=0}^{r-m_{12}} \frac{q^{\frac{t}{2}(r_1-m_{12}-m_{22})}}{\Gamma_q(r_1+m_{11}-m_{12}-m_{22}+1)} \frac{(-m_{22})_t^q\,(r_1-m_{22}+1)_t^q}{[t]_q!\,(m_{12}-m_{22}+2)_t^q} \times$$

$$\times\ {}_3F_2{}^q\,(r+m_{11}-m_{12}-m_{22}+1,\ r_1-m_{12}-m_{22},\ m_{11}-m_{12}-t;$$
$$r_1+m_{11}-m_{12}-m_{22}+1,\ m_{11}-m_{12}-m_{22};\ q^{\frac{1}{2}(t+m_{12}-r-1)}) =$$

$$= (-1)^{m_{12}+r_1}\, q^{\frac{1}{2}\left(r_1+(m_{11}+r-m_{12})(r_1-m_{12}-m_{22})+m_{12}(m_{22}-1)\right)} \times$$

$$\times \frac{[r+1]_q!\,[r-r_1]_q!\,[r+m_{11}-m_{12}-m_{22}]_q!}{[r+1-m_{22}]_q!\,[r-m_{12}]_q!\,[m_{12}-r_1]_q!} \times$$

$$\times \frac{[r_1-m_{22}]_q!\,[m_{12}-m_{11}]_q!}{[m_{12}+1]_q!\,[m_{11}-m_{22}]_q!\,[m_{22}]_q!} \tag{6.224}$$

In order to show better the properties of the above formula, we will rewrite it in representation independent parameters:

$$\begin{aligned} b_1 &= r_1-m_{12}+m_{11}-m_{22}+1\,, & b_2 &= m_{11}-m_{12}-m_{22} \\ m_1 &= m_{12}-r_1\,, \quad m_2 = m_{22}\,, & N &= r-m_{12} \end{aligned} \tag{6.225}$$

Now we rewrite (6.224) in the new variables using also the q-Pochhammer symbol:

$$\frac{1}{\Gamma_q(b_1)} \sum_{t=0}^{N} q^{-\frac{t}{2}(m_1+m_2)} \frac{(-m_2)_t^q\,(b_1-b_2-m_2)_t^q}{(b_1-b_2+m_1-m_2+1)_t^q\,(1)_t^q} \times$$
$$\times\ {}_3F_2{}^q(-(m_1+m_2), b_1+m_1+N, b_2+m_2-t; b_1, b_2; q^{\frac{1}{2}(t-N-1)}) =$$
$$= (-1)^{m_1+m_2}\, q^{-\frac{1}{2}(\,m_1(m_1+b_1+N)+m_2(m_2+b_2+N)+m_1m_2)} \times$$
$$\times \frac{(b_1-b_2+m_1+1)_N^q\,(m_1+1)_N^q}{(b_1-b_2+m_1-m_2+1)_N^q\,(1)_N^q} \frac{(1)_{m_1+m_2}^q}{[b_1+m_1-1]_q!\,(b_2)_{m_2}^q} \tag{6.226}$$

For the better comparison with the literature on q-summation formulae we rewrite our formula using notation from Gasper–Rahman [311]. We shall use the definition (1.2.15) for the *q-shifted factorial*:

$$(a;q)_n = \begin{cases} 1, & n=0 \\ (1-a)(1-aq)\ldots(1-aq^{n-1}), & n=1,2,\ldots \end{cases} \tag{6.227}$$

and (1.2.22) for the basic hypergeometric series:

$$\begin{aligned} {}_r\phi_s\,(a_1,\ldots,a_r;\,b_1,\ldots,b_s;\,q\,,z) &= \\ &= \sum_{n=0}^{\infty} \frac{(a_1;q)_n\,\ldots\,(a_r;q)_n}{(q;q)_n\,(b_1;q)_n\,\ldots\,(b_s;q)_n} \left[(-1)^n\, q^{\binom{n}{2}}\right]^{1+s-r} z^n \end{aligned} \tag{6.228}$$

For completeness we mention also the relation between our q-Pochhammer symbol and the notation of [311]:

$$(a)_n^q = (-\lambda)^{-n} q^{-(n-1)n/4} q^{-na/2} (q^a;q)_n\,, \quad \lambda \equiv q^{1/2} - q^{-1/2} \tag{6.229}$$

Now our summation formula (6.224) or (6.226) can be rewritten as (when $b_1 > 0$):

$$\begin{aligned} &\sum_{t=0}^{N} q^t\, \frac{(q^{-m_2};q)_t\,(q^{b_1-b_2-m_2};q)_t}{(q^{b_1-b_2+m_1-m_2+1};q)_t\,(q;q)_t} \times \\ &\times {}_3\phi_2(q^{-(m_1+m_2)}, q^{b_1+m_1+N}, q^{b_2+m_2-t}; q^{b_1}, q^{b_2}; q; q^{t-N}) = \\ &= (-1)^{m_1+m_2}\, q^{-\frac{1}{2}(m_1+m_2)(m_1+m_2+1+2N)} \times \\ &\times \frac{(q^{b_1-b_2+m_1+1};q)_N\,(q^{m_1+1};q)_N}{(q^{b_1-b_2+m_1-m_2+1};q)_N\,(q;q)_N}\, \frac{(q;q)_{m_1+m_2}}{(q^{b_1};q)_{m_1}\,(q^{b_2};q)_{m_2}} \end{aligned} \tag{6.230}$$

This new summation formula seems unknown also for the classical case $q = 1$. Partial cases can be found in the literature. For instance, the case $N = 0$; that is, $m_{12} = r$, reduces to a q-Karlsson–Minton formula (cf. (1.9.8) of [311]):

$$\begin{aligned} &{}_3\phi_2(q^{-(m_1+m_2)}, q^{b_1+m_1}, q^{b_2+m_2}; q^{b_1}, q^{b_2}; q; 1) = \\ &= (-1)^{m_1+m_2}\, q^{-\frac{1}{2}(\,m_1+m_2)(m_1+m_2+1)}\, \frac{(q;q)_{m_1+m_2}}{(q^{b_1};q)_{m_1}\,(q^{b_2};q)_{m_2}}\,. \end{aligned} \tag{6.231}$$

It corresponds to a 0-balanced ${}_3\phi_2$ [311].

6.4.7 Weight Pyramid of the *SU*(3) UIRs

6.4.7.1 Geometrical Construction of the Weight Pyramid

First let us recall some well-known facts about the UIRs of $SU(3)$ which hold also for the (anti)holomorphic representations of $SL(3)$, also for the Lie algebras and quantum groups. Fix such a representation, that is, the non-negative integers r_1, r_2, so that we have a representation of dimension d_{r_1+1,r_2+1}. It is customary to depict the weight lattice of every such irrep in the (I_z, Y) plane. We recall that the notation comes from the popular application in which I_z is the third component of isospin, and Y is the hypercharge. The points of the weight diagram form a hexagon, the sides of the hexagon containing alternatively $r_1 + 1$, $r_2 + 1$ points. (Thus, the hexagon degenerates into a triangle if $r_1 r_2 = 0$.) Each point of the weight diagram represents all states with the same weight and differing only by the values of isospin I, for which the corresponding I_z is admissible. It is also customary to connect all points with the same multiplicity. Then the resulting figure consists of nested hexagons if $r_1 r_2 \neq 0$, the most outward one containing the states with multiplicity one, the next inwards – the states with multiplicity two, and so on. When $r_1 r_2 = 0$ the resulting figure consists of nested triangles; moreover, each weight has multiplicity one and that is why such representations are called *flat representations*.

Now for our purposes we shall replace this customary weight diagram with a *hexagonal pyramid* (when $r_1 r_2 \neq 0$) in the following way. We consider now a three-dimensional picture adding also the direction perpendicular to the (I_z, Y) plane. The points in that plane have coordinates, say $(i_z, y, 0)$. Next we replace each point of the weight lattice of multiplicity m and coordinate $(i_z, y, 0)$ by m equally spaced points in direction perpendicular to the (I_z, Y) plane which points have coordinates: (i_z, y, k), $k = 0, 1, \ldots, m-1$. We consider now each point of the so formed pyramid as one state; that is, each point has also fixed value of isospin I and there is no multiplicity. From the algebraic formulae given in next section we shall see that for fixed (I_z, Y) the value of isospin I diminishes as k increases.

Thus, we obtain a pyramid of height $r_0 \equiv \min(r_1, r_2)$.

Consider now the states with coordinates (i_z, y, k) for a fixed k. We shall say that these states form a *layer*. We note now that by construction each such layer is actually a weight diagram in the I_z and Y axis and has the form of a hexagon. Moreover, this hexagon has exactly the form of a standard $SU(3)$ weight diagram – the difference is that we put only one GWZ state at each site. Of course, it is important how we distribute the states with the same weight and this is what we explain next.

Let us agree, in order to save space, to omit the first row of the standard $SU(3)$ GWZ pattern $(\mathbf{m})$ since we shall work with fixed representation parameters r_1, r_2. Namely, we set:

$$\begin{bmatrix} m_{12} & & m_{22} \\ & m_{11} & \end{bmatrix} \equiv \begin{pmatrix} r & & r_1 & & 0 \\ & m_{12} & & m_{22} & \\ & & m_{11} & & \end{pmatrix} \tag{6.232}$$

We place the GWZ states on our pyramid in the following manner. The bottom, or zeroth, layer contains both the lowest-weight state and the highest-weight state of our representation. Overall it contains the following states:

$$\begin{bmatrix} r & & r_1 \\ & r_1 & \end{bmatrix}\begin{bmatrix} r & & r_1 \\ & r_1{+}1 & \end{bmatrix} \quad \cdots \quad \begin{bmatrix} r & & r_1 \\ & r & \end{bmatrix}$$

$$\cdots$$

$$\begin{bmatrix} r & & 1 \\ & 1 & \end{bmatrix}\begin{bmatrix} r & & 1 \\ & 2 & \end{bmatrix} \quad \cdots \qquad\qquad \begin{bmatrix} r & & 1 \\ & r & \end{bmatrix}$$

$$\begin{bmatrix} r & & 0 \\ & 0 & \end{bmatrix}\begin{bmatrix} r & & 0 \\ & 1 & \end{bmatrix} \quad \cdots \qquad\qquad\qquad \begin{bmatrix} r & & 0 \\ & r & \end{bmatrix}$$

$$\cdots$$

$$\begin{bmatrix} r_1{+}1 & & 0 \\ & 0 & \end{bmatrix}\begin{bmatrix} r_1{+}1 & & 0 \\ & 1 & \end{bmatrix} \quad \cdots \quad \begin{bmatrix} r_1{+}1 & & 0 \\ & r_1{+}1 & \end{bmatrix}$$

$$\begin{bmatrix} r_1 & & 0 \\ & 0 & \end{bmatrix}\begin{bmatrix} r_1 & & 0 \\ & 1 & \end{bmatrix} \quad \cdots \quad \begin{bmatrix} r_1 & & 0 \\ & r_1 & \end{bmatrix} \tag{6.233}$$

The lowest-weight state $\begin{bmatrix} r_1 & & 0 \\ & 0 & \end{bmatrix}$ is in the bottom-left corner, the highest-weight state $\begin{bmatrix} r & & r_1 \\ & r & \end{bmatrix}$ is in the top-right corner, of this hexagon. (Of course, these states and the others on the edges of this initial hexagon are with no multiplicity, so their placement is more or less standard.) Analogously, we put the following states on the k-th layer, $k \le r_1, r_2$:

$$\begin{bmatrix} r{-}k & & r_1 \\ & r_1 & \end{bmatrix}\begin{bmatrix} r{-}k & & r_1 \\ & r_1{+}1 & \end{bmatrix} \cdots \begin{bmatrix} r{-}k & & r_1 \\ & r{-}k & \end{bmatrix}$$

$$\cdots$$

$$\begin{bmatrix} r{-}k & & k{+}1 \\ & k{+}1 & \end{bmatrix}\begin{bmatrix} r{-}k & & k{+}1 \\ & k{+}2 & \end{bmatrix} \quad \cdots \quad \begin{bmatrix} r{-}k & & k{+}1 \\ & r{-}k & \end{bmatrix}$$

$$\begin{bmatrix} r{-}k & & k \\ & k & \end{bmatrix}\begin{bmatrix} r{-}k & & k \\ & k{+}1 & \end{bmatrix} \quad \cdots \qquad\qquad \begin{bmatrix} r{-}k & & k \\ & r{-}k & \end{bmatrix}$$

$$\cdots$$

$$\begin{bmatrix} r_1{+}1 & & k \\ & k & \end{bmatrix}\begin{bmatrix} r_1{+}1 & & k \\ & k{+}1 & \end{bmatrix} \quad \cdots \quad \begin{bmatrix} r_1{+}1 & & k \\ & r_1{+}1 & \end{bmatrix}$$

$$\begin{bmatrix} r_1 & & k \\ & k & \end{bmatrix}\begin{bmatrix} r_1 & & k \\ & k{+}1 & \end{bmatrix} \quad \cdots \quad \begin{bmatrix} r_1 & & k \\ & r_1 & \end{bmatrix} \tag{6.234}$$

(Note that $k = 0$ represents the bottom layer.) Clearly, there are $r_1 - k + 1$ states on the bottom row of the above hexagon, $r_1 - k + 2$ states on the next row, and so on, and $r - 2k + 1$ states on the middle (longest) row, then the number of states decreases by one, the top row having $r - k - r_1 + 1 = r_2 - k + 1$ states. If we sum these we obtain that the number of states in the k-th layer presented in (6.234) is:

$$N^k_{r_1,r_2} = \tfrac{1}{2}(r+1)(r+2) + r_1 r_2 + 3k^2 - 3k(r+1) \tag{6.235}$$

From this it is easy to see that the number of states on the first k layers is:

$$\sum_{s=0}^{k-1} N^s_{r_1,r_2} = \frac{k}{2}\left(r^2 + 6r + 2r_1 r_2 + 2k^2 - 3k(r+2)\right) \tag{6.236}$$

We make now the observation that the latter number is equal to the difference of two $SU(3)$ dimensions:

$$\sum_{s=0}^{k-1} N^s_{r_1,r_2} = d_{r_1+1,r_2+1} - d_{r_1+1-k,r_2+1-k} \tag{6.237}$$

that is, the dimension of the irrep we are considering minus the dimension of an irrep with each representation parameter r_i decreased by k. This seems natural since the latter representation has a weight pyramid with bottom layer the $(k+1)$-th layer of our pyramid.

6.4.7.2 Algebraic Description of the Weight Pyramid

Now we explain the placement of the GWZ states on our pyramid. This is related to a procedure to obtain all GWZ states starting from the lowest weight state. (A similar procedure starting from the highest-weight state was used in the previous section.) To derive the necessary for the procedure relations between the GWZ states, up to normalization constants, it is enough to use only the fact that the GWZ states are eigenvectors of the operators $\hat{I}_z, \hat{Y}, \hat{I}^2$. Note that $\hat{I}^2$ is the Casimir of the $U_q(sl(2))$ quantum subgroup generated by $X_1^\pm, H_1$. We recall the relation of these eigenvalues to the parameters of the GWZ pattern:

$$\begin{bmatrix} m_{12} & & m_{22} \\ & m_{11} & \end{bmatrix} = \begin{bmatrix} I + \frac{1}{2}Y + \frac{1}{3}(r+r_1) & & -I + \frac{1}{2}Y + \frac{1}{3}(r+r_1) \\ & I_z + \frac{1}{2}Y + \frac{1}{3}(r+r_1) & \end{bmatrix} \tag{6.238}$$

with I_z, Y, I denoting the eigenvalues of the corresponding operators.

Before giving the explicit formulae we mention some general facts: The states on a fixed row of a fixed layer (6.234) are states with the same value of Y and I, while I_z varies between $-I$ and I. On a fixed layer the value of Y increases by 1 from the bottom to the top row. The states which have the same weight and differ only by the value of I are one above the other in the pyramid, the value I decreasing from the bottom up.

First, we describe the states on a fixed layer (hexagon), say, the k-th one.

Starting from the state in low-left corner of the hexagon, that is, $\left[\begin{smallmatrix} r_1 & & k \\ & k & \end{smallmatrix}\right]$, we first obtain the states on the south-west edge of the hexagon:

$$\begin{gathered} (X_2^+)^s \begin{bmatrix} r_1 & & k \\ & k & \end{bmatrix} = \mathcal{N}_2(s,k) \begin{bmatrix} r_1+s & & k \\ & k & \end{bmatrix}, s = 0,1,\dots r_2 - k, \\ \mathcal{N}_2(s,k) = \left(\frac{[s]_q!\,[r_1+1+s]_q!\,[r_2]_q!\,[r_1+1-k]_q!}{[r_1+1]_q!\,[r_2-s]_q!\,[r_1+1-k+s]_q!} \right)^{1/2} \end{gathered} \tag{6.239}$$

Now we prove the following lemma which is our main technical tool for the procedure.

Lemma: Let ψ be an eigenstate of $\hat{I}^2$, $\hat{I}_z$ and η with eigenvalues $\mu(\mu+1)$, $-\mu$ and κ, respectively, and let

$$\psi^+ = \widehat{C}\,\psi$$

where $\widehat{C}$ is the following operator:

$$\begin{aligned}\widehat{C} &\equiv X_3^+\,[H_1]_q + X_2^+\,X_1^+\,q^{\frac{H_1}{2}} = \\ &= X_1^+\,X_2^+\,[H_1]_q + X_2^+\,X_1^+\,[1-H_1]_q\end{aligned} \tag{6.240}$$

Then either $\psi^+ = 0$ or ψ^+ is an eigenstate of $\hat{I}^2$, $\hat{I}_z$ and η with eigenvalues $(\mu-\frac{1}{2})(\mu+\frac{1}{2})$, $-\mu+\frac{1}{2}$ and $\kappa+1$, respectively. In terms of GWZ pattern: if $\psi \leftrightarrow \begin{bmatrix} m_{12} & & k \\ & k & \end{bmatrix}$ then $\psi^+ \leftrightarrow \begin{bmatrix} m_{12} & & k+1 \\ & k+1 & \end{bmatrix}$ unless $k = r_1$.

The proof is given in [245]. ◊

Using the above lemma we obtain the states on the north-west edge of the hexagon:

$$\begin{aligned}&\widehat{C}^t (X_2^+)^{r_2-k} \begin{bmatrix} r_1 & & k \\ & k & \end{bmatrix} = \mathcal{N}_2(r_2-k,k)\,\widehat{C}^t \begin{bmatrix} r-k & & k \\ & k & \end{bmatrix} = \\ &= \mathcal{N}_2(r_2-k,k)\,\mathcal{N}_3(t) \begin{bmatrix} r-k & & k+t \\ & k+t & \end{bmatrix}, \qquad t = 0,1,\ldots r_1-k, \\ &\mathcal{N}_3(t) = \left(\frac{[r-k+1]_q!\,[r-2k+1]_q!\,[r_1-k]_q!\,[k+t]_q!}{[r-k+1-t]_q!\,[r-2k+1-t]_q!\,[r_1-k-t]_q!\,[k]_q!} \right)^{1/2}\end{aligned} \tag{6.241}$$

Now all other states of the k-th layer are obtained by the action of the operator X_1^+ to the states on the edges (6.239),(6.241):

$$\begin{aligned}&(X_1^+)^u (X_2^+)^s \begin{bmatrix} r_1 & & k \\ & k & \end{bmatrix} = \mathcal{N}_2(s,k)(X_1^+)^u \begin{bmatrix} r_1+s & & k \\ & k & \end{bmatrix} = \\ &= \mathcal{N}_1(u,s,k)\,\mathcal{N}_2(s,k) \begin{bmatrix} r_1+s & & k \\ & k+u & \end{bmatrix}, \\ &\qquad s = 0,1,\ldots r_2-k, \quad u = 0,1,\ldots r_1-k+s \\ &\mathcal{N}_1(u,s,k) = \left(\frac{[r_1+s-k]_q!\,[u]_q!}{[r_1+s-k-u]_q!} \right)^{1/2}\end{aligned} \tag{6.242a}$$

$$\begin{aligned}&(X_1^+)^u \widehat{C}^t (X_2^+)^{r_2-k} \begin{bmatrix} r_1 & & k \\ & k & \end{bmatrix} = \\ &\mathcal{N}_2(r_2-k,k)\,\mathcal{N}_3(t)(X_1^+)^u \begin{bmatrix} r-k & & k+t \\ & k+t & \end{bmatrix} =\end{aligned} \tag{6.242b}$$

$$= \mathcal{N}_1'(u)\, \mathcal{N}_2(r_2-k,k)\, \mathcal{N}_3(t) \begin{bmatrix} r-k & & k+t \\ & k+t+u & \end{bmatrix},$$

$$t = 0,1,\dots r_1-k, \quad u = 0,1,\dots r-2k-t,$$

$$\mathcal{N}_1'(u) = \left(\frac{[r-2k-t]_q![u]_q!}{[r-2k-t-u]_q!}\right)^{1/2}$$

Finally we explain how to obtain the lower-left-corner states $\begin{bmatrix} r_1 & & k \\ & k & \end{bmatrix}$ starting from the lowest-weight state $\begin{bmatrix} r_1 & & 0 \\ & 0 & \end{bmatrix}$. This is achieved by using again the lemma above:

$$\widehat{C}^k \begin{bmatrix} r_1 & & 0 \\ & 0 & \end{bmatrix} = \mathcal{N}_3'(k,r) \begin{bmatrix} r_1 & & k \\ & k & \end{bmatrix}, \quad k = 0,1,\dots,r_0 = \min(r_1,r_2)$$

$$\mathcal{N}_3'(k,r) = \left(\frac{[r+1]_q!\,[r_1+1]_q!\,[r_1]_q!\,[k]_q!}{[r+1-k]_q!\,[r_1+1-k]_q!\,[r_1-k]_q!}\right)^{1/2} \tag{6.243}$$

For further use we note that relation (6.243) may be rewritten in two alternative ways:

$$\mathcal{N}_3'(k,r) \begin{bmatrix} r_1 & & k \\ & k & \end{bmatrix} = \prod_{s=1}^{k} \widehat{C}_s \begin{bmatrix} r_1 & & 0 \\ & 0 & \end{bmatrix} = \tag{6.244a}$$

$$= \sum_{j=0}^{k} (-1)^{k-j}\, q^{\frac{j}{2}(j-r_1-1)} \binom{k}{j}_q \frac{[r_1-j]_q!}{[r_1-k]_q!} \times$$

$$\times (X_2^+)^j (X_3^+)^{k-j} (X_1^+)^j \begin{bmatrix} r_1 & & 0 \\ & 0 & \end{bmatrix} \tag{6.244b}$$

$$\widehat{C}_s \equiv X_3^+\,[s-1-r_1]_q + X_2^+\,X_1^+\,q^{\frac{1}{2}(s-1-r_1)} =$$

$$= X_1^+X_2^+[s-1-r_1]_q + X_2^+X_1^+\,[r_1-s+2]_q$$

The proof of (6.244) is given in [245]. ◊

We should mention that similar formulae to (6.239) and (6.242a) for the relation between GWZ states may be found the literature (cf., e. g., [47, 65, 143]). However, at the time we could not find in the literature formulae involving the operator $\widehat{C}$.

In this subsection we have not specified any realization of $U_q(sl(3))$. If we want to have the GWZ states realized as polynomials then we first identify the lowest-weight state $\begin{bmatrix} r_1 & & 0 \\ & 0 & \end{bmatrix}$ with the function 1 and then use the representation (6.112).

Finally we note the similarity of formula (6.244b) with the formula giving the singular vector in (2.37) for A_2. It is this similarity that will be exploited in the next section in order to prove the explicit realization of the irregular irreps in terms of GWZ states.

6.4.8 The Irregular Irreps in Terms of GWZ States

In the present section we combine the results of the previous sections to derive our main result. We set $q = e^{2\pi i/N}$, so that $[x]_q = \sin(\pi x/N)/\sin(\pi/N)$. We consider the irregular representations characterized by (2.179), and we restrict the representation parameters $r_i = m_i - 1$ as needed in the current situation:

$$1 < r_1 + 1, r_2 + 1 \; < N < r_1 + r_2 + 2 = r + 2 < 2N \tag{6.245}$$

With this the relevant singular vectors are (cf. (2.39) and (2.61) for A_2):

$$v_i = (X_i^+)^{r_i+1} v_0, \quad i = 1, 2, \tag{6.246}$$

$$\begin{aligned}
v_s^{\bar m} &= \mathscr{P}^{\bar m}(X_1^+, X_2^+, X_3^+) v_0 \,, \\
\mathscr{P}^{\bar m} &= \sum_{j=0}^{\bar m} (-1)^{\bar m - \bar m_1} q^{\frac{j}{2}(j-r_1-1)} \binom{\bar m}{j}_q \frac{[r_1+1-\bar m]_q!}{[r_1+1-j]_q!} \times \\
&\quad \times (X_2^+)^j (X_3^+)^{\bar m - j} (X_1^+)^j \,, \\
&\quad \bar m = r + 2 - N \,.
\end{aligned}$$

As we know to obtain an irreducible representations we have to factor out the Verma submodule built on these singular vectors, or, in a function space realization of the lowest-weight representations, impose corresponding vanishing conditions using the corresponding invariant differential operators. In the GWZ basis the lowest-weight vector is $\left[\begin{smallmatrix} r_1 & & 0 \\ & 0 & \end{smallmatrix}\right]$, while the vanishing conditions following from above are:

$$(X_i^+)^{r_i+1} \begin{bmatrix} r_1 & & 0 \\ & 0 & \end{bmatrix} = 0 \tag{6.247}$$

$$\mathscr{P}^{\bar m}(X_1^+, X_2^+, X_3^+) \begin{bmatrix} r_1 & & 0 \\ & 0 & \end{bmatrix} = 0 \tag{6.248}$$

Actually, the restrictions from (6.247) are valid in the GWZ basis by construction, e. g., from (6.239) one would obtain:

$$\begin{aligned}
(X_2^+)^{r_2+1} \begin{bmatrix} r_1 & & 0 \\ & 0 & \end{bmatrix} &= \mathscr{N}_2(s,0)|_{s=r_2+1} \begin{bmatrix} r+1 & & 0 \\ & 0 & \end{bmatrix} = \\
&= \mathscr{N}_2(s,0)|_{s=r_2+1} \begin{pmatrix} r & & r_1 & & 0 \\ & r+1 & & 0 & \\ & & 0 & & \end{pmatrix} = 0
\end{aligned} \tag{6.249}$$

since the latter is an impossible GWZ state (the betweenness constraint (6.155) is violated), and $\mathscr{N}_2(s,0)|_{s=r_2+1} \sim \left(\Gamma_q(r_2+1-s)\right)^{-1}|_{s=r_2+1} = 0$. Analogously from (6.242a) one would obtain:

$$\begin{aligned}
(X_1^+)^{r_1+1} \begin{bmatrix} r_1 & & 0 \\ & 0 & \end{bmatrix} &= \mathscr{N}_1(u,0,0)|_{u=r_1+1} \begin{bmatrix} r_1 & & 0 \\ & r_1+1 & \end{bmatrix} = \\
&= \mathscr{N}_1(u,0,0)|_{u=r_1+1} \begin{pmatrix} r & & r_1 & & 0 \\ & r_1 & & 0 & \\ & & r_1+1 & & \end{pmatrix} = 0
\end{aligned} \tag{6.250}$$

again the latter is an impossible GWZ state, and $\mathcal{N}_1(u,0,0)|_{u=r_1+1} \sim\sim (\Gamma_q(r_1+1-u))^{-1}|_{u=r_1+1} = 0$.

Thus the only new condition is (6.248). Indeed, it means that the lower-left-corner state $\left[\begin{smallmatrix} r_1 & \bar{m} \\ & \bar{m} \end{smallmatrix}\right]$ on the $\bar{m}$-th layer of our pyramid decouples from the irrep. This is clear from (6.244b) since the expression giving $\left[\begin{smallmatrix} r_1 & \bar{m} \\ & \bar{m} \end{smallmatrix}\right]$ is just $\mathscr{P}^{\bar{m}}(X_1^+, X_2^+, X_3^+)$. The decoupling of this state follows also from the explicit normalization factor in (6.243) with $k = m$ and $r = N + \bar{m} - 2$ since:

$$\mathcal{N}_3'(\bar{m}, N + \bar{m} - 2) = 0 \tag{6.251}$$

which follows from:

$$\frac{[r+1]_q!}{[r+1-k]_q!}\Big|_{\substack{r=N+\bar{m}-2\\ k=\bar{m}}} = \frac{[N+\bar{m}-1]_q!}{[N-1]_q!} =$$

$$= [N+\bar{m}-1]_q\,[N+\bar{m}-2]_q \dots [N]_q = 0$$

since $[N]_q = \sin(\pi N/N)/\sin(\pi/N) = 0$ when $q = e^{2\pi i/N}$.

The decoupling of the state $\left[\begin{smallmatrix} r_1 & \bar{m} \\ & \bar{m} \end{smallmatrix}\right]$ implies the decoupling of the lower-left-corner states on the higher layers, that is, the states $\left[\begin{smallmatrix} r_1 & k \\ & k \end{smallmatrix}\right]$ with $k > \bar{m}$. This follows by noting that because of the factorization formula (6.244b) can be written also as:

$$\mathcal{N}_3''(k,\bar{m}) \begin{bmatrix} r_1 & k \\ & k \end{bmatrix} = \prod_{s=\bar{m}+1}^{k} \hat{C}_s \begin{bmatrix} r_1 & \bar{m} \\ & \bar{m} \end{bmatrix} \tag{6.252}$$

$$\mathcal{N}_3''(k,\bar{m}) = \left(\frac{[k]_q!\,[r+1-\bar{m}]_q!\,[r_1+1-\bar{m}]_q!\,[r_1-\bar{m}]_q!}{[\bar{m}]_q!\,[r+1-k]_q!\,[r_1+1-k]_q!\,[r_1-k]_q!}\right)^{1/2}$$

$$k = \bar{m}+1, \dots, r_0 = \min(r_1, r_2)$$

that is, these states are descendants of $\left[\begin{smallmatrix} r_1 & \bar{m} \\ & \bar{m} \end{smallmatrix}\right]$. For consistency we note also that:

$$\mathcal{N}_3'(k, N+\bar{m}-2) \begin{cases} \neq 0 & \text{for k} < \bar{\text{m}} \\ 0 & \text{for k} \geq \bar{\text{m}} \end{cases} \tag{6.253}$$

Clearly, together with the lower-left-corner states decouple also the states on their layers; that is, all states on layers $k = \bar{m}, \bar{m}+1, \dots, r_0$. Thus, we are left with the states on the first $\bar{m}$ layers. Their number is given by (6.236) and (6.237), with $k = \bar{m}$.

Thus, we have obtained the **explicit description of the irregular representations** of $U_q(sl(3))$ in terms of the GWZ basis. These are the states displayed in (6.234) for $k = 0, 1, \dots, \bar{m} - 1 = r + 1 - N$.

We note that in the case when $\bar{m} = 1$; that is, $N = r + 1$, the irregular irrep is flat.

Finally, we discuss the representation action of $U_q(sl(3))$ in our irregular irreps. First we stress that when we consider *unnormalized* GWZ states the $U_q(sl(3))$ action is given straightforwardly as action on a truncated Verma module basis and there is no need even to display it explicitly. A little more care is needed when we consider the *normalized* GWZ basis. First we recall the standard action of $U_q(sl(3))$ on the normalized GWZ basis [47, 65], when q is not a nontrivial root of 1:

$$H_1 \begin{bmatrix} m_{12} & & m_{22} \\ & m_{11} & \end{bmatrix} = (2m_{11} - m_{12} - m_{22}) \begin{bmatrix} m_{12} & & m_{22} \\ & m_{11} & \end{bmatrix} \tag{6.254a}$$

$$X_1^+ \begin{bmatrix} m_{12} & & m_{22} \\ & m_{11} & \end{bmatrix} = \left([m_{12} - m_{11}]_q\right)^{\xi} \left([m_{11} - m_{22} + 1]_q\right)^{\xi'} \times \\ \times \begin{bmatrix} m_{12} & & m_{22} \\ & m_{11}+1 & \end{bmatrix} \tag{6.254b}$$

$$X_1^- \begin{bmatrix} m_{12} & & m_{22} \\ & m_{11} & \end{bmatrix} = \left([m_{12} - m_{11} + 1]_q\right)^{1-\xi} \left([m_{11} - m_{22}]_q\right)^{1-\xi'} \times \\ \times \begin{bmatrix} m_{12} & & m_{22} \\ & m_{11}-1 & \end{bmatrix} \tag{6.254c}$$

$$H_2 \begin{bmatrix} m_{12} & & m_{22} \\ & m_{11} & \end{bmatrix} = (2(m_{12} + m_{22}) - m_{11} - r - r_1) \times \\ \times \begin{bmatrix} m_{12} & & m_{22} \\ & m_{11} & \end{bmatrix} \tag{6.254d}$$

$$X_2^+ \begin{bmatrix} m_{12} & & m_{22} \\ & m_{11} & \end{bmatrix} = a_1^+ \begin{bmatrix} m_{12}+1 & & m_{22} \\ & m_{11} & \end{bmatrix} + \\ + a_2^+ \begin{bmatrix} m_{12} & & m_{22}+1 \\ & m_{11} & \end{bmatrix} \tag{6.254e}$$

$$X_2^- \begin{bmatrix} m_{12} & & m_{22} \\ & m_{11} & \end{bmatrix} = a_1^- \begin{bmatrix} m_{12}-1 & & m_{22} \\ & m_{11} & \end{bmatrix} + \\ + a_2^- \begin{bmatrix} m_{12} & & m_{22}-1 \\ & m_{11} & \end{bmatrix} \tag{6.254f}$$

$$a_1^+ = \frac{\left([r - m_{12}]_q\right)^{\eta_1} \left([m_{12} - r_1 + 1]_q\right)^{\eta_2}}{[m_{12} - m_{22} + 1]_q^{1/2}} \times \\ \times \frac{\left([m_{12} + 2]_q\right)^{\eta_3} \left([m_{12} - m_{11} + 1]_q\right)^{1-\xi}}{[m_{12} - m_{22} + 2]_q^{1/2}},$$

$$a_2^+ = \frac{\left([r - m_{22} + 1]_q\right)^{\zeta_1} \left([r_1 - m_{22}]_q\right)^{\zeta_2}}{[m_{12} - m_{22}]_q^{1/2}} \times \\ \times \frac{\left([m_{22} + 1]_q\right)^{\zeta_3} \left([m_{11} - m_{22}]_q\right)^{1-\xi'}}{[m_{12} - m_{22} + 1]_q^{1/2}},$$

$$
\begin{aligned}
a_1^- &= \frac{\left([r-m_{12}+1]_q\right)^{1-\eta_1}\left([m_{12}-r_1]_q\right)^{1-\eta_2}}{[m_{12}-m_{22}]_q^{1/2}} \times \\
&\quad\times \frac{\left([m_{12}+1]_q\right)^{1-\eta_3}\left([m_{12}-m_{11}]_q\right)^{\xi}}{[m_{12}-m_{22}+1]_q^{1/2}}, \\
a_2^- &= \frac{\left([r-m_{22}+2]_q\right)^{1-\zeta_1}\left([r_1-m_{22}+1]_q\right)^{1-\zeta_2}}{[m_{12}-m_{22}+1]_q^{1/2}} \times \\
&\quad\times \frac{\left([m_{22}]_q\right)^{1-\zeta_3}\left([m_{11}-m_{22}+1]_q\right)^{\xi'}}{[m_{12}-m_{22}+2]_q^{1/2}},
\end{aligned}
\tag{6.255}
$$

where the parameters $\xi, \xi', \eta_1, \eta_2, \eta_3, \zeta_1, \zeta_2, \zeta_3$ (introduced in [47]) take independently the values $0, \frac{1}{2}, 1$, the value $\frac{1}{2}$ for all of them being the classical choice. Note, however, that some of the nonclassical choices have to be excluded if we want that the coefficients would automatically become zero for impossible GWZ states. Thus, there are the following exclusions: $\xi \neq 0$, $\xi' \neq 1$, $\eta_1 \neq 0$, $\eta_2 \neq 1$, $\zeta_2 \neq 0$, $\zeta_3 \neq 1$. Note that partial cases of (6.254) were actually used in the algebraic description of the pyramid above (with all extra parameters equal to $\frac{1}{2}$). Note also that for the unnormalized GWZ basis (6.254) would also hold, however, the coefficients $a_i^\pm$ would be different; in particular, they will not contain any denominators.

For our purposes below we comment the action of the generators in relation to our pyramid structure (still in the generic q case). The action of the generators $X_1^\pm$ is confined on fixed rows, which is expected since these rows form irreps of the $U_q(sl(2))$ quantum subgroup generated by $X_1^\pm$, H_1. The action of the generators $X_2^\pm$ is more interesting. Consider the k-th layer. Then under the action the operator X_2^+ the states on the middle row (starting on the left with $\left[\begin{smallmatrix} r-k & & k \\ & k & \end{smallmatrix}\right]$) and the rows above it are mapped into a state on the same layer (cf. the second term in (6.254e)) and a state on the layer $k-1$ (cf. the first term in (6.254e)), while the states below the middle row are mapped into a state on the same layer (cf. the first term in (6.254e)) and a state on the layer $k+1$ (cf. the second term in (6.254e)). Analogously, under the action the operator X_2^- the states on the middle row and the rows below it are mapped into a state on the same layer (cf. the first term in (6.254f)) and a state on the layer $k-1$ (cf. the second term in (6.254f)), while the states above the middle row are mapped into a state on the same layer (cf. the second term in (6.254f)) and a state on the layer $k+1$ (cf. the first term in (6.254f)). Certainly, in all cases the two resulting states are one above the other since they have the same weights (eigenvalues of H_i). Note also that in some cases one of the two resulting states may miss when the initial state is on some of the sides or edges of the pyramid.

When q is a root of unity, as specified in the beginning of this section, there are *two possible problems* when using formulae (6.254). The first problem is that the action of the generators $X_2^\pm$ is mixing in general neighbouring layers and thus we have to ensure that formulae (6.254) will respect our factorization of the upper layers of the pyramid (which is so by construction if we use unnormalized GWZ states). This problem was cleared in [245].

The second possible problem, which is not specific for our approach and which was discussed in [4], is that there may arise zeros in the denominators of the coefficients (6.255). This necessitates modifications of (6.254) which were partially given in [4], and then in [245] where we also have checked that these modified formulae do not contradict our factorization. The modifications in (6.254) are as follows. First, we make the choice:

$$\xi = 1, \quad \xi' = 0, \quad \eta_1 = 1, \quad \zeta_3 = 0 \tag{6.256}$$

and then we set all remaining parameters equal to their classical value $\frac{1}{2}$. Thus we have instead of (6.254b,c) and (6.255):

$$X_1^+ \begin{bmatrix} m_{12} & & m_{22} \\ & m_{11} & \end{bmatrix} = [m_{12} - m_{11}]_q \begin{bmatrix} m_{12} & & m_{22} \\ & m_{11}+1 & \end{bmatrix} \tag{6.254b'}$$

$$X_1^- \begin{bmatrix} m_{12} & & m_{22} \\ & m_{11} & \end{bmatrix} = [m_{11} - m_{22}]_q \begin{bmatrix} m_{12} & & m_{22} \\ & m_{11}-1 & \end{bmatrix} \tag{6.254c'}$$

$$a_1^+ = [r - m_{12}]_q \left(\frac{[m_{12} - r_1 + 1]_q \, [m_{12} + 2]_q}{[m_{12} - m_{22} + 1]_q \, [m_{12} - m_{22} + 2]_q} \right)^{1/2} \tag{6.255$'$}$$

$$a_2^+ = [m_{11} - m_{22}]_q \left(\frac{[r - m_{22} + 1]_q \, [r_1 - m_{22}]_q}{[m_{12} - m_{22}]_q \, [m_{12} - m_{22} + 1]_q} \right)^{1/2}$$

$$a_1^- = [m_{12} - m_{11}]_q \left(\frac{[m_{12} - r_1]_q \, [m_{12} + 1]_q}{[m_{12} - m_{22}]_q \, [m_{12} - m_{22} + 1]_q} \right)^{1/2}$$

$$a_2^- = [m_{22}]_q \left(\frac{[r - m_{22} + 2]_q \, [r_1 - m_{22} + 1]_q}{[m_{12} - m_{22} + 1]_q \, [m_{12} - m_{22} + 2]_q} \right)^{1/2}$$

6.5 The Case of $U_q(sl(4))$

6.5.1 Elementary Representations

In this section following [216, 217] we consider in more detail the case $n = 4$. It is convenient (also for the comparison with the $q = 1$ case) to make the following change of variables:

$$\begin{aligned} &Y_{31} = \tilde{Y}_{31} - q\tilde{Y}_{21}\tilde{Y}_{32}\,, \quad Y_{41} = \tilde{Y}_{41} - q\tilde{Y}_{21}\tilde{Y}_{42}\,, \\ &Y_{21} = -q\tilde{Y}_{21}\,, \quad Y_{43} = q\tilde{Y}_{43}\,, \\ &Y_{ij} = \tilde{Y}_{ij}\,, \quad \text{for } (ij) = (32), (42)\,. \end{aligned} \tag{6.257}$$

Using (6.51) we have:

$$
\begin{aligned}
Y_{i\ell} Y_{ij} &= q^{1-2\delta_{\ell 2}} Y_{ij} Y_{i\ell}\,, \quad 4 \geq i > \ell > j \geq 1\,, && (6.258a)\\
Y_{kj} Y_{ij} &= q^{1-2\delta_{i2}} Y_{ij} Y_{kj}\,, \quad 4 \geq k > i > j \geq 1\,, && (6.258b)\\
Y_{41} Y_{32} &= Y_{32} Y_{41} + \lambda Y_{31} Y_{42}\,, && (6.258c)\\
Y_{4i} Y_{j1} &= Y_{j1} Y_{4i}\,, \quad (ij) = (23), (32)\,, && (6.258d)\\
Y_{ki} Y_{ij} &= q^{1-2\delta_{i3}} Y_{ij} Y_{ki} - (-1)^{\delta_{i3}} \lambda Y_{kj}\,, \quad 4 \geq k > i > j \geq 1 && (6.258e)
\end{aligned}
$$

(each of (6.258a,b,e) has four cases). Note that (6.51g,h,i) holds also for $Y_{j\ell}$ replacing $\tilde{Y}_{j\ell}$.

Note that for q a phase ($|q| = 1$) the q-coset in the Y coordinates is invariant under the anti-linear anti-involution ω acting as $\tilde{\omega}$ (cf. (6.52)) with $n = 4$:

$$\omega(Y_{j\ell}) = Y_{5-\ell,5-j}\,. \qquad (6.259)$$

Thus it can be considered as a q-coset of the conformal quantum group $SU_q(2,2)$.

The reduced functions for the $\mathcal{U}$ action are (cf. (6.50)):

$$
\begin{aligned}
\tilde{\varphi}(\bar{Y}, \bar{\mathcal{D}}) &= \sum_{i,j,k,\ell,m,n \in \mathbb{Z}_+} \mu_{ijk\ell mn} \tilde{\varphi}_{ijk\ell mn} && (6.260a)\\
\tilde{\varphi}_{ijk\ell mn} &= (Y_{21})^i (Y_{31})^j (Y_{32})^k (Y_{41})^\ell (Y_{42})^m (Y_{43})^n \times\\
&\quad \times (\mathcal{D}_1)^{r_1} (\mathcal{D}_2)^{r_2} (\mathcal{D}_3)^{r_3} && (6.260b)
\end{aligned}
$$

Now the action of $U_q(sl(4))$ on (6.260) is given explicitly by:

$$
\begin{aligned}
\hat{\pi}_{\bar{r}}(k_1) \tilde{\varphi}_{ijk\ell mn} &= q^{i+(j-k+\ell-m-r_1)/2} \tilde{\varphi}_{ijk\ell mn}\,, && (6.261)\\
\hat{\pi}_{\bar{r}}(k_2) \tilde{\varphi}_{ijk\ell mn} &= q^{k+(-i+j+m-n-r_2)/2} \tilde{\varphi}_{ijk\ell mn}\,,\\
\hat{\pi}_{\bar{r}}(k_3) \tilde{\varphi}_{ijk\ell mn} &= q^{n+(-j-k+\ell+m-r_3)/2} \tilde{\varphi}_{ijk\ell mn}\,, && (6.262)
\end{aligned}
$$

$$
\begin{aligned}
\hat{\pi}_{\bar{r}}(X_1^+) \tilde{\varphi}_{ijk\ell mn} &= q^{-1+(-j+k-\ell+m)/2} [r_1 - i]_q \tilde{\varphi}_{i+1,jk\ell mn} +\\
&\quad + q^{i-r_1-1+(j-k-\ell+m)/2} [k]_q \tilde{\varphi}_{i,j+1,k-1,\ell mn} +\\
&\quad + q^{i-r_1-1+(j-k+\ell-m)/2} [m]_q \tilde{\varphi}_{ijk,\ell+1,m-1,n}\,,\\
\hat{\pi}_{\bar{r}}(X_2^+) \tilde{\varphi}_{ijk\ell mn} &= q^{r_2-k+(i-j-m+n)/2} [i]_q \tilde{\varphi}_{i-1,j+1,k\ell mn} +\\
&\quad + q^{(i-j+m-n)/2} [j - i + k + m - n - r_2]_q \tilde{\varphi}_{ij,k+1,\ell mn} +\\
&\quad + q^{-r_2+(-i+j+k+3m-3n)/2} [\ell]_q \tilde{\varphi}_{i,j+1,k,\ell-1,m+1,n} +\\
&\quad + q^{k-r_2+(-i+j+m-n)/2} [n]_q \tilde{\varphi}_{ijk\ell,m+1,n-1}\,,
\end{aligned}
$$

$$\begin{aligned}
\hat{\pi}_{\bar{r}}(X_3^+)\tilde{\varphi}_{ijk\ell mn} &= -q^{r_3-1-n+(j+k-\ell-m)/2}[j]_q\tilde{\varphi}_{i,j-1,k,\ell+1,mn} - \\
&\quad -q^{r_3-1-n+(3j+k-3\ell-m)/2}[k]_q\tilde{\varphi}_{ij,k-1,\ell,m+1,n} + \\
&\quad +q^{-1+(-j-k+\ell+m)/2}[n-r_3]_q\tilde{\varphi}_{ijk\ell m,n+1}, \qquad (6.263)\\
\hat{\pi}_{\bar{r}}(X_1^-)\tilde{\varphi}_{ijk\ell mn} &= q^{1+(-j+k-\ell+m)/2}[i]_q\tilde{\varphi}_{i-1,jk\ell mn} + \\
&\quad +q^{i+2+(-j+k-\ell+m)/2}[j]_q\tilde{\varphi}_{i,j-1,k+1,\ell mn} + \\
&\quad +q^{i+2+(j-k-\ell+m)/2}[\ell]_q\tilde{\varphi}_{ijk,\ell-1,m+1,n}, \\
\hat{\pi}_{\bar{r}}(X_2^-)\tilde{\varphi}_{ijk\ell mn} &= -q^{(-i+j-m+n)/2}[k]_q\tilde{\varphi}_{ij,k-1,\ell mn}, \\
\hat{\pi}_{\bar{r}}(X_3^-)\tilde{\varphi}_{ijk\ell mn} &= -q^{-n+(-j-3k+\ell+3m)/2}[\ell]_q\tilde{\varphi}_{i,j+1,k,\ell-1,mn} - \\
&\quad -q^{-n+(-j-k+\ell+m)/2}[m]_q\tilde{\varphi}_{ij,k+1,\ell,m-1,n} - \\
&\quad -q^{1+(-j-k+\ell+m)/2}[n]_q\tilde{\varphi}_{ijk\ell m,n-1}. \qquad (6.264)
\end{aligned}$$

It is easy to check that $\hat{\pi}_{\bar{r}}(k_i)$, $\hat{\pi}_{\bar{r}}(X_i^\pm)$ satisfy (6.10).

From (6.263) and (6.264) one can easily write down the explicit action of the non-simple root generators. These are defined as follows [198, 360]:

$$\begin{aligned}
X_{ab}^\pm &= \pm q^{\mp 1/2}\left(q^{1/2}X_a^\pm X_b^\pm - q^{-1/2}X_b^\pm X_a^\pm\right), (ab) = (12), (23),\\
X_{13}^\pm &= \pm q^{\mp 1/2}\left(q^{1/2}X_1^\pm X_{23}^\pm - q^{-1/2}X_{23}^\pm X_1^\pm\right) = \\
&= \pm q^{\mp 1/2}\left(q^{1/2}X_{12}^\pm X_3^\pm - q^{-1/2}X_3^\pm X_{12}^\pm\right). \qquad (6.265)
\end{aligned}$$

We give only the negative roots action, since these formulae will be used below:

$$\begin{aligned}
\hat{\pi}_{\bar{r}}(X_{12}^-)\tilde{\varphi}_{ijk\ell mn} &= -q^{(i-k-\ell+n+3)/2}[j]_q\tilde{\varphi}_{i,j-1,k\ell mn} + \\
&\quad + q^{j+(i-k-\ell+n+5)/2}\lambda\,[k]_q[\ell]_q\tilde{\varphi}_{ij,k-1,\ell-1,m+1,n}\\
\hat{\pi}_{\bar{r}}(X_{23}^-)\tilde{\varphi}_{ijk\ell mn} &= -q^{(-i+k+\ell-n+3)/2}[m]_q\tilde{\varphi}_{ijk\ell,m-1,n}, \\
\hat{\pi}_{\bar{r}}(X_{13}^-)\tilde{\varphi}_{ijk\ell mn} &= -q^{3+(i+j-m-n)/2}[\ell]_q\tilde{\varphi}_{ijk,\ell-1,mn}. \qquad (6.266)
\end{aligned}$$

Further we consider the restricted functions (cf. (6.59)):

$$\begin{aligned}
\hat{\varphi}(\bar{Y}) &= \sum_{i,j,k,\ell,m,n\in\mathbb{Z}_+} \mu_{ijk\ell mn}\hat{\varphi}_{ijk\ell mn}, \qquad (6.267)\\
\hat{\varphi}_{ijk\ell mn} &= (Y_{21})^i(Y_{31})^j(Y_{32})^k(Y_{41})^\ell(Y_{42})^m\,(Y_{43})^n.
\end{aligned}$$

As a consequence of the intertwining property (6.60), we obtain that $\hat{\varphi}_{ijk\ell mn}$ obey the same transformation rules (6.261), (6.263), (6.264), and (6.266), as $\tilde{\varphi}_{ijk\ell mn}$.

Recall that we consider the representations $\hat{\pi}_{\bar{r}}$ for arbitrary complex r_i and we expect as in the $q = 1$ case (cf. Section I.4.) that whenever some $m_i = r_i + 1$ or $m_{ij} = m_i + \cdots + m_j$ $(i < j)$ is a positive integer the representations are reducible and there exist invariant subspaces. We give now two simple examples.

Let $m_1 = r_1 + 1 \in \mathbb{N}$. Then it is clear that functions $\tilde{\varphi}$ with $\mu_{ijk\ell mn} = 0$ if $i \geq m_1$ form an invariant subspace since:

$$\hat{\pi}_{\tilde{r}}(X_1^+)\tilde{\varphi}_{r_1,jk\ell mn} = q^{(j+m-\ell-2-k)/2}[k]_q\tilde{\varphi}_{r_1,j+1,k-1,\ell mn} + \\ +q^{(j+\ell-k-2-m)/2}[m]_q\tilde{\varphi}_{r_1,jk,\ell+1,m-1,n} \tag{6.268}$$

and all other operators in (6.261), (6.263) and (6.264) either preserve or lower the index i. The same is true for the functions $\hat{\varphi}$. In particular, for $r_1 = 0$ the functions in the invariant subspace do not depend on the variable Y_{21}.

Analogously if $m_3 = r_3 + 1 \in \mathbb{N}$ the functions $\tilde{\varphi}$ with $\mu_{ijk\ell mn} = 0$ if $n \geq m_3$ form an invariant subspace since:

$$\hat{\pi}_{\tilde{r}}(X_3^+)\tilde{\varphi}_{ijk\ell m,r_3} = -q^{(k+j+m-\ell-2)/2}[j]_q\tilde{\varphi}_{i,j-1,k,\ell+1,m,r_3} - \\ -q^{(k+3j+m-3\ell-2)/2}[m]_q\tilde{\varphi}_{ij,k-1,\ell,m+1,r_3}, \tag{6.269}$$

and all other operators in (6.261), (6.263) and (6.264) either preserve or lower the index n, the same holding for the functions $\hat{\varphi}$. In particular, for $r_3 = 0$ the functions in the invariant subspace do not depend on the variable Y_{43}.

It is an useful exercise to rewrite the transformation rules (6.261), (6.263), (6.264), and (6.266) for the functions $\hat{\varphi}$ using the operators (6.61), (6.62), and (6.85).

6.5.2 Intertwining Operators

The general prescription for finding the intertwining operators was already discussed in detail. In order to apply this procedure here we need the explicit action of $\pi_R(X_i^-)$ on our functions. First we have to calculate the action on the new basis $Y_{j\ell}$. We have instead of (6.67b):

$$\pi_R(X_i^-)(Y_{j\ell})^n = (-1)^{\delta_{i1}}\delta_{i\ell}\delta_{i+1,j}q^{n-1/2}[n]_q(Y_{i+1,i})^{n-1}\mathcal{D}_{i+1}\mathcal{D}_i^{-2}\mathcal{D}_{i-1}, \quad i = 1,3 \\ \pi_R(X_2^-)(Y_{j\ell})^n = q^{-\delta_{\ell1}-\delta_{j4}}q^{(n-2)(\ell-1)+1/2}[n]_qY_{2\ell}(Y_{j\ell})^{n-1}Y_{j3}\mathcal{D}_3\mathcal{D}_2^{-2}\mathcal{D}_1 \tag{6.270}$$

where we again use $\mathcal{D}_4 = \mathcal{D}_0 = Y_{jj} = 1_{\mathcal{A}}$, $Y_{j\ell} = 0$ for $j < \ell$.

Using (6.270) and (6.67a) we obtain:

$$\pi_R(X_1^-)\tilde{\varphi}^{r_1,r_2,r_3}_{ijk\ell mn} = -q^{i+j-k+\ell-m+(r_1-1)/2}[i]_q\tilde{\varphi}^{r_1-2,r_2+1,r_3}_{i-1,jk\ell mn} + \\ +q^{(r_1-1)/2}[r_1]_q\tilde{\varphi}^{r_1,r_2,r_3}_{ijk\ell mn}Z_{12}, \tag{6.271a}$$

$$\begin{aligned}
\pi_R(X_2^-)\tilde{\varphi}^{r_1,r_2,r_3}_{ijk\ell mn} &= q^{2k+m-n+(r_2-1)/2}\,[j]_q\tilde{\varphi}^{r_1+1,r_2-2,r_3+1}_{i+1,j-1,k\ell mn} + \\
&+q^{k+m-n+(r_2-3)/2}\,[k]_q\tilde{\varphi}^{r_1+1,r_2-2,r_3+1}_{ij,k-1,\ell mn} + \\
&+ q^{k-j+2m-n+(r_2-3)/2}\,[\ell]_q\tilde{\varphi}^{r_1+1,r_2-2,r_3+1}_{i+1,jk,\ell-1,m,n+1} + \\
&+ q^{m-n+(r_2-5)/2}\,[m]_q\tilde{\varphi}^{r_1+1,r_2-2,r_3+1}_{ijk\ell,m-1,n+1} - \\
&-q^{2m-n+(r_2-3)/2}\lambda[k]_q[\ell]_q\tilde{\varphi}^{r_1+1,r_2-2,r_3+1}_{i,j+1,k-1,\ell-1,m,n+1} + \\
&+ q^{(r_2-1)/2}\,[r_2]_q\tilde{\varphi}^{r_1,r_2,r_3}_{ijk\ell mn}Z_{23}\,,
\end{aligned} \tag{6.271b}$$

$$\begin{aligned}
\pi_R(X_3^-)\tilde{\varphi}^{r_1,r_2,r_3}_{ijk\ell mn} &= q^{n+(r_3-1)/2}\,[n]_q\tilde{\varphi}^{r_1,r_2+1,r_3-2}_{ijk\ell m,n-1} + \\
&+ q^{(r_3-1)/2}\,[r_3]_q\tilde{\varphi}^{r_1,r_2,r_3}_{ijk\ell mn}Z_{34}\,,
\end{aligned} \tag{6.271c}$$

where we have labelled the functions also with the representation parameters r_s. As in the classical case [197] the right action is taking out from the representation space $\mathcal{C}_{\bar{r}}$, and while some of the terms are functions from other representation spaces (depending on which X_s^- is acting), there are terms involving the $Z_{j\ell}$ variables which do not belong to any of our representation spaces. These terms vanish only when the respective r_s is equal to zero, and in these cases (6.271) describe three different intertwining operators corresponding to the simple roots of the root system of $sl(4)$. If $r_s \in \mathbb{N}$ then the terms with $Z_{j\ell}$ vanish exactly when we take $(\pi_R(X_s^-))^{m_s}$ [197], [211], $m_s = r_s + 1$.

Indeed, we know from the general prescription that if $m_s \in \mathbb{N}$ then there exist an intertwining operator $I_s^{m_s} = (\pi_R(X_s^-))^{m_s}$. We have the following intertwining properties (cf. (6.73)):

$$I_1^{m_1} \circ \pi_{m_1,m_2,m_3} = \pi_{-m_1,m_{12},m_3} \circ I_1^{m_1}\,, \quad m_1 \in \mathbb{N}\,, \tag{6.272a}$$

$$I_2^{m_2} \circ \pi_{m_1,m_2,m_3} = \pi_{m_{12},-m_2,m_{23}} \circ I_2^{m_2}\,, \quad m_2 \in \mathbb{N}\,, \tag{6.272b}$$

$$I_3^{m_3} \circ \pi_{m_1,m_2,m_3} = \pi_{m_1,m_{23},-m_3} \circ I_3^{m_3}\,, \quad m_3 \in \mathbb{N}\,, \tag{6.272c}$$

where we label the representations with the numbers m_s instead of $r_s = m_s - 1$ to simplify the notation. The expressions for two of these operators (up to $q^{\cdots}$ factors) are:

$$(\pi_R(X_1^-))^{m_1}\tilde{\varphi}^{m_1,m_2,m_3}_{ijk\ell mn} = (-1)^{m_1}\frac{[i]_q!}{[i-m_1]_q!}\tilde{\varphi}^{-m_1,m_{12},m_3}_{i-m_1,jk\ell mn} \tag{6.273}$$

$$(\pi_R(X_3^-))^{m_3}\tilde{\varphi}^{m_1,m_2,m_3}_{ijk\ell mn} = \frac{[n]_q!}{[n-m_3]_q!}\tilde{\varphi}^{m_1,m_{23},-m_3}_{ijk\ell m,n-m_3} \tag{6.274}$$

It will be convenient to use also the following notation for the coordinates of the coset:

$$\xi = Y_{21}\,, \quad x = Y_{31}\,, \quad u = Y_{32}\,, \quad w = Y_{41}\,, \quad y = Y_{42}\,, \quad \eta = Y_{43}. \tag{6.275}$$

Having in mind the preceding discussion let us introduce the following q-difference operators (using notation (6.61), (6.62), (6.85), and (6.275)):

$$\hat{I}_1 \equiv -\,q^{(r_1-1)/2}\hat{\mathscr{D}}_\xi T_\xi\, T_x\, T_w\,(T_u T_y)^{-1} \tag{6.276a}$$

$$\begin{aligned}\hat{I}_2 \equiv\ & q^{(r_2-3)/2}\Big(q\hat{M}_\xi\hat{\mathscr{D}}_x T_u^2 + \hat{\mathscr{D}}_u T_u + \\ & +\hat{M}_\xi\hat{M}_\eta\hat{\mathscr{D}}_w T_x^{-1}\, T_y\, T_u + q^{-1}\hat{M}_\eta\hat{\mathscr{D}}_y - \\ & -\lambda\hat{M}_x\,\hat{M}_\eta\hat{\mathscr{D}}_u\hat{\mathscr{D}}_w T_y\Big)T_y T_\eta^{-1}\end{aligned} \tag{6.276b}$$

$$\hat{I}_3 \equiv q^{(r_3-1)/2}\hat{\mathscr{D}}_\eta T_\eta \tag{6.276c}$$

It is not difficult to see that if $m_s \in \mathbb{N}$ we have (cf. (6.76)):

$$\hat{I}_s^{m_s} = I_s^{m_s} = (\pi_R(X_s^-))^{m_s} \tag{6.277}$$

We go back to the general situation. There are altogether six different operators corresponding to the positive roots of Δ which exist when the respective number from the set $m_1, m_2, m_3, m_{12}, m_{23}, m_{13}$ is a positive integer. We have considered the three simple roots. To obtain the remaining three operators it is enough to substitute in (6.74) the expressions in the $sl(4)$ case given in (2.37) for the singular vectors corresponding to the three nonsimple roots $\alpha_{12}, \alpha_{23}, \alpha_{13}$, realized when $m_{12} \in \mathbb{N}$, $m_{23} \in \mathbb{N}$, $m_{13} \in \mathbb{N}$, respectively. We shall give explicitly the cases we need in the next chapter.

7 q-Maxwell Equations Hierarchies

Summary

In this chapter we start by using q-conformal invariance to propose a new q-Minkowski space–time and q-Maxwell equations. We are using an indexless formulation in which the spin properties are expressed not through Lorentz indices but through polynomial dependence on two conjugate variables, $z, \bar{z}$. The proposed new q-Minkowski coordinates together with $z, \bar{z}$ can be interpreted as the six local coordinates of a $SU_q(2,2)$ flag manifold. The new q-Maxwell equations are q-conformal invariant and are the first members of an infinite new hierarchy of q-difference equations parametrized by an integer $n \in \mathbb{Z}_+$. We also present a generalized q-Maxwell equations hierarchy indexed by two integers which includes the initial q-Maxwell equations hierarchy as a subfamily. Another subfamily of the generalized q-Maxwell equations hierarchy is the potential q-Maxwell equations hierarchy. Yet another subfamily of the generalized q-Maxwell equations hierarchy is the q-d'Alembert equations hierarchy with first member the q-d'Alembert equation. The latter hierarchy intersects the initial q-Maxwell equations hierarchy exactly with the q-Maxwell equations. Further, we present polynomial solutions and q-plane-wave solutions of the q-d'Alembert equation. Next, we present q-plane-wave solutions of the potential q-Maxwell hierarchy. Then we present q-plane-wave solutions of the full q-Maxwell equations. We also consider the q-Weyl gravity equations hierarchy and present q-plane-wave solutions of the lowest member which is q-deformation of linear conformal gravity. As a small detour we present a multiparameter deformation of quantum Minkowski space–time. This chapter is based mainly on [214, 215, 221, 226, 229, 237–240, 247].

7.1 Maxwell Equations Hierarchy

The present section follows mostly [214]. It is well known that Maxwell equations

$$\partial^\mu F_{\mu\nu} = J_\nu \tag{7.1a}$$

$$\partial^\mu \, {}^*F_{\mu\nu} = 0, \tag{7.1b}$$

(where ${}^*F_{\mu\nu} \equiv \epsilon_{\mu\nu\rho\sigma} F^{\rho\sigma}$, $\epsilon_{\mu\nu\rho\sigma}$ being totally antisymmetric with $\epsilon_{0123} = 1$), or, equivalently

$$\begin{aligned} &\partial_k E_k = J_0 (= 4\pi\rho), \quad \partial_0 E_k - \varepsilon_{k\ell m} \partial_\ell H_m = J_k (= -4\pi j_k), \\ &\partial_k H_k = 0\,, \quad \partial_0 H_k + \varepsilon_{k\ell m} \partial_\ell E_m = 0\,, \end{aligned} \tag{7.2}$$

where $E_k \equiv F_{k0}$, $H_k \equiv (1/2)\varepsilon_{k\ell m} F_{\ell m}$, can be rewritten in the following manner:

$$\partial_k F_k^\pm = J_0\,, \quad \partial_0 F_k^\pm \pm i \varepsilon_{k\ell m} \partial_\ell F_m^\pm = J_k\,, \tag{7.3}$$

where

$$F_k^\pm \equiv E_k \pm i H_k. \tag{7.4}$$

DOI 10.1515/9783110427707-007

Not so well known is the fact that the eight equations in (7.3) can be rewritten as two conjugate scalar equations in the following way:

$$I^+ F^+(z) = J(z, \bar{z}) , \tag{7.5a}$$
$$I^- F^-(\bar{z}) = J(z, \bar{z}) , \tag{7.5b}$$

where

$$I^+ = \bar{z}\partial_+ + \partial_v - \tfrac{1}{2}\left(\bar{z}z\partial_+ + z\partial_v + \bar{z}\partial_{\bar{v}} + \partial_-\right)\partial_z , \tag{7.6a}$$
$$I^- = z\partial_+ + \partial_{\bar{v}} - \tfrac{1}{2}\left(\bar{z}z\partial_+ + z\partial_v + \bar{z}\partial_{\bar{v}} + \partial_-\right)\partial_{\bar{z}} , \tag{7.6b}$$

$$x_\pm \equiv x_0 \pm x_3, \quad v \equiv x_1 - ix_2, \quad \bar{v} \equiv x_1 + ix_2, \tag{7.7a}$$
$$\partial_\pm \equiv \partial/\partial x_\pm, \quad \partial_v \equiv \partial/\partial v, \quad \partial_{\bar{v}} \equiv \partial/\partial \bar{v}, \tag{7.7b}$$

$$F^+(z) \equiv z^2(F_1^+ + iF_2^+) - 2zF_3^+ - (F_1^+ - iF_2^+) , \tag{7.8a}$$
$$F^-(\bar{z}) \equiv \bar{z}^2(F_1^- - iF_2^-) - 2\bar{z}F_3^- - (F_1^- + iF_2^-) , \tag{7.8b}$$
$$J(z, \bar{z}) \equiv \bar{z}z(J_0 + J_3) + z(J_1 + iJ_2) + \bar{z}(J_1 - iJ_2) + (J_0 - J_3) = \equiv \bar{z}zJ_+ + zJ_v + \bar{z}J_{\bar{v}} + J_- \tag{7.8c}$$

where we continue to suppress the x_μ, respectively, $x_\pm, v, \bar{v}$, dependence in F and J. (The conjugation mentioned above is standard and in our terms it is: $I^+ \longleftrightarrow I^-$, $F^+(z) \longleftrightarrow F^-(\bar{z})$.)

It is easy to recover (7.3) from (7.5) – just note that both sides of each equation are first-order polynomials in each of the two variables z and $\bar{z}$; then comparing the independent terms in (7.5) one gets at once (7.3).

Writing the Maxwell equations in the simple form (7.5) has also important conceptual meaning. The point is that each of the two scalar operators I^+, I^- is indeed a single object, namely, it is an intertwiner of the conformal group, while the individual components in (7.1)–(7.3) do not have this interpretation. This is also the simplest way to see that the Maxwell equations are conformally invariant, since this is equivalent to the intertwining property.

Let us be more explicit. The physically relevant representations T^χ of the four-dimensional conformal algebra $su(2,2)$ may be labelled by $\chi = [n_1, n_2; d]$, where n_1, n_2 are non-negative integers fixing finite-dimensional irreducible representations of the Lorentz subalgebra, (the dimension being $(n_1 + 1)(n_2 + 1)$), and d is the conformal dimension (or energy). (In the literature these Lorentz representations are labelled also by $(j_1, j_2) = (n_1/2, n_2/2)$.) Then the intertwining properties of the operators in (7.6) are given by:

$$I^+ : C^+ \longrightarrow C^0\,, \quad I^+ \circ T^+ = T^0 \circ I^+\,, \tag{7.9a}$$
$$I^- : C^- \longrightarrow C^0\,, \quad I^- \circ T^- = T^0 \circ I^-\,, \tag{7.9b}$$

where $T^a = T^{\chi^a}$, $a = 0, +, -$, $C^a = C^{\chi^a}$ are the representation spaces, and the signatures are given explicitly by:

$$\chi^+ = [2, 0; 2]\,, \quad \chi^- = [0, 2; 2]\,, \quad \chi^0 = [1, 1; 3]\,, \tag{7.10}$$

as anticipated. Indeed, $(n_1, n_2) = (1, 1)$ is the four-dimensional Lorentz representation (carried by J_μ above), and $(n_1, n_2) = (2, 0), (0, 2)$ are the two conjugate three-dimensional Lorentz representations (carried by $F_k^\pm$ above), while the conformal dimensions are the canonical dimensions of a current ($d = 3$), and of the Maxwell field ($d = 2$). We see that the variables $z, \bar{z}$ are related to the spin properties, and we shall call them "spin variables". More explicitly, a Lorentz spin-tensor $G(z, \bar{z})$ with signature (n_1, n_2) is a polynomial in $z, \bar{z}$ of order n_1, n_2, respectively.

Formulae (7.9) and (7.10) are part of an infinite hierarchy of couples of first-order intertwiners given already in [235] for the Euclidean conformal group $SU^*(4)$, and then for the conformal group $SU(2, 2)$ in [194, 503]. (Note that [235, 503] use a different approach, while [194] already uses the essential features of [197] in the context of the conformal group; see also Volume 1.)

Explicitly, instead of (7.9) and (7.10) we have [194]:

$$I_n^+ : C_n^+ \longrightarrow C_n^0\,, \quad I_n^+ \circ T_n^+ = T_n^0 \circ I_n^+\,, \tag{7.11a}$$
$$I_n^- : C_n^- \longrightarrow C_n^0\,, \quad I_n^- \circ T_n^- = T_n^0 \circ I_n^-\,, \tag{7.11b}$$

where $T_n^a = T^{\chi_n^a}$, $C_n^a = C^{\chi_n^a}$, and the signatures are:

$$\chi_n^+ = [n + 2, n; 2]\,, \quad \chi_n^- = [n, n + 2; 2]\,, \quad \chi_n^0 = [n + 1, n + 1; 3]\,, \quad n \in \mathbb{Z}_+\,, \tag{7.12}$$

while instead of (7.5) we have:

$$I_n^+ F_n^+(z, \bar{z}) = J_n(z, \bar{z})\,, \tag{7.13a}$$
$$I_n^- F_n^-(z, \bar{z}) = J_n(z, \bar{z})\,, \tag{7.13b}$$

where

$$I_n^+ = \frac{n+2}{2}\Big(\bar{z}\partial_+ + \partial_v\Big) - \tfrac{1}{2}\Big(\bar{z}z\partial_+ + z\partial_v + \bar{z}\partial_{\bar{v}} + \partial_-\Big)\partial_z\,, \tag{7.14a}$$
$$I_n^- = \frac{n+2}{2}\Big(z\partial_+ + \partial_{\bar{v}}\Big) - \tfrac{1}{2}\Big(\bar{z}z\partial_+ + z\partial_v + \bar{z}\partial_{\bar{v}} + \partial_-\Big)\partial_{\bar{z}}\,, \tag{7.14b}$$

($n \in \mathbb{Z}_+$), while $F_n^+(z, \bar{z})$, $F_n^-(z, \bar{z})$, $J_n(z, \bar{z})$, are polynomials in $z, \bar{z}$ of degrees $(n + 2, n)$, $(n, n + 2)$, $(n + 1, n + 1)$, respectively, as explained above. If we want to use the notation with indices as in (7.1), then $F_n^+(z,\bar{z})$ and $F_n^-(z,\bar{z})$ correspond to $F_{\mu\nu,\alpha_1,\dots,\alpha_n}$, which is antisymmetric in the indices μ,ν, symmetric in $\alpha_1,\dots,\alpha_n$, and traceless in every pair of indices, while $J_n(z,\bar{z})$ corresponds to $J_{\mu,\alpha_1,\dots,\alpha_n}$, which is symmetric and traceless in every pair of indices. Note, however, that the analogues of (7.1) would be much more complicated if one wants to write explicitly all components. The crucial advantage of (7.13) is that the operators $I_n^\pm$ are given just by a slight generalization of $I^\pm = I_0^\pm$.

We shall call the hierarchy of equations (7.13) the **Maxwell hierarchy**. The Maxwell equations are the zero member of this hierarchy.

To proceed further we rewrite (7.14) in the following form:

$$I_n^+ = \tfrac{1}{2}\Big((n+2)I_1I_2 - (n+3)I_2I_1\Big)\,, \tag{7.15a}$$

$$I_n^- = \tfrac{1}{2}\Big((n+2)I_3I_2 - (n+3)I_2I_3\Big)\,, \tag{7.15b}$$

where

$$I_1 \equiv \partial_z\,, \quad I_2 \equiv \bar{z}z\partial_+ + z\partial_v + \bar{z}\partial_{\bar{v}} + \partial_-\,, \quad I_3 \equiv \partial_{\bar{z}}\,. \tag{7.16}$$

We note in passing that group-theoretically the operators I_a correspond to the three simple roots of the root system of $sl(4)$, while the operators $I_n^\pm$ correspond to the two nonsimple nonhighest roots [194, 197].

Remark 7.1. If we use induction from the ten-dimensional parabolic $P_1 = M_1A_1N_1$ (cf. [232]), the variables there are $x_+, v, \bar{v}, z, \bar{z}$ (from N_1), y (from M_1). The relation between the variables of the P_0 (or P_2) induction and P_1 induction is:

$$x_+^0 = x_+^1 + y(z^1)^2\,, \quad x_-^0 = y\,, \quad v^0 = v^1 + yz^1\,, \quad z^0 = z^1$$

$$x_+^1 = x_+^0 - x_-^0(z^0)^2\,, \quad y = x_-^0\,, \quad v^1 = v^0 - x_-^0z^0\,, \quad z^1 = z^0\,.$$

From this change of variables follow:

$$\partial_+^0 = \partial_+^1\,, \partial_v^0 = \partial_v^1\,, \partial_-^0 = \partial_y - z^1\partial_v^1 - \bar{z}^1\partial_{\bar{v}}^1 - z\bar{z}\partial_+^1$$

$$\partial_z^0 = \partial_z^1 - y\bar{z}^1\partial_+^1 - y\partial_v^1\,, \partial_{\bar{z}}^0 = \partial_{\bar{z}}^1 - yz^1\partial_+^1 - y\partial_{\bar{v}}^1\,.$$

Correspondingly, the operators I_k from (7.16) in the P_1 variables are:

$$I_1 = \partial_z - y\bar{z}\partial_+ - y\partial_v\,, I_2 = \partial_y\,, I_3 = \partial_{\bar{z}} - yz\partial_+ - y\partial_{\bar{v}}$$

where we have omitted the supscript 1. ◊

This is the form - (7.15) that we generalize for the q-deformed case. In fact, we can write at once the general form, which follows from the expressions for the singular vectors corresponding to those nonsimple nonhighest roots given by (2.37) with $u = 1$, $m = 1$, $n_{i_1} = 1$, $q_{i_1} = 1$:

$$ {}_qI_n^+ = \frac{1}{2}\Big([n+2]_q I_1^q I_2^q - [n+3]_q I_2^q I_1^q \Big) , \tag{7.17a} $$

$$ {}_qI_n^- = \frac{1}{2}\Big([n+2]_q I_3^q I_2^q - [n+3]_q I_2^q I_3^q \Big) . \tag{7.17b} $$

It is our task (using the previous sections) to make this form explicit by first generalizing the variables and then the functions and the operators.

7.2 Quantum Minkowski Space–Time

7.2.1 q-Minkowski Space–Time

The variables $x_\pm, v, \bar{v}, z, \bar{z}$ have definite group-theoretical meaning, namely, they are six local coordinates on the coset $\mathscr{Y} = SL(4)/B$, where B is the Borel subgroup of $SL(4)$ consisting of all upper diagonal matrices. (Equally well one may take the coset $SL(4)/B^-$, where B^- is the Borel subgroup of lower diagonal matrices.) Under the natural conjugation (cf. also below), this is also a coset of the conformal group $SU(2,2)$.

We know from Section 4.5 what are the properties of the noncommutative coordinates on the $SL_q(4)$ coset. We make the following identification (compare with (6.275)):

$$ \begin{aligned} x_+ &= w = Y_{41} , \quad & x_- &= u = Y_{32} \\ v &= x = Y_{31} , \quad & \bar{v} &= y = Y_{42} \\ z &= \xi = Y_{21} , \quad & \bar{z} &= \eta = Y_{43} \end{aligned} \tag{7.18} $$

for the q-Minkowski space–time coordinates and for the spin coordinates, which we denote as their classical counterparts. Thus, we obtain for the commutation rules of the q-Minkowski space–time coordinates (cf. (6.258)):

$$ \begin{aligned} & x_\pm v = q^{\pm 1} v x_\pm , \quad x_\pm \bar{v} = q^{\pm 1} \bar{v} x_\pm , \\ & x_+ x_- - x_- x_+ = \lambda v \bar{v} , \quad \bar{v} v = v \bar{v} . \end{aligned} \tag{7.19} $$

It is easy to notice that these relations are as the $GL_q(2)$ commutation relations [462], if we identify our coordinates with the standard a, b, c, d generators of $GL_q(2)$ as follows:

$$M = \begin{pmatrix} a & b \\ c & d \end{pmatrix} = \begin{pmatrix} x_+ & v \\ \bar{v} & x_- \end{pmatrix}. \tag{7.20}$$

The q-Minkowski length is defined as the $GL_q(2)$ q-determinant:

$$\ell_q \doteq \det_q M = ad - qbc = x_+x_- - q\bar{v}v\,, \tag{7.21}$$

and hence it commutes with the q-Minkowski coordinates. It has the correct classical limit $\ell_{q=1} = x_0^2 - \vec{x}^2$.

We know from (5.183) that for q phase ($|q| = 1$) the commutation relations (7.19) are preserved by an antilinear anti-involution ω acting as (cf. (5.183)):

$$\omega(x_\pm) = x_\pm\,, \quad \omega(v) = \bar{v}\,, \tag{7.22}$$

from which follows also that $\omega(\ell_q) = \ell_q$.

Remark 7.2. Note that relations (7.19) are different from the commutation relations of q-Minkowski space–time (with q real) in [123, 455, 544], (cf. also [173, 174], and references therein). Later, Majid [456] has shown that the latter q-Minkowski space of [123, 455, 544] can be obtained by a quantum Wick rotation (twisting) from a q-Euclidean space. The latter is also related to $GL_q(2)$, as our q-Minkowski space; however, for q real and under a different anti-linear anti-involution: $\tilde{\omega}_E(a) = d$, $\tilde{\omega}_E(b) = -q^{-1}c$; that is, for the matrix M (cf. (7.20)) this is the unitary * while with our conjugation (7.22) M is hermitian. ◊

The commutation rules of the spin variables $\bar{z}, z$ between themselves, with the q-Minkowski coordinates and with the q-Minkowski length are (cf. (6.258)):

$$\begin{aligned}
&\bar{z}z = z\bar{z}\,,\\
&x_+z = q^{-1}zx_+\,, \quad x_-z = qzx_- - \lambda v\,,\\
&vz = q^{-1}zv\,, \quad \bar{v}z = qz\bar{v} - \lambda x_+\,,\\
&\bar{z}x_+ = qx_+\bar{z}\,, \quad \bar{z}x_- = q^{-1}x_-\bar{z} + \lambda\bar{v}\,,\\
&\bar{z}v = q^{-1}v\bar{z} + \lambda x_+\,, \quad \bar{z}\bar{v} = q\bar{v}\bar{z}\,,\\
&z\ell_q = \ell_q z\,, \quad \bar{z}\ell_q = \ell_q\bar{z}.
\end{aligned} \tag{7.23}$$

Certainly, the commutation relations (7.23) are also preserved (for q phase) by the conjugation ω – supplementing (7.22) by $\omega(z) = \bar{z}$ (all follow from (5.183)). Thus, with this conjugation $\mathscr{Y}_q$ becomes a coset of $SU_q(2,2)$.

From (6.260) we know the normally ordered basis of the q-coset $\mathcal{Y}_q$ considered as an associative algebra:

$$\hat{\varphi}_{ijk\ell mn} = z^i v^j x_-^k x_+^\ell \bar{v}^m \bar{z}^n, \quad i, j, k, \ell, m, n \in \mathbb{Z}_+. \tag{7.24}$$

Let us denote by $\mathcal{Z}$, $\bar{\mathcal{Z}}$, and $\mathcal{M}_q$ the associative algebras with unity generated by $z, \bar{z}$ and $x_\pm, v, \bar{v}$, respectively. These three algebras are subalgebras of $\mathcal{Y}_q$, and we notice the following structure of $\mathcal{Y}_q$:

$$\mathcal{Y}_q \cong \mathcal{Z} \ltimes \mathcal{M}_q \rtimes \bar{\mathcal{Z}}, \tag{7.25}$$

where $A \ltimes B$ denotes the tensor product of A and B with A acting on B.

7.2.2 Multiparameter Quantum Minkowski Space–Time

In this subsection following [221], we shall present the multiparameter version of our quantum Minkowski space–time. We start from the case $n = 4$ of the multiparameter deformation $GL_{q,\mathbf{q}}(n)$ of $GL(n)$, which we discussed in Section 4.5. The flag manifold $\tilde{\mathcal{Y}}_{q,\mathbf{q}} = GL_{q,\mathbf{q}}(n)/\tilde{B}_{q,\mathbf{q}}(n)$ depends on the same number of parameters $(n^2 - n + 2)/2$. For $n = 4$ we pass from the variables Y_{ij} to the variables on the above coset in the manner of (7.18), keeping the same notation as in the one-parameter case of Section 6.5 and the previous subsection. Thus, we obtain the explicit multiparameter commutation relations (instead of (7.19) and (7.23), $\lambda \equiv q - q^{-1}$):

$$\begin{aligned}
x_+ v &= \frac{q_{23}q_{34}}{q_{24}} v x_+ , & \bar{v} x_+ &= \frac{q_{14}}{q_{12}q_{24}} x_+ \bar{v} , \\
x_- v &= \frac{q_{13}}{q_{12}q_{23}} v x_- , & \bar{v} x_- &= \frac{q_{13}q_{34}}{q_{14}} x_- \bar{v} , \\
\bar{v} v &= \frac{q_{13}q_{34}}{q_{12}q_{24}} v \bar{v} , \\
\frac{q\, q_{24}}{q_{23}q_{34}} x_+ x_- &= \frac{q_{12}q_{24}}{q\, q_{14}} x_- x_+ + \lambda v \bar{v} ,
\end{aligned} \tag{7.26}$$

$$\begin{aligned}
\bar{z} z &= \frac{q_{13}q_{24}}{q_{14}q_{23}} z \bar{z} , \\
\bar{z} x_+ &= \frac{q_{13}q_{34}}{q_{14}} x_+ \bar{z} , & \bar{z} x_- &= \frac{q_{23}q_{34}}{q^2 q_{24}} x_- \bar{z} + \lambda \bar{v} , \\
\bar{z} \bar{v} &= \frac{q_{23}q_{34}}{q_{24}} \bar{v} \bar{z} , & \bar{z} v &= \frac{q_{13}q_{34}}{q^2 q_{14}} v \bar{z} + \lambda x_+ , \\
x_+ z &= \frac{q_{14}}{q_{12}q_{24}} z x_+ , & x_- z &= \frac{q^2 q_{13}}{q_{12}q_{23}} z x_- - \lambda v , \\
v z &= \frac{q_{13}}{q_{12}q_{23}} z v , & \bar{v} z &= \frac{q^2 q_{14}}{q_{12}q_{24}} z \bar{v} - \lambda x_+ .
\end{aligned} \tag{7.27}$$

Thus, in (7.26) we have the expected seven-parameter quantum Minkowski space–time.

We note that when all deformation parameter are phases; that is, $|q| = 1$, $|q_{ij}| = 1$, and in addition hold the following relations:

$$q_{13} = \frac{q_{12}q_{24}}{q_{34}}, \quad q_{14} = \frac{q_{12}q_{24}^2}{q_{23}q_{34}}, \tag{7.28}$$

then the commutation relations (7.26) and (7.27) are preserved by the antilinear anti-involution ω acting as in the previous subsection.

Further, we recall from Section 4.5.5 that the dual quantum algebra $U_{q,\mathbf{q}}(gl(n))$ has the quantum algebra $U_{q,\mathbf{q}}(sl(n))$ as a commutation subalgebra but not as a co-subalgebra. In order to achieve the complete splitting of $U_{q,\mathbf{q}}(sl(n))$ we have to impose some relations between the parameters; thus, the genuine multiparameter deformation $U_{q,\mathbf{q}}(sl(n))$ depends on $(n^2 - 3n + 4)/2$ parameters. Using the same conditions we also ensure that we can restrict from $GL_{q,\mathbf{q}}(n)$ to $SL_{q,\mathbf{q}}(n)$.

Thus, in the case of $n = 4$ for the genuine $U_{q,\mathbf{q}}(sl(4))$ we have four parameters. Explicitly, we achieve this by imposing that the parameters $q_{i,i+1}$ are expressed through the rest as:

$$q_{12} = \frac{q^3}{q_{13}q_{14}}, \quad q_{23} = \frac{q^4}{q_{13}q_{14}q_{24}}, \quad q_{34} = \frac{q^3}{q_{14}q_{24}}. \tag{7.29}$$

Thus, the four-parameter quantum Minkowski space–time and the embedding quantum flag manifold $\mathscr{Y}_{q,\mathbf{q}}$ are given by (7.26) and (7.27) with (7.29) enforced.

If we would like to enforce also the conjugation ω, then there are more relations between the deformation parameters, namely, we get:

$$q_{12} = q_{23} = q_{34} = \frac{q^2}{q_{14}}, \quad q_{13} = q_{24} = q, \tag{7.30}$$

and all deformation parameter are phases.

Thus, in this case we have a two-parameter deformation and using the above relations (7.26) and (7.27) simplify as follows:

$$\begin{aligned} &x_+v = p\,vx_+, \qquad \bar{v}x_+ = p^{-1}x_+\bar{v}, \\ &x_-v = p^{-1}vx_-, \qquad \bar{v}x_- = p\,x_-\bar{v}, \\ &\bar{v}v = v\bar{v}, \\ &\frac{q}{p}x_+x_- = \frac{p}{q}x_-x_+ + \lambda v\bar{v}, \end{aligned} \tag{7.31}$$

$$\begin{aligned} &\bar{z}z = z\bar{z}, \\ &\bar{z}x_+ = p\,x_+\bar{z}, \qquad \bar{z}x_- = \frac{p}{q^2}x_-\bar{z} + \lambda\bar{v}, \end{aligned} \tag{7.32}$$

$$\bar{z}\bar{v} = p\,\bar{v}\bar{z}\,, \qquad \bar{z}v = \frac{p}{q^2}\,v\bar{z} + \lambda x_+\,,$$

$$x_+z = p^{-1}\,zx_+\,, \qquad x_-z = \frac{q^2}{p}\,zx_- - \lambda v\,,$$

$$vz = p^{-1}\,zv\,, \qquad \bar{v}z = \frac{q^2}{p}\,z\bar{v} - \lambda x_+\,,$$

where $p \equiv q^3/q_{14}^2$.

7.3 q-Maxwell Equations Hierarchy

We return to the one-parameter setting of Section 7.2.1. We introduce now the representation spaces C^χ, $\chi = [n_1, n_2; d]$. The elements of C^χ, which we shall call (abusing the notion) functions, are polynomials in $z, \bar{z}$ of degrees n_1, n_2, respectively, and formal power series in the q-Minkowski variables. (In the general $U_q(sl(n))$ situation the signatures n_1, n_2 are complex numbers and the functions are formal power series in $z, \bar{z}$ too, cf. (5.38b).) Namely, these functions are given by:

$$\hat{\varphi}_{n_1,n_2}(\bar{Y}) = \sum_{\substack{i,j,k,\ell,m,n\in\mathbb{Z}_+\\ i\le n_1,\ n\le n_2}} \mu^{n_1,n_2}_{ijk\ell mn}\,\hat{\varphi}_{ijk\ell mn}\,, \tag{7.33}$$

where $\bar{Y}$ denotes the set of the six coordinates on $\mathscr{Y}_q$. Thus the analogues of $F_n^\pm$, J_n, cf. (7.13), are:

$${}_qF_n^+ = \hat{\varphi}_{n+2,n}(\bar{Y})\,, \quad {}_qF_n^- = \hat{\varphi}_{n,n+2}(\bar{Y})\,, \quad {}_qJ_n = \hat{\varphi}_{n+1,n+1}(\bar{Y}). \tag{7.34}$$

Using the above we now present explicitly a q version of the Maxwell hierarchy of equations. We recall that the explicit form of the operators I_a in (7.16) is obtained by the infinitesimal right action of the three simple root generators of $sl(4)$ on the coset $\mathscr{Y}$ (cf. (5.150)). Adapting this to our notation we have for the q-analogues of I_a (cf. (6.276)):

$${}_qI_1 = \hat{\mathscr{D}}_z T_z\, T_v\, T_+\,(T_-T_{\bar{v}})^{-1} \tag{7.35a}$$

$$\begin{aligned}{}_qI_2 = \Big(&q\hat{M}_z\hat{\mathscr{D}}_v T_-^2 + \hat{\mathscr{D}}_- T_- + \\ &+\hat{M}_z\hat{M}_{\bar{z}}\hat{\mathscr{D}}_+ T_-\, T_{\bar{v}}\, T_v^{-1} + q^{-1}\hat{M}_{\bar{z}}\hat{\mathscr{D}}_{\bar{v}} - \\ &-\lambda\hat{M}_v\hat{M}_{\bar{z}}\hat{\mathscr{D}}_-\hat{\mathscr{D}}_+ T_{\bar{v}}\Big)T_{\bar{v}}T_{\bar{z}}^{-1}\end{aligned} \tag{7.35b}$$

$${}_qI_3 = \hat{\mathscr{D}}_{\bar{z}} T_{\bar{z}}. \tag{7.35c}$$

With this we have now the q-Maxwell hierarchy of equations – it remains just to substitute the operators of (7.35) in (7.17). In fact, we can also rewrite these in the q-analog of (7.13). We have:

$$
\begin{aligned}
{}_qI_n^+ = \tfrac{1}{2}\Bigg(&\left(q\,\hat{\mathscr{D}}_v + \hat{M}_{\bar z}\,\hat{\mathscr{D}}_+\,(T_-\,T_v)^{-1}T_{\bar v}\right)[n+2-N_z]_q - \\
&-q^{-n-2}\Big(\hat{\mathscr{D}}_-\,T_- + q^{-1}\hat{M}_{\bar z}\hat{\mathscr{D}}_{\bar v} - \\
&-\lambda\,\hat{M}_v\,\hat{M}_{\bar z}\,\hat{\mathscr{D}}_-\,\hat{\mathscr{D}}_+\,T_{\bar v}\Big)\,\hat{\mathscr{D}}_z\Bigg)\,T_+\,T_-\,T_v\,T_z\,T_{\bar z}^{-1}
\end{aligned} \tag{7.36a}
$$

$$
\begin{aligned}
{}_qI_n^- = \tfrac{1}{2}\Big(&\hat{\mathscr{D}}_{\bar v} + q\,\hat{M}_z\,\hat{\mathscr{D}}_+\,T_{\bar v}\,T_-\,T_v^{-1} - \\
&-q\lambda\,\hat{M}_v\,\hat{\mathscr{D}}_-\,\hat{\mathscr{D}}_+\,T_{\bar v}\Big)\,T_{\bar v}\,[n+2-N_{\bar z}]_q - \\
&-\tfrac{1}{2}\,q^{n+3}\Big(\hat{\mathscr{D}}_- + q\,\hat{M}_z\,\hat{\mathscr{D}}_v T_-\Big)\,\hat{\mathscr{D}}_{\bar z}\,T_-\,T_{\bar v}.
\end{aligned} \tag{7.36b}
$$

Clearly, for $q = 1$ the operators in (7.35) and (7.36) coincide with (7.15) and (7.16), respectively.

With this the final result for the q-Maxwell hierarchy of equations is (cf. (7.34)):

$$ {}_qI_n^+\,{}_qF_n^+ = {}_qJ_n\,, \tag{7.37a} $$

$$ {}_qI_n^-\,{}_qF_n^- = {}_qJ_n. \tag{7.37b} $$

Remark 7.3. Note that our free q-Maxwell equations, obtained from (7.37) for $n = 0$, and ${}_qJ_0 = 0$, are different from the free q-Maxwell equations of [472, 508]. The advantages of our equations are (1) they have simple indexless form; (2) we have a whole hierarchy of equations; (3) we have the full equations, and not only their free counterparts; (4) our equations are q-conformal invariant, not only q-Lorentz [472], or q-Poincaré [508], invariant. ◊

Formulae (7.13), (7.11), and (7.12) are part of a much more general classification scheme (mentioned above, cf. [194, 198]) involving also other intertwining operators, and of arbitrary order. A subset of this scheme are two infinite two-parameter families of representations which are intertwined by the same operators (7.14) (cf. [213]). The latter set was called generalized q-Maxwell hierarchy, the q-Maxwell hierarchy being just a one-parameter subhierarchy. Explicitly, instead of (7.11), (7.12) we have:

$$
\begin{aligned}
&I^+_{n_1^+,n_2^+} : C^+_{n_1^+,n_2^+} \longrightarrow C^{0+}_{n_1^+,n_2^+}, \\
&I^+_{n_1^+,n_2^+} \circ T^+_{n_1^+,n_2^+} = T^{0+}_{n_1^+,n_2^+} \circ I^+_{n_1^+,n_2^+},
\end{aligned} \tag{7.38a}
$$

$$
\begin{aligned}
&I^-_{n_1^-,n_2^-} : C^-_{n_1^-,n_2^-} \longrightarrow C^{0-}_{n_1^-,n_2^-}, \\
&I^-_{n_1^-,n_2^-} \circ T^-_{n_1^-,n_2^-} = T^{0-}_{n_1^-,n_2^-} \circ I^-_{n_1^-,n_2^-},
\end{aligned} \tag{7.38b}
$$

where $T^a_{n_1^\pm,n_2^\pm} = T^{\chi^a_{n_1^\pm,n_2^\pm}}$, $C^a_{n_1^\pm,n_2^\pm} = C^{\chi^a_{n_1^\pm,n_2^\pm}}$, $a = \pm$, or $a = 0\pm$, and

$$\chi^{+}_{n_1^+,n_2^+} = [n_1^+, n_2^+; \frac{n_1^+ - n_2^+}{2} + 1] \tag{7.39a}$$
$$\chi^{0+}_{n_1^+,n_2^+} = [n_1^+ - 1, n_2^+ + 1; \frac{n_1^+ - n_2^+}{2} + 2], \qquad n_1^+ \in \mathbb{N}, n_2^+ \in \mathbb{Z}_+,$$
$$\chi^{-}_{n_1^-,n_2^-} = [n_1^-, n_2^-; \frac{n_2^- - n_1^-}{2} + 1] \tag{7.39b}$$
$$\chi^{0-}_{n_1^-,n_2^-} = [n_1^- + 1, n_2^- - 1; \frac{n_2^- - n_1^-}{2} + 2], \qquad n_1^- \in \mathbb{Z}_+, n_2^- \in \mathbb{N},$$

while instead of (7.13) in the $q = 1$ case and (7.37) in the q-deformed case, we have:

$$_qI^+_{n_1^+} F^+_{n_1^+,n_2^+}(z,\bar{z}) = J^+_{n_1^+,n_2^+}(z,\bar{z}), \tag{7.40a}$$
$$_qI^-_{n_2^-} F^-_{n_1^-,n_2^-}(z,\bar{z}) = J^-_{n_1^-,n_2^-}(z,\bar{z}), \tag{7.40b}$$

where $_qI^+_{n_1^+}$, $_qI^-_{n_2^-}$ are given by (7.36) (or (7.14) for $q = 1$), while $F^\pm_{n_1^\pm,n_2^\pm}(z,\bar{z})$, $J^\pm_{n_1^\pm,n_2^\pm}(z,\bar{z})$ are polynomials in $z, \bar{z}$ of degrees $(n_1^\pm, n_2^\pm)$, $(n_1^\pm \mp 1, n_2^\pm \pm 1)$, respectively.

The crucial feature which unifies these representations is the form of the operators $_qI^\pm_n$, which is not generalized anymore in equations (7.40).

We call the hierarchy of equations (7.40) the **generalized q-Maxwell hierarchy**. The q-Maxwell hierarchy is obtained in the partial case when $\chi^{0+}_{n_1^+,n_2^+} = \chi^{0-}_{n_1^-,n_2^-} = \chi^0_n$ which fixes three of the four parameters: $n_1^+ - 2 = n_2^+ = n_1^- = n_2^- - 2 = n$.

Another one-parameter subhierarchy of the generalized q-Maxwell hierarchy involves the two signatures of $\chi^+_n = [n+2, n; 2]$, $\chi^-_n = [n, n+2; 2]$, and in addition

$$\chi^{00}_n = [n+1, n+1; 1] = \{n+2, -1-n, n+2\}, \qquad n \in \mathbb{Z}_+ \tag{7.41}$$

The intertwining relations are:

$$I^+_{n-1} : C^{00}_n \longrightarrow C^-_n, \quad I^+_{n-1} \circ T^{00}_n = T^-_n \circ I^+_{n-1}, \tag{7.42a}$$
$$I^-_{n-1} : C^{00}_n \longrightarrow C^+_n, \quad I^-_{n-1} \circ T^{00}_n = T^+_n \circ I^-_{n-1}, \tag{7.42b}$$

where $T^{00}_n = T^{\chi^{00}_n}$, $C^{00}_n = C^{\chi^{00}_n}$: Thus, instead of (7.13) in the $q = 1$ case and (7.37) in the q-deformed case, we have:

$$_qI^+_{n-1}\,{}_qA_n = {}_qF^-_n, \tag{7.43a}$$
$$_qI^-_{n-1}\,{}_qA_n = {}_qF^+_n, \tag{7.43b}$$

where $_qI^\pm_n$ are given by (7.36) (or (7.14) for $q = 1$), $_qA_n$ has the signature χ^{00}_n.

This hierarchy will be called the **potential q-Maxwell hierarchy**. The reason is that the lowest member obtained for $n = 0$ and $q = 1$ is just:

$$\partial_{[\mu}A_{\nu]} = F_{\mu\nu}. \tag{7.44}$$

Of course, as in the classical case these equations have auxiliary character w.r.t. (7.1). One of the reasons for their introduction is to make transparent the gauge invariance of the Maxwell equations. We recall that substituting (7.44) in (7.1b) gives an identity, while from (7.1a) one gets:

$$\Box A_\mu - \partial_\mu \partial^\sigma A_\sigma = J_\mu \tag{7.45a}$$

$$\Box \equiv \partial^\sigma \partial_\sigma \tag{7.45b}$$

Thus the eight equations (7.1) are reduced to the four equations (7.45). The lost equations are actually traded for *gauge symmetry*:

$$A_\mu \mapsto A'_\mu = A_\mu + \partial_\mu \phi \tag{7.46}$$

since the last substitution leave $F_{\mu\nu}$ and (7.45) unchanged. One uses (7.46) to simplify (7.45) by setting:

$$\partial^\sigma A_\sigma = 0 \tag{7.47}$$

This is called the *Lorentz gauge condition*, and it is equivalent to find suitable ϕ. Indeed, if $\partial^\sigma A_\sigma \neq 0$, take ϕ so that $\Box\phi = -\partial^\sigma A_\sigma$; then it follows that $\partial^\sigma A'_\sigma = 0$. Thus, one may assume that (7.47) holds, and then from (7.45) follows:

$$\Box A_\mu = J_\mu\,. \tag{7.48}$$

Further we note a special gauge symmetry now also of (7.47):

$$\begin{aligned} A_\mu \mapsto A'_\mu &= A_\mu + \partial_\mu \phi_0 \\ \Box\phi_0 &= 0 \end{aligned} \tag{7.49}$$

This may be used to get rid of one component of A_μ, if the same component of J_μ is zero; for example, if $J_0 = 0$, take ϕ_0 so that $\partial_0\phi_0 = -A_0$, $A'_0 = 0$. Thus in this case we have:

$$\Box A_k = J_k\,, \quad k = 1, 2, 3, A_0 = J_0 = 0 \tag{7.50a}$$

$$\partial_k A_k = 0. \tag{7.50b}$$

The last condition (7.50)b is called *Coulomb gauge condition*. This gauge also used when $\partial_0 A_0 = 0$.

Let us see how these things are related to representation theory. The fact that using (7.44) Maxwell equations reduces to four equations is expressed group-theoretically for the whole hierarchy by the fact that the two possible composition maps intertwining the "potential" representations χ_n^{00} and "current" representations χ_n^0 (via χ_n^+ or χ_n^-)

coincide; that is,

$$ {}_qI^-_n \circ {}_qI^+_{n-1} = {}_qI^+_n \circ {}_qI^-_{n-1} \equiv {}_q\tilde{\Box}_n \tag{7.51a} $$

$$ {}_q\tilde{\Box}_n : C^{00}_n \longrightarrow C^0_n , \quad {}_q\tilde{\Box}_n \circ {}_qT^{00}_n = {}_qT^0_n \circ {}_q\tilde{\Box}_n . \tag{7.51b} $$

and the equation is:

$$ \begin{aligned} {}_q\tilde{\Box}_n \, {}_qA_n &= {}_qI^-_n \circ {}_qI^+_{n-1} \, {}_qA_n = {}_qI^-_n \, {}_qF^-_n = \\ &= {}_qI^+_n \circ {}_qI^-_{n-1} \, {}_qA_n = {}_qI^+_n \, {}_qF^+_n = \\ &= {}_qJ_n . \end{aligned} \tag{7.52} $$

Further, as an example we consider the Maxwell case; that is, $n = 0$, setting also $\tilde{\Box}_q \equiv_q \tilde{\Box}_0$, $A \equiv {}_qA_0$. After a short calculation we find first for $q = 1$:

$$ \tilde{\Box} A = \Box A - I_2 (\partial \cdot A) = J \tag{7.53a} $$

$$ (\partial \cdot A) \equiv \tfrac{1}{2}(\partial_- A_+ + \partial_+ A_- - \partial_\nu A_{\bar\nu} - \partial_{\bar\nu} A_\nu) = \partial^\mu A_\mu \tag{7.53b} $$

Thus, also in this language the suitable gauge condition is the Lorentz one (7.47), while for the elimination of one further degree of freedom here it is more convenient to set $A_+ = 0$ or $A_- = 0$. This is called a *light-front gauge condition.*

In the q-deformed case we have instead of (7.53):

$$ \tilde{\Box}_q \, A = \Box_q \, A - I^q_2 \, (\partial \cdot A)_q = J \tag{7.54a} $$

$$ \Box_q \equiv \left(\hat{\mathscr{D}}_{\bar\nu} \, \hat{\mathscr{D}}_\nu - q \, \hat{\mathscr{D}}_- \, \hat{\mathscr{D}}_+ \, T_\nu \, T_{\bar\nu} \right) T_\nu \, T_{\bar\nu} \, T_+ \, T_- \tag{7.54b} $$

$$ \begin{aligned} (\partial \cdot A)_q \equiv \tfrac{1}{2}\Big(& q^2 \, \hat{\mathscr{D}}_- \; T_\nu \, T_+ \, A_+ + q \, \hat{\mathscr{D}}_+ \, T_{\bar\nu} \, T_+ \, A_- - \\ & - q^3 \, \hat{\mathscr{D}}_\nu \, T_- \, T_\nu \, T_+ \, A_{\bar\nu} - \hat{\mathscr{D}}_{\bar\nu} \, T_-^{-1} \, T_\nu \, T_+ \, A_\nu + \\ & + q\lambda \, \hat{M}_\nu \, \hat{\mathscr{D}}_- \, \hat{\mathscr{D}}_+ \, T_-^{-1} \, T_{\bar\nu} \, T_\nu \, T_+ \, A_\nu \Big). \end{aligned} \tag{7.54c} $$

Further we consider the free equations, that is, $J = 0$, in the q-Lorentz gauge:

$$ \Box_q \, A = 0 \, , \tag{7.55a} $$

$$ (\partial \cdot A)_q = 0 \, . \tag{7.55b} $$

Since the first equation is valid component-wise, we can use its A_ν component to simplify the gauge condition. Thus, finally we have:

$$ \begin{aligned} \Box_q \, A &= \left(\hat{\mathscr{D}}_{\bar\nu} \, \hat{\mathscr{D}}_\nu - q \, \hat{\mathscr{D}}_- \, \hat{\mathscr{D}}_+ \, T_\nu \, T_{\bar\nu} \right) T_\nu \, T_{\bar\nu} \, T_+ \, T_- \, A = 0 \\ 2(\partial \cdot A)_q &= q^2 \, \hat{\mathscr{D}}_- \; T_\nu \, T_+ \, A_+ + q \, \hat{\mathscr{D}}_+ \, T_{\bar\nu} \, T_+ \, A_- - \\ & \quad - q^3 \, \hat{\mathscr{D}}_\nu \, T_- \, T_\nu \, T_+ \, A_{\bar\nu} - \hat{\mathscr{D}}_{\bar\nu} \, T_-^{-1} \, T_\nu^{-1} \, T_+ \, A_\nu = 0. \end{aligned} \tag{7.56} $$

7.4 q-d'Alembert Equations Hierarchy

Here we consider another one-parameter subhierarchy of the generalized q-Maxwell hierarchy which is obtained from (7.39) for $n_1^+ = n_2^- = r \in \mathbb{N}$, $n_1^- = n_2^+ = 0$; that is,

$$\chi_r^{d+} = [r, 0; \frac{r}{2} + 1],$$
$$\chi_r^{d0+} = [r-1, 1; \frac{r}{2} + 2], \quad r \in \mathbb{N} \tag{7.57a}$$
$$\chi_r^{d-} = [0, r; \frac{r}{2} + 1],$$
$$\chi_r^{d0-} = [1, r-1; \frac{r}{2} + 2], \quad r \in \mathbb{N}, \tag{7.57b}$$

where the two conjugated equations follow from (7.40):

$${}_qI_r^+ F_r^{d+} = J_r^{d+}, \tag{7.58a}$$
$${}_qI_r^- F_r^{d-} = J_r^{d-}, \tag{7.58b}$$

where ${}_qI_r^\pm$ is given by (7.36).

For the minimal possible value of the parameter $r = 1$, we obtain the two conjugate *q-Weyl equations*.

The case $r = 2$ gives the q-Maxwell equations (note that $J_2^{d+} = J_2^{d-}$). This is the only intersection of the present hierarchy with the q-Maxwell hierarchy.

We call this hierarchy *q-d'Alembert hierarchy* following the classical case (cf. [215] and Volume 1), due to the following. We consider the representations $\chi_a^{d\pm}$ for the excluded above value $r = 0$, when they coincide. Thus, we set: $\chi^d \equiv \chi_0^{d\pm} = [0, 0; 1]$, $F^d \equiv F_0^{d\pm}$. Furthermore, the relevant equation is the q-d'Alembert equation [215]:

$$\Box_q F^d = J^d \tag{7.59}$$

where the signature of J^d is $\chi^{d0} = [0, 0; 3]$, and $\Box_q$ is as in (7.56).

Finally, we recall [215] that the solutions of the free equations (7.58) satisfy also the q-d'Alembert equation.

7.4.1 Solutions of the q-d'Alembert Equation

Here and in the next Subsection we follow [226] to find solutions of the q-d'Alembert equation (7.59) with trivial RHS:

$$\Box_q F^d = 0\,. \tag{7.60}$$

We recall that the elements of our representation spaces are formal power series in the variables $x_\pm, v, \bar{v}, z, \bar{z}$ of the coset $\mathscr{Y}$. But here there is no dependence on the spin variables $z, \bar{z}$ and our solutions will be power series in the q-Minkowski variables $x_{\pm}, v, \bar{v}$:

$$\hat{\varphi} = \sum_{j,n,\ell,m\in\mathbb{Z}_+} \mu_{jn\ell m}\, \hat{\varphi}_{jn\ell m}, \qquad \hat{\varphi}_{jn\ell m} = v^j\, x_-^n\, x_+^\ell\, \bar{v}^m. \tag{7.61}$$

We substitute the above in $\Box_q\, \hat{\varphi} = 0$ to obtain:

$$\Box_q\, \hat{\varphi} = \sum_{j,n,\ell,m\in\mathbb{Z}_+} \mu_{jn\ell m} \Box_q\, \hat{\varphi}_{jn\ell m} = 0, \tag{7.62a}$$

$$\begin{aligned} \Box_q\, \hat{\varphi}_{jn\ell m} &= q^{1+n+2m+2j+\ell} [n]_q [\ell]_q\, \hat{\varphi}_{j,n-1,\ell-1,m} - \\ &-q^{n+j+\ell+m} [j]_q [m]_q\, \hat{\varphi}_{j-1,n,\ell,m-1}. \end{aligned} \tag{7.62b}$$

We first show two polynomial solutions (with$a, b, c, d \in \mathbb{Z}_+$):

$$\hat{\varphi} = \sum_{n=0}^{n_{a,b}} q^{n(c+d+n)} \frac{(-a)_n^q\, (-b)_n^q}{(c+1)_n^q\, (d+1)_n^q}\, v^{n+d}\, x_-^{a-n}\, x_+^{b-n}\, \bar{v}^{c+n}, \tag{7.63}$$

where$(\alpha)_n^q = \Gamma_q(\alpha+n)/\Gamma_q(\alpha)$ is the q-Pochhammer symbol,

$$\hat{\varphi}_{a,b,c} = \sum_{n=0}^{n_{a,b}} q^{n(c+n)} \frac{(-a)_n^q\, (-b)_n^q}{(c+1)_n^q\, [n]_q!}\, v^n\, x_-^{a-n}\, x_+^{b-n}\, \bar{v}^{c+n}, \tag{7.64}$$

where $n_{a,b} = \min(a, b)$.

7.4.2 q-Plane-Wave Solutions

Next we look for solutions of the q-d'Alembert equation in terms of a q-deformation of the classical plane wave exp $(k \cdot x)$, where

$$(k \cdot x) = k^\mu x_\mu = \tfrac{1}{2}(k_- x_+ + k_+ x_- - k_v \bar{v} - k_{\bar{v}} v), \tag{7.65}$$

and $(k_v, k_-, k_+, k_{\bar{v}})$ are related to the components k_μ of the four-momentum as the variables $(v, x_-, x_+, \bar{v})$ are related to x_μ. Clearly, the natural q-deformation of the plane wave is:

$$(\exp(k \cdot x))_q = \sum_{s=0}^{\infty} \frac{1}{[s]_q!} f_s(v, x_-, x_+, \bar{v}), \tag{7.66}$$

where f_s is a homogeneous polynomial of degrees in both sets of variables $(k_v, k_-, k_+, k_{\bar{v}})$ and $(v, x_-, x_+, \bar{v})$, such that $(f_s)|_{q=1} = (k \cdot x)^s$. Thus, we set $f_0 = 1$. One may expect that f_s for $s > 1$ would be equal or at least proportional to $(f_1)^s$, but it turns out that this is not the case. In order to proceed systematically, we have to impose the conditions of q-Lorentz covariance and the q-d'Alembert equation.

The complexification of the q-Lorentz subalgebra of the q-conformal algebra is generated by $k_j^{\pm}, X_j^{\pm}, j = 1, 3$. Using (6.263a,c) and (6.264a,c) it is easy to check that:

$$\pi(X_j^{\pm})\,\mathscr{L}_q = 0, \quad \Longrightarrow \quad \pi(X_j^{\pm})\,(\mathscr{L}_q)^s = 0, \qquad j = 1, 3. \tag{7.67}$$

Since $(k \cdot x)^s$ is a scalar as $(\mathscr{L}_q)^s$, then also the q-deformations f_s should be scalars, and thus also should obey (7.67). In order to implement this we suppose that the momentum components are also noncommutative obeying the same rules (7.19) as the q-Minkowski coordinates, and that they commute with the coordinates. Also the ordering of the momentum basis will be the same for the coordinates. Taking all this into account we can see that a natural expression for f_s is:

$$\begin{aligned} f_s = & \sum_{a,b,n\,\in\mathbb{Z}_+} \beta^s_{a,b,n} \frac{(-1)^{s-a-b}}{\Gamma_q(a-n+1)\,\Gamma_q(b-n+1)\,[n]_q!} \times \\ & \times \frac{k_v^{s-a-b+n}\,k_-^{b-n}\,k_+^{a-n}\,k_{\bar v}^n\,v^n\,x_-^{a-n}\,x_+^{b-n}\,\bar v^{s-a-b+n}}{\Gamma_q(s-a-b+n+1)}, \end{aligned} \tag{7.68}$$

where we have introduced some factors that are obvious from the correspondence with the case $q = 1$. (The expression in (7.68) does not involve terms that would vanish for $q = 1$. Actually, we shall see that such expressions would lead to noncovariant momenta light cone.) In order to implement q-Lorentz covariance we impose the conditions:

$$\pi(X_j^{\pm})\,f_s = 0, \qquad j = 1, 3. \tag{7.69}$$

For this calculation we suppose that the q-Lorentz action on the noncommutative momenta is given by (6.79a,c), (6.263a,c), and (6.264a,c). We also have to use the twisted derivation rule which here is:

$$\begin{aligned} &\pi(X_j^{\pm})\,\psi\cdot\psi' = \pi(X_j^{\pm})\,\psi\,\cdot\,\pi(k_j^{-1})\,\psi' + \pi(k_j)\,\psi\,\cdot\,\pi(X_j^{\pm})\,\psi', \\ &\psi = k_v^{s-a-b+n}\,k_-^{b-n}\,k_+^{a-n}\,k_{\bar v}^n\,, \psi' = v^n\,x_-^{a-n}\,x_+^{b-n}\,\bar v^{s-a-b+n}. \end{aligned} \tag{7.70}$$

The four conditions (7.69) bring eight relations between the coefficientsβ; however, only three are independent, namely, the relations:

$$\beta^s_{a,b,n} = q^{-s-2n+a+2b}\,\beta^s_{a,b-1,n}, \tag{7.71a}$$

$$\beta^s_{a,b,n} = q^{s-2n-2a+b}\,\beta^s_{a-1,b,n}, \tag{7.71b}$$

$$\beta^s_{a,b,n} = q^{s+4n-2a-2b-2}\,\beta^s_{a,b,n-1}, \tag{7.71c}$$

solving which we find the following solution:

$$\beta^s_{a,b,n} \;=\; q^{n(s-2a-2b+2n)\,+\,a(s-a-1)\,+\,b(-s+a+b+1)}\;\beta^s_{0,0,0}, \tag{7.72}$$

that is, for each $s \geq 1$ only one constant remains to be fixed.

Next we impose the q-d'Alembert equation on f_s:

$$\Box_q f_s = 0, \tag{7.73}$$

which holds trivially for $s = 0, 1$. For $s \geq 2$ we substitute (7.68) to obtain (for details see [226]):

$$\begin{aligned}
\Box_q f_s &= (q\,k_-\,k_+ \;-\; k_\nu\,k_{\bar\nu}) \times \\
&\times \sum_{a,b,n\,\in\mathbb{Z}_+} \frac{(-1)^{s-a-b}\,\beta^s_{a,b,n}\,q^{2s+2n-a-b}}{\Gamma_q(a-n)\,\Gamma_q(b-n)\,\Gamma_q(s-a-b+n+1)\,[n]_q!} \times \\
&\times\; k_\nu^{s-a-b+n}\,k_-^{b-n-1}\,k_+^{a-n-1}\,k_{\bar\nu}^n\,\hat\varphi_{n,a-n-1,b-n-1,s-a-b+n} = \\
&= (k_-\,k_+ \;-\; q^{-1}k_\nu\,k_{\bar\nu})\,\frac{q^{2s}\,\beta^s_{0,0,0}}{\beta^{s-2}_{0,0,0}}\,f_{s-2}.
\end{aligned} \tag{7.74}$$

If (7.73) holds then for every $s \geq 2$ we obtain (as for q=1) the condition that the momentum operators are on the q-Lorentz covariant *q-light cone* (cf. (7.21)):

$$\mathscr{L}^k_q = k_- k_+ - q^{-1}\,k_\nu k_{\bar\nu} = 0. \tag{7.75}$$

Now it remains only to fix the coefficient $\beta^s_{0,0,0}$. We note that for q=1 it holds:

$$(k\cdot x)|_{k\,\to\,x} = (x\cdot x) = \mathscr{L}, \tag{7.76}$$

and thus we shall impose the conditions:

$$(f_s)|_{k\,\to\,x} = (\mathscr{L}_q)^s. \tag{7.77}$$

Next we note that:

$$(\mathscr{L}_q)^s = \sum_{n=0}^{s} (-1)^n \binom{s}{n}_q q^{n(n-s-1)}\,\nu^n\,x_-^{s-n}\,x_+^{s-n}\,\bar\nu^n. \tag{7.78}$$

A tedious calculation shows that:

$$(f_s)|_{k\,\to\,x} = \beta^s_{0,0,0}\,(\mathscr{L}_q)^s \sum_{p=0}^{s} \frac{q^{(s-p)(p-1)+p}}{[p]_q!\,[s-p]_q!}, \tag{7.79}$$

and comparing (7.79) with (7.77) we finally obtain:

$$\left(\beta^s_{0,0,0}\right)^{-1} = \sum_{p=0}^{s} \frac{q^{(s-p)(p-1)+p}}{[p]_q!\,[s-p]_q!}. \tag{7.80}$$

Note that $\left(\beta^s_{0,0,0}\right)^{-1}|_{q=1} = 2^s/s!$, as expected.

Finally, we note that our f_s for $s > 1$ is not equal, and not even proportional, to $(f_1)^s$. Actually, imposing the q-d'Alembert equation on $(f_1)^s$ will bring a s-dependent relation between the momenta, which is not q-Lorentz covariant. For instance, for $s = 2$ imposing: $\Box_q\,(f_1)^2 = 0$ results in the following condition on the momenta: $[2]_q\,k_-k_+ = (3-q^2)\,k_v\,k_{\bar{v}}$ instead of (7.75) (cf. more details in [226]).

Thus, though our q-plane wave has some properties analogous to the classical one, it is not an exponent or q-exponent. Thus, it differs conceptually from the classical plane wave and may serve as a regularization of the latter.

7.4.3 q-Plane-Wave Solutions for Non-Zero Spin

Here we follow [239] looking for solutions of the free equations (7.58).

$$ {}_qI^+_r\,\tilde{\varphi} = 0, \tag{7.81a}$$

$$ {}_qI^-_r\,\hat{\varphi} = 0. \tag{7.81b}$$

We start with (7.81b). As we know from [215] since it depends only on one spin-variable $\bar{z}$ that equation becomes a couple of equations:

$$\left([r-N_{\bar{z}}]_q\hat{\mathscr{D}}_+T_{\bar{v}}T_v^{-1} - q^{r+1}\hat{\mathscr{D}}_v\hat{\mathscr{D}}_{\bar{z}}T_-\right)\,T_-T_{\bar{v}}\,\hat{\varphi} = 0\,, \tag{7.82a}$$

$$\left([r-N_{\bar{z}}]_q\hat{\mathscr{D}}_{\bar{v}} - q^{r+1}\hat{\mathscr{D}}_-\hat{\mathscr{D}}_{\bar{z}}T_v^2T_-\right)\,T_{\bar{v}}\,\hat{\varphi} = 0\,. \tag{7.82b}$$

The spin dependence is encoded in the spin variable $\bar{z}$ in which the solutions depend polynomially of degree $r \in \mathbb{N}$. As it was shown in [215], if a function satisfies (7.82) then it satisfies also the q-d'Alambert equation (7.60). Thus, it is justified to look for solutions in terms of q-deformation of the plane wave:

$$\widehat{\exp}_q(k\cdot x) = \sum_{s=0}^{\infty} \frac{1}{[s]_q!}\,h_s\,, \tag{7.83}$$

$$h_s = \beta^s \sum_{a,b,n\in\mathbb{Z}_+} \frac{(-1)^{s-a-b}\,q^{n(s-2a-2b+2n)\,+\,a(s-a-1)\,+\,b(-s+a+b+1)}\,q^{P_s(a,b)}}{\Gamma_q(a-n+1)\Gamma_q(b-n+1)\Gamma_q(s-a-b+n+1)[n]_q!}\times$$

$$\times\, k_v^{s-a-b+n}k_-^{b-n}k_+^{a-n}k_{\bar{v}}^{n}v^{n}x_-^{a-n}x_+^{b-n}\bar{v}^{s-a-b+n}\,, \tag{7.84}$$

$$(\beta^s)^{-1} = \sum_{p=0}^{s} \frac{q^{(s-p)(p-1)+p}}{[p]_q!\,[s-p]_q!}\,,$$

This deformation of the plane wave generalizes the one from the previous subsection. To obtain the latter one has to replace $P_s(a,b)$ by 0. Each h_s satisfies the q-d'Alembert equation (7.60) on the momentum q-cone (7.75).

We look for the solutions of (7.82) in a form analogous to (7.83):

$$\hat{\varphi} = \sum_{s=0}^{\infty} \frac{1}{[s]_q!} \hat{\varphi}_s . \tag{7.85}$$

The solutions are constructed component-wise; that is, we solve (7.82) separately for each $\hat{\varphi}_s$ and we find that

$$\begin{aligned} \hat{\varphi}_s &= \sum_{m=0}^{r} \hat{\gamma}_m^{rs} \left(\prod_{i=-r+1}^{-m} (k_+ - q^{i-A_s} k_v \bar{z}) \right) \times \\ &\quad \times \left(\prod_{j=-m+1}^{0} (k_{\bar{v}} - q^{j-A_s} k_- \bar{z}) \right) h_s , \end{aligned} \tag{7.86a}$$

$$P_s(a,b) = A_s a + P_s(b), \tag{7.86b}$$

where $\hat{\gamma}_m^{rs}$ are $r+1$ independent constants, A_s is an arbitrary constant, and $P_s(b)$ is an arbitrary polynomial in b.

In order to be able to write the general solution of the system (7.82) in terms of the deformed plane wave we have to suppose that the $\hat{\gamma}_m^{rs}$ and A_s for different s coincide: $\hat{\gamma}_m^{rs} = \hat{\gamma}_m^{r}$, $A_s = A$. Then we have:

$$\begin{aligned} \hat{\varphi} &= \sum_{m=0}^{r} \hat{\gamma}_m^{r} \left(\prod_{i=-r+1}^{-m} (k_+ - q^{i-A} k_v \bar{z}) \right) \times \\ &\quad \times \left(\prod_{j=-m+1}^{0} (k_{\bar{v}} - q^{j-A} k_- \bar{z}) \right) \widehat{\exp}_q(k \cdot x) . \end{aligned} \tag{7.87}$$

We pass now to equation (7.81a). As in the first case it produces a couple of equations:

$$\left([r - N_z]_q D_v - q^r D_z D_- T_- \right) T_v \tilde{\varphi} = 0 , \tag{7.88a}$$

$$\left([r - N_z]_q D_+ T_{\bar{v}}^{-1} - q^r D_z D_{\bar{v}} T_- T_v \right) T_- \tilde{\varphi} = 0 . \tag{7.88b}$$

As found in [247] for these equations we need to use a basis conjugate to the basis in (7.61); that is,

$$\begin{aligned} \tilde{\varphi} &= \sum_{j,n,\ell,m \in \mathbb{Z}_+} \mu_{jn\ell m} \tilde{\varphi}_{jn\ell m} , \\ \tilde{\varphi}_{jn\ell m} &= \bar{v}^m x_+^\ell x_-^n v^j = \omega(\hat{\varphi}_{jn\ell m}). \end{aligned} \tag{7.89}$$

We also recall that here the q-d'Alembert equation is slightly different [239]:

$$\left(\hat{\mathscr{D}}_- \hat{\mathscr{D}}_+ - q\, \hat{\mathscr{D}}_\nu \hat{\mathscr{D}}_{\bar{\nu}} T_\nu T_{\bar{\nu}}\right) T_- T_+ \tilde{\varphi} = 0 \tag{7.90}$$

though it coincides with (7.56) when $q = 1$.

Analogously to the first case we use the expansion

$$\tilde{\varphi} = \sum_{s=0}^{\infty} \frac{1}{[s]_q!} \tilde{\varphi}_s , \tag{7.91}$$

and we solve it again component-wise. Here we shall use another deformation of the plane wave:

$$\widetilde{\exp}_q(k \cdot x) = \sum_{s=0}^{\infty} \frac{1}{[s]_q!} \tilde{h}_s , \tag{7.92}$$

$$\begin{aligned} \tilde{h}_s &= \tilde{\beta}^s \sum_{a,b,n} \frac{(-1)^{s-a-b}\, q^{n(2a+2b-2n-s) + a(a-s-1) + b(s-a-b+1)}\, q^{Q_s(a,b)}}{\Gamma_q(a-n+1)\, \Gamma_q(b-n+1)\, \Gamma_q(s-a-b+n+1)\, [n]_q!} \times \\ &\quad \times k_{\bar{\nu}}^n k_+^{a-n} k_-^{b-n} k_\nu^{s-a-b+n}\, \bar{\nu}^{s-a-b+n} x_+^{b-n} x_-^{a-n} \nu^n , \\ \left(\tilde{\beta}^s\right)^{-1} &= \sum_{p=0}^{s} \frac{q^{(p-s)(p-1)+p}}{[p]_q!\, [s-p]_q!} , \end{aligned} \tag{7.93}$$

where $Q_s(a, b)$ are arbitrary polynomials. The $\tilde{h}_s$ has the same properties as the h_s, but the conjugated basis is used; in particular, they satisfy the q-d'Alembert equation (7.90) on the momentum q-cone (7.75). The solutions of (7.88) are polynomials of degree r in the spin variable z. Explicitly they are given by:

$$\tilde{\varphi}_s = \sum_{m=0}^{r} \tilde{\gamma}_m^{rs} \left(\prod_{i=-r+2}^{-m+1} (k_+ - q^{i+B_s} k_{\bar{\nu}} z) \right) \times$$
$$\times \left(\prod_{j=-m+2}^{1} (k_\nu - q^{j+B_s} k_- z) \right) \tilde{h}_s , \tag{7.94a}$$
$$Q_s(a, b) = Q_s(a) + B_s b, \tag{7.94b}$$

where $\tilde{\gamma}_m^{rs}$ are $r+1$ independent constants, $Q_s(a)$ is an arbitrary polynomial in a, and B_s is an arbitrary constant. In order to be able to write the general solution of the system (7.88) in terms of the deformed plane wave, we have to suppose that the $\tilde{\gamma}_m^{rs}$ and A_s for different s coincide: $\tilde{\gamma}_m^{rs} = \hat{\gamma}_m^r$ and $B_s = B$. Then we have:

$$\tilde{\varphi} = \sum_{m=0}^{r} \tilde{\gamma}_m^r \left(\prod_{i=-r+2}^{-m+1} (k_+ - q^{i+B} k_{\bar{v}} z) \right) \times$$
$$\times \left(\prod_{j=-m+2}^{1} (k_v - q^{j+B} k_- z) \right) \widetilde{\exp}_q(k \cdot x) . \tag{7.95}$$

7.5 q-Plane-Wave Solutions of the Potential q-Maxwell Hierarchy

Here we use results of [240]. We mentioned that the q-d'Alembert hierarchy for $r = 2$ intersects with the q-Maxwell hierarchy for $n = 0$. Thus, we shall identify $\hat{\varphi}, \tilde{\varphi}$ at $r = 2$ from the previous subsection with ${}_qF_0^{\pm}$

$$\hat{\varphi}_{r=2} = {}_qF_0^- , \qquad \tilde{\varphi}_{r=2} = {}_qF_0^+ \tag{7.96}$$

Accordingly, we would like to use the solutions for $\hat{\varphi}, \tilde{\varphi}$ in equations (7.43):

$${}_qI_{-1}^+ \, {}_qA_0 = {}_qF_0^- = \hat{\varphi}_{r=2} , \tag{7.97a}$$
$${}_qI_{-1}^- \, {}_qA_0 = {}_qF_0^+ = \tilde{\varphi}_{r=2} . \tag{7.97b}$$

We start with solving (7.97a) for ${}_qA^0$, with ${}_qF_0^- = \hat{\varphi}$ given by (7.87). We write:

$${}_qA^0 = \bar{z}zA_+ + zA_v + \bar{z}A_{\bar{v}} + A_- = \sum_{s=o}^{\infty} \frac{1}{[s]_q!} \, {}_qA_s^0 \, h_{s+1}^- \tag{7.98}$$

$$A_\kappa = A_\kappa(k,x) = \sum_{s=o}^{\infty} \frac{1}{[s]_q!} A_\kappa^s(k) \, h_{s+1}^- , \qquad \kappa = \pm, v, \bar{v} . \tag{7.99}$$

Substituting, we take into account that the action of ${}_qI_{-1}^+$ converts h_{s+1}^- into h_s^-, but this requires $P_{s+1}^-(a,b) = P_s^-(a,b) = P^-(a,b) = Ca + Q(b) = Ca + Bb$ (in the last step we use the fact that $Q(b)$ has to be linear in b and any constant term would be absorbed in the constant $\hat{\gamma}$). Then comparing the coefficients of 1, $\bar{z}$, $\bar{z}^2$ we obtain, respectively:

$$\begin{aligned}
&(q^{s+2} A_-^s(k) k_{\bar{v}} + q^{-1-B} A_v^s(k) k_+) \, h_s^- = \\
&= -2d_s \, (\hat{\gamma}_0^{s-} k_+^2 + \hat{\gamma}_1^{s-} k_+ k_{\bar{v}} + \hat{\gamma}_2^{s-} k_{\bar{v}}^2) \, h_s^- , \\
&\quad (q^{s+2+B} A_-^s(k) k_- - q^{s+1+B+C} A_{\bar{v}}^s k_{\bar{v}} - q^{C-2} A_+^s(k) k_+ + q^{-1} A_v^s(k) k_v) \, h_s^- = \\
&= -2d_s \, [2]_q q^{-C} (\hat{\gamma}_0^{s-} k_v k_+ + \hat{\gamma}_1^{s-} k_v k_{\bar{v}} + \hat{\gamma}_2^{s-} k_- k_{\bar{v}}) \, h_s^- \\
&\quad (q^{s+1+B} A_{\bar{v}}^s(k) k_- + q^{-2} A_+^s(k) k_v) \, h_s^- = \\
&\quad 2d_s \, q^{-2C-1} (\hat{\gamma}_0^{s-} k_v^2 + \hat{\gamma}_1^{s-} k_v k_- + \hat{\gamma}_2^{s-} k_-^2) \, h_s^- \\
&\quad d_s \equiv \beta^s / \beta^{s+1} .
\end{aligned} \tag{7.100}$$

Note, however, that only two of these three equations are independent when they are compatible (see below). Furthermore, we see that $A_\kappa^s(k)$ should be linear in k and in fact should be given as follows:

$$A_+^s(k) = \lambda_+^s k_V + \nu_+^s k_- , \qquad A_-^s(k) = \lambda_-^s k_{\bar{V}} + \nu_-^s k_+ , \tag{7.101}$$
$$A_V^s(k) = \lambda_V^s k_+ + \nu_V^s k_{\bar{V}} , \qquad A_{\bar{V}}^s(k) = \lambda_{\bar{V}}^s k_- + \nu_{\bar{V}}^s k_V$$

where for the constants we have:

$$\begin{aligned} \lambda_V^s &= -2d_s q^{1+B} \hat{\gamma}_0^{s-} , \quad \lambda_-^s = -2d_s q^{-s-2} \hat{\gamma}_2^{s-} , \\ \nu_V^s &= -q^{s+4+B} \nu_-^s - 2d_s q^{2+B} \hat{\gamma}_1^{s-} \\ \lambda_+^s &= 2d_s q^{1-2C} \hat{\gamma}_0^{s-} , \quad \lambda_{\bar{V}}^s = 2d_s q^{-2C-s-2-B} \hat{\gamma}_2^{s-} , \\ \nu_+^s &= -q^{s+4+B} \nu_{\bar{V}}^s + 2d_s q^{2-2C} \hat{\gamma}_1^{s-} \\ C &= -B , \end{aligned} \tag{7.102}$$

where the last condition arises from compatibility between the equations (7.100).

Now we substitute this result for ${}_qA^0$ in (7.97b). It turns out that we obtain a result compatible with the general solution above only when $B = C = 0$. Thus, in fact ${}_qA^0$ is given in terms of the original components f_{s+1} (cf. (7.84)). Furthermore, the action of ${}_qI_{-1}^-$ converts f_{s+1} into h_s^+, with $P_s^+(a, b) = -2b$, $B_s = 2 + s$. The result is:

$$\begin{aligned} \tilde{\varphi}_s &= ({}_qF_0^+)_s = {}_qI_{-1}^- \, {}_qA_s^0 = \\ &= -q^{s+1} \frac{(\nu_{\bar{V}}^s + \nu_-^s)}{2d_s} (k_+ - q^{s+2} z k_{\bar{V}})(k_V - q^{s+3} z k_-) \, h_s^+ , \end{aligned} \tag{7.103}$$

which is a special case of the general solution (7.94), with $\tilde{\gamma}_0^{2,s} = \tilde{\gamma}_2^{2,s} = 0$, $\tilde{\gamma}_1^{2,s} = -q^{s+1} (\nu_{\bar{V}}^s + \nu_-^s)/2d_s$. Thus, the resulting ${}_qF_0^+$ is not given in terms of the q-plane wave (only componentwise).

Let us now repeat the calculations in the other order, namely, we solve (7.97b) for ${}_qA^0$ with ${}_qF_0^+$ given by (7.94), but since we want this to be compatible with what we obtained above we take: $P_s^+(a, b) = -2b$. We use again the decomposition (7.98) but with f_{s+1} instead of h_{s+1}. Substituting and comparing the coefficients of 1, z, z^2, we obtain, respectively:

$$\begin{aligned} &(q^{s+1} A_-^s(k) k_V + q^{s+2} A_{\bar{V}}^s(k) k_+) \, h_s^+ = \\ &= -2d_s (\hat{\gamma}_0^{s+} k_V^2 + \hat{\gamma}_1^{s+} k_V k_+ + \hat{\gamma}_2^{s+} k_+^2) \, h_s^+ , \\ &(q^s A_-^s(k) k_- + q^{s+1} A_{\bar{V}}^s(k) k_{\bar{V}} - q A_+^s(k) k_+ - q^{-1} A_V^s(k) k_V) \, h_s^+ = \\ &= -2d_s [2]_q (\hat{\gamma}_0^{s+} k_V k_- + \hat{\gamma}_1^{s+} k_V k_{\bar{V}} + \hat{\gamma}_2^{s+} k_+ k_{\bar{V}}) \, h_s^+ , \\ &(q^{-2} A_+^s(k) k_{\bar{V}} + q^{-3} A_V^s(k) k_-) \, h_s^+ = \\ &= 2d_s \, (\hat{\gamma}_0^{s+} k_-^2 + \hat{\gamma}_1^{s+} k_- k_{\bar{V}} + \hat{\gamma}_2^{s+} k_{\bar{V}}^2) \, h_s^+ . \end{aligned} \tag{7.104}$$

Now instead of (7.101) we have:

$$A_+^s(k) = \mu_+^s k_{\bar{V}} + \nu_+^s k_- , \qquad A_-^s(k) = \mu_-^s k_V + \nu_-^s k_+ , \tag{7.105}$$
$$A_V^s(k) = \mu_V^s k_- + \nu_V^s k_{\bar{V}} , \qquad A_{\bar{V}}^s(k) = \mu_{\bar{V}}^s k_+ + \nu_{\bar{V}}^s k_V ,$$

where from the constants μ^s, ν^s only six can be determined (due to the gauge freedom). Making some choice we find:

$$\begin{aligned}
&\mu_-^s = -2d_s q^{-s-1}\hat{\gamma}_0^{s+}\,, \quad \mu_{\bar{v}}^s = -2d_s q^{-s-2}\hat{\gamma}_2^{s+}\,,\\
&\nu_{\bar{v}}^s = -\nu_-^s - 2d_s q^{-s-2}\hat{\gamma}_1^{s+}
\end{aligned} \tag{7.106}$$

$$\begin{aligned}
&\mu_v^s = 2d_s q^3\,\hat{\gamma}_0^{s+}\,, \quad \mu_+^s = 2d_s q^2\,\hat{\gamma}_2^{s+}\,,\\
&\nu_+^s = -\nu_v^s + 2d_s q^2\,\hat{\gamma}_1^{s+}.
\end{aligned} \tag{7.107}$$

Now we can substitute this result for ${}_qA^0$ in (7.97a). The action of ${}_qI_{-1}^+$ converts f_{s+1} into f_s, and we obtain for the components:

$$\hat{F}_s^- = -\frac{(\nu_v^s q^{-2} + \nu_-^s q^{s+2})}{2d_s}(k_+ - q^{-1}k_v\bar{z})(k_{\bar{v}} - k_-\bar{z})\,f_s, \tag{7.108}$$

which is consistent with the solution (7.86), with $\hat{\gamma}_0^{2,s} = \hat{\gamma}_2^{2,s} = 0$, $\hat{\gamma}_1^{2,s} = -(\nu_v^s q^{-2} + \nu_-^s q^{s+2})/2d_s$. Thus, the resulting ${}_qF_0^-$ is not given in terms of the q-plane wave (only componentwise).

Finally, we impose that we use the same ${}_qA^0$ for ${}_qF_0^+ = \tilde{\varphi}_{r=2}$ and ${}_qF_0^- = \hat{\varphi}_{r=2}$. Then instead of (7.101) and (7.105) we have:

$$A_+^s(k) = \nu_+^s k_-\,, \quad A_-^s(k) = \nu_-^s k_+\,, \quad A_v^s(k) = \nu_v^s k_{\bar{v}}\,, \quad A_{\bar{v}}^s(k) = \nu_{\bar{v}}^s k_v, \tag{7.109}$$

where from the four constants in (7.109) only three can be determined since their sum is zero:

$$\nu_+^s + \nu_-^s + \nu_v^s + \nu_{\bar{v}}^s = 0 \tag{7.110}$$

and using (7.102) and (7.106) we have:

$$\begin{aligned}
&\nu_v^s = -q^{s+4}\,\nu_-^s - 2d_s q^2\hat{\gamma}_1^{s-}\,, \quad \nu_{\bar{v}}^s = -\nu_-^s - 2d_s q^{-s-2}\hat{\gamma}_1^{s+}\,,\\
&\nu_+^s = q^{s+4}\,\nu_-^s + 2d_s q^2\,(\hat{\gamma}_1^{s+} + \hat{\gamma}_1^{s-}).
\end{aligned} \tag{7.111}$$

The disappearance of the constants λ^s, μ^s is consistent with $\hat{\gamma}_0^{s\pm} = \hat{\gamma}_2^{s\pm} = 0$. Substituting (7.106) in (7.103) and (7.108) we obtain, respectively:

$$\hat{F}_s^+ = \hat{\gamma}_1^{s+}\,q^{-1}\,(k_+ - q^{s+2}zk_{\bar{v}})(k_v - q^{s+3}zk_-)\,h_s^+\,, \tag{7.112}$$

$$\hat{F}_s^- = \hat{\gamma}_1^{s-}\,(k_+ - q^{-1}k_v\bar{z})(k_{\bar{v}} - k_-\bar{z})\,f_s. \tag{7.113}$$

We stress that for each s there are only three independent constants: $\hat{\gamma}_1^{s\pm}$, ν_-^s, the latter entering only the expressions for the *q-potentials* (7.109) and being a manifestation of the gauge freedom. We can eliminate the A_- components by setting $\nu_-^s = 0$ and/or the A_+ components by setting $\hat{\gamma}_1^{s+} = -\hat{\gamma}_1^{s-} - q^{s+2}\nu_-^s/2d_s$.

Finally we note that we can write ${}_qF_0^-$ in terms of $\widehat{\exp}_q(k,x)$ but not ${}_qF_0^+$ because of the s dependence in the prefactors. If we use the basis (7.89) the roles of ${}_qF_0^-$ and ${}_qF_0^+$ would be exchanged.

If we want ${}_qF_0^\pm$ on an equal footing then one should consider ${}_qF_0^-$ on the basis (7.61) and ${}_qF_0^+$ on the basis (7.89). However, then one should use two different q-potentials and furthermore should ensure that the two are not mixing because of the equations (7.40); that is, the q-potential obtained from solving from one of the equations (7.40) should give zero contribution after substitution in the other. This is easy to ensure through the gauge-freedom constants in the q-potentials, for example, setting $v_{\bar{v}}^s + v_-^s = 0$ we obtain that $\hat{F}_s^+ = 0$ in (7.103). Thus, the fields ${}_qF_0^+$ and ${}_qF_0^-$ may be seen as living on different copies of q-Minkowski space–time, similarly to the two four-dimensional sheets in the Connes–Lott model [153].

7.6 q-Plane-Wave Solutions of the Full q-Maxwell Equations

Here we use results from [237, 238]. First we shall use the basis (7.61). The general solutions of (7.37) for $n = 0$ in the homogeneous case ($J = 0$) are:

$$\hat{F}^{h\pm} \doteq \left({}_qF_0^\pm\right)_{J=0} = \sum_{m,s=0}^{\infty} \frac{1}{[s]_q!} \hat{F}_{ms}^{h\pm}(k) f_s\,, \tag{7.114}$$

$$\begin{aligned} \hat{F}_{ms}^{h+}(k) = \sum_{i=0}^{m} \Big(& \sum_{j=0}^{m-i} \hat{p}_{ij}^{ms1}\, k_v^i k_-^{m-i-j} k_{\bar{v}}^j\, (k_v - q^{s+6} z k_-)(k_v - q^{s+3} z k_-) + \\ & + \hat{p}_i^{ms2}\, k_v^i k_{\bar{v}}^{m-i}\, (k_v - q^{s+6} z k_-)(k_+ - q^{s+3} z k_{\bar{v}}) + \\ & + \sum_{j=0}^{m-i} \hat{p}_{ij}^{ms3}\, k_v^i k_+^{m-i-j} k_{\bar{v}}^j\, (k_+ - q^{s+6} z k_{\bar{v}})(k_+ - q^{s+3} z k_{\bar{v}}) \Big)\,, \end{aligned} \tag{7.115}$$

$$\begin{aligned} \hat{F}_{ms}^{h-}(k) = \sum_{i=0}^{m} \Big(& \sum_{j=0}^{m-i} \hat{r}_{ij}^{ms1}\, k_v^i k_-^{m-i-j} k_{\bar{v}}^j\, (k_{\bar{v}} - q^{-1} \bar{z} k_-)(k_{\bar{v}} - \bar{z} k_-) + \\ & + \hat{r}_i^{ms2}\, k_v^i k_{\bar{v}}^{m-i}\, (k_+ - q^{-1} \bar{z} k_v)(k_{\bar{v}} - \bar{z} k_-) + \\ & + \sum_{j=0}^{m-i} \hat{r}_{ij}^{ms3}\, k_v^i k_+^{m-i-j} k_{\bar{v}}^j\, (k_+ - q^{-1} \bar{z} k_v)(k_+ - \bar{z} k_v) \Big)\,, \end{aligned} \tag{7.116}$$

where $\hat{p}_{i(j)}^{msa}$, $\hat{r}_{i(j)}^{msa}$ are independent constants. The check that these are solutions is done as in the previous sections. Actually, the solution for $\hat{\varphi}_{r=2}$ given in (7.86) is obtained here for $m = 0$. As for (7.86) going to (7.87) the solution (7.117) can be written in terms of the deformed plane wave if we suppose that the $\hat{r}_{i(j)}^{msa}$ for different s coincide: $\hat{r}_{i(j)}^{msa} = \hat{r}_{i(j)}^{ma}$. Then we have:

$$\hat{F}^{h-} = \sum_{m=0}^{\infty} \hat{F}_m^{h-}(k)\exp_q(k,x)\,, \quad \hat{F}_m^{h-}(k) = \hat{F}_{ms}^{h-}(k)\,. \tag{7.117}$$

Also as before the solution (7.115) cannot be written in terms of the same deformed plane wave.

In the inhomogeneous case the solutions of (7.37) for $n = 0$ are:

$$_qJ^0 = \bar{z}z\hat{J}_+ + z\hat{J}_\nu + \bar{z}\hat{J}_{\bar{\nu}} + \hat{J}_-\,, \tag{7.118}$$

$$\hat{J}_\kappa = \sum_{m,s=o}^{\infty} \frac{1}{[s]_q!}\hat{J}_\kappa^{ms}(k) f_{s-1}\,, \quad \kappa = \pm, \nu, \bar{\nu}\,, \tag{7.119}$$

$$\begin{aligned}
\hat{J}_+^{ms}(k) &= -\hat{K}_m^s(k)\,k_-\,, \\
\hat{J}_-^{ms}(k) &= -q^{-s-2}\hat{K}_m^s(k)\,k_+\,, \\
\hat{J}_\nu^{ms}(k) &= \hat{K}_m^s(k)\,k_{\bar{\nu}}\,, \\
\hat{J}_{\bar{\nu}}^{ms}(k) &= q^{-s-2}\hat{K}_m^s(k)\,k_\nu\,, \\
\hat{K}_m^s(k) &\doteq \hat{\gamma}_\nu^s k_\nu^{m+1} + \hat{\gamma}_-^s k_-^{m+1} + \hat{\gamma}_+^s k_+^{m+1} + \hat{\gamma}_{\bar{\nu}}^s k_{\bar{\nu}}^{m+1}\,,
\end{aligned} \tag{7.120}$$

$$_qF_0^\pm = \hat{F}^\pm + \hat{F}^{h\pm}\,, \tag{7.121}$$

$$\hat{F}^\pm = \sum_{m,s=0}^{\infty} \frac{1}{[s]_q!}\hat{F}_{ms}^\pm(k) f_s\,, \tag{7.122}$$

$$\begin{aligned}
\hat{F}_{ms}^+(k) &= 2d_s q^{-s}\big((q^{-s-5}\hat{\gamma}_-^s k_-^m + z\hat{\gamma}_\nu^s k_\nu^m)(k_\nu - q^{s+3}zk_-) + \\
&\quad + (q^{-s-5}\hat{\gamma}_{\bar{\nu}}^s k_{\bar{\nu}}^m + z\hat{\gamma}_+^s k_+^m)(k_+ - q^{s+3}zk_{\bar{\nu}})\big)\,, \\
\hat{F}_{ms}^-(k) &= 2d_s q^{-2s-2}\big((\hat{\gamma}_-^s k_-^m + q^{-2}\bar{z}\hat{\gamma}_{\bar{\nu}}^s k_{\bar{\nu}}^m)(k_{\bar{\nu}} - \bar{z}k_-) + \\
&\quad + (\hat{\gamma}_\nu^s k_\nu^m + q^{-2}\bar{z}\hat{\gamma}_+^s k_+^m)(k_+ - \bar{z}k_\nu)\big)\,,
\end{aligned}$$

where $d_s = \beta^s/\beta^{s+1}$. As in the homogeneous case we can make $\hat{F}_{ms}^-(k)$ independent of s by choosing $\hat{\gamma}_\kappa^s \sim q^{2s}\,d_s^{-1}$, but we cannot make $\hat{F}_{ms}^+(k)$ or $\hat{J}_\kappa^{ms}(k)$ independent of s.

Since we work with the full Maxwell equations, we have also to check the q-deformation of the current conservation $\partial^\nu J_\nu = 0$:

$$I_{13}J = 0\,, \tag{7.123}$$

$$\begin{aligned}
I_{13} = {} & q^3\,[N_z - 1]_q\,T_z\,\hat{d}_{\bar{z}}\,\hat{d}_\nu\,T_\nu\,T_-\,T_+ + q\,\hat{d}_z\,T_z\,\hat{d}_{\bar{z}}\,\hat{d}_-\,T_\nu\,T_+ + \\
& + q\,[N_z - 1]_q\,T_z\,[N_{\bar{z}} - 1]_q\,\hat{d}_+\,T_+\,T_{\bar{\nu}} + \\
& + q^{-1}\,[N_{\bar{z}} - 1]_q\,\hat{d}_z\,T_z\,\hat{d}_{\bar{\nu}}\,T_\nu\,T_-^{-1}\,T_+ - \\
& - \lambda\,\hat{M}_\nu\,[N_{\bar{z}} - 1]_q\,\hat{d}_z\,T_z\,\hat{d}_-\,\hat{d}_+\,T_\nu\,T_-^{-1}\,T_+\,T_{\bar{\nu}}.
\end{aligned} \tag{7.124}$$

Substituting (7.118 and 7.119) in the above we get:

$$q\,J^s_+(k)\,k_+ + J^s_V(k)\,k_V + q^{s+2}\,J^s_{\bar V}\,k_{\bar V} + q^{s+1}\,J^s_-(k)\,k_- = 0. \tag{7.125}$$

The latter is fulfilled by the explicit expressions in (7.120), but we should note that these expressions fulfil also the following splittings of (7.125):

$$\begin{aligned} &q\,J^s_+(k)\,k_+ + J^s_V(k)\,k_V = 0\,, \quad q\,J^s_{\bar V}(k)\,k_{\bar V} + J^s_-(k)\,k_- = 0\,,\\ &J^s_+(k)\,k_+ + q^{s+1}\,J^s_{\bar V}(k)\,k_{\bar V} = 0\,, \quad J^s_V(k)\,k_V + q^{s+1}\,J^s_-(k)\,k_- = 0\,. \end{aligned} \tag{7.126}$$

Furthermore the expressions from (7.120) fulfil also:

$$\begin{aligned} &q\,J^s_+(k)\,k_{\bar V} + J^s_V(k)\,k_- = 0\,, \quad q\,J^s_{\bar V}(k)\,k_+ + J^s_-(k)\,k_V = 0\,,\\ &J^s_+(k)\,k_V + q^{s+1}\,J^s_{\bar V}(k)\,k_- = 0\,, \quad J^s_V(k)\,k_+ + q^{s+1}\,J^s_-(k)\,k_{\bar V} = 0\,. \end{aligned} \tag{7.127}$$

Now we shall use the basis (7.89). Then solutions of (7.37) for $n = 0$ in the homogeneous case ($J = 0$) are:

$$\tilde{F}^{h\pm} \doteq \left({}_qF^{\pm}_0\right)_{J=0} = \sum_{m,s=0}^{\infty} \frac{1}{[s]_q!}\,\tilde{F}^{h\pm}_{ms}(k)\,\tilde{h}_s\,, \tag{7.128}$$

$$\begin{aligned} \tilde{F}^{h+}_{ms}(k) = \sum_{i=0}^{m}\Big(&\sum_{j=0}^{m-i} \tilde{p}^{ms1}_{ij}\,k^i_{\bar V}k^{m-i-j}_- k^j_V\,(k_V - zk_-)(k_V - qzk_-) + \\ &+ \tilde{p}^{ms2}_i\,k^i_{\bar V}k^{m-i}_V\,(k_+ - zk_{\bar V})(k_V - qzk_-) + \\ &+ \sum_{j=0}^{m-i} \tilde{p}^{ms3}_{ij}\,k^i_{\bar V}k^{m-i-j}_+ k^j_V\,(k_+ - zk_{\bar V})(k_+ - qzk_{\bar V})\Big)\,, \end{aligned} \tag{7.129}$$

$$\begin{aligned} \tilde{F}^{h-}_{ms}(k) = \sum_{i=0}^{m}\Big(&\sum_{j=0}^{m-i} \mathrm{tr}^{ms1}_{ij}\,k^i_{\bar V}k^{m-i-j}_- k^j_V\,(k_{\bar V} - q^{s+1}\bar{z}k_-)(k_{\bar V} - q^{s+2}\bar{z}k_-) + \\ &+ \mathrm{tr}^{ms2}_i\,k^i_V k^{m-i}_{\bar V}\,(k_{\bar V} - q^{s+1}\bar{z}k_-)(k_+ - q^{s+2}\bar{z}k_V) + \\ &+ \sum_{j=0}^{m-i} \mathrm{tr}^{ms3}_{ij}\,k^i_V k^{m-i-j}_+ k^j_{\bar V}\,(k_+ - q^{s+1}\bar{z}k_V)(k_+ - q^{s+2}\bar{z}k_V)\Big), \end{aligned} \tag{7.130}$$

where $\tilde{p}^{msa}_{i(j)}$, $\mathrm{tr}^{msa}_{i(j)}$ are independent constants, $Q_s(a,b) = 0$ in $\tilde{h}_s$. (The solution for $\tilde{\varphi}_{r=2}$ given in (7.94) is obtained here for $m = 0$.) The solution (7.129) can be written in terms of the deformed plane wave if we suppose that the $\tilde{p}^{msa}_{i(j)}$ for different s coincide: $\tilde{p}^{msa}_{i(j)} = \tilde{p}^{ma}_{i(j)}$. Then we have:

$$\tilde{F}^{h+} = \sum_{m=0}^{\infty} \tilde{F}_m^{h+}(k)\, \widetilde{\exp}_q(k,x)\,, \quad \tilde{F}_m^{h+}(k) = \tilde{F}_{ms}^{h+}(k)\,. \tag{7.131}$$

In the inhomogeneous case the solutions of (7.37) for $n = 0$ are:

$$ {}_qJ^0 = \bar{z}z\tilde{J}_+ + z\tilde{J}_\nu + \bar{z}\tilde{J}_{\bar{\nu}} + \tilde{J}_-\,, \tag{7.132}$$

$$\tilde{J}_\kappa = \sum_{m,s=0}^{\infty} \frac{1}{[s]_q!} \tilde{J}_\kappa^{ms}(k)\, \tilde{h}_{s-1}\,, \quad \kappa = \pm, \nu, \bar{\nu}\,, \tag{7.133}$$

$$\begin{aligned}
\tilde{J}_+^{ms}(k) &= -q^{s+1}\tilde{K}_m^s(k)\, k_-\,, \\
\tilde{J}_-^{ms}(k) &= -q^{-1}\tilde{K}_m^s(k)\, k_+\,, \\
\tilde{J}_\nu^{ms}(k) &= \tilde{K}_m^s(k)\, k_{\bar{\nu}}\,, \\
\tilde{J}_{\bar{\nu}}^{ms}(k) &= q^s\tilde{K}_m^s(k)\, k_\nu\,, \\
\tilde{K}_m^s(k) &\doteq \tilde{\gamma}_\nu^s k_\nu^{m+1} + \tilde{\gamma}_-^s k_-^{m+1} + \tilde{\gamma}_+^s k_+^{m+1} + \tilde{\gamma}_{\bar{\nu}}^s k_{\bar{\nu}}^{m+1}\,,
\end{aligned} \tag{7.134}$$

$$ {}_qF_0^\pm = \tilde{F}^\pm + \tilde{F}^{h\pm}\,, \tag{7.135}$$

$$\tilde{F}^\pm = \sum_{m,s=0}^{\infty} \frac{1}{[s]_q!} \tilde{F}_{ms}^\pm(k)\, \tilde{h}_s\,, \tag{7.136}$$

$$\begin{aligned}
\tilde{F}_{ms}^+(k) &= 2\tilde{d}_s q^{s-2}\big((\tilde{\gamma}_-^s k_-^m + q^{-1}z\tilde{\gamma}_\nu^s k_\nu^m)(k_\nu - qzk_-) + \\
&\quad + (\tilde{\gamma}_{\bar{\nu}}^s k_{\bar{\nu}}^m + q^{-1}z\tilde{\gamma}_+^s k_+^m)(k_+ - qzk_{\bar{\nu}})\big)\,, \\
\tilde{F}_{ms}^-(k) &= 2\tilde{d}_s\big((q^{-s-3}\tilde{\gamma}_-^s k_-^m + q\bar{z}\tilde{\gamma}_{\bar{\nu}}^s k_{\bar{\nu}}^m)(k_{\bar{\nu}} - q^{s+2}\bar{z}k_-) + \\
&\quad + (q^{-s-3}\tilde{\gamma}_\nu^s k_\nu^m + q\bar{z}\tilde{\gamma}_+^s k_+^m)(k_+ - q^{s+2}\bar{z}k_\nu)\big)\,,
\end{aligned}$$

where $\tilde{d}_s = \tilde{b}^s/\tilde{b}^{s+1}$, $Q_s(a,b) = 0$ in $\tilde{h}_s$. We can make $\tilde{F}_{ms}^+(k)$ independent of s by choosing $\tilde{\gamma}_\kappa^s \sim q^{-s}\,\tilde{d}_s^{-1}$, but we cannot make $\tilde{F}_{ms}^-(k)$ or $\tilde{J}_\kappa^{ms}(k)$ independent of s.

Also here we shall check whether the q-deformation of the current conservation (7.123) is fulfilled. The analog of (7.124) in the basis (7.89) is:

$$\begin{aligned}
I_{13} = {}& [N_z - 1]_q\, \hat{\mathcal{D}}_{\bar{z}}\, T_{\bar{z}}\, \hat{\mathcal{D}}_\nu\, T_{\bar{\nu}}\, T_+\, T_-^{-1} + q\, \hat{\mathcal{D}}_{\bar{z}}\, T_{\bar{z}}\, \hat{\mathcal{D}}_z\, \hat{\mathcal{D}}_-\, T_{\bar{\nu}}\, T_+ + \\
& + q\,[N_{\bar{z}} - 1]_q\, T_{\bar{z}}\,[N_z - 1]_q\, \hat{\mathcal{D}}_+\, T_+\, T_\nu + \\
& + q^2\,[N_{\bar{z}} - 1]_q\, \hat{\mathcal{D}}_z\, T_{\bar{z}}\, \hat{\mathcal{D}}_{\bar{\nu}}\, T_{\bar{\nu}}\, T_-\, T_+ - \\
& - \lambda\, q\, \hat{M}_\nu\,[N_{\bar{z}} - 1]_q\, \hat{\mathcal{D}}_z\, T_{\bar{z}}\, \hat{\mathcal{D}}_-\, \hat{\mathcal{D}}_+\, T_-\, T_+.
\end{aligned} \tag{7.137}$$

Then the analog of (7.125) is:

$$J^s_+(k)\,k_+ + q^s J^s_\nu(k)\,k_\nu + J^s_{\bar\nu}\,k_{\bar\nu} + q^s J^s_-(k)\,k_- = 0. \tag{7.138}$$

The latter is fulfilled by the explicit expressions in (7.134), but we should note that these expressions fulfil also the following splittings of (7.138):

$$\begin{aligned} &J^s_+(k)\,k_+ + q^s J^s_\nu(k)\,k_\nu = 0\,, \quad J^s_{\bar\nu}(k)\,k_{\bar\nu} + q^s J^s_-(k)\,k_- = 0\,,\\ &J^s_+(k)\,k_+ + J^s_{\bar\nu}(k)\,k_{\bar\nu} = 0\,, \quad J^s_\nu(k)\,k_\nu + J^s_-(k)\,k_- = 0\,. \end{aligned} \tag{7.139}$$

Furthermore the expressions from (7.134) fulfil also:

$$\begin{aligned} &J^s_+(k)\,k_{\bar\nu} + q^s J^s_\nu(k)\,k_- = 0\,, \quad J^s_{\bar\nu}(k)\,k_+ + q^s J^s_-(k)\,k_\nu = 0\,,\\ &J^s_+(k)\,k_\nu + J^s_{\bar\nu}(k)\,k_- = 0\,, \quad J^s_\nu(k)\,k_+ + J^s_-(k)\,k_{\bar\nu} = 0\,. \end{aligned} \tag{7.140}$$

Summarizing, we have given solutions of the full q-Maxwell equations in two conjugated bases (7.61) and (7.89). The solutions of the homogeneous equations are also more general than the solutions for $\hat\varphi$ and $\tilde\varphi$ for general r. As before we see that the roles of the solutions F^+ and F^- are exchanged in the two conjugated bases. We note also that the current components are different: $\hat J^{ms}_\kappa \neq \tilde J^{ms}_\kappa$ (for $q \neq 1$, $\kappa \neq \nu$), and in both cases they cannot be made independent of s. Thus, there is no advantage of choosing either of the bases (7.61) or (7.89). It may be also possible to use both in a Connes-Lott type model [153].

7.7 q-Weyl Gravity Equations Hierarchy

In this section we follow [229, 238]. Here we study another hierarchy which is given as follows:

$$\begin{array}{ccccc} & & C^+_m & & \\ & \nearrow & & \searrow & \\ C^h_m & & & & C^T_m \\ & \searrow & & \nearrow & \\ & & C^-_m & & \end{array} \tag{7.141}$$

where $m \in \mathbb{N}$, and the corresponding signatures are:

$$\begin{aligned} \chi^+_m &= [2m, 0; 2], \quad \chi^-_m = [0, 2m; 2],\\ \chi^h_m &= [m, m; 2-m], \quad \chi^T_m = [m, m; 2+m]. \end{aligned} \tag{7.142}$$

For future reference we also give the *Dynkin labels* $\chi = \{m_1, m_2, m_3\}$ of these representations:

$$\begin{aligned}
\chi_m^+ &= \{2m+1, -m-1, 1\}, \quad \chi_m^- = \{1, -m-1, 2m+1\}, \\
\chi_m^h &= \{m+1, -1, m+1\}, \quad \chi_m^T = \{m+1, -2m-1, m+1\}.
\end{aligned} \tag{7.143}$$

The arrows on (7.141) represent invariant differential operators of order m. It is a partial case of the general conformal scheme parametrized by three natural numbers p, ν, n (cf. formula (6.170) and figure (6.171) of Volume 1), setting here: $\nu = 1$, $p = n = m$. This hierarchy intersects with the Maxwell hierarchy for the lowest value $m = 1$. Here we consider the linear Weyl gravity which is obtained for $m = 2$.

7.7.1 Linear Conformal Gravity

We start with the $q = 1$ situation, and we first write the linear conformal gravity equations, or Weyl gravity equations in our indexless formulation, trading the indices for two conjugate variables $z, \bar{z}$.

Weyl gravity is governed by the Weyl tensor $C_{\mu\nu\sigma\tau}$, which is given in terms of the Riemann curvature tensor $R_{\mu\nu\sigma\tau}$, Ricci curvature tensor $R_{\mu\nu}$, scalar curvature R:

$$C_{\mu\nu\sigma\tau} = R_{\mu\nu\sigma\tau} - \tfrac{1}{2}(g_{\mu\sigma}R_{\nu\tau} + g_{\nu\tau}R_{\mu\sigma} - g_{\mu\tau}R_{\nu\sigma} - g_{\nu\sigma}R_{\mu\tau}) + \tfrac{1}{6}(g_{\mu\sigma}g_{\nu\tau} - g_{\mu\tau}g_{\nu\sigma})R\,, \tag{7.144}$$

where $g_{\mu\nu}$ is the metric tensor. Linear conformal gravity is obtained when the metric tensor is written as: $g_{\mu\nu} = \eta_{\mu\nu} + h_{\mu\nu}$, where $\eta_{\mu\nu}$ is the flat Minkowski metric, $h_{\mu\nu}$ are small so that all quadratic and higher-order terms are neglected. In particular: $R_{\mu\nu\sigma\tau} = \frac{1}{2}(\partial_\mu\partial_\tau h_{\nu\sigma} + \partial_\nu\partial_\sigma h_{\mu\tau} - \partial_\mu\partial_\sigma h_{\nu\tau} - \partial_\nu\partial_\tau h_{\mu\sigma})$. The equations of linear conformal gravity are:

$$\partial^\nu\partial^\tau C_{\mu\nu\sigma\tau} = T_{\mu\sigma}\,, \tag{7.145}$$

where $T_{\mu\nu}$ is the energy-momentum tensor. From the symmetry properties of the Weyl tensor it follows that it has ten independent components. These may be chosen as follows (introducing notation for future use):

$$\begin{aligned}
C_0 = C_{0123}\,, \quad C_1 = C_{2121}\,, \quad C_2 = C_{0202}\,, \quad C_3 = C_{3012}\,, \\
C_4 = C_{2021}\,, \quad C_5 = C_{1012}\,, \quad C_6 = C_{2023}\,, \\
C_7 = C_{3132}\,, \quad C_8 = C_{2123}\,, \quad C_9 = C_{1213}\,.
\end{aligned} \tag{7.146}$$

Furthermore, the Weyl tensor transforms as the direct sum of two conjugate Lorentz irreps, which we shall denote as $C^\pm$ (cf. (7.142) for $m = 2$). The tensors $T_{\mu\nu}$ and $h_{\mu\nu}$ are symmetric and traceless with nine independent components.

Further, we shall use again the fact that a Lorentz irrep (spin-tensor) with signature (n_1, n_2) may be represented by a polynomial $G(z, \bar{z})$ in $z, \bar{z}$ of order n_1, n_2, respectively. More explicitly, for the Weyl gravity representations mentioned above we use:

$$C^+(z) = z^4 C_4^+ + z^3 C_3^+ + z^2 C_2^+ + zC_1^+ + C_0^+ \,, \qquad (7.147)$$
$$C^-(\bar{z}) = \bar{z}^4 C_4^- + \bar{z}^3 C_3^- + \bar{z}^2 C_2^- + \bar{z}C_1^- + C_0^- \,,$$

$$\begin{aligned} T(z,\bar{z}) &= z^2\bar{z}^2 T'_{22} + z^2\bar{z}T'_{21} + z^2 T'_{20} + \\ &+ z\bar{z}^2 T'_{12} + z\bar{z}T'_{11} + zT'_{10} + \\ &+ \bar{z}^2 T'_{02} + \bar{z}T'_{01} + T'_{00} \,, \end{aligned} \qquad (7.148)$$

$$\begin{aligned} h(z,\bar{z}) &= z^2\bar{z}^2 h'_{22} + z^2\bar{z}h'_{21} + z^2 h'_{20} + \\ &+ z\bar{z}^2 h'_{12} + z\bar{z}h'_{11} + zh'_{10} + \\ &+ \bar{z}^2 h'_{02} + \bar{z}h'_{01} + h'_{00} \,, \end{aligned} \qquad (7.149)$$

where the indices on the RHS are not Lorentz-covariance indices, they just indicate the powers of $z, \bar{z}$. The components $C_k^\pm$ are given in terms of the Weyl tensor components as follows:

$$\begin{aligned} C_0^+ &= C_2 - \tfrac{1}{2}C_1 - C_6 + i(C_0 + \tfrac{1}{2}C_3 + C_7) \\ C_1^+ &= 2(C_4 - C_8 + i(C_9 - C_5)) \\ C_2^+ &= 3(C_1 - iC_3) \\ C_3^+ &= 8(C_4 + C_8 + i(C_9 + C_5)) \\ C_4^+ &= C_2 - \tfrac{1}{2}C_1 + C_6 + i(C_0 + \tfrac{1}{2}C_3 - C_7) \\ C_0^- &= C_2 - \tfrac{1}{2}C_1 - C_6 - i(C_0 + \tfrac{1}{2}C_3 + C_7) \\ C_1^- &= 2(C_4 - C_8 - i(C_9 - C_5)) \\ C_2^- &= 3(C_1 + iC_3) \\ C_3^- &= 2(C_4 + C_8 - i(C_9 + C_5)) \\ C_4^- &= C_2 - \tfrac{1}{2}C_1 + C_6 - i(C_0 + \tfrac{1}{2}C_3 - C_7). \end{aligned} \qquad (7.150)$$

while the components T'_{ij} are given in terms of $T_{\mu\nu}$ as follows:

$$\begin{aligned} T'_{22} &= T_{00} + 2T_{03} + T_{33} \\ T'_{11} &= T_{00} - T_{33} \\ T'_{00} &= T_{00} - 2T_{03} + T_{33} \\ T'_{21} &= T_{01} + iT_{02} + T_{13} + iT_{23} \\ T'_{12} &= T_{01} - iT_{02} + T_{13} - iT_{23} \\ T'_{10} &= T_{01} + iT_{02} - T_{13} - iT_{23} \\ T'_{01} &= T_{01} - iT_{02} - T_{13} + iT_{23} \\ T'_{20} &= T_{11} + 2iT_{12} - T_{22} \\ T'_{02} &= T_{11} - 2iT_{12} - T_{22} \end{aligned} \qquad (7.151)$$

and similarly for h'_{ij} in terms of $h_{\mu\nu}$.

In these terms all linear conformal Weyl gravity equations (7.145) (cf. also (7.141)) may be written in compact form as the following pair of equations:

$$I^+C^+(z) = T(z,\bar{z}), \qquad I^-C^-(\bar{z}) = T(z,\bar{z}), \tag{7.152}$$

where the operators $I^\pm$ are given as follows:

$$\begin{aligned}
I^+ = \Big(& z^2\bar{z}^2\partial_+^2 + z^2\partial_v^2 + \bar{z}^2\partial_{\bar{v}}^2 + \partial_-^2 + \\
&+2z^2\bar{z}\partial_v\partial_+ + 2z\bar{z}^2\partial_+\partial_{\bar{v}} + 2z\bar{z}(\partial_-\partial_+ + \partial_v\partial_{\bar{v}}) + \\
&+2\bar{z}\partial_-\partial_{\bar{v}} + 2z\partial_v\partial_-\Big)\partial_z^2 - \\
&-6\Big(z\bar{z}^2\partial_+^2 + z\partial_v^2 + 2z\bar{z}\partial_v\partial_+ + \bar{z}^2\partial_+\partial_{\bar{v}} + \\
&+\bar{z}(\partial_-\partial_+ + \partial_v\partial_{\bar{v}}) + \partial_v\partial_-\Big)\partial_z + \\
&+12\Big(\bar{z}^2\partial_+^2 + \partial_v^2 + 2\bar{z}\partial_v\partial_+\Big),
\end{aligned} \tag{7.153}$$

$$\begin{aligned}
I^- = \Big(& z^2\bar{z}^2\partial_+^2 + z^2\partial_v^2 + \bar{z}^2\partial_{\bar{v}}^2 + \partial_-^2 + \\
&+2z^2\bar{z}\partial_v\partial_+ + 2z\bar{z}^2\partial_+\partial_{\bar{v}} + 2z\bar{z}(\partial_-\partial_+ + \partial_v\partial_{\bar{v}}) + \\
&+2\bar{z}\partial_-\partial_{\bar{v}} + 2z\partial_v\partial_-\Big)\partial_{\bar{z}}^2 - \\
&-6\Big(z^2\bar{z}\partial_+^2 + \bar{z}\partial_{\bar{v}}^2 + 2z\bar{z}\partial_+\partial_{\bar{v}} + z^2\partial_v\partial_+ + \\
&+z(\partial_-\partial_+ + \partial_v\partial_{\bar{v}}) + \partial_-\partial_{\bar{v}}\Big)\partial_{\bar{z}} + \\
&+12\Big(z^2\partial_+^2 + \partial_{\bar{v}}^2 + 2z\partial_+\partial_{\bar{v}}\Big)
\end{aligned}$$

using the Minkowski conformal variables. We recall that in terms of these variables the d'Alembert equation is:

$$\Box\,\varphi = (\partial_-\partial_+ - \partial_v\,\partial_{\bar{v}})\,\varphi = 0\,. \tag{7.154}$$

To make more transparent the origin of (7.152) and in the same time to derive the quantum group deformation of (7.152) and (7.153) we first introduce the following parameter-dependent operators:

$$\begin{aligned}
I_n^+ &= \tfrac{1}{2}\Big(n(n-1)I_1^2I_2^2 - 2(n^2-1)I_1I_2^2I_1 + n(n+1)I_2^2I_1^2\Big), \\
I_n^- &= \tfrac{1}{2}\Big(n(n-1)I_3^2I_2^2 - 2(n^2-1)I_3I_2^2I_3 + n(n+1)I_2^2I_3^2\Big),
\end{aligned} \tag{7.155}$$

where $I_1 = \partial_z$, $I_2 = \bar{z}z\partial_+ + z\partial_v + \bar{z}\partial_{\bar{v}} + \partial_-$, and $I_3 = \partial_{\bar{z}}$ are from (7.16). We recall that group-theoretically the operators I_a correspond to the three simple roots of the root system of $sl(4)$, while the operators $I_n^\pm$ correspond to the singular vectors for the two nonsimple nonhighest roots. More precisely, the operator I_n^+ is obtained from the $sl(4)$ formula for the singular vector given by (2.37) of weight $m_{12}\alpha_{12} = 2\alpha_{12}$. Analogously, the operator I_n^- is obtained from the same formula for weight $m_{23}\alpha_{23} = 2\alpha_{23}$. The parameter $n = \max(2j_1, 2j_2)$.

It is easy to check that we have the following relation:

$$I^{\pm} = I_4^{\pm}\,, \tag{7.156}$$

that is, (7.152) are written as:

$$I_4^+ \, C^+(z) = T(z,\bar{z}), \qquad I_4^- \, C^-(\bar{z}) = T(z,\bar{z}). \tag{7.157}$$

This is the form that is immediately generalizable to the q-deformed case in next subsection.

Using the same operators we can write down the pair of equations which give the Weyl tensor components in terms of the metric tensor:

$$I_2^+ h(z,\bar{z}) = C^-(\bar{z}), \qquad I_2^- h(z,\bar{z}) = C^+(z). \tag{7.158}$$

We stress again the advantage of the indexless formalism due to which two different pairs of equations – (7.157) and (7.158) – may be written using the same parameter-dependent operator expressions by just specializing the values of the parameter.

7.7.2 q-Plane-Wave Solutions of q-Weyl Gravity

We consider now the q-deformed setting supposing that q is not a nontrivial root of unity.

Using the $U_q(sl(4))$ formula for the singular vector given in (2.37), we obtain for the q-analogue of (7.15):

$$\begin{aligned}
{}_qI_n^+ &= \tfrac{1}{2}\Big([n]_q \, [n-1]_q \, {}_qI_1^2 \, {}_qI_2^2 - [2]_q \, [n-1]_q \, [n+1]_q \, {}_qI_1 \, {}_qI_2^2 \, {}_qI_1 + \\
&\quad + [n]_q \, [n+1]_q \, {}_qI_2^2 \, {}_qI_1^2 \Big), \\
{}_qI_n^- &= \tfrac{1}{2}\Big([n]_q \, [n-1]_q \, {}_qI_3^2 \, {}_qI_2^2 - [2]_q \, [n-1]_q \, [n+1]_q \, {}_qI_3 \, {}_qI_2^2 \, {}_qI_3 + \\
&\quad + [n]_q \, [n+1]_q \, {}_qI_2^2 \, {}_qI_3^2 \Big),
\end{aligned} \tag{7.159}$$

where the q-deformed versions ${}_qI_a$ of (7.16) are given in (7.35).

Then the q-Weyl gravity equations are (cf. (7.157)):

$${}_qI_4^+ C^+(z) = T(z,\bar{z}), \qquad {}_qI_4^- C^-(\bar{z}) = T(z,\bar{z}), \tag{7.160}$$

while q-analogues of (7.158) are:

$${}_qI_2^+ h(z,\bar{z}) = C^-(\bar{z}), \qquad {}_qI_2^- h(z,\bar{z}) = C^+(z). \tag{7.161}$$

For the solutions we shall use the basis (7.61). The solutions of the first equation in (7.160) in the homogeneous case ($T = 0$) are:

$$ {}_qC_0^+ = \sum_{s=0}^{\infty} \frac{1}{[s]_q!} \hat{C}_s^+ , \tag{7.162} $$

$$ \begin{aligned} \hat{C}_s^+ &= \sum_{m=0}^{4} \hat{\gamma}_m^{s+} \left(\prod_{i=0}^{-m+3} (k_+ - q^{i+B_s+s+4} k_{\bar{v}} z) \right) \times \\ &\times \left(\prod_{j=-m+4}^{3} (k_v - q^{j+B_s+s+4} k_{-} z) \right) \hat{h}_s^+ , \end{aligned} \tag{7.163} $$

where h_s^+ is h_s with:

$$ P_s(a,b) = P_s^+(a,b) \equiv R_s(a) + B_s b, \tag{7.164} $$

$\hat{\gamma}_m^{s+}$, B_s are arbitrary constants and $R_s(a)$ is an arbitrary polynomial in a. In order to be able to write the above solution in terms of the deformed plane wave, we have to suppose that the $\hat{\gamma}_m^{s+}$, $B_s + s$ for different s coincide: $\hat{\gamma}_m^{s+} = \tilde{\gamma}_m^+$, for example, we can make the choice $B_s = B' - s - 4$. Then we have:

$$ \begin{aligned} {}_qC_0^+ &= \sum_{m=0}^{4} \tilde{\gamma}_m^+ \left(\prod_{i=0}^{-m+3} (k_+ - q^{i+B'} k_{\bar{v}} z) \right) \times \\ &\times \left(\prod_{j=-m+4}^{3} (k_v - q^{j+B'} k_{-} z) \right) \widehat{\exp}_q^+(k,x) , \end{aligned} \tag{7.165} $$

where $\widehat{\exp}_q^+(k,x)$ is $\widehat{\exp}_q(k,x)$ with the choice (7.164).

The solutions of the second equation in (7.160) are:

$$ {}_qC_0^- = \sum_{s=0}^{\infty} \frac{1}{[s]_q!} \hat{C}_s^- \tag{7.166} $$

$$ \begin{aligned} \hat{C}_s^- &= \sum_{m=0}^{4} \hat{\gamma}_m^{s-} \left(\prod_{i=-1}^{-m+2} (k_+ - q^{i-D_s} k_v \bar{z}) \right) \times \\ &\times \left(\prod_{j=-m+3}^{2} (k_{\bar{v}} - q^{j-D_s} k_{-} \bar{z}) \right) h_s^- \end{aligned} \tag{7.167} $$

where h_s^- is h_s with:

$$ P_s(a,b) = P_s^-(a,b) \equiv D_s a + Q_s(b), \tag{7.168} $$

$\hat{\gamma}_m^{s-}$, D_s are arbitrary constants and $Q_s(b)$ is an arbitrary polynomial. In order to be able to write this solution in terms of the deformed plane wave, we have to suppose that the $\hat{\gamma}_m^{s-}$, D_s for different s coincide: $\hat{\gamma}_m^{s-} = \hat{\gamma}_m^-$, $D_s = D$. Then we have:

$$\begin{aligned} {}_qC_0^- &= \sum_{m=0}^{4} \hat{\gamma}_m^- \left(\prod_{i=-1}^{-m+2} (k_+ - q^{i-D} k_\nu \bar{z}) \right) \times \\ &\times \left(\prod_{j=-m+3}^{2} (k_{\bar{\nu}} - q^{j-D} k_- \bar{z}) \right) \widehat{\exp}_q^-(k, x), \end{aligned} \tag{7.169}$$

where $\widehat{\exp}_q^-(k, x)$ is $\widehat{\exp}_q(k, x)$ with the choice (7.168).

Bibliography

[1] M.C. Abbott and D.A. Lowe, "Statistical entropy of three-dimensional q-deformed Kerr-de Sitter space," Phys. Lett. **B646** (2007) 275–278.

[2] B. Abdesselam, "Special Representations of U_q(sl(N)) at roots of unity," J. Phys. **A29** (1996) 1201–1214.

[3] B. Abdesselam, D. Arnaudon and M. Bauer, "Centre and representations of U_q(sl(2|1)) at roots of unity," J. Phys. **A30** (1997) 867–880.

[4] B. Abdesselam, D. Arnaudon and A. Chakrabarti, "Representations of U_q(sl(N)) at roots of unity," J. Phys. **A28** (1995) 5495–5508.

[5] B. Abdesselam, A. Chakrabarti and R. Chakrabarti, "On U_h(sl(2)), U_h(e(3)) and their representations," Int. J. Mod. Phys. **A12** (1997) 2301–2319; "Irreducible representations of Jordanian quantum algebra U_h(sl(2)) via a nonlinear map," Mod. Phys. Lett. **A11** (1996) 2883–2891; "Towards a general construction of nonstandard R_h-matrices as contraction limits of R_q-matrices: The $\mathcal{U}_h(sl(N))$ algebra case," Mod. Phys. Lett. **A13** (1998) 779–790.

[6] B. Abdesselam, A. Chakrabarti, R. Chakrabarti and J. Segar, "Maps and twists relating $\mathcal{U}(sl(2))$ and the nonstandard $\mathcal{U}_h(sl(2))$: Unified construction," Mod. Phys. Lett. **A14** (1999) 765–777.

[7] B. Abdesselam, A. Chakrabarti, V.K. Dobrev and S.G. Mihov, "Higher dimensional unitary braid matrices: Construction, associated structures and entanglements," J. Math. Phys. **48**, 053508 (2007).

[8] B. Abdesselam, A. Chakrabarti, V.K. Dobrev and S.G. Mihov, "Higher dimensional multiparameter unitary and nonunitary braid matrices: Even dimensions," J. Math. Phys. **48** (2007) 103505.

[9] B. Abdesselam, A. Chakrabarti, V.K. Dobrev and S.G. Mihov, "Exotic bialgebras from 9x9 unitary braid matrices," Phys. At. Nucl. **74**, 6 (2011) 824–831 (851–857 Russian edition).

[10] B. Abdesselam, A. Chakrabarti, V.K. Dobrev and S.G. Mihov, "Exotic bialgebras from 9×9 braid matrices," in: "Quantum Groups, Quantum Foundations, and Quantum Information: A Festschrift for Tony Sudbery," J. Phys.: Conf. Ser. Vol. 254 (2010) 012001.

[11] E. Abe, *Hopf Algebras*, Cambridge Tracts in Math. N. 74 (Cambridge University Press, Cambridge, 1980).

[12] M. Adivar, "Quadratic pencil of difference equations: Jost solutions, spectrum, and principal vectors," Quaest. Math. **33** (2010) 305–323.

[13] A. Aghamohammadi, "The two-parametric extension of h deformation of GL(2), and the differential calculus on its quantum plane," Mod. Phys. Lett. **A8** (1993) 2607–2613.

[14] A. Aghamohammadi, M. Khorrami and A. Shariati, "Jordanian deformation of SL(2) as a contraction of its Drinfeld-Jimbo deformation," J. Phys. **A28** (1995) L225–L232.

[15] A. Agostini, G. Amelino-Camelia and F. D'Andrea, "Hopf algebra description of noncommutative space-time symmetries," Int. J. Mod. Phys. **A19** (2004) 5187–5220.

[16] N. Aizawa, "Representation functions for Jordanian quantum group $SL_h(2)$ and Jacobi polynomials," J. Phys. **A33** (2000) 3735–3752; "Tensor Operators for U_h(sl(2))," J. Phys. **A31** (1998) 5467–5475; "Irreducible decomposition for tensor product representations of Jordanian quantum algebras," J. Phys. **A30** (1997) 5981–5992; "Symplecton for $U_h(sl(2))$ and representations of $SL_h(2)$," J. Math. Phys. **40** (1999) 5921–5938; "Drinfeld twist for two-parametric deformation of gl(2) and sl(1/2)," Czech. J. Phys. **48** (1998) 1273–1278.

[17] N. Aizawa, R. Chakrabarti, S.S. Naina Mohammed, J. Segar, "Universal T-matrix, representations of $osp_q(1/2)$ and little q-Jacobi polynomials," J. Math. Phys. **47** (2006) art. no. 123511.

[18] N. Aizawa, R. Chandrashekar and J. Segar, "Lowest weight representations, singular vectors and invariant equations for a class of conformal Galilei algebras," SIGMA **11** (2015) 002.

[19] N. Aizawa, F.J. Herranz, J. Negro and M. A. del Olmo, "Twisted conformal algebra so(4,2)," J. Phys. **A35** (2002) 8179–8196.

[20] F.C. Alcaraz, D. Arnaudon, V. Rittenberg and M. Scheunert, "Finite chains with quantum affine symmetries," Int. J. Mod. Phys. **A9** (1994) 3473–3496.

[21] F.C. Alcaraz and V. Rittenberg, "Reaction-diffusion processes as physical realizations of Hecke algebras," Phys. Lett. **B314** (1993) 377–380.

[22] F.C. Alcaraz, M. Droz, M. Henkel and V. Rittenberg, "Reaction-diffusion processes, critical dynamics and quantum chains," Ann. Phys. **230** (1994) 250–302.

[23] A.N.F. Aleixo and A.B. Balantekin, "Algebraic construction of coherent states for nonlinear quantum deformed systems," J. Phys. **A45** (2012) art. no. 165302; "Accidental degeneracies in nonlinear quantum deformed systems," J. Phys. **A44** (2011) 365303; "Entropy and entanglement dynamics in a quantum deformed coupled system," J. Phys. **A41** (2008) art. no. 315302; "An algebraic construction of generalized coherent states associated with q-deformed models for primary shape-invariant systems," J. Phys. **A40** (2007) 3463–3480.

[24] A.N.F. Aleixo, A.B. Balantekin and M.A. Candido-Ribeiro, "An algebraic q-deformed form for shape-invariant systems," J. Phys. **A36** (2003) 11631.

[25] A. Alekseev, L.D. Faddeev and M. Semenov-Tian-Shansky, "Hidden quantum groups inside Kac-Moody algebra," Commun. Math. Phys. **149** (1992) 335–345.

[26] A. Alekseev and S.L. Shatashvili, "Quantum groups and WZW models," Commun. Math. Phys. **133** (1990) 353–368.

[27] A.Y. Alekseev and I.T. Todorov, "Quadratic brackets from symplectic forms," Nucl. Phys. **B421** (1994) 413–428.

[28] A. Algin, "Quantum groups $GL_{pq}(2)$ – and $SU_{q1/q2}(2)$ – invariant bosonic gases: A comparative study," Int. J. Theor. Phys. 46, 71–84 (2009); A. Algin and M. Arik, "A two-parameter deformed susy algebra for $SU_{p/q}$(n)-covariant (p,q)-deformed fermionic oscillators," Mod. Phys. Lett. **A20** (2005) 613–622.

[29] S. Alisauskas, "The multiple sum formulas for 12j coefficients of SU(2) and $u_q(2)$", J. Math. Phys. **43** (2002) 1547–1568.

[30] L. Alvarez-Gaume, C. Gomez and G. Sierra, "Quantum group interpretation of some conformal field theories," Phys. Lett. **220B** (1989) 142–152.

[31] L. Alvarez-Gaume, C. Gomez and G. Sierra, "Hidden quantum symmetries in rational conformal field theories," Nucl. Phys. **B319** (1989) 155–186.

[32] L. Alvarez-Gaume, C. Gomez and G. Sierra, "Duality and quantum groups," Nucl. Phys. **B330** (1990) 347–398.

[33] R. Alvarez-Nodarse and J. Arvesu, "On the q-polynomials in the exponential lattice x(s) = c(1)q(s)+c(3)," Integr. Trans. Spec. **F8** (1999) 299–324.

[34] R. Alvarez-Nodarse and Y.F. Smirnov, "The dual Hahn q-polynomials in the Lattice x(s)=(s)(q)(s+1)(q) and the q-algebras $SU_q(2)$ and $SU_q(1,1)$," J. Phys. **A29** (1996) 1435–1452.

[35] G. Amelino-Camelia and S. Majid, "Waves on noncommutative space-time and gamma-ray bursts," Int. J. Mod. Phys. **A15** (2000) 4301–4324; G. Amelino-Camelia, "Enlarged bound on the measurability of distances and quantum Kappa Poincare group," Phys. Lett. **B392** (1997) 283–286.

[36] H.H. Andersen, P. Polo and K. Wen, "Representations of quantum algebras," Invent. Math. **104** (1991) 1–59.

[37] G.E. Andrews, Regional Conference Series in Math. No. 66 (AMS, Providence, 1986).

[38] G.E. Andrews and R. Askey, "Enumeration of partitions: The role of Eulerian series and q-orthogonal polynomials," in: *Higher Combinatorics*, ed. M. Aigner (Reide., 1977) pp. 3–26.

[39] B.L. Aneva, V.K. Dobrev and S.G. Mihov, "Duality for the Jordanian matrix quantum group $GL_{g,h}(2)$," J. Phys. **A30** (1997) 6769–6781.

[40] A. Appel and V. Toledano-Laredo, “Quasi-Coxeter categories and quantum groups,” arXiv:1610.09741 [math.QA] (2016).
[41] I.Y. Arefeva and I.V. Volovich, “Quantum group gauge fields,” Mod. Phys. Lett. **A6** (1991) 893–908.
[42] M. Arik and B.T. Kaynak, “Non-commutative phase and the unitarization of $GL_{p,q}(2)$,” J. Math. Phys. **44** (2003) 1730–1734.
[43] D. Arnaudon, “Periodic and flat irreducible representations of $SU(3)_q$,” Commun. Math. Phys. **134** (1990) 523–537; Nonintegrable “Representations of the restricted quantum analog of sl(3),” J. Phys. **A30** (1997) 3527–3542; “Representations of U_q(sl(3)) at roots of 1 (in the restricted specialization),” Czech. J. Phys. **47** No 11 (1997) 1075–1082; “A note on the generalised lie algebra $sl(2)_q$,” J. Phys. **A31** (1998) 6647–6652.
[44] D. Arnaudon, “New fusion rules and r-matrices for sl(n)(q) at roots of unity,” Phys. Lett. **B280** (1992) 31–38.
[45] D. Arnaudon and A. Chakrabarti, “q-analog of IU(n) for q a root of unity,” Phys. Lett. **255B** (1991) 242–248.
[46] D. Arnaudon and A. Chakrabarti, “Periodic representations of $SO(5)_q$,” Phys. Lett. **B262** (1991) 68–70.
[47] D. Arnaudon and A. Chakrabarti, “Periodic and partially periodic representations of $SU(N)_q$”, Commun. Math. Phys. **139** (1991) 461–478; “Flat periodic representations of $U_q(g)$,” Commun. Math. Phys. **139** (1991) 605–618.
[48] D. Arnaudon, A. Chakrabarti, V.K. Dobrev et al. “On Combined Standard-Nonstandard or Hybrid (q,h)-Deformations,” J. Math. Phys. **42** (2001) 1236–1249.
[49] D. Arnaudon, A. Chakrabarti, V.K. Dobrev and S.G. Mihov, “Duality for exotic bialgebras,” J. Phys. **A34** (2001) 4065–82.
[50] D. Arnaudon, A. Chakrabarti, V.K. Dobrev and S.G. Mihov, “Duality and representations for new exotic bialgebras,” J. Math. Phys. **43** (2002) 6238–6264.
[51] D. Arnaudon, A. Chakrabarti, V.K. Dobrev and S.G. Mihov, “Spectral decomposition and Baxterisation of exotic bialgebras and associated noncommutative geometries,” Int. J. Mod. Phys. **A18** (2003) 4201–4213.
[52] D. Arnaudon, A. Chakrabarti, V.K. Dobrev and S.G. Mihov, “Exotic bialgebra S03: Representations, baxterisation and applications,” Ann. Henri Poincare, **7** (2006) 1351–1373, [volume dedicated to D. Arnaudon].
[53] D. Arnaudon, A. Sedrakyan and T. Sedrakyan, “Multi-leg integrable ladder models,” Nucl. Phys. **B676** (2004) 615–636; “Integrable N-leg ladder models,” Int J. Mod. Phys. **A19** Suppl. S (2004) 16–33.
[54] D. Arnaudon, A. Sedrakyan, T. Sedrakyan and P. Sorba, “Generalization of the U_q(gl(N)) algebra and staggered models,” Lett. Math. Phys. **58** (2001) 209–222.
[55] J. Arvesu, “Quantum algebras $su_q(2)$ and $su_q(1,1)$ associated with certain q-Hahn polynomials: A revisited approach,” Electron Trans. Numer. Anal. **24** (2006) 24–44.
[56] P. Aschieri, “Real forms of quantum orthogonal groups, q Lorentz groups in any dimension,” Lett. Math. Phys. **49** (1999) 1–15.
[57] P. Aschieri and F. Bonechi, “On the noncommutative geometry of twisted spheres,” Lett. Math. Phys. **59** (2002) 133–156.
[58] P. Aschieri and L. Castellani, “Universal enveloping algebra and differential calculi on inhomogeneous orthogonal q groups,” J. Geom. Phys. **26** (1998) 247–271; “R matrix formulation of the quantum inhomogeneous group $ISO_{q,r}(N)$ and $ISp_{q,r}(N)$,” Lett. Math. Phys. **36** (1996) 197–211. “Bicovariant calculus on twisted ISO(N), quantum Poincare group and quantum Minkowski space,” Int. J. Mod. Phys. **A11** (1996) 4513–4550; “Inhomogeneous quantum groups $IGL_{q,r}(N)$ – universal enveloping algebra and differential-calculus”, Int. J. Mod. Phys. **A11** (1996) 1019–1056.

[59] P. Aschieri and L. Castellani, "An Introduction to noncommutative differential geometry on quantum groups," Int. J. Mod. Phys. **A8** (1993) 1667–1706.

[60] P. Aschieri, L. Castellani and A.M. Scarfone, "Quantum orthogonal planes: $ISO_{q,r}(N)$ and $SO_{q,r}(N)$ bicovariant calculi and differential geometry on quantum Minkowski space," Eur. Phys. J. **C7** (1999) 159–175.

[61] R. Askey and J. Wilson, "Some basic hypergeometric orthogonal polynomials that generalize Jacobi polynomials," Mem. AMS **54** (1985) No. 319.

[62] H. Awata and Y. Yamada, "Five-dimensional AGT relation and the deformed beta-ensemble," Prog. Theor. Phys. **124** (2010) 227–262.

[63] H. Awata, M. Noumi and S. Odake, "Heisenberg realization for $U_q(sl(n))$ on the flag manifold," Lett. Math. Phys. **30** (1993) 35–44.

[64] F. Bachmaier and C. Blohmann, "Separation of noncommutative differential calculus on quantum Minkowski space," J. Math. Phys. **47** (2006) Art. No. 023501.

[65] G.E. Baird and L.C. Biedenharn, "On the representations of the semisimple lie groups. II," J. Math. Phys. **4** (1963) 1449–1466.

[66] T.H. Baker and P.J. Forrester, "The Calogero-Sutherland model and generalized classical polynomials," Commun. Math. Phys. **188** (1997) 175–216.

[67] A. Ballesteros, N.R. Bruno and F.J. Herranz, "Quantum (anti)de Sitter algebras and generalizations of the kappa-Minkowski space," in: *Symmetry Methods in Physics*, eds. C. Burdik, O. Navratil and S. Posta (Joint Institute for Nuclear Research, Dubna, Russia, 2004), pp. 1–20.

[68] A. Ballesteros and F.J. Herranz, "Universal R-matrix for non-standard quantum sl(2,R)," J. Phys. **A29** (1996) L311–L316.

[69] A. Ballesteros, F.J. Herranz and C. Meusburger, "Drinfel'd doubles for (2+1)-gravity," Class. Quantum Grav. **30** (2013) 155012; "Three-dimensional gravity and Drinfel'd doubles: Spacetimes and symmetries from quantum deformations," Phys. Lett. **B687** (2010) 375–381.

[70] A. Ballesteros and F.J. Herranz, F. Musso and P. Naranjo, "The kappa-(A)dS quantum algebra in (3+1) dimensions," arXiv:1612.03169 [hep-th] (2016).

[71] A. Ballesteros, F.J. Herranz and P. Naranjo, "Towards 3+1 gravity through Drinfel'd doubles with cosmological constant," Phys. Lett. **B746** (2015) 37–43.

[72] A. Ballesteros, F.J. Herranz and J. Negro, "Boson representations, non-standard quantum algebras and contractions," J. Phys. **A30** (1997) 6797–6810.

[73] A. Ballesteros, F.J. Herranz, J. Negro and L.M. Nieto, "Twist maps for non-standard quantum algebras and discrete Schrodinger symmetries," J. Phys. **A33** (2000) 4859–4870.

[74] A. Ballesteros, F.J. Herranz, et al., "Non-standard quantum so(2,2) and beyond," J. Phys. **A28** (1995) 941–956.

[75] A. Ballesteros, F.J. Herranz and P. Parashar, "(1+1) Schrodinger Lie bialgebras and their Poisson–Lie groups," J. Phys. **A33** (2000) 3445–3465; "Multiparametric quantum gl(2): Lie bialgebras, quantum R-matrices and non-relativistic limits," J. Phys. **A32** (1999) 2369–2385; "A Jordanian quantum two-photon/Schrödinger algebra," J. Phys. **A30** (1997) 8587–8597; "Quantum harmonic oscillator algebras as non-relativistic limits multiparametric gl(2) quantizations, Czech. J. Phys. **48** (1998) 1465–1470.

[76] J. Balog, L. Dabrowski and L. Feher, "A new quantum deformation of SL(3)," Phys. Lett. **B257** (1991) 74–78.

[77] R. Barbier and M. Kibler, "Application of a two-parameter quantum algebra to rotational spectroscopy of nuclei," Rept. Math. Phys. **38** (1996) 221–226.

[78] R. Barbier, J. Meyer and M. Kibler, "An $U_{qp}(u_2)$ rotor model for rotational bands of superdeformed nuclei," Int. J. Mod. Phys. **E4** (1995) 385–410.

[79] M.T. Batchelor, L. Mezincescu, R.I. Nepomechie and V. Rittenberg, "q-deformations of the O(3) symmetric spin-1 Heisenberg chain," J. Phys. **A23** (1990) L141.

[80] C. Bauer and H. Wachter, "Operator representations on quantum spaces," Eur. Phys. J. **C31** (2003) 261–275.

[81] P.F. Baum, K. De Commer and P.M. Hajac, "Free actions of compact quantum group on unital C*-algebras," arXiv:1304.2812 (2013).

[82] V.V. Bazhanov, "Trigonometric solutions of triangle equations and classical lie algebras" Phys. Lett. **B159** (1985) 321–324; "Integrable quantum systems and classical Lie algebras," Commun. Math. Phys. **113** (1987) 471–503.

[83] J. Beck and V.G. Kac, "Finite-dimensional representations of quantum affine algebras at roots of unity," J. Am. Math. Soc. **9** (1996) 391–423.

[84] N. Beisert and P. Koroteev, "Quantum deformations of the one-dimensional Hubbard model," J. Phys. **A41** (2008) 255204.

[85] A.A. Belavin, A.M. Polyakov and A.B. Zamolodchikov, "Infinite conformal symmetry in two-dimensional quantum field theory," Nucl. Phys. **B241** (1984) 333–380.

[86] A. Beliakova, K. Habiro, A.D. Lauda and B. Webster, "Current algebras and categorified quantum groups," arXiv:1412.1417 [math.QA] (2014).

[87] G. Benkart, S.J. Kang and H.H. Lee, "On the centre of two-parameter quantum groups," Proc. R. Soc. Edinburgh, A-Math **136** (2006) 445–472.

[88] G. Benkart, M. Pereira and S. Witherspoon, "Yetter-Drinfeld modules under cocycle twists," J. Algebra **324** (2010) 2990–3006.

[89] G. Benkart and S. Witherspoon, "Two-parameter quantum groups and Drinfeld doubles," Alg. Represent. Th. **7** (2004) 261–286.

[90] D. Berenstein and S.A. Cherkis, "Deformations of N=4 SYM and integrable spin chain models," Nucl. Phys. **B702** (2004) 49–85.

[91] N. Bergeron, Y. Gao and N.-H. Hu, "Drinfel'd doubles and Lusztig's symmetries of two-parameter quantum groups," J. Algebra, **301** (2006) 378–405.

[92] D. Bernard, "Vertex operator representations of the quantum affine algebra $U_q(B_r^{(1)})$," Lett. Math. Phys. **17** (1989) 239–245.

[93] D. Bernard, "Quantum Lie algebras and differential calculus on quantum groups," Prog. Theor. Phys. Suppl. **102** (1990) 49–66, eds. T. Eguchi, T. Inami and T. Miwa.

[94] D. Bernard and A. Le Clair, "q-deformation of SU(1,1) conformal ward identities and q-strings," Phys. Lett. **B227** (1990) 417–423.

[95] D. Bernard and A. Le Clair, "Quantum group symmetries and nonlocal currents in 2D QFT," Commun. Math. Phys. **142** (1991) 99–138.

[96] I.N. Bernstein, I.M. Gel'fand and S.I. Gel'fand, "Structure of representations generated by highest weight vectors," Funkts. Anal. Prilozh. **5**(1) (1971) 1–9; English translation: Funct. Anal. Appl. **5** (1971) 1–8.

[97] V.B. Bezerra, E.M.F. Curado and M.A. Rego-Monteiro, "Perturbative computation in a generalized quantum field theory," Phys. Rev **D66** (2002) 085013.

[98] A. Bhattacharya and S. Wang, "Kirchberg's factorization property for discrete quantum groups," arXiv:1512.08737 [math.OA] (2015).

[99] L.C. Biedenharn, "The quantum group $SU_q(2)$ and a q-analogue of the boson operators," J. Phys. **A22** (1989) L873–L878; "A q-boson realization of the quantum group $SU_q(2)$ and the theory of q-tensor operators," in: Proceedings of the Quantum Groups Workshop (Clausthal, 1989) eds. H.D. Doebner and J.D. Hennig, Lecture Notes in Physics, Vol. 370 (Springer-Verlag, Berlin, 1990) pp. 67–88.

[100] L.C. Biedenharn and M.A. Lohe, "An extension of the Borel-Weil construction to the quantum group $U_q(n)$," Commun. Math. Phys. **146** (1992) 483–504.

[101] L.C. Biedenharn and M.A. Lohe, *Quantum Group Symmetry and q-Tensor Algebras* (World Scientific, Singapore, 1995) 293 pages.

[102] L.C. Biedenharn and J.D. Louck, *Angular Momentum in Quantum Physics*, Vol. 8, *Encyclopedia of Mathematics and Its Applications* (Addison-Wesley, Reading, MA, 1981).

[103] L.C. Biedenharn and M. Tarlini, "On q-tensor operators for quantum groups," Lett. Math. Phys. **20** (1990) 271–278.

[104] C. Blohmann, "Free q-deformed relativistic wave equations by representation theory," Eur. Phys. J. **C30** (2003) 435–445.

[105] D. Bonatsos, C. Daskaloyannis and K. Kokkotas, "WKB equivalent potentials for q-deformed harmonic and anharmonic-oscillators," J. Math. Phys. **33** (1992) 2958–2965.

[106] A. Borowiec, J. Lukierski and A. Pachol, "Twisting and kappa-Poincare," J. Phys. **A47** (2014) 405203.

[107] A. Borowiec, J. Lukierski and V.N. Tolstoy, "Quantum deformations of D=4 Lorentz algebra revisited: Twistings of q-deformation," Eur. Phys. J. **C57** (2008) 601–611.

[108] G. Bosnjak and V.V. Mangazeev, "Construction of R-matrices for symmetric tensor representations related to $U_q(\widehat{sl_n})$," arXiv:1607.07968 [math-ph] (2016).

[109] N. Bourbaki, *Groupes at algèbres de Lie, Chapitres 4,5 et 6* (Hermann, Paris, 1968).

[110] P. Bouwknegt, J. McCarthy and K. Pilch, "Free field realizations of WZNW Models: BRST complex and its quantum group structure," Phys. Lett. **B234** (1990) 297–303; "Quantum group structure in the Fock space resolutions of sl(n) representations," Commun. Math. Phys. **131** (1990) 125–156; "Free field approach to two-dimensional conformal field theories," Progr. Theor. Phys. Suppl. **102** (1990) 67–135.

[111] M. Brannan, "Approximation properties for free orthogonal and free unitary quantum groups," J. Reine Ang. Math. **672** (2012) 223–251.

[112] T. Bridgeland, "Quantum groups via Hall algebras of complexes," Ann. Math. **177** (2013) 739–759.

[113] B. Bruegmann, R. Gambini and J. Pullin, "Jones polynomials for intersecting knots as physical states of quantum gravity," Nucl. Phys. **B385** (1992) 5877–603.

[114] T. Brzezinski and S. Majid, "Quantum group gauge theory on quantum spaces," Commun. Math. Phys. **157** (1993) 591–638; Erratum: [Commun. Math. Phys. **167** (1995) 235].

[115] F. Bruhat, "Sur les representations induites des groups de Lie," Bull. Soc. Math. France, **84** (1956) 97–205.

[116] J. Brundan and A. Kleshchev, "Parabolic presentations of the Yangian," Commun. Math. Phys. **254** (2005) 191–220.

[117] C. Burdik and P. Hellinger, "Universal R-matrix for a two-parametric quantization of gl(2)," J. Phys. **A25** (1992) L629–L631; "The universal R-matrix and the Yang-Baxter equation with parameters," *ibid.* L1023–L1027.

[118] C. Burdik and O. Navratil, "Extremal vectors of the Verma modules of the Lie algebra B_2 in Poincare-Birkhoff-Witt basis," Phys. Part. Nucl. Lett. **11** (2014) 938–940.

[119] C. Burdik and R. Tomasek, "The two parameter deformation of the supergroup GL(1|1), its differential calculus and its lie algebra," Lett. Math. Phys. **26** (1992) 97–103.

[120] N. Burroughs, "The universal R-matrix for $U_q(sl(3))$ and beyond!," Commun. Math. Phys. **127** (1990) 109–128; "Relating the approaches to quantised algebras and quantum groups," **133** (1990) 91–117.

[121] A.G. Bytsko and J. Teschner, "Quantization of models with non-compact quantum group symmetry: Modular XXZ magnet and lattice sinh-Gordon model," J. Phys. **A39** (2006) 12927–12981.

[122] G. Carnovale and I.I. Simion, "On small modules for quantum groups at roots of unity," Boll. d'Unione Mat. Ital. (2016) doi:10.1007/s40574-016-0094-9, arXiv:1512.04724 [math.RT].

[123] U. Carow-Watamura, M. Schlieker, M. Scholl and S. Watamura, "Quantum Lorentz group," Int. J. Mod. Phys. **A6** (1991) 3081–3108; "Tensor representation of the quantum group $SL_q(2, \mathbb{C})$ and quantum Minkowski space," Zeit. f. Physik **C48** (1990) 159–166.
[124] U. Carow-Watamura, M. Schlieker and S. Watamura, "$SO_q(N)$ covariant differential calculus on quantum space and quantum deformation of Schrodinger equation," Zeit. f. Physik **C49** (1991) 439–446.
[125] U. Carow-Watamura, M. Schlieker, S. Watamura and W. Weich, "Bicowariant differential calculus on quantum groups $SU_q(N)$ and $SO_q(N)$," Commun. Math. Phys. **142** (1991) 605–641.
[126] U. Carow-Watamura and S. Watamura, "The quantum group as a symmetry – the Schrödinger equation of the N-dimensional q-deformed harmonic oscillator," Prog. Theor. Phys. Sup. **118** (1995) 375–389; "Harmonic oscillators on quantum space," in: *Nonlinear, Deformed and Irreversible Quantum Systems*, Proceedings Clausthal Symposium 1994, eds. H.-D. Doebner, V.K. Dobrev and P. Nattermann (World Scientific, Singapore, 1995) pp. 416–427.
[127] R. Carroll, "Aspects of quantum groups and integrable systems," in: Proc. 5th Int. Conf. Symm. Nonl. Math. Phys., Kiev, Ukraine, 2003, Vol. 50 Proc. Inst. Math. NAS Ukraine, eds. A.G. Nikitin et al. (2004) pp. 356–367.
[128] L. Castellani, "Differential calculus on $ISO_q(N)$, quantum Poincare algebra and q gravity," Commun. Math. Phys. **171** (1995) 383–404.
[129] S. Celik, "Z_3-graded differential geometry of the quantum plane," J. Phys. **A35** (2002) 6307–6318; S. Celik, E.M. Özkan and E. Yaşar, "Cartan calculus on the quantum space R_q^3," Turk. J. Math. **33** (2009) 75–83.
[130] B.L. Cerchiai, G. Fiore and J. Madore, "Geometrical tools for quantum euclidean spaces," Commun. Math. Phys. **217** (2001) 521–554.
[131] G. Chaco'n-Acosta, E. Manrique, L. Dagdug and H.A. Morales-Te'cotl, "Statistical thermodynamics of polymer quantum systems," SIGMA, **7** (2011) Art. no. 110.
[132] M. Chaichian, J.A. de Azcarraga, P. Presnajder and F. Rodenas, "Oscillator realization of the q-deformed anti-de Sitter algebra," Phys. Lett. **B291** (1992) 411–417.
[133] M. Chaichian, J. A. de Azcarraga and F. Rodenas, "q-Fock representations of the q-Lorentz algebra and irreducible tensors," in: *Symmetries in Science VI*, eds. B. Gruber (Plenum Press, 1993) pp. 157–170.
[134] M. Chaichian and A. Demichev, *Introduction to Quantum Groups* (World Scientific, 1996).
[135] M. Chaichian and A. Demichev, "Quantum Poincare group without dilatation and twisted classical algebra," J. Math. Phys. **36** (1995) 398–413; "Inhomogeneous quantum groups without dilatations," Phys. Lett. **A188** (1994) 205–209.
[136] M. Chaichian, A. Demichev and P. Presnajder, "Quantum field theory on noncommutative space-times and the persistence of ultraviolet divergences," Nucl. Phys. **B567** (2000) 360–390. M. Chaichian, A. Demichev and P. Presnajder, "Quantum field theory on the noncommutative plane with $E_q(2)$ symmetry," J. Math. Phys. **41** (2000) 1647–1671.
[137] M. Chaichian, A. Demichev, P. Presnajder, M.M. Sheikh-Jabbari and A. Tureanu, "Aharonov-Bohm effect in noncommutative spaces," Phys. Lett. **B527** (2002) 149–154.
[138] M. Chaichian, A. Demichev, P. Presnajder and A. Tureanu, "Space-time noncommutativity, discreteness of time and unitarity," Eur. Phys. J. **C20** (2001) 767–772.
[139] M. Chaichian and D. Ellinas, "On the polar decomposition of the quantum SU(2) algebra," J. Phys. **A23** (1990) L291–L296.
[140] M. Chaichian, D. Ellinas and P. Kulish, "Quantum algebra as the dynamical symmetry of the deformed. Jaynes-Cumminis model," Phys. Rev. Lett. **65** (1990) 980–983.
[141] M. Chaichian, D. Ellinas and Z. Popowicz, "Quantum conformal algebra with central extension," Phys. Lett. **B248** (1990) 95–99.

[142] M. Chaichian, P.P. Kulish and J. Lukierski, “q-deformed Jacobi identity, q-oscillators and q-deformed infinite-dimensional algebras,” Phys. Lett. **B237** (1990) 401–406.

[143] A. Chakrabarti, “q Analogs of Iu(n) and U(n,1),” J. Math. Phys. **32** (1991) 1227–1234.

[144] A. Chakrabarti, V.K. Dobrev and S.G. Mihov, “On a ‘New’ deformation of GL(2),” Eur. Phys. J. **B58** (2007) 135–136.

[145] R. Chakrabarti and R. Jagannathan, “On a nonstandard two parametric quantum algebra and its connections with $U_{p,q}$(gl(2)) and $U_{p,q}$(gl(1/1)),” Z. Phys. **C66** (1995) 523–528; “On the Hopf structure of $U_{p,q}$(gl(1|1)) and the universal $\mathcal{T}$-matrix of $Fun_{p,q}$(GL(1|1)),” Lett. Math. Phys. **37** (1996) 191–199; “On $U_{p,q}$(gl(2)) and a (p,q)-Virasoro algebra,” J. Phys. **A27** (1994) 2023–2036; “Parabosons, Virasoro-type algebras and their deformations,” J. Phys. **A27** (1994) L277–L284; “A (p,q) oscillator realization of two parameter quantum algebras,” J. Phys. **A24** (1991) L711–L718.

[146] R. Chakrabarti and C. Quesne, “On Jordanian $\mathcal{U}_{q,h}(gl(2))$ algebra and its T matrices via a contraction method,” Int. J. Mod. Phys. **A14** (1999) 2511–2529.

[147] V. Chari and A. Pressley, *A Guide to Quantum Groups* (Cambridge University Press, 1994).

[148] I. Cherednik, “Double affine Hecke algebras and Macdonald’s conjectures,” Ann. Math. Second Series, **141** (1995) 191–216.

[149] D. Chicherin and R. Kirschner, “Yangian symmetric correlators,” Nucl. Phys. **B877** (2013) 484–505; “Monodromy operators and symmetric correlators,” J. Phys. Conf. Ser. **474** (2013) 012014.

[150] S. Cho, S.-J. Kang, C.-H. Kim, K.S. Park, “Two-parameter quantum groups and quantum planes,” J. Kor. Phys. Soc. **29** (1996) 279–283.

[151] M. Cohen and S. Westreich, “Characters and a Verlinde-type formula for symmetric Hopf algebras,” J. Algebra, **320** (2008) 4300–4316.

[152] A. Connes, *Noncommutative Geometry* (Academic Press, San Diego, CA, 1994, ISBN 0-12-185860-X) 661 pages.

[153] A. Connes and J. Lott, “Particle models and noncommutative geometry,” Nucl. Phys. (Proc. Suppl.) **B18** (1991) 29–47.

[154] J.F. Cornwell, “A 2-parameter deformation of the universal enveloping algebra of sl(3,C),” Z. f. Phys. **C56** (1992) 245–252; “Multiparameter deformations of the universal enveloping-algebras of the simple Lie-algebras A_l for all $l \geq 2$ and the Yang-Baxter equation,” J. Math. Phys. **33** (1992) 3963–3977.

[155] E. Corrigan, D.B. Fairlie, P. Fletcher and R. Sasaki, “Some aspects of quantum groups and supergroups,” J. Math. Phys. **31** (1990) 776–780.

[156] M. Couture, “Braided GL_q(2,C) and its dual-space,” Can. J. Phys. **72** (1994) 336–341.

[157] L. Crane and I. Frenkel, “Four-dimensional topological field theory, Hopf categories, and the canonical bases,” J. Math. Phys. **35** (1994) 5136–5154.

[158] E. Cremmer and G.-L. Gervais, “The quantum group structure associated with non-linearly extended Virasoro algebras,” Commun. Math. Phys. **134** (1990) 619–632.

[159] W. Cui, “On crystal bases of two-parameter (v, t)-quantum groups,” arXiv:1411.4727 [math.QA] (2014).

[160] E.M.F. Curado, Y. Hassouni, M.A. Rego-Monteiro and L.M.C.S. Rodrigues, “Generalized Heisenberg algebra and algebraic method: The example of an infinite square-well potential,” Phys. Lett. **A372** (2008) 3350–3355.

[161] T. Curtright, “Quantum Backlund transformations and conformal algebras,” in: *Physics and Geometry*, eds. L.L. Chau and W. Nahm (Plenum Press, 1990) pp. 279–289.

[162] T. Curtright, G. Ghandour and C. Zachos, “Quantum algebra deforming maps, Clebsch-Gordan coefficients, coproducts, U and R matrices,” J. Math. Phys. **32** (1991) 676–688.

[163] T. Curtright and C. Zachos, “Deforming maps for quantum algebras,” Phys. Lett. **B243** (1990) 237–244.

[164] L. Dabrowski, V.K. Dobrev and R. Floreanini, "q-difference intertwining operators for a Lorentz quantum algebra," J. Math. Phys. **35** (1994) 971–985.

[165] L. Dabrowski, V.K. Dobrev, R. Floreanini and V. Husain, "Positive energy representations of the conformal quantum algebra," Phys. Lett. **B302** (1993) 215–222.

[166] L. Dabrowski and L. Wang, "Two parameter quantum deformation of GL(1|1)," Phys. Lett. **B266** (1991) 51–54.

[167] I. Damiani, "La R-matrice pour les algèbres quantiques de type affine non tordu," Ann. Sci. Ecole Norm. Sup. **31** (1998) 493–523.

[168] R. Da Rocha, A.E. Bernardini and J. Vaz Jr., "kappa-deformed Poincaré algebras and quantum Clifford-Hopf algebras," Int. J. Geom. Meth. Mod. Phys. **7** (2010) 821–836.

[169] E. Date, M. Jimbo and T. Miwa, "Representations of U(q)(gl(n,C)) at q=0 and the Robinson-Schensted correspondence," in: *Vadim Knizhnik Memorial Volume*, eds. L. Brink, D. Friedan and A.M. Polyakov (World Scientific, 1990) pp. 185–211.

[170] E. Date, M. Jimbo, K. Miki and T. Miwa, R matrix for cyclic representations of $U_q(sl(3,\mathbb{C})^{(1)})$ at $q^3 = 1$, Phys. Lett. **A148**, 45–49 (1990).

[171] E. Date, M. Jimbo, K. Miki and T. Miwa, Cyclic representations of $U_q(sl(n+1,C))$ at $q^n = 1$. Publ. Res. I. Math. Sci. **27**, 347–366 (1991).

[172] E. Date, M. Jimbo, K. Miki and T. Miwa, New R matrices associated with cyclic representations of $U_q(a(2)^2)$, Kyoto preprint RIMS-706 (1990); "Generalized chiral Potts models and minimal cyclic representations of $U_q(gl(n,C))$," Commun. Math. Phys. **137** (1991) 133–147.

[173] J.A. de Azcarraga, P.P. Kulish and F. Rodenas, "Reflection equations and q-Minkowski space algebras," Lett. Math. Phys. **32** (1994) 173–182; "Quantum groups and deformed special relativity," Fortschr. Phys. **44** (1996) 1–40.

[174] J.A. de Azcarraga and F. Rodenas, "Deformed Minkowski spaces: Classification and properties," J. Phys. **A29** (1996) 1215–1226.

[175] C. De Concini and V.G. Kac, "Representations of quantum groups at roots of 1," in: *Colloque Dixmier*, Progress Math. **92** (Birkhäuser, 1990) pp. 471–506.

[176] C. De Concini and V.G. Kac, "Representations of quantum groups at roots of 1: reduction to the exceptional case," Proceedings of "Infinite analysis, Part A,B (Kyoto, 1991)," Adv. Ser. Math. Phys. **16A** (World Scientific, River Edge, NJ, 1992) pp. 141–149.

[177] C. De Concini, V.G. Kac and C. Procesi, "Quantum coadjoint action," J. AMS, **5** (1992) 151–189; "Some remarkable degenerations of quantum groups," Commun. Math. Phys. **157** (1993) 405–427.

[178] W.A. de Graaf, "Constructing homomorphisms between Verma modules," J. Lie Theory **15** (2005) 415–428.

[179] F. Delduc, M. Magro and B. Vicedo, "On classical q-deformations of integrable sigma-models," JHEP 1311 (2013) 192.

[180] G.W. Delius and A. Hüffmann, "On quantum Lie algebras and quantum root systems," J. Phys. **A29** (1996) 1703–1722.

[181] G.W. Delius and N.J. MacKay, "Quantum group symmetry in sine-Gordon and affine Toda field theories on the half line," Commun. Math. Phys. **233** (2003) 173–190.

[182] G.W. Delius, A. Hüffmann, M. D. Gould and Y.-Z. Zhang, "Quantum Lie algebras associated to $U_q(gl_n)$ and $U_q(sl_n)$," J. Phys. **A29** (1996) 5611–5617.

[183] E.E. Demidov, Yu.I. Manin, E.E. Mukhin and D.V. Zhdanovich, "Non-standard quantum deformations of GL(n) and constant solutions of the Yang-Baxtr equation," Progr. Theor. Phys. Suppl. **102** (1990) 203–218.

[184] S. Derkachov, D. Karakhanyan, R. Kirschner and P. Valinevich, "Iterative construction of U_q(sl(n+1)) representations and Lax matrix factorisation," Lett. Math. Phys. **85** (2008) 221–234.

[185] J. de Souza, E.M.F. Curado and M.A. Rego-Monteiro, "Generalized Heisenberg algebras and Fibonacci series," J. Phys. **A39** (2006) 10415–10425.
[186] J. de Souza, N.M. Oliveira-Neto and C.I. Ribeiro-Silva, "A method based on a nonlinear generalized Heisenberg algebra to study the molecular vibrational spectrum," Eur. Phys. J. **D40** (2006) 205–210.
[187] H.J. De Vega, "Yang-Baxter algebras, integrable theories and quantum groups," Int. J. Mod. Phys. **A4** (1989) 2371–2463 [Nucl. Phys. Proc. Suppl. **18A** (1990) 229–282].
[188] P. Di Francesco and R. Kedem, "Quantum Q systems: From cluster algebras to quantum current algebras," Lett. Math. Phys. (2016) (2016) doi:10.1007/s11005-016-0902-2.
[189] P. Di Francesco and P. Mathieu, "Singular vectors and conservation laws of quantum KdV type equations," Phys. Lett. **B278**(1992) 79–84.
[190] M. Dimitrijevic, L. Jonke, L. Moller, E. Tsouchnika, J. Wess and M. Wohlgenannt, "Deformed field theory on kappa space-time," Eur. Phys. J. **C31** (2003) 129–138.
[191] P.A.M. Dirac, "A remarkable representation of the 3+2 de Sitter group," J. Math. Phys. **4** (1963) 901–909.
[192] J. Dixmier, *Enveloping Algebras* (North Holland, New York, 1977).
[193] V.K. Dobrev, "Multiplet classification of the reducible elementary representations of real semisimple Lie groups: The $SO_e(p,q)$ example," Lett. Math. Phys. **9** (1985) 205–211.
[194] V.K. Dobrev, "Elementary representations and intertwining operators for SU(2,2): I," J. Math. Phys. **26** (1985) 235–251.
[195] V.K. Dobrev, "Multiplet classification of the indecomposable highest weight modules over affine Lie algebras and invariant differential operators: The $A_\ell^{(1)}$ example," Talk at the Conference on Algebraic Geometry and Integrable Systems, Oberwolfach, July 1984 and ICTP, Trieste, preprint, IC/85/9 (1985).
[196] V.K. Dobrev, "Multiplet classification of the indecomposable highest weight modules over the Neveu-Schwarz and Ramond superalgebras," Lett. Math. Phys. **11** (1986) 225–234.
[197] V.K. Dobrev, "Canonical construction of intertwining differential operators associated with representations of real semisimple Lie groups," Rep. Math. Phys. **25** (1988) 159–181; first as ICTP Trieste preprint IC/86/393 (1986).
[198] V.K. Dobrev, "Multiplet classification of highest weight modules over quantum universal enveloping algebras: The $U_q(sl(3,\mathbb{C}))$ example," in: Proceedings of the International Group Theory Conference (St. Andrews, 1989), eds. C.M. Campbell and E.F. Robertson, Vol. 1, London Math. Soc. Lecture Note Series 159 (Cambridge University Press, 1991) pp. 87–104; first as ICTP Trieste internal report IC/89/142 (June 1989).
[199] V.K. Dobrev, "Character formulae for $U_q(sl(3,\mathbb{C}))$ representations," Proceedings, XIII Johns–Hopkins Workshop "Knots, Topology and Field Theory" (Florence, June 1989), ed. L. Lusanna (World Scientific, Singapore, 1989) pp. 539–547.
[200] V.K. Dobrev, "Classification and characters of $U_q(sl(3,\mathbb{C}))$ representations," Proceedings of the Quantum Groups Workshop (Clausthal, 1989) eds. H.-D. Doebner and J.-D. Hennig, Lecture Notes in Physics, Vol. 370 (Springer-Verlag, Berlin, 1990) pp. 107–117.
[201] V.K. Dobrev, "Characters of the $U_q(sl(3,\mathbb{C}))$ highest weight modules," Progr. Theor. Phys. Suppl. **102** (1990) 137–158; Proceedings, "Symmetries and Algebraic Structures in Physics, Pt. 2 – Integral Systems, Solid State Physics and Theory of Phase Transitions," Proc. Lebedev Institute vol. 188, eds. V.V. Dodonov & V.I. Manko (Nova Science Publishers, New York, 1991) pp. 207–211.
[202] V.K. Dobrev, "Representations of quantum groups," Invited review talk, Proceedings of the International Symposium "Symmetries in Science V: Algebraic Structures, their Representations, Realizations and Physical Applications" (Schloss Hofen, Vorarlberg, Austria, 30.7.-3.8.1990), eds. B. Gruber et al (Plenum Press, NY, 1991) pp. 93–135.

[203] V.K. Dobrev, "Representations of quantum groups for roots of 1," Invited review talk, Proceedings of the 14th Johns Hopkins Workshop "Nonperturbative Methods in Low Dimensional Field Theories" (Debrecen, August 1990) eds. G. Domokos et al. (World Scientific, Singapore, 1991) pp. 69–106.

[204] V.K. Dobrev, "Canonical q – deformations of noncompact Lie (super-) algebras," J. Phys. **A26** (1993) 1317–1334; first as Göttingen University preprint (July 1991).

[205] V.K. Dobrev, "Singular vectors of representations of quantum groups," J. Phys. **A25** (1992) 149–160.

[206] V.K. Dobrev, "Singular vectors of quantum groups representations for straight Lie algebra roots," Lett. Math. Phys. **22** (1991) 251–266.

[207] V.K. Dobrev, "Introduction to quantum groups," Göttingen University preprint (April 1991), Invited plenary lecture at the 22nd Annual Iranian Mathematics Conference (March 13–16, 1991, Mashhad, Iran), Proceedings, eds. M.R.R. Moghaddam and M.A. Pourabdollah (Ferdowsi University of Mashhad, 1995) pp. 91–121.

[208] V.K. Dobrev, "Singular vectors of representations of quantum groups and B-G-G resolution," in: Proceedings II Symposium on Topological and Geometrical Methods in Field Theory (May 26 – June 1, 1991, Turku, Finland), eds. J. Mickelsson and O. Pekonen (World Scientific, Singapore, 1992) pp. 104–120.

[209] V.K. Dobrev, "Duality for the matrix quantum group $GL_{p,q}(2, \mathbb{C})$," J. Math. Phys. **33** (1992) 3419–3430.

[210] V.K. Dobrev, unpublished (May 1993).

[211] V.K. Dobrev, "q – difference intertwining operators for U_q(sl(n)): General setting and the case n=3," J. Phys. **A27** (1994) 4841–4857 & 6633–6634.

[212] V.K. Dobrev, "Positive energy representations of noncompact quantum algebras," Invited talk at the Workshop on Generalized Symmetries in Physics (Clausthal, July 1993), Proceedings, eds. H.-D. Doebner et al. (World Scientific, Singapore, 1994) pp. 90–110.

[213] V.K. Dobrev, *Representations of Quantum Groups and q – Deformed Invariant Wave Equations*, Dr. Habil. Thesis, Tech. Univ. Clausthal 1994 (Papierflieger Verlag, Clausthal-Zellerfeld, 1995) ISBN 3-930697-59-9.

[214] V.K. Dobrev, "New q – Minkowski space-time and q – Maxwell equations hierarchy from q - conformal invariance," Phys. Lett. **341B** (1994) 133–138; **346B** (1995) 427.

[215] V.K. Dobrev, "Subsingular vectors and conditionally invariant (q-deformed) equations," J. Phys. **A28** (1995) 7135–7155.

[216] V.K. Dobrev, "q-difference intertwining operators for U_q(sl(4)) and q-conformal invariant equations," in: "Symmetries in Science VIII," Proceedings of the International Symposium (August 1994, Bregenz, Austria), ed. B. Gruber (Plenum Press, NY, 1995) p. 55–84.

[217] V.K. Dobrev, "q – difference intertwining operators and q – conformal invariant equations," Acta Appl. Math. **44** (1996) 81–116.

[218] V.K. Dobrev, "Representations of the Jordanian quantum algebra U_h(sl(2))," Proceedings of the 10th International Conference 'Problems of Quantum Field Theory' (Alushta, Crimea, 13–18.5.1996), eds. D. Shirkov et al. (Publishing Department of JINR, Dubna, 1996) pp. 104–110.
[218]′ Preprint version with more details: ICTP preprint IC/97/163 (1997).

[219] V.K. Dobrev, "Invariant differential operators and characters of the AdS_4 algebra," J. Phys. **A39** (2006) 5995–6020; hep-th/0512354.

[220] V.K. Dobrev, "Invariant differential operators in noncommutative quantum group setting," Talk at Workshop on Noncommutative Field Theory and Gravity 21–27.9.2015, Corfu; Proceedings, PoS CORFU **2015** (2016) 120.

[221] V.K. Dobrev, "Multiparameter quantum Minkowski space-time and quantum Maxwell equations hierarchy," Invited Talk at the III National Congress of Bulgarian Physicists (Sofia, 2016), Bulg. J. Phys. **44** no.1 (2017); arXiv:1612.05034.

[222] V.K. Dobrev, H.-D. Doebner, U. Franz and R. Schott, "Lévy Processes on $U_q(g)$ as Infinitely Divisible Representations," in: *Probability on Algebraic Structures*, eds. G. Budzban, Ph. Feinsilver and A. Mukherjea, Contemp. Math. vol. 261 (American Math. Society, 2000) pp. 181–192.

[223] V.K. Dobrev and M. El Falaki, "Quantum group $U_q(A_\ell)$ singular vectors in Poincare-Birkhoff-Witt basis," Lett. Math. Phys. **49** (1999) 47–57.

[224] V.K. Dobrev and M. El Falaki, "Quantum group $U_q(D_\ell)$ singular vectors in Poincare-Birkhoff-Witt basis," J. Phys. **A33** (2000) 6321–6332.

[225] V.K. Dobrev and R. Floreanini, "The massless representations of the conformal quantum algebra," J. Phys. **A27** (1994) 4831–4840.

[226] V.K. Dobrev and B.S. Kostadinov, "Noncommutative quantum conformal plane wave deformation," Phys. Lett. **B439** (1998) 337–344.

[227] V.K. Dobrev, G. Mack, V.B. Petkova, S.G. Petrova and I.T. Todorov, *Harmonic Analysis on the n-Dimensional Lorentz Group and Its Applications to Conformal Quantum Field Theory*, Lecture Notes in Physics, Vol. 63 (Springer-Verlag, Berlin-Heidelberg-New York, 1977).

[228] V.K. Dobrev and S.G. Mihov, "Induced representations for duals of two-parameter GL(2) deformations," Invited talk (given by V.K.D.) at the XIV Max Born Symposium, Karpacz (1999), Proceedings, "New Symmetries and Integrable Models," eds. A. Frydryszak, J. Lukierski and Z.Popowicz (World Scientific, 2000) pp. 39–61.

[229] V.K. Dobrev and S.G. Mihov, "q-plane wave solutions of q-Weyl gravity," Phys. At. Nucl. **71** (5) (2008) 862–867.

[230] V.K. Dobrev, A.D. Mitov and P. Truini, "Normalized $U_q(sl(3))$ Gelfand-(Weyl)-Zetlin basis and new summation formulae for q-hypergeometric functions," J. Math. Phys. **41** (2000) 7752–7768.

[231] V.K. Dobrev and P. Moylan, "Finite-dimensional singletons of the quantum anti de Sitter Phys. Lett. **B315** (1993) 292–298.

[232] V.K. Dobrev and P. Moylan, "Induced representations and invariant integral operators for SU(2,2)," Fortschr. Phys. **42** (1994) 339–392.

[233] V.K. Dobrev and P. Parashar, "Duality for multiparametric quantum GL(n), J. Phys. **A26** (1993) 6991–7002.

[234] V.K. Dobrev and P. Parashar, "Duality for a Lorentz quantum group," Lett. Math. Phys. **29** (1993) 259–269.

[235] V.K. Dobrev and V.B. Petkova, "Elementary representations and intertwining operators for the group $SU^*(4)$," Rep. Math. Phys. **13** (1978) 233–277.

[236] V.K. Dobrev and V.B. Petkova, "All positive energy unitary irreducible representations of extended conformal supersymmetry," Phys. Lett. **162B** (1985) 127–132;

[237] V.K. Dobrev and S.T. Petrov, "q-plane wave solutions of the full q-Maxwell equations," Phys. Part. Nucl. **33** (7) (2002) 43–52.

[238] V.K. Dobrev and S.T. Petrov, q-Conformal Invariant Equations and q-Plane Wave Solutions, J. Math. Phys. **45** (2004) 3788–3799.

[239] V.K. Dobrev, S.T. Petrov and B.S. Zlatev, "Solutions of q-deformed equations with quantum conformal symmetry," Talk by V.K.D. at PDE2000 Conference (Clausthal, Germany, 2000), Proceedings "Partial Differential Equations and Spectral Theory," ed. M. Demuth et al, *Operator Theory: Advances and Applications, Vol. 126* (Birkhäuser, Berlin, 2001) pp. 113–118.

[240] V.K. Dobrev, S.T. Petrov and B.S. Zlatev, "Helicity Asymmetry of q-Plane Wave Solutions of q-Maxwell Equations," Phys. At. Nucl., **64** (2001) 2110–2115.

[241] V.K. Dobrev, R.P. Roussev and P.A. Terziev, “Induced representations of $U_q(so(3))$ with subrepresentations of integer spin only,” J. Phys. **A31** (1998) 9269–9277.
[242] V.K. Dobrev and E. Sezgin, “Spectrum and character formulae of $so(3, 2)$ unitary representations,” Lect. Notes Phys. **379** (Springer-Verlag, Berlin, 1990) 227–238.
[243] V.K. Dobrev and A. Sudbery, “Representations of the generalized Lie algebra $sl(2)_q$,” J. Phys. **A31** (1998) 6635–6645.
[244] V.K. Dobrev and P. Truini, “Polynomial realization of the $U_q(sl(3))$ Gelfand-(Weyl)-Zetlin basis,” J. Math. Phys. **38** (1997) 3750–3767.
[245] V.K. Dobrev and P. Truini, “Irregular $U_q(sl(3))$ representations at roots of unity via Gelfand-(Weyl)-Zetlin basis,” J. Math. Phys. **38** (1997) 2631–2651.
[246] V.K. Dobrev, P. Truini and L.C. Biedenharn, “Representation theory approach to the polynomial solutions of q-difference equations: $U_q(sl(3))$ and beyond,” J. Math. Phys. **35** (1994) 6058–6075.
[247] V.K. Dobrev and B.S. Zlatev, “q-plane wave solutions of q-deformed equations with quantum conformal symmetry,” Czech. J. Phys. **50** (2000) 53–58.
[248] H.D. Doebner, J.D. Hennig and W. Lücke, “Mathematical guide to quantum groups,” in: *Quantum Groups*, Proceedings 8th International Workshop on Mathematical Physics (Clausthal, 1989), eds. H.D. Doebner and J.D. Hennig, Lecture Notes in Physics, Vol. 370 (Springer-Verlag, Berlin, 1990) pp. 29–63.
[249] S. Doplicher, K. Fredenhagen and J.E. Roberts, “The quantum structure of spacetime at the Planck scale and quantum fields,” Commun. Math. Phys. **172** (1995) 187–220.
[250] B. Drabant, M. Schlieker, W. Weich and B. Zumino, “Complex quantum groups and their quantum enveloping algebras,” Commun. Math. Phys. **147** (1992) 625–634.
[251] V.G. Drinfeld, “Hopf algebras and the quantum ang-Baxter equation,” Dokl. Akad. Nauk SSSR **283** (1985) 1060–1064 (in Russian); English translation: Soviet. Math. Dokl. **32** (1985) 254–258.
[252] V.G. Drinfeld, “Degenerate affine Hecke algebras and Yangians,” Funct. Anal. Appl. **20** (1986) 58–60. [Funkts. Anal. Ego Pril. (in Russian) **20** (1986) 69–70.]
[253] V.G. Drinfeld, “Quantum groups,” in: *Proceedings of the International Congress of Mathematicians (MSRI, Berkeley, 1986)*, ed. A.M. Gleason (American Mathematical Society, Providence, RI, 1987) pp. 798–820.
[254] V.G. Drinfeld, “A New realization of Yangians and quantized affine algebras,” Sov. Math. Dokl. **36** (1988) 212–216.
[255] V.G. Drinfeld, “Quasi-Hopf algebras,” Leningrad Math J. (Alg. Anal.) **1** (1989) 1419–1457.
[256] V.G. Drinfeld, “Quasi-Hopf algebras and Knizhnik-Zamolodchikov equations,” in: *Problems of Modern Quantum Field Theory*, Invited Lectures of the Spring School held in Alushta, USSR, 24.4–5.5.1989, eds. A.A. Belavin, A.U. Klimyk and A.B. Zamolodchikov (Springer-Verlag, Berlin, 1989, ISBN 3-540-51833-9; 0-387-51833-9) pp. 1–13.
[257] V.G. Drinfeld, “On quasi-triangular quasi-Hopf algebras and a group closely connected to $Gal(\bar{Q}/Q)$,” Leningrad Math J. (Alg. Anal.) **2** (1991) 829–860.
[258] M.S. Dijkhuizen, “The double covering of the quantum group $SO_q(3)$,” Rend. Circ. Math. Palermo Serie II **37** (1994) 47–57.
[259] F. Dumas and L. Rigal, “Prime spectrum and automorphisms for 2x2 Jordanian matrices,” Commun. Algebra, **30** (2002) 2805–2828.
[260] A. Dutriaux, “Braided vector fields on a quantum hyperboloid via the quantum group $U_q(sl(2))$, J. Phys. **A42** (2009) art. no. 445203.
[261] A. Dutriaux and D. Gurevich, “Maxwell operator on q-Minkowski space and q-hyperboloid algebras,” J. Phys. **A41** (2008) art. no. 315203.

[262] M. Eastwood, Suppl. Rendiconti Circolo Matematici di Palermo, Serie II, Numero 43 (1996) 57–76.

[263] D. Ellinas and J. Sobczyk, “Quantum Heisenberg group and algebra: Contraction, left and right regular representations,” J. Math. Phys. **36** (1995) 1404–1412; “On the representation theory of quantum Heisenberg group and algebra,” Czech. J. Phys. 44 (1994) 1019–1027.

[264] T. Enright, R. Howe and N. Wallach, “A classification of unitary highest weight modules,” in: *Representations of Reductive Groups*, ed. P. Trombi (Birkhäuser, Boston, 1983) pp. 97–143.

[265] C.J. Efthimiou and D.A. Spector, “A Collection of exercises in two-dimensional physics. Part 1,” hep-th/0003190, 232 pages.

[266] K. Erdmann, “Type A Hecke algebras and related algebras,” Archiv d. Math. **82** (2004) 385–390.

[267] N.T. Evans, “Discrete series for the universal covering group of the $3+2$ de Sitter group,” J. Math. Phys. **8** (1967) 170–184.

[268] H. Ewen, O. Ogievetsky and J. Wess, “Quantum matrices in two-dimensions,” Lett. Math. Phys. **22** (1991) 297.

[269] L.D. Faddeev, “Integrable models in 1+1 dimensional quantum field theory,” in: *Les Houches Lectures 1982* (Elsevier, Amsterdam, 1984) pp. 561–607.

[270] L.D. Faddeev, “On the exchange matrix for WZNW model,” Commun. Math. Phys. **132** (1990) 131–138.

[271] L.D. Faddeev, “Modular double of the quantum group $SL_q(2,R)$,” in: *Lie Theory and Its Applications in Physics*, “Springer Proceedings in Mathematics and Statistics,” Vol. 111, ed. V. Dobrev (Springer, Tokyo-Heidelberg, 2014) pp. 21–31.

[272] L.D. Faddeev, N.Yu. Reshetikhin and L.A. Takhtajan, “Quantization of Lie groups and Lie algebras,” Leningrad Math. J. **1** (1990) 193–225, Alg. Anal. **1** (1989) no.1, 178–206,

[273] L.D. Faddeev, E.K. Sklyanin and L.A. Takhtajan, “The quantum inverse problem method. 1,” Theor. Math. Phys. **40** (1979) 688–706.

[274] L.D. Faddeev and L.A. Takhtajan, “The quantum method of the inverse problem and the Heisenberg XYZ model,” Uspekhi Mat. Nauk **34** (1979) 13–63 (in Russian).

[275] L.D. Faddeev and L.A. Takhtajan, “Liouville model on the lattice,” in: Proceedings of a seminar series held at DAPHE & LPTHE, Univ. Pierre et Marie Curie (Paris, 1984–1985), Lect. Notes Phys., **246** (1986) pp. 166–179.

[276] D. Fairlie, “Quantum deformations of SU(2),” J. Phys. A: Math. Gen. **23** (1990) L183–L187.

[277] D.B. Fairlie and C.K. Zachos, “Multiparameter associative generalizations of canonical commutation relations and quantized planes,” Phys. Lett. **256B** (1991) 43–49.

[278] F. Falceto and K. Gawedzki, “Lattice Wess-Zumino-Witten model and quantum groups,” J. Geom. Phys. **11** (1993) 251–279.

[279] Z. Fan and Y. Li, “Two-parameter quantum algebras, canonical bases and categorifications,” Int. Math. Res. Notices, **16** (2015) 7016–7062.

[280] V.A. Fateev and S.L. Lukyanov, “Poisson Lie groups and classical W algebras,” Int. J. Mod. Phys. **A7** (1992) 853–876; “Vertex operators and representations of quantum universal enveloping algebras,” Int. J. Mod. Phys. **A7** (1992) 1325–1360.

[281] B.L. Feigin, A.M. Gainutdinov, A.M. Semikhatov and I.Y. Tipunin, “Kazhdan-Lusztig-dual quantum group for logarithmic extensions of Virasoro minimal models,” J. Math. Phys. **48** (2007) 032303; “Modular group representations and fusion in logarithmic conformal field theories and in the quantum group center,” Commun. Math. Phys. **265** (2006) 47–93.

[282] B. Feigin and E. Frenkel, “Quantum W algebras and elliptic algebras,” Commun. Math. Phys. **178** (1996) 653–678.

[283] P. Feinsilver, U. Franz and R. Schott, “Feynman-Kac formula and Appell systems on quantum groups,” Compt. Rend. Acad. Sci. Ser. I - Math. **321** (1995) 1615–1619; “Duality and multiplicative stochastic processes on quantum groups,” J. Theor. Probab. **10** (1997) 795–818.

[284] G. Felder, J. Frohlich and G. Keller, "Braid matrices and structure constants for minimal conformal models," Commun. Math. Phys. **124** (1989) 647–664.

[285] A.T. Filipov, A.P. Isaev and A.B. Kurdikov, "Paragrassmann differential-calculus," Theor. Math. Phys. **94** (1993) 150–165; "ParaGrassmann analysis and quantum groups," Mod. Phys. Lett. A **A7** (1992) 2129–2142.

[286] J.R. Finkelstein, "q-Gravity," Lett. Math. Phys. **38** (1996) 53–62; q-deformation of the Lorentz group, J. Math. Phys. **37** (1996) 953–964.

[287] G. Fiore, M. Maceda and J. Madore, "Metrics on the real quantum plane," J. Math. Phys. **43** (2002) 6307–6324.

[288] R. Fioresi, "Commutation relations among generic quantum minors in $\mathcal{O}_q(M_n(k))$," J. Algebra, **280** (2004) 655–682.

[289] M. Flato and C. Fronsdal, "One massless particle equals two Dirac singletons," Lett. Math. Phys. **2** (1978) 421–426.

[290] M. Flato, L.K. Hadjiivanov and I.T. Todorov, "Quantum deformations of singletons and of free zero mass fields," Found. Phys. **23** (1993) 571–586.

[291] E.G. Floratos, "Representations of the quantum group $GL_q(2)$ for values of q on the unit circle," Phys. Lett. **B233** (1989) 395–399.

[292] A. Francis, "Centralizers of Iwahori-Hecke algebras II: The general case," Alg. Colloq., **10** (2003) 95–100.

[293] J.M. Franko, "Braid group representations arising from the Yang-Baxter equation," J. Knot Theory Ramif. **19** (2010) 525–538.

[294] U. Franz, A. Kula and A. Skalski, "Levy processes on quantum permutation groups," arXiv:1510.08321 [math.QA] (2015).

[295] L. Frappat, V. Hussin and G. Rideau, "Classification of the quantum deformations of the superalgebra GL(1/1)," J. Phys. **A31** (1998) 4049–4072.

[296] L. Frappat and A. Sciarrino, "Lattice space-time from Poincare and kappa Poincare algebras," Phys. Lett. **B347** (1995) 28–32.

[297] D.S. Freed, "Higher algebraic structures and quantization," Commun. Math. Phys. **159** (1994) 343–398.

[298] L. Freidel and J.M. Maillet, "Quadratic algebras and integrable systems," Phys. Lett. **B262** (1991) 278–284.

[299] I.B. Frenkel and N. Jing, "Vertex representations of quantum affine algebras," Proc. NAS USA **85** (1988) 9373–9377.

[300] A. Freslon, "Examples of weakly amenable discrete quantum groups," J. Funct. Anal. **265** (2013) 2164–2187.

[301] J. Fröhlich and T. Kerler, *Quantum Groups, Quantum Categories, and Quantum Field Theory*, Lecture Notes in Math. No 1542 (Springer-Verlag, Berlin, 1993).

[302] C. Fronsdal, "Elementary particles in a curved space," Rev. Mod. Phys. **37**, 221–224 (1965).

[303] J. Fuchs, *Affine Lie Algebras and Quantum Groups. An Introduction with Applications in Conformal Field Theory* (Cambridge University Press, 1992).

[304] P. Furlan, A.Ch. Ganchev and V.B. Petkova, "Quantum groups and fusion rules multiplicities," Nucl. Phys. **B343** (1990) 205–227.

[305] P. Furlan, A.Ch. Ganchev and V.B. Petkova, "Remarks on the quantum group structure of the rational $c < 1$ conformal theories," Int. J. Mod. Phys. **A6** (1991) 4859–4884.

[306] P. Furlan, L.K. Hadjiivanov and I.T. Todorov, "Indecomposable $U_q(sl(n))$ modules for q(h) =-1 and BRS intertwiners," J. Phys. **A34** (2001) 4857–4880.

[307] P. Furlan, Y.S. Stanev and I.T. Todorov, "Coherent state operators and n point invariants for $U_q(sl(2))$," Lett. Math. Phys. **22** (1991) 307–320.

[308] V.M. Futorny and D.J. Melville, “Quantum deformations of α-stratified modules,” Alg. Repres. Theory, **1** (1998) 135–153.

[309] A.Ch. Ganchev and V.B. Petkova, “U_q(s1(2)) invariant operators and minimal theories fusion matrices,” Phys. Lett **B233** (1989) 374–382.

[310] S. Garibaldi, R.M. Guralnick and D.K. Nakano, “Globally irreducible weyl modules for quantum groups,” arXiv:1612.03118 [math.RT] (2016).

[311] G. Gasper and M. Rahman, *Basic Hypergeometric Series* (Cambridge University Press, 1990).

[312] K. Gawedzki, “Classical origin of quantum group symmetries in Wess-Zumino-Witten conformal field theory,” Commun. Math. Phys. **139** (1991) 201–214.

[313] N. Geer and B. Patureau-Mirand, “The trace on projective representations of quantum groups,” arXiv:1610.09129 [math.QA] (2016).

[314] I.M. Gelfand, M.I. Graev and N.Y. Vilenkin, *Generalised Functions*, vol. 5 (Academic Press, New York, 1966).

[315] I.M. Gel’fand and M.L. Zetlin, “Finite-dimensional representations of the group of unimodular matrices,” Dokl. Akad. Nauk. SSSR **71** 5 (1950) 825–828.

[316] M. Gerstenhaber and A. Giaquinto, “Boundary solutions of the classical Yang–Baxter equation,” Lett. Math. Phys. **40** (1997) 337–353.

[317] J.-L. Gervais, “The quantum group structure of 2D gravity and minimal models. I” Commun. Math. Phys. **130** (1990) 257–283; “Solving the strongly coupled 2-D gravity: 1. Unitary truncation and quantum group structure,” Commun. Math. Phys. **138** (1991) 301–338.

[318] P.I. Golod and A.U. Klimyk, *Matematicheskie Osnovy Teorii Symetrii* (Naukova dumka, Kiev, 1992) (in Russian).

[319] C. Gomez and G. Sierra, “Quantum group meaning of the Coulomb gas,” Phys. Lett. **B240** (1990) 149–157; “The quantum symmetry of rational conformal field theories,” Nucl. Phys. **B352** (1991) 791–828.

[320] C. Gomez and G. Sierra, “Integrability and uniformization in Liouville theory: The Geometrical origin of quantized symmetries,” Phys. Lett. **B255** (1991) 51–60.

[321] K.R. Goodearl and M.T. Yakimov, “Quantum cluster algebras and quantum nilpotent algebras,” Proc. Nat. Acad. Sci. **111** (2014) 9696–9703.

[322] F.M. Goodman and H. Wenzl, “Littlewood-Richardson coefficients for Hecke algebras at roots of unity,” Adv. Math. 82 (1990) 244–265.

[323] M.D. Gould, R.B. Zhang and A.J. Bracken, “Generalized Gelfand invariants and characteristic identities for quantum groups,” J. Math. Phys. **32** (1991) 2298–2303r.

[324] M.D. Gould and L.C. Biedenharn, “The pattern calculus for tensor operators in quantum groups,” J. Math. Phys. **33** (1992) 3613–3635.

[325] A.R. Gover and R.B. Zhang, “Geometry of quantum homogeneous vector bundles and representation theory of quantum groups I,” Rev. Math. Phys. **11** (1999) 533–552.

[326] N.A. Gromov, D.B. Efimov, I.V. Kostyakov and V.V. Kuratov, “Non-commutative low dimension spaces and superspaces associated with contracted quantum groups and supergroups,” Czech. J. Phys. **54** (2004) 1297–1303.

[327] V.A. Groza, “Degenerate series representations of the q-deformed algebra so'_q(r,s), SIGMA **3** (2007) 064.

[328] V.A. Groza, I.I. Kachurik and A.U. Klimyk, “On Clebsch-Gordan coefficients and matrix-elements of representations of the quantum algebra U_q(su2),” J. Math. Phys **31** (1990) 2769–2780.

[329] B. Gruber, Y.F. Smirnov and Y.I. Kharitonov, “Quantum algebra U_q(gl(3)) and nonlinear optics,” J. Russ. Laser Res. **24** (2003) 56–68.

[330] E. Guadagnini, M. Martellini and M. Mintchev, “Braids and quantum group symmetry in Chern-Simons theory,” Nucl. Phys. **B336** (1990) 581–609; “Chern-Simons holonomies and the appearance of quantum groups,” Phys. Lett. **B235** (1990) 275–281.

[331] D. Gurevich and P. Saponov, "Braided affine geometry and q-analogs of wave operators," J. Phys. **A42** (2009) art. no. 313001.

[332] T. Hakioglu, "Admissible cyclic representations and an algebraic approach to quantum phase," J. Phys. **A31** 707–721 (1998).

[333] Harish-Chandra, "Representations of semisimple Lie groups: IV,V," Am. J. Math. **77** (1955) 743–777; **78** (1956) 1–41.

[334] Harish-Chandra, "Discrete series for semisimple Lie groups: II," Ann. Math. **116** (1966) 1–111.

[335] T. Hayashi, "Q-analogues of Clifford and Weyl algebras-spinor and oscillator representations of quantum enveloping algebras," Commun. Math. Phys. **127** (1990) 129–144; "Quantum groups and quantum determinants," J. Algebra **152** (1992) 146–165.

[336] M. Hazewinkel, "Witt vectors. Part 1," *Handbook of Algebra, Vol. 6*, ed. M. Hazewinkel (Elsevier B.V., Amsterdam, 2009, ISBN: 978-044453257-2) pp. 319–472.

[337] L. Hellström and S.D. Silvestrov, *Commuting Elements in q-Deformed Heisenberg Algebras* (World Scientific, Singapore, 2000) 257 pages.

[338] F.J. Herranz, "New quantum conformal algebras and discrete symmetries," Phys. Lett. **B543** (2002) 89–97; "Non-standard quantum $so(3,2)$ and its contractions," J. Phys. **A30** (1997) 6123–6129; "Non-standard 2+1 conformal bialgebras and their quantum deformations," Czech. J. Phys. **47** (1997) 1171–1178.

[339] J. Hietarinta, "Solving the two-dimensional constant quantum Yang–Baxter equation," J. Math. Phys. **34** (1993) 1725–1756.

[340] H. Hinrichsen, P.P. Martin, V. Rittenberg and M. Scheunert, "On the two-point correlation functions for the U_q(su(2)) invariant spin one-half Heisenberg chain at roots of unity," Nucl. Phys. **B415** (1994) 533–556.

[341] H. Hinrichsen and V. Rittenberg, "A two parameter deformation of the SU(1/1) superalgebra and the XY quantum chain in a magnetic field," Phys. Lett. **B275** (1992) 350–354.

[342] C.-L. Ho, A.I. Solomon and C.-H. Oh, "Quantum entanglement, unitary braid representation and Temperley-Lieb algebra," Europhys. Lett. (EPL) **92** (2010) art. no. 30002.

[343] T.J. Hollowood, "Quantizing SL(N) solitons and the Hecke algebra," Int. J. Mod. Phys. **A8** (1993) 947–982.

[344] M.N. Hounkonnou, J. Desire and B. Kyemba, "R(p,q)-calculus: Differentiation and integration," SUT J. Math. **49** (2013) 145–167.

[345] J.E. Humphreys, *Reflection Groups and Coxeter Groups*, Cambridge Studies on Advanced Math. Vol. 29 (Cambridge University Press, 1990).

[346] E. Ilievski and B. Zunkovic, "Quantum group approach to steady states of boundary-driven open quantum systems," J. Stat. Mech.: Theory Exp. **2014**. 1, Art. no P01001.

[347] R. Iordanescu, F.F. Nichita and I.M. Nichita, "Non-associative algebras, Yang-Baxter equations, and quantum computers," Bulg. J. Phys. **41** (2014) 71–76.

[348] R. Iordanescu and P. Truini, "Quantum groups and Jordan structures," Bull. Univ. Politechnica Bucharest, Ser. A: App. Math. Phys. **57–58** (1996) 43–60.

[349] I.C.H. Ip, "On tensor product decomposition of positive representations of $U_{q\tilde{q}}$(sl(2,R))," arXiv:1511.07970 [math.RT] (2015).

[350] A.P. Isaev, "Paragrassmann integral, discrete systems and quantum groups," Int. J. Mod. Phys. **A12** (1997) 201–206.

[351] A.P. Isaev and R.P. Malik, "Deformed traces and covariant quantum algebras for quantum groups $GL_{qp}(2)$ and $GL_{qp}(1|1)$," Phys. Lett. **B280** (1992) 219–226.

[352] A.P. Isaev and Z. Popowicz, "q trace for the quantum groups and q deformed Yang-Mills theory," Phys. Lett. **B281** (1992) 271–278; "Quantum group gauge theories and covariant quantum algebras," Phys. Lett. **B307** (1993) 353–361.

[353] U.N. Iyer and D.A. Jordan, "Noetherian algebras of quantum differential operators," Alg. Repr. Theory, **18** (2015) 1593–1622.

[354] A.D. Jacobs and J.F. Cornwell, "Classification of bicovariant differential calculi on the Jordanian quantum groups GL(h,g)(2) and $SL_h(2)$ and quantum Lie algebras," J. Phys. **A31 44** (1998) 8869–8904.

[355] R. Jagannathan and J. Van der Jeugt, "Finite dimensional representations of the quantum group $GL_{p,q}(2)$ using the exponential map from $U_{p,q}(gl(2))$," J. Phys. **A28** (1995) 2819–2832; "The exponential map for representations of $U_{p,q}(gl(2))$," Czech. J. Phys. **46** (1996) 269–275.

[356] H.P. Jakobsen, "Quantized Dirac operators," Czech. J. Phys. **50** (2000) 1265–1270.

[357] P. Jara and J. Jodar, "Finite-dimensional duality on the generalized Lie algebra $s1(2)_q$," J. Phys. **A35** (2002) 3683–3696.

[358] A. Jevicki, M. Mihailescu and S. Ramgoolam, "Noncommutative spheres and the AdS / CFT correspondence," JHEP **0010** (2000) 008.

[359] A. Jevicki and S. Ramgoolam, "Noncommutative gravity from the AdS/CFT correspondence," JHEP **9904** (1999) 032.

[360] M. Jimbo, "A q-difference analog of U(g) and the Yang-Baxter equation," Lett. Math. Phys. **10** (1985) 63–69.

[361] M. Jimbo, "A q-analog of U(gl(n+1)), Hecke algebra and the Yang-Baxter equation," Lett. Math. Phys. **11** (1986) 247–252.

[362] M. Jimbo, "Quantum R matrix for the generalized Toda system," Commun. Math. Phys. **102** (1986) 537–547.

[363] M. Jimbo, "Introduction to the Yang-Baxter equation," Int. J. Mod. Phys. **A4** (1989) 3759–3777.

[364] M. Jimbo, K. Misra, T. Miwa and M. Okado, "Combinatorics of representations of $U_q(sl(n)^{(1)})$ at q=0," Commun. Math. Phys. **136** (1991) 543–566.

[365] N. Jing, L. Zhang and M. Liu, "Wedge modules for two-parameter quantum groups," Contemp. Math. **602** (2013) 115–121; N. Jing, H. Zhang, "On finite-dimensional representations of two-parameter quantum affine algebras," J. Algebra Appl. **15** (2016) 1650054.

[366] R.S. Johal and R.K. Gupta, "Two parameter quantum deformation of $U(2) \supset U(1)$ dynamical symmetry and the vibrational spectra of diatomic molecules," Int. J. Mod. Phys. **E7** (1998) 553–557.

[367] A. Joseph, *Quantum Groups and Their Primitive Ideals, Vol. 29* (Springer Science & Business Media, 2012).

[368] A. Joseph and D. Todoric, "On the quantum KPRV determinants for semisimple and affine Lie algebras," Algebra. Rep. Theory **5** (2002) 57–99.

[369] B. Jurco, "Differential calculus on quantized simple Lie groups," Lett. Math. Phys. **22** (1991) 177–186.

[370] B. Jurco and M. Schlieker, "On Fock space representations of quantized enveloping algebras related to non-commutative differential geometry," J. Math. Phys. **36** (1995) 3814–3821.

[371] C. Juszczak, "D=3 real quantum conformal algebra," J. Phys. **A27** (1994) 385–392.

[372] V.G. Kac, *Infinite–Dimensional Lie Algebras. An Introduction*, Progr. Math. Vol. **44** (Birkhäuser, Boston, 1983).

[373] V.G. Kac and D. Kazhdan, "Structure of representations with highest weight of infinite–dimensional Lie algebras," Adv. Math. **34** (1979) 97–108.

[374] M. Kalantar, P. Kasprzak and A. Skalski, "Open quantum subgroups of locally compact quantum groups," arXiv:1511.03952 [math.OA], to appear in Adv. Math.

[375] S.-J. Kang, M. Kashiwara, M. Kim and S. Oh, "Symmetric quiver Hecke algebras and R-matrices of quantum affine algebras III," Proc. London Math. Soc. **111** (2015) 420–444.

[376] V. Karimipour, "The quantum de Rham complexes associated with $SL_h(2)$," Lett. Math. Phys. **30** (1994) 87–98.
[377] E. Karolinsky, A. Stolin and V. Tarasov, "From dynamical to non-dynamical twists," Lett. Math. Phys. **71** (2005) 173–178.
[378] R.M. Kashaev, "The Hyperbolic volume of knots from quantum dilogarithm," Lett. Math. Phys. **39** (1997) 269–275.
[379] M. Kashiwara, "Crystalizing the q-analogue of universal enveloping algebras," Commun. Math. Phys. **133** (1990) 249–260.
[380] C. Kassel, *Quantum Groups*, Vol. 155 (Springer Science & Business Media, 2012).
[381] D. Kazhdan and G. Lusztig, "Representations of Coxeter groups and Hecke algebras," Inv. Math. **53** (1979) 165–184.
[382] D. Kazhdan and G. Lusztig, "Schubert varieties and Poincare duality," Proc. Symp. Pure Math. **34** (AMS, Providence, RI, 1980) pp. 185–203.
[383] A.A. Kehagias, P.A.A. Meessen and G. Zoupanos, "Deformed Poincaré algebra and field theory," Phys. Lett. **B346** (1995) 262–268.
[384] M. Khorrami and A. Shariati, M.R. Abolhassani and A. Aghamohammadi, "A triangular deformation of the two-dimensional Poincaré algebra," Mod. Phys. Lett. **A10** (1995) 873–883.
[385] S.M. Khoroshkin and V.N. Tolstoy, "Universal R-matrix for quantized (super)algebras," Commun. Math. Phys. **141** (1991) 599–617; "The Cartan-Weyl basis and the universal R-matrix for quantum KacŰMoody algebras and superalgebras," in: *Quantum Symmetries*, Proceedings, Workshop on Quantum Groups of the II Wigner Symposium (Goslar, July 1991), eds. H.-D. Doebner and V.K. Dobrev (World Scientific, Singapore, 1993) pp. 336–351.
[386] S.M. Khoroshkin and V.N. Tolstoy, "The uniqueness theorem for the universal R-matrix," Lett. Math. Phys. **24** (1992) 231–244.
[387] M.R. Kibler, R.M. Asherova and Yu.F. Smirnov, "Some aspects of q and qp boson calculus," in: Proceedings *Symmetries in Science VIII*, ed. B. Gruber (Plenum Press, NY, 1995) pp. 241–254.
[388] Y. Kimura and Fan Qin, "Graded quiver varieties, quantum cluster algebras and dual canonical basis," Adv. Math. **262** (2014) 261–312.
[389] A.N. Kirillov and N.Yu. Reshetikhin, "Representations of the algebra $U_q(sl(2))$, q-orthogonal polynomials and invariants of links," in: *New Developments in the Theory of Knots*, ed. T. Kohno, Adv. Ser. Math. Phys. vol. 11 (World Scientific, Singapore, 1990) pp. 202–256; "q-Weyl group and multiplicative formula for universal R-matrices," Commun. Math. Phys. **134** (1990) 421–431.
[390] R. Kirschner, "Integrable chains with Jordan-Schwinger representations," J. Phys. Conf. Ser. **411** (2013) 012018.
[391] M. Klimek, "D=2 and D=4 realization of κ-conformal algebra," Czech. J. Phys. **46** (1996) 187–194; M. Klimek and J. Lukierski, "Kappa deformed realization of D=4 conformal algebra," Acta Phys. Polon. **B26** (1995) 1209–1216.
[392] A.U. Klimyk and S. Pakuliak, "Infinite-dimensional representations of quantum algebras," in: *Symmetries in Science VI*, ed. B. Gruber (Plenum Press, 1993) pp. 389–414.
[393] A.U. Klimyk and K. Schmüdgen, *Quantum Groups and Their Representations* (Springer-Verlag, Berlin, 1997).
[394] A.W. Knapp and G.J. Zuckerman, "Classification theorems for representations of semisimple groups," Lecture Notes in Math. **587** (Springer-Verlag, Berlin, 1977) 138–159; "Classification of irreducible tempered representations of semisimple groups," Ann. Math. **116** (1982) 389–501.
[395] V.G. Knizhnik and A.B. Zamolodchikov, "Current algebra and Wess–Zumino model in two-dimensions," Nucl. Phys. **B247** (1984) 83–103.
[396] T.H. Koornwinder, "Orthogonal polynomials in connection with quantum groups," in: *Orthogonal Polynomials: Theory and Practice*, NATO ASI Series C, Vol. 294 (Kluwer Academic Publishers, 1990) pp. 257–292.

[397] P. Kosinski, J. Lukierski and P. Maslanka, "Local D=4 field theory on kappa deformed Minkowski space," Phys. Rev. **D62** (2000) 025004.

[398] P. Kosinski and M. Majewski, "The bicovariant differential calculus on the three-dimensional kappa-Poincaré group," Acta Phys. Polon. **B27** (1996) 2137–2153.

[399] P. Kosinski, P. Maslanka, J. Lukierski and A. Sitarz, "Generalized kappa-deformations and deformed relativistic scalar fields on noncommutative Minkowski space," in: Proceedings Conference "Topics in Mathematical Physics, General Relativity and Cosmology," 2002, Mexico City, eds. H. Garcia-Compean et al. (World Scientific, 2006) pp. 255–277.

[400] Y. Kosmann-Schwarzbach, "From 'Quantum Groups' to 'Quasi-Quantum Groups'," in: Proceedings, International Symposium "Symmetries in Science V: Algebraic Structures, their Representations, Realizations and Physical Applications" (Schloss Hofen, Vorarlberg, Austria, 30.7.-3.8.1990), eds. B. Gruber et al. (Plenum Press, New York, 1991) pp. 369–393.

[401] J. Kowalski-Glikman, "Testing dispersion relations of quantum Kappa-Poincare algebra on cosmological ground," Phys. Lett. **B499** (2001) 1–8.

[402] C. Krishnan and E. di Napoli, "Can quantum de Sitter space have finite entropy?," Class. Quantum Grav. **24** (2007) 3457–3463.

[403] P.P. Kulish, "A two-parameter quantum group and a gauge transform" (in Russian), Zapiski Nauch. Semin. LOMI **180** (1990) 89–93.

[404] P.P. Kulish and E.V. Damashinsky, "On the q oscillator and the quantum algebra SU(q)(1,1)," J. Phys. **A23** (1990) L415–L420.

[405] P.P. Kulish and N.Yu. Reshetikhin, "Quantum linear problem for the Sine-Gordon equation and higher representation," Zap. Nauch. Semin. LOMI **101** (1981) 101–110 (in Russian); English translation: J. Soviet. Math. **23** (1983) 2435–2441.

[406] P.P. Kulish, N.Yu. Reshetikhin and E.K. Sklyanin, "Yang-Baxter equation and representation theory. 1." Lett. Math. Phys. **5** (1981) 393–403.

[407] P.P. Kulish and R. Sasaki, "Covariance properties of reflection equation algebras," Prog. Theor. Phys. **89** (1993) 741–762.

[408] P.P. Kulish and E.K. Sklyanin, "Quantum inverse scattering method and the Heisenberg ferromagnet," Phys. Lett. **A70** (1979) 461–463.

[409] P.P. Kulish and E.K. Sklyanin, "On the solution of the Yang-Baxter equation," J. Sov. Math. **19** (1982) 1596–1620 [Zap. Nauchn. Semin. **95** (1980) 129–160].

[410] P.P. Kulish and E.K. Sklyanin, "Quantum spectral transform method. Recent developments," Lect. Notes Phys. **151** (1982) 61–119.

[411] P.P. Kulish and E.K. Sklyanin, "The general $U_q(sl(2))$ invariant XXZ integrable quantum spin chain," J. Phys. **A24** (1991) L435–L439.

[412] P.P. Kulish and E.K. Sklyanin, "Algebraic structures related to the reflection equations," J. Phys. **A25** (1992) 5963–5976.

[413] A. Kundu, "YangŰ-Baxter algebra and generation of quantum integrable models," Theor. Math. Phys. **151** (2007) 831–842; "Algebraic approach in unifying quantum integrable models," Phys. Rev. Lett. **82** (1999) 3936–3939; "Algebraic construction of quantum integrable models including inhomogeneous models," Rep. Math. Phys. **46** (2000) 125–136.

[414] A. Kundu and P. Truini, "Universal R-matrix of reductive Lie-algebras and quantum integrable systems from its color representation," J. Phys. **A28** (1995) 4089–4108.

[415] A. Kuniba, "Quantum R matrix for G_2 and a solvable 175-vertex model," J. Phys. **A23** (1990) 1349–1362.

[416] B.A. Kupershmidt, "Classification of quantum group structures on the group GL(2)," J. Phys. **A27** (1994) L47–L52.

[417] J. Kustermans and S. Vaes, "Locally compact quantum groups," Ann. Sci. l'Ecole Normale Superieure **33**.6 (2000) 837–934.

[418] M. Lagraa and N. Touhami, "The inhomogeneous quantum groups from differential calculi with classical dimension," J. Math. Phys. **40** (1999) 6052–6070; "Lie algebra of noncommutative inhomogeneous Hopf algebra," J. Math. Phys. **39** (1998) 5007–5014.

[419] R.P. Langlands, "On the classification of irreducible representations of real algebraic groups," in: *Representation Theory and Harmonic Analysis on Semi-simple Lie Groups*, eds. P. Sally and D. Vogan, Math. Surveys and Monographs, Vol. 31 (AMS, 1989) pp. 101–170; (first as IAS Princeton preprint, 1973).

[420] F. Lemeux, "The fusion rules of some free wreath product quantum groups and applications," J. Funct. Anal. **267** (2014) 2507–2550.

[421] A. Lerda and S. Sciuto, "Anyons and quantum groups," Nucl. Phys. **B401** (1993) 613–643.

[422] S.Z. Levendorsky, "Twisted function algebras on a compact quantum group and their representations," St. Petersburg (Leningrad) Math. J. **3** (1992) 405–423.

[423] S.Z. Levendorsky and Ya.S. Soibelman, *Algebras of Functions on Quantum Groups, Part 1*, Math. Surveys and Monographs, vol. 56 (American Mathematical Soc., 1998).

[424] S.Z. Levendorsky and Ya.S. Soibelman, "Some applications of the quantum Weyl groups," J. Geom. Phys. **7** (1990) 24–254.

[425] S.Z. Levendorsky and Ya.S. Soibelman, "Algebras of functions on compact quantum groups, Schubert cells and quantum tori," Commun. Math. Phys. **139** (1991) 141–170.

[426] S. Levenderovskii, Ya. Soibelman and V. Stukopin, "The quantum Weyl group and the universal quantum R-matrix for affine Lie algebra $A_1^{(1)}$," Lett. Math. Phys. **27** (1993) 253–264.

[427] D.A. Lowe, "Statistical entropy of two-dimensional dilaton de Sitter space," Phys. Lett. **B624** (2005) 275–280.

[428] R. Lü and K. Zhao, "Finite-dimensional simple modules over generalized Heisenberg algebras," Linear Algebra Appl. **475** (2015) pp. 276–291.

[429] J. Lukierski, From quantum deformations of relativistic symmetries to modified kinematics and dynamics, Acta Phys. Polon. **41** (2010) 2937–2965; "Fundamental masses from quantum symmetries," Acta Phys. Polon. **B27** (1996) 849–864.

[430] J. Lukierski and V.D. Lyakhovsky, "Two-parameter extensions of the kappa-Poincare quantum deformation," Contemp. Math. **391** (2005) 281–288.

[431] J. Lukierski, V. Lyakhovsky and M. Mozrzymas, "Kappa deformations of D=4 Weyl and conformal symmetries," Phys. Lett. **B538** (3–4) (2002) 375–384; κ-deformations of D=3 conformal versus deformations of D=4 AdS symmetries," Mod. Phys. Lett. **A18** (2003) 753–770.

[432] J. Lukierski, P. Minnaert and M. Mozrzymas, "New quantum deformations of D=4 conformal algebra," Czech. J. Phys. **46** (1996) 217–225; "Quantum deformations of conformal algebras introducing fundamental mass parameters," Phys. Lett. **B371** (1996) 215–222.

[433] J. Lukierski, P. Minnaert and A. Nowicki, "D=4 quantum Poincaré-Heisenberg algebra," in: *Symmetries in Science VI*, ed. B. Gruber (Plenum Press, 1993), p. 469–475.

[434] J. Lukierski and A. Nowicki, "Quantum deformations of D=4 Poincare and Weyl algebra from q-deformed D=4 conformal algebra," Phys. Lett. **B279** (1992) 299–307.

[435] J. Lukierski, A. Nowicki and H. Ruegg, "New quantum Poincare algebra and k deformed field theory," Phys. Lett. **B293** (1992) 344–352; "Real forms of complex quantum anti-De Sitter algebra $U_q(Sp(4;C))$ and their contraction schemes," Phys. Lett. **B271** (1991) 321–328; "κ-deformed Poincaré algebra and some physical consequences," in: *Symmetries in Science VI*, B. Gruber (ed.), (Plenum Press, 1993) pp. 477–487; "Quantum deformations of nonsemisimple algebras: The example of D=4 inhomogeneous rotations," J. Math. Phys. **35** (1994) 2607–2616.

[436] J. Lukierski and H. Ruegg, "Quantum kappa Poincare in any dimension," Phys. Lett. **B329** (1994) 189

[437] J. Lukierski, H. Ruegg, A. Nowicki and V.N. Tolstoi, "q-deformation of Poincare algebra," Phys. Lett. **B264** (1991) 331–338.

[438] J. Lukierski, H. Ruegg and W. Ruehl, "From κ-Poincaré algebra to κ-Lorentz quasigroup. A deformation of relativistic symmetry," Phys. Lett. **B313** (1993) 357–366.

[439] J. Lukierski, A. Nowicki and J. Sobczyk, "All real forms of $U_q(sl(4,C))$ and D=4 conformal quantum algebras," J. Phys. **A26** (1993) 4047–4058.

[440] J. Lukierski and M. Woronowicz, "New Lie-algebraic and quadratic deformations of Minkowski space from twisted Poincare symmetries," Phys. Lett. **B633** (2006) 116–124.

[441] G. Lusztig, "Quantum deformations of certain simple modules over enveloping algebras," Adv. Math. **70** (1988) 237–249.

[442] G. Lusztig, "Modular representations and quantum groups," Contemp. Math. **82** (1988) 59–77.

[443] G. Lusztig, "On quantum groups," J. Algebra **131** (1990) 466–475.

[444] G. Lusztig, "Canonical bases arising from quantized enveloping algebras," J. AMS **3**.2 (1990) 447–498.

[445] G. Lusztig, "Canonical bases arising from quantized enveloping algebras. II," Progr. Theor. Phys. Suppl. **102** (1990) 175–201.

[446] G. Lusztig, *Introduction to Quantum Groups*, Progress Math. 110 (Birkäuser, Boston-Basel-Berlin, 1993).

[447] V. Lyubashenko and A. Sudbery, "Quantum Lie algebras of type A_n," J. Math. Phys. **39** (1998) 3487–3504.

[448] A.J. MacFarlane, "On q-analogues of the quantum harmonic oscillator and the quantum group $SU(2)_q$, J. Phys. **A22** (1989) 4581–4588.

[449] G. Mack, "All unitary ray representations of the conformal group SU(2,2) with positive energy," Commun. Math. Phys. **55** (1977) 1–28.

[450] G. Mack and V. Schomerus, "Conformal field algebras with quantum symmetry from the theory of superselection sectors," Commun. Math. Phys. **134** (1990) 139–196; "Quasi-Hopf quantum symmetry in quantum theory," Nucl. Phys. **B370** (1992) 185–230.

[451] N.J. MacKay, "Introduction to Yangian symmetry in integrable field theory," Int. J. Mod. Phys. **A20** (2005) 7189–7218.

[452] J. Madore, H. Steinacker, "Propagator on the h-deformed Lobachevsky plane," J. Phys. **A33** (2000) 327–342.

[453] M. Maggiore, "Quantum groups, gravity, and the generalized uncertainty principle," Phys. Rev. **D49** (1994) 5182–5187.

[454] S. Majid, "Quasitriangular Hopf algebras and Yang-Baxter equations," Int. J. Mod. Phys. **A5** (1990) 1–91.

[455] S. Majid, "Examples of braided groups and braided matrices," J. Math. Phys. **32** (1991) 3246–3253; "Quantum and braided linear algebra," J. Math. Phys. **34** (1993) 1176–1196; "Braided momentum in the q-Poincare group," J. Math. Phys. **34** (1993) 2045–2058.

[456] S. Majid, "q-Euclidean space and quantum group wick rotation by twisting," J. Math. Phys. **35** (1994) 5025–5034.

[457] S. Majid, *Foundations of Quantum Group Theory* (Cambridge University Press, 2000).

[458] S. Majid and H. Ruegg, "Bicrossproduct structure of kappa Poincare group and noncommutative geometry," Phys. Lett. **B334** (1994) 348–354.

[459] F. Malikov, "Quantum groups: Singular vectors and BGG resolution," in: *Infinite Analysis*, Proceedings, RIMS Research Project 1991 part. B, Adv. Series in Math. Phys. vol. 16, eds. A. Tsuchiya, T. Eguchi, M. Jimbo (World Scientific, Singapore, 1992) pp. 623–643.

[460] F.G. Malikov, B.L. Feigin and D.B. Fuchs, "Singular vectors in Verma modules over Kac–Moody algebras," Funkts. Anal. Prilozh. **20**.2 (1986) 25–37; English translation: Funct. Anal. Appl. **20** (1986) 103–113.

[461] Yu.I. Manin, Some remarks on Koszul algebras and quantum groups, Ann. Inst. Fourier **37** (1987) 191–205.
[462] Yu.I. Manin, "Multiparametric quantum deformation of the general linear supergroup," Commun. Math. Phys. **123** (1989) 163–175.
[463] Yu.I. Manin, "Quantized theta functions," Prog. Theor. Phys. Suppl. **102** (1990) 219–228.
[464] Yu.I. Manin, *Topics in Noncommutative Geometry* (Princeton University Press, 1991).
[465] T. Masuda, K. Mimachi, Y. Nakagami, M. Noumi, Y. Sabuti and K. Ueno, "Unitary representations of the quantum group $SU_q(1,1)$: I. Structure of the dual space of $U_q(sl(2))$," Lett. Math. Phys. **19** (1990) 187–194; "II. Matrix elements of unitary representations and the basic hypergeometric functions," Lett. Math. Phys. **19** (1990) 195–204.
[466] T. Masuda, K. Mimachi, Y. Nakagami, M. Noumi and K. Ueno, "Representations of the quantum groups and a q-analogue of orthogonal polynomials," C.R. Acad. Sci. Paris **307** Ser I, (1988) 559–564.
[467] T. Masuda, Y. Nakagami and S.L. Woronowicz, "A C*-algebraic framework for quantum groups," Int. J. Math. **14** (2003) 903–1001.
[468] D. Maulik and A. Okounkov, "Quantum groups and quantum cohomology," arXiv:1211.1287 [math.AG].
[469] R.J. McDermott and D. Parashar, "Contraction of the $G_{r,s}$ quantum group to its nonstandard analogue and corresponding coloured quantum groups," J. Math. Phys. **41** (2000) 2403–2416; "Realizations of quantum $GL_{p,q}(2)$ and Jordanian $GL_{h,h1}(2)$," Czech. J. Phys. **50** (2000) 157–162.
[470] L. Mesref, "q-deformed conformal correlation functions," Int. J. Mod. Phys. **A20** (2005) 1471–1479; "Quantum field theories," Int. J. Mod. Phys. **A20** (2005) 5317–5351; "Maps between deformed and ordinary gauge fields," Int. J. Theor. Phys. **44** (2005) 1549–1557.
[471] R. Meyer, S. Roy and S.L. Woronowicz, "Semidirect products of C*-quantum groups: Multiplicative unitaries approach," Commun. Math. Phys. (2016) 1-34doi:10.1007/s00220-016-2727-3; "Quantum group-twisted tensor products of C*-algebras," Int. J. Math. **25** (2014) Art. no. 1450019.
[472] U. Meyer, "Wave equations on q-Minkowski space," Commun. Math. Phys. **174** (1995) 457–475.
[473] L. Mezincescu and R.I. Nepomechie, "Integrability of open spin chains with quantum algebra symmetry," Int. J. Mod. Phys. **A6** (1991) 5231–5248; Addendum: [Int. J. Mod. Phys. **A7** (1992) 5657–5659]
[474] L. Mezincescu, R.I. Nepomechie and V. Rittenberg, "Bethe ansatz solution of the Fateev-Zamolodchikov quantum spin chain with boundary terms," Phys. Lett. **A147** (1990) 70–78.
[475] C. Micu and E. Papp, "Non-polynomial solutions to the q-difference form of the Harper equation," J. Phys. **A31** (1998) 2881–2887.
[476] D. Mikulovic, A. Schmidt and H. Wachter, "Grassmann variables on quantum spaces," Eur. Phys. J. **C45** (2006) 529–544.
[477] K.C. Misra and T. Miwa, "Crystal base for the basic representation of $U_q(sl(n)^{(1)})$," Commun. Math. Phys. **134** (1990) 79–88.
[478] R.L. Mkrtchyan and L.A. Zurabyan, Casimir operators for quantum sl_n groups, Yerevan preprint YERPHI – 1149 (26) – 89 (1989).
[479] A.I. Molev, *Yangians and Classical Lie Algebras* (Providence, RI, AMS (2007) ISBN 978-0-8218-4374-1); *Yangians and Classical Lie Algebras* (in Russian) (Moscow, MCNMO, 2009; ISBN 978-5-94057-498-9).
[480] A.I. Molev, "Gelfand-Tsetlin bases for classical Lie algebras," in: *Handbook of Algebra*, Vol. 4, ed. M. Hazewinkel (Elsevier, 2006) pp. 109–170.
[481] A. Molev, M. Nazarov and G. Olshansky, "Yangians and classical Lie algebras," Russ. Math. Surv. **51** (1996) 205 (83 pp.).

[482] G. Moore and N.Yu. Reshetikhin, "A comment on quantum group symmetry in conformal field theory," Nucl. Phys. **B328** (1989) 557–574.

[483] G. Moore and N. Seiberg, "Classical and quantum conformal field theory," Commun. Math. Phys. **123** (1989) 177–254.

[484] P. Moylan, "Representations of classical Lie algebras from their quantum deformations," Inst. Phys. Conf. Ser. No 173 (2003) pp. 683–686; "Embedding Euclidean Lie algebras into quantum structures," Czech. J. Phys. **47** (1997) 1251–1258.

[485] A. Mudrov, "Regularization of Mickelsson generators for non-exceptional quantum groups," arXiv:1512.08666 [math.QA] (2015).

[486] F.J. Narganes-Quijano, "Cyclic representations of a q-deformation of the Virasoro algebra," J. Phys. **A24** (1991) 593–601. "The quantum deformation of SU(1,1) as the dynamical symmetry of the anharmonic oscillator," pp. 1699–1707.

[487] M.L. Nazarov, "Quantum Berezinian and the classical Capelli identity," Lett. Math. Phys. **21** (1991) 123–131.

[488] M. Nazarov and V. Tarasov, "Yangians and Gelfand-Zetlin bases," Publ. Res. Inst. Math. Sci. Kyoto **30** (1994) 459–478.

[489] J. Negro, F.J. Herranz and A. Ballesteros, "On a class of non-standard quantum algebras and their representations," Czech. J. Phys. **47** (1997) 1259–1266.

[490] N.A. Nekrasov, "Milne universe, tachyons, and quantum group," Surv. High Energy Phys. **17** (2002) 115–124.

[491] N. Nekrasov, V. Pestun and S. Shatashvili, "Quantum geometry and quiver gauge theories," arXiv:1312.6689 [hep-th].

[492] S. Neshveyev and L. Tuset, "Deformation of C^*-algebras by cocycles on locally compact quantum groups," Adv. Math. **254** (2014) 454–496.

[493] M.V. Ngu, N.G. Vinh, N.T. Lan and V.N. Ai, "Cooper pair of superconductivity in the coordinate representation and q-deformed harmonic oscillator," J. Phys.: Conf. Ser. **726** (2016) 012017.

[494] M.R. Niedermaier, "Irrational free field resolutions for W(sl(n)) and extended Sugawara construction," Commun. Math. Phys. **148** (1992) 249–281; "W(sl(n)): Existence, Cartan basis and infinite Abelian subalgebras," in: Cargese 1991, Proceedings, *New Symmetry Principles in QFT* (NATO Sci.Ser.B v. 295 1992) pp. 493–503.

[495] M. Noumi and K. Mimachi, "Quantum 2-spheres and big q-Jacobi polynomials," Commun. Math. Phys. **128** (1990) 521–531.

[496] R. Oeckl, "Braided quantum field theory," Commun. Math. Phys. **217** (2001) 451–473; "Untwisting noncommutative R**d and the equivalence of quantum field theories," Nucl. Phys. **B581** (2000) 559–574.

[497] O. Ogievetsky and J. Wess, "Relations between GL(p,q)(2)'s," Z. Physik **C50** (1991) 123–131.

[498] O. Ogievetsky, W.B. Schmidke, J. Wess and B. Zumino, "q-deformed Poincare algebra," Commun. Math. Phys. **150** (1992) 495–518.

[499] C. Ohn, "A *-product on SL(2) and the corresponding nonstandard sl(2)," Lett. Math. Phys. **25** (1992) 85–88.

[500] M. Okado, "Quantum R matrices related to the spin representations of B_n and D_n," Commun. Math. Phys. **134** (1990) 467–486.

[501] M.A. Olshanetsky and V.B.K. Rogov, "dS-AdS structures in the non-commutative Minkowski spaces," Theor. Math. Phys. **144** (2005) 1315–1343; "Unitary representations of quantum Lorentz group and quantum relativistic Toda chain," Theor. Math. Phys. **130** (2002) 299–322.

[502] P. Parashar, "Jordanian $U_{h,s}gl(2)$ and its colored realization," Lett. Math. Phys. **45** (1998) 105–112.

[503] V.B. Petkova and G.M. Sotkov, "The six-point families of exceptional representations of the conformal group," Lett. Math. Phys. **8** (1984) 217–226.

[504] B. Parshall and J.-P. Wang, *Quantum Linear Groups*, Mem. Am. Math. Soc. 439 (1991).
[505] V. Pasquier, "Continuum limit of lattice models built on quantum groups," Nucl. Phys. **B295** [**FS21**] (1989) 491–510.
[506] V. Pasquier and H. Saleur, "Common structures between finite systems and conformal field theories through quantum groups," Nucl. Phys. **B330** [**FS21**] (1990) 523–556.
[507] M.K. Patra, "A realization of quantum algebras – some applications," J. Phys. **A30** (1997) 1259–1271.
[508] M. Pillin, "q deformed relativistic wave equations," J. Math. Phys. **35** (1994) 2804–2817.
[509] L. Pittner, *Algebraic Foundations of Non-Commutative Differential Geometry and Quantum Groups*, Lecture Notes in Physics Monographs, M. 39 (Springer, Berlin-Heidelberg, 1996).
[510] P. Podles, "Solutions of Klein–Gordon and Dirac equations on quantum Minkowski spaces," Commun. Math. Phys. **181** (1996) 569–586; "Symmetries of quantum spaces. Subgroups and quotient spaces of quantum SU(2) and SO(3) groups," Commun. Math. Phys. **170** (1995) 1–20; "Differential calculus on quantum spheres," Lett. Math. Phys. **18** (1989) 107–119; "Quantum Minkowski spaces," in: *Particles, Fields, and Gravitation*, ed. J. Rembielinski, Lodz, AIP Conference Proceedings 453 (1998) pp. 97–106.
[511] P. Podles and S.L. Woronowicz, "On the structure of inhomogeneous quantum groups," Commun. Math. Phys. **185** (1997) 325–358.; "On the classification of quantum Poincaré groups," Commun. Math. Phys. **178** (1996) 61–82; "Quantum deformation of Lorentz group," Commun. Math. Phys. **130** (1990) 381–431.
[512] A. Polychronakos, "A classical realization of quantum algebras," Mod. Phys. Lett. **A5** (1990) 2325–2333; "Consistency conditions and representations of a q deformed Virasoro algebra," Phys. Lett. **B256** (1991) 35–40.
[513] T. Popov, "Homogeneous algebras, parastatistics and combinatorics," Inst. Phys. Conf. Ser. **173** (2003) 463–467.
[514] K. Przanowski, "The isomorphism Hopf *-algebras between kappa-Poincare algebra in case g_{00}=0 and "null plane" quantum Poincare algebra, Acta Phys. Pol. **B28** (1997) 1635–1641.
[515] C. Quesne, "Coherent states, K-matrix theory and q-boson realizations of the quantum algebra $su_q(2)$," Phys. Lett. **A153** (1991) 303–307.
[516] C. Quesne, "Covariant (hh')-deformed bosonic and fermionic algebras as contraction limits of q-deformed ones," Int. J. Theor Phys. **38** (1999) 1905–1923; "Duals of coloured quantum universal enveloping algebras and coloured universal $\mathscr{T}$-matrices," J. Math. Phys. **39** (1998) 1199–1222; "Colored quantum universal enveloping algebras," J. Math. Phys. **38** (1997) 6018–6040; "Complementarity of $su_q(3)$ and $u_q(2)$ and q-boson realization of the $su_q(3)$ irreducible representations" J. Phys. **A25** (1992) 5977–5998; "Raising and lowering operators for $u_q(n)$," J. Phys. **A26** (1993) 357–372.
[517] D.E. Radford and J. Towber, "Yetter-Drinfel'd categories associated to an arbitrary bialgebra," J. Pure Appl. Algebra **87** (1993) 259–279.
[518] A. Ram, M. Lanini and P. Sobaje, "A Fock space model for decomposition numbers for quantum groups at roots of unity," arXiv:1612.03120 [math.RT] (2016).
[519] C. Ramirez, H. Ruegg and M. Ruiz-Altaba, "Explicit quantum symmetries of WZNW theories," Phys. Lett. **B247** (1990) 499–508; "The Contour picture of quantum groups: Conformal field theories," Nucl. Phys. **B364** (1991) 195–253.
[520] S. Raum and M. Weber, "The full classification of orthogonal easy quantum groups," Commun. Math. Phys. **341** (2016) 751–779.
[521] N.Yu. Reshetikhin, "Quantized universal enveloping algebras, the Yang-Baxter equation and invariants of links. I & II," LOMI Leningrad preprints E-4-87, E-17-87 (1987).
[522] N.Yu. Reshetikhin, "Multiparametric quantum groups and twisted quasitriangular Hopf algebras," Lett. Math. Phys. **20** (1990) 331–335.

[523] N.Yu. Reshetikhin and M.A. Semenov-Tian-Shansky, "Central extensions of quantum current groups," Lett. Math. Phys. **19** (1990) 133–142.

[524] N.Yu. Reshetikhin and F.A. Smirnov, "Hidden quantum group symmetry and integrable perturbations of conformal field theories," Commun. Math. Phys. **131** (1990) 157–177.

[525] N.Yu. Reshetikhin and V.G. Turaev, "Invariants of 3-manifolds via link polynomials and quantum groups," Inv. Math. **103** (1991) 547–597; "Ribbon graphs and their invaraints derived from quantum groups," Commun. Math. Phys. **127** (1990) 1–26.

[526] C.I. Ribeiro-Silva, E.M.F. Curado and M.A. Rego-Monteiro, "Deformed scalar quantum electrodynamics as a phenomenological model for composite scalar particles," J. Phys. **A41** (2008) art. no. 145404.

[527] V. Rittenberg and M. Scheunert, "Tensor operators for quantum groups and applications," J. Math. Phys. **33** (1992) 436–445.

[528] A. Ritz and G.C. Joshi, "A q-Lorentz algebra from q-deformed harmonic oscillators," Chaos Soliton Fract. **8** (1997) 823–834.

[529] P. Roche and D. Arnaudon, "Irreducible representations of the quantum analog of SU(2)," Lett. Math. Phys. **17** (1989) 295–300.

[530] L.J. Romans, "Realizations of classical and quantum W(3) symmetry," Nucl. Phys. **B352** (1991) 829–848.

[531] M. Rosso, "Repre'sentation irre'ductibles de dimension finie du q-analogue de l'alge'bre enveloppante d'une algebre de Lie simple," C.R. Acad. Sci. Paris **305** Serie I (1987) 587–593.

[532] M. Rosso, "Finite dimensional representations of the quantum analog of the enveloping algebra of a complex simple lie algebra," Commun. Math. Phys. **117** (1987) 581–593.

[533] M. Rosso, "An analogue of P.B.W. theorem and the universal R-matrix for $U_h(sl(N+1))$," Commun. Math. Phys. **124** (1989) 307–318.

[534] H.C. Rosu, "Hybridizing the Skyrmion with an anti-de Sitter bag," Nuovo Cim. **B108** (1993) 313–329.

[535] E.C. Rowell, Y. Zhang, Y.S. Wu and M.-L. Ge, "Extraspecial two-groups, generalized Yang-Baxter equations and Braiding quantum gates," Quantum Inf. Comput. **10** (2010) 685–702.

[536] S. Roy and B. Ghosh, "Study of controlled dense coding with some discrete tripartite and quadripartite states," Int. J. Quantum Inf. **13** (2015) Art. no. 1550033.

[537] S. Roy, S.L. Woronowicz, "Landstad-Vaes theory for locally compact quantum groups," arXiv:1606.03728 [math.OA] (2016).

[538] H. Ruegg, "A simple derivation of the quantum Clebsch-Gordan coefficients for $SU(2)_q$," J. Math. Phys. **31** (1990) 1085–1087.

[539] H. Saleur, "Quantum osp(1,2) and solutions of the graded Yang-Baxter equation," Nucl. Phys. **B336** (1990) 363–376.

[540] H. Saleur and D. Altschüler, "Level rank duality in quantum groups," Nucl. Phys. **B354** (1991) 579–613.

[541] N.A. Sang, D.T.T. Thuy, N.T.H. Loan and V.N. Ai, "Energy spectrum inverse problem of q-deformed harmonic oscillator and WBK approximation," J. Phys.: Conf. Ser. **726** (2016)

[542] A. Schirrmacher, "The multiparametric deformation of GL(n) and the differential calculus on the quantum vector space," Zeit. f. Physik **C50** (1991) 321–328.

[543] A. Schirrmacher, J. Wess and B. Zumino, "The two parameter deformation of GL(2), its differential calculus, and Lie algebra," Zeit. f. Physik **C49** (1991) 317–324.

[544] W.B. Schmidke, J. Wess and B. Zumino, "A q-deformed Lorentz algebra," Zeit. f. Physik **C52** (1991) 471–476.

[545] A. Schmidt and H. Wachter, "q-Deformed quantum Lie algebras," J. Geom. Phys. **56** (2006) 2289–2325.

[546] K. Schmüdgen and A. Schüler, "Bicovariant differential calculi on quantum groups A_{n-1}, B_n, C_n and D_n," in: Proceedings, Workshop on Generalized Symmetries in Physics (Clausthal, July 1993), eds. H.-D. Doebner, V.K. Dobrev and A.G. Ushveridze (World Scientific, Singapore, 1994), pp. 185–195.

[547] P. Schupp, P. Watts and B. Zumino, "Differential geometry on linear quantum groups," Lett. Math. Phys. **25** (1992) 139–148.

[548] J. Schwenk and J. Wess, "A q deformed quantum mechanical toy model," Phys. Lett. **B291** (1992) 273–277.

[549] A. Shafiekhani, "U_q(sl(n)) difference operator realization," Mod. Phys. Lett. **A9** (1994) 3273–3283.

[550] N.N. Shapovalov, "On a bilinear form on the universal enveloping algebra of a complex semisimple Lie algebra," Funkts. Anal. Prilozh. **6** (4) 65–70 (1972); English translation: Funct. Anal. Appl. **6**, 307–312 (1972).

[551] A. Shariati and A. Aghamohammadi, "A simple method for constructing the inhomogeneous quantum group IGL_q(n) and its universal enveloping algebra U_q(IGL(n))," J. Math. Phys. **36** (1995) 7103–7108.

[552] D. Shklyarov, S. Sinelshchikov, A. Stolin and L. Vaksman, "Non-compact quantum groups and quantum Harish-Chandra modules," Nucl. Phys. Proc. Suppl. **102** (2001) 334–337.

[553] D. Shklyarov, S. Sinelshchikov and L. Vaksman, "Geometric realizations for some series of representations of the quantum group $SU_{2,2}$," Matem. Fiz. Analiz, Geom., **8** (2001) 90–110.

[554] D. Shklyarov and G.K. Zhang, "Covariant q-differential operators and unitary highest weight representations for $U_q$su(n,n)," J. Math. Phys. **46** (2005) art. no. 062307.

[555] S. Shnider and S. Sternberg, *Quantum Groups (From coalgebras to Drinfeld algebras)* (International Press, Boston, 1993).

[556] S. Sinel'shchikov, A. Stolin and L. Vaksman, "Differential calculi on some quantum prehomogeneous vector spaces," J. Math. Phys. **48** (2007) 073514.

[557] A. Sitarz, "Noncommutative differential calculus on the kappa Minkowski space," Phys. Lett. **B349** (1995) 42–48.

[558] E.K. Sklyanin, "Quantum inverse scattering method. Selected topics," in: *Quantum Group and Quantum Integrable Systems*, Nankai Lectures on Mathematical Physics: Nankai Institute of Mathematics, 1991 (World Scientific, 1992) pp. 63–97; "Some algebraic structures connected with the Yang-Baxter equation," Funkts. Anal. Prilozh. **16** (1982) 27–34, (in Russian); English translation: Funct. Anal. Appl. **16** (1982) 263–270; "Some algebraic structures connected with the Yang-Baxter equation. Representations of quantum algebras," Funkts. Anal. Prilozh. **17** (1983) 34–48 (in Russian); English translation: Funct. Anal. Appl. **17** (1983) 274–288; "On an algebra generated by quadratic relations," Uspekhi Mat. Nauk **40** (1985) 214 (in Russian).

[559] K. Sogo, Y. Akutsu and T. Abe, "New factorized S-matrix and its application to exactly solvable q-state model. I, Prog. Theor. Phys. **70** (1983) 730–738.

[560] Ya.S. Soibelman, "The algebra of functions on a compact quantum group, and its representations," Leningrad (St. Petersburg) Math. J. **2** (1991) 161–178.

[561] M.A. Sokolov, "Exponential representation of Jordanian matrix quantum group $GL_h(2)$," J. Phys. **A33** (2000) 417–425.

[562] H. Steinacker, "Unitary representations of noncompact quantum groups at roots of unity," Rev. Math. Phys. **13** (2001) 1035–1054; "Finite–dimensional unitary representations of quantum Anti–de Sitter groups at roots of unity," Commun. Math. Phys. **192** (1998) 687–706; "Quantum anti-de Sitter space and sphere at roots of unity," Adv. Theor. Math. Phys. **4** (2000) 1–44; "Quantum anti-De Sitter space at roots of unity," Int. J. Mod. Phys. **B14** (2000) 2499–2501.

[563] P. Stovicek, "Some remarks on U_q(sl(2,R)) at root of unity," Czech. J. Phys. **50** (2000) 1353–1358; "A construction of representations for quantum groups: An example of U_q(so(5))," Czech. J. Phys. **48** (1998) 1501–1506; "A construction of representations and quantum homogeneous spaces," Lett. Math. Phys. **47** (1999) 125–138.

[564] A. Sudbery, "Non-commuting coordinates and differential operators," Proceedings of Workshop on Quantum Groups, Argonne National Lab (1990), eds. T. Curtright, D. Fairlie and C. Zachos (World Scientific, 1991) p. 33.

[565] A. Sudbery, "Consistent multiparameter quantization of GL(n)," J. Phys. **A23** (1990) L697–L704.

[566] A. Sudbery, "The quantum orthogonal mystery," in: *Quantum Groups: Formalism and Applications*, eds. J. Lukierski et al. (Polish Scientific Publishers PWN, 1995), pp. 303–316.

[567] A. Sudbery, "SU_q(n) gauge theory," Phys. Lett. **B375** (1996) 75–80.

[568] A. Sudbery, "Quantum-group gauge theory," in: *Quantum Group Symposium at Group21*, Proceedings of a Symposium at the XXI Intern. Colloquium on Group Theoretical Methods in Physics (Goslar, July 1996), eds. H.-D. Doebner and V.K. Dobrev (Heron Press, Sofia, 1997) pp. 45–52.

[569] C.-P. Sun and H.-C. Fu, "The q-deformed boson realisation of the quantum group SU(n)q and its representations," J. Phys. **A22** (1989) L983–L986; "The q analog of the boson algebra, its representation on the Fock space, and applications to the quantum group," J. Math. Phys. **32** (1991) 595–598.

[570] M.E. Sweedler, *Hopf Algebras* (Benjamin, New York, 1969).

[571] E.H. Tahri, "Quantum group structure on the quantum plane," J. Math. Phys. **39** (1998) 2983–2992.

[572] M. Takeuchi, "Matrix bialgebras and quantum groups," Israel J. Math. **72** (1990) 232–251.

[573] L.A. Takhtajan, "Introduction to quantum groups," Proceedings of the Quantum Groups Workshop (Clausthal, 1989) eds. H.-D. Doebner and J.-D. Hennig, Lecture Notes in Physics, Vol. 370 (Springer-Verlag, Berlin, 1990) pp. 3–28; Adv. Stud. Pure Math. **19** (1989) 435–457.

[574] X. Tang, "Ringel-Hall algebras and two-parameter quantized enveloping algebras," Pac. J. Math. **247** (2010) 213–240.

[575] P. Terwilliger, "Finite-dimensional irreducible U_q(sl2)-modules from the equitable point of view," Linear Algebra Appl. **439** (2013) 358–400.

[576] V.N. Tolstoy, "Extremal projectors for quantized Kac-Moody superalgebras and some of their applications," in: Proceedings, Quantum Groups Workshop (Clausthal, Germany, 1989), eds. H.D. Doebner and J.D. Hennig, Lect. Notes in Physics, V. 370 (Springer-Verlag, Berlin, 1990) 118–125.

[577] V.N. Tolstoy, "Drinfeldians," Proceedings, II International Workshop "Lie Theory and Its Applications in Physics" (Clausthal, August 1997); eds. H.-D. Doebner, V.K. Dobrev and J. Hilgert (World Scientific, Singapore, 1998, ISBN 981-02-3539-9) pp. 325–337.

[578] V.N. Tolstoy and S.M. Khoroshkin, "The universal R-matrix for quantum untwisted affine Lie algebras," Funct. Anal. Appl. **26** (1992) 69–71.

[579] P. Truini and V.S. Varadarajan, "The concept of a quantum semisimple group," Lett. Math. Phys. **21** (1991) 287–292; "Universal deformations of reductive Lie-algebras," Lett. Math. Phys. **26** (1992) 53–65; "Quantization of reductive Lie-algebras - construction and universality," Rev. Math. Phys. **5** (1993) 363–416; "Aspects of Lie algebra quantization," in: *Symmetries in Science VI*, ed. B. Gruber (Plenum Press, 1993) pp. 731–740.

[580] A.V. Turbiner, "Lie algebras and linear operators with invariant subspace," in: *Lie Algebras, Cohomologies and New Applications in Quantum Mechanics*, eds. N. Kamran and P. Olver *Contemporary Mathematics*, vol. 160 (1994) pp. 263–310.

[581] U.N. Iyer and T.C. McCune, "Quantum differential operators on the quantum plane," J. Algebra **260** (2003) 577–591.
[582] K. Ueno, T. Takebayashi and Y. Shibukawa, "Gelfand-Zetlin basis for U_q(gl(N+1)) modules," Lett. Math. Phys. **18** (1989) 215–221.
[583] L.L. Vaksman and L.I. Korogodsky, "An algebra of bounded functions on the quantum group of the motions of the plane, and q-analogues of Bessel functions," Sov. Math. Dokl. **39** (1989) 173–177.
[584] L.L. Vaksman and Ya.S. Soibelman, "Algebra of functions on the quantum group SU(2)," Funkts. Anal. Prilozh. **22**.3 (1988) 1–14; English translation: Funct. Anal. Appl. **22** (1988) 170–181.
[585] P.A. Valinevich, "Factorization of the R-matrix for the quantum algebra U_q(sln), J. Math. Sci. **168** (2010) 811–819.
[586] A. Van Daele and S.L. Woronowicz, "Duality for the quantum E(2) group," Pac. J. Math. **173** (1996) 375–385.
[587] J. Van der Jeugt, "Representations and Clebsch-Gordan coefficients for the Jordanian quantum algebra U_h(sl(2))," J. Phys. **A31** (1998) 1495–1508; "Quantum algebra embeddings - deforming functionals and algebraic approach," Can. J. Phys. **72** (1994) 519–526; "The Jordanian deformation of su(2) and Clebsch-Gordan coefficients," Czech. J. Phys. **47** (1997) 1283–1289.
[588] H.J. de Vega, "Yang-Baxter algebras, integrable theories and quantum groups," Int. J. Mod. Phys. **A4** (1989) 2371–2465.
[589] J.-L. Verdier, "Groupes quantiques (dáprès V.G. Drinfel'd)," in: *Séminaire Bourbaki*, No. 685, *Astérisque*, **152–153** (1987) 305–319.
[590] N.Ya. Vilenkin and A.U. Klimyk, *Representation of Lie Groups and Special Functions*, Vol. 3: *Classical and Quantum Groups and Special Functions* (Klumer Academic Publishers, Dordrecht, 1992) 629 pages; Vol. 4: *Recent Advances* (ibid. 1995) 497 pages.
[591] N.G. Vinh, M.V. Ngu, N.T. Lan, L.T.K. Thanh and N.A. Viet, "The possible conection between q-defomed hamornic oscillator formation and anharmonicity," J. Phys.: Conf. Ser. **726** (2016) 012018; "Q-deformed harmonic oscillator and Morse-like anharmonic potential," J. Phys.: Conf. Ser. **627** (2015) 012022.
[592] A.A. Vladimirov, "A closed expression for the universal R-matrix in a non-standard quantum double," Mod. Phys. Lett. **A8** (1993) 2573–2578.
[593] H. Wachter, "Analysis on q-deformed quantum spaces," Int. J. Mod. Phys. **A22** (2007) 95–164; "q-integration on quantum spaces," Eur. Phys. J. **C32** (2004) 281–297; "q-exponentials on quantum spaces," Eur. Phys. J. **C37** (2004) 379–389.
[594] G. Wang, K. Xue, C. Sun and G. Du, "Yang-Baxter R matrix, entanglement and Yangian," Quantum Inf. Process **11** (2012) 1775–1784.
[595] M. Weber, "On the classification of easy quantum groups," Adv. Math. **245** (2013) 500–533.
[596] J. Wess and B. Zumino, "Covariant differential calculus on the quantum hyperplane," Nucl. Phys. (Proc. Suppl.) **B18** (1990) 302–312; "Differential calculus on quantum planes and applications," talk by J. Wess at the XXIV International Symposium (Gosen, 1990), *Proceedings, Theory of Elementary Particles* (Zeuthen, Germany, Inst. fur Hochenergiephysik, 1991) pp. 1–11.
[597] E. Witten, "Gauge theories, vertex models and quantum groups," Nucl. Phys. **B330** (1990) 285–346.
[598] C. Worawannotai, "Dual polar graphs, the quantum algebra U_q(sl2), and Leonard systems of dual q-Krawtchouk type," Linear Algebra Appl. **438** (2013) 443–497.
[599] S.L. Woronowicz, "Quantum exponential function," Rev. Math. Phys. **12** (2000) 873–920; "From multiplicative unitaries to quantum groups," Int. J. Math. **7** (1996) 127–149; "Quantum SU(2) and E(2) groups. Contraction procedure," Commun. Math. Phys. **149** (1992) 637–652;

"Operator equalities related to the quantum E(2) group," Commun. Math. Phys. **144** (1992) 417–428; "Quantum E(2) group and its Pontryagin dual," Lett. Math. Phys. **23** (1991) 251–263; "Compact matrix pseudogroups," Commun. Math. Phys. **111** (1987) 613–665; "A remark on compact matrix quantum groups," Lett. Math. Phys. **21** (1991) 35–39; "New quantum deformation of SL(2,C). Hopf algebra level," Rep. Math. Phys. **30** (1991) 259–269; "Differential calculus on compact matrix pseudogroups (quantum groups),"Commun. Math. Phys. **122** (1989) 125–170; "Unbounded elements affiliated with C^*-algebras and non-compact quantum groups," Commun. Math. Phys. **136** (1990) 399–432.

[600] S.L. Woronowicz and K. Napiorkowski, "Operator theory in the C*-algebra framework," Rep. Math. Phys. **31** (1992) 353–371.

[601] S.L. Woronowicz and S. Zakrzewski, "Quantum Lorentz group having Gauss decomposition property," Publ. RIMS, Kyoto Univ. **28** (1992) 809–824.

[602] S.L. Woronowicz and S. Zakrzewski, "Quantum deformations of the Lorentz group. The Hopf *-algebra level," Compos. Math. **90** (1994) 211–243.

[603] S.L. Woronowicz and S. Zakrzewski, "Quantum 'ax + b' group," Rev. Math. Phys. **14** (2002) 797–828.

[604] Wei Xiao, "Differential equations and singular vectors in Verma modules over sl(n,C)," Acta Math. Sinica, English Ser. **31** (2015) 1057–1066.

[605] D. Yau,. "Hom-quantum groups: I. Quasi-triangular Hom-bialgebras," J. Phys. **A45** (2012) 065203.

[606] C.K. Zachos, "Quantum deformations," Proceedings Workshop on Quantum Groups, Argonne Nat. Lab (1990); eds. T. Curtright, D. Fairlie and C. Zachos (Singapore, World Scientific, 1991) 10 pages; "Quantum maps for deformed algebras," Proc. Int. Symp. "Symmetries in Science V: Algebraic Structures, their Representations, Realizations and Physical Applications" (Schloss Hofen, Vorarlberg, Austria, 30.7.-3.8.1990), eds. B. Gruber et al (Plenum Press, NY, 1991) pp. 593–610.

[607] C.K. Zachos,"Paradigms of quantum algebras," Contemp. Math. **134** (1992) 351–377.

[608] S. Zakrzewski, "A Hopf star-algebra of polynomials on the quantum SL(2,R) for a 'unitary' R-matrix," Lett. Math. Phys. **22** (1991) 287–289.

[609] R.B. Zhang, A.J. Bracken and M.D. Gould, "Solution of the graded Yang-Baxter equation associated with the vector representation of Uq(osp(M2n))," Phys. Lett. **B257** (1991) 133–139; "Generalized Gelfand invariants of quantum groups" J. Phys. **A24** (1991) 937–944; "From representations of the Braid group to solutions of the Yang-Baxter equation," Nucl. Phys. **B354** (1991) 625–652.

[610] Y. Zhang and M.-L. Ge, "GHZ states, almost-complex structure and Yang–Baxter equation (I)," Quantum Inf. Process **6** (2007) 363–379.

[611] Y. Zhang, N. Jing and M.-L. Ge, "Quantum algebras associated with Bell states," J. Phys. **A41** (2008) art. no. 055310.

[612] S.Y. Zhou, "On the irreducible module of quantum group $U_q(B_2)$ at a root of 1," J. Phys. **A27** (1994) 2401–2406; "Verma module of $U_q(B_2)$," Chin. Phys. Lett. **11** (1994) 1–3.

Author Index

Abe 1
Arnaudon 187

Biedenharn 281
Bourbaki 22

Chari–Pressley 11
Connes 111

De Concini 37
Dirac 89
Dixmier 79
Drinfeld 1, 7, 118

Faddeev 1
Faddeev–Reshetikhin–Takhtajan 19, 109, 140

Gasper–Rahman 310
Gelfand–Graev–Vilenkin 234

Jimbo 1, 7

Kac 37, 79
Kazhdan–Lusztig 84
Kulish 1
Kulish–Reshetikhin 1

Lusztig 34, 84

Majid 332
Manin 107, 111, 120

Reshetikhin 1

Shapovalov 300
Sklyanin 1
Sudbery 115
Sweedler 2

Takhtajan 1

Wess–Zumino 111
Woronowicz 109, 111
Woronowicz–Zakrzewski 223

Subject Index

E_β 6
H_β 6
I^Λ 35
L_Λ 35
R-matrix 109
$U_q(\mathscr{G})$ 4
V^Λ 34
$Y_\eta(\mathfrak{g})$ 17
Δ 5
Γ_+ 37
Γ_q 42
$e(\mu)$ 79
q-Gamma function 42
q-Karlsson–Minton formula 311
q-Pochhammer symbol 299
q-Weyl equations 340
q-Weyl group 8
q-binomial coefficients $C_n^s(q)$ 15
q-commutes 121
q-coset 256
q-d'Alembert hierarchy 340
q-light cone 343
q-number 5
q-potentials 349
q-Serre relations 4
q-shifted factorial 310
q-special functions 42
$[m]_q$ 5
$\binom{n}{k}_q$ 5
ad_q 6
$\det_\mu^\lambda$ 37

antipode 2

baxterized braid equation 188
baxterized Yang–Baxter equation 188
bialgebra 1
Borel subalgebra 34
Borel subgroup 205
braid matrix 188
Bruhat decompositions 20

Cartan decomposition 21
Cartan involution 16
Cartan matrix 4
Cartan subalgebra 5
Cartan–Weyl 6
Casimir operator 10
centre 10
character 79
Chevalley generators 5
coalgebra 2
coassociativity 1
cocommutative 2
compact roots 22
complex roots 22
comultiplication 1
contravariant Hermitian form 37
coproduct 1
Coulomb gauge condition 338
counit 1
cyclic representations 76

descendant 89
Di 89
dual Coxeter number 84
Dynkin labels 355

essentially quasi-triangular Hopf algebras 7

flat representations 312
formal exponent 79
FRT 109

gauge symmetry 338
Gauss decomposition 204
generalized partition function, $P(\mu)$ 37
graded algebra 185
graded Hopf algebra 219
Grassmannian quantum plane 107

highest weight 34
highest-weight modules (HWMs) 34
highest-weight vector v_0 34

imaginary root 15
in duality 113
integral dominant elements 37
irregular representations 82

Jimbo quantum algebra 11

K_β 12
Kazhdan–Lusztig polynomials 84

left and right inverse 262
left and right quantum cofactor matrix 261
left regular representation 170, 201, 263
light-front gauge condition 339
Lorentz gauge condition 338
Lorentz spin-tensor 329

massless representations of so(3,2) 89
massless representations of so(4,2) 98
maximal parabolic subalgebra 29
modular representations 82
multiplet 58

normal ordering 6
number operators 38

opposite antipode 2
opposite comultiplication 2

parabolic subalgebras 20
permutation matrix 156
pseudo quasi-triangular Hopf algebras 7

quantum algebras 5
quantum cofactor matrices 121
quantum determinant 108
quantum group 2, 3
quantum Lie algebras 238
quantum matrices 107
quantum monodromy matrix 109
quantum plane 107
quantum pseudogroups 109
quantum signature 108
quantum universal enveloping algebras 5
quasi-bialgebras 4
quasi-Hopf algebras 4
quasi-triangular Hopf algebra 2

Rac 89
rational form 11
real rank 21
real roots 21
reduced Weyl group 96
restricted root space 21
restricted root system 21
restricted roots 21
right covariance 203
right regular representation (RRR) 173
RTT relations 167

Shapovalov bilinear form 243
singleton representations 89
singular root 24
singular vector 35, 47
split rank 21
straight root 40
subroot 41
subsingular vector 99

triangular decomposition 5
triangular Hopf algebra 4
twisted (right) derivation rule 202
twisted derivation rule 201

universal R-matrix 2

Verma module 34

weight representations 172
weight vector 172
Weyl character formula 80
Weyl group W 8
Weyl subalgebra 33
Weyl tensor 355

Yang–Baxter equation (YBE) 3

De Gruyter Studies in Mathematical Physics

Volume 38
Alexander N. Petrov, Sergei M. Kopeikin, Robert R. Lompay, Bayram Tekin
Metric theories of gravity: Perturbations and conservation laws, 2017
ISBN 978-3-11-035173-6, e-ISBN (PDF) 978-3-11-035178-1,
e-ISBN (EPUB) 978-3-11-038340-9, Set-ISBN 978-3-11-035179-8

Volume 37
Igor Olegovich Cherednikov, Frederik F. Van der Veken
Parton Densities in Quantum Chromodynamics, Gauge invariance, path-dependence and Wilson lines, 2016
ISBN 978-3-11-043939-7, e-ISBN (PDF) 978-3-11-043060-8,
e-ISBN (EPUB) 978-3-11-043068-4, Set-ISBN 978-3-11-043061-5

Volume 36
Alexander B. Borisov, Vladimir V. Zverev
Nonlinear Dynamics: Non-Integrable Systems and Chaotic Dynamics, 2016
ISBN 978-3-11-043938-0, e-ISBN (PDF) 978-3-11-043058-5,
e-ISBN (EPUB) 978-3-11-043067-7, Set-ISBN 978-3-11-043059-2

Volume 35
Vladimir K. Dobrev
Invariant Differential Operators: Volume 1 Noncompact Semisimple Lie Algebras and Groups, 2016
ISBN 978-3-11-043542-9, e-ISBN (PDF) 978-3-11-042764-6,
e-ISBN (EPUB) 978-3-11-042780-6, Set-ISBN 978-3-11-042765-3

Volume 34
Abram I. Fet
Group Theory of Chemical Elements, 2016
ISBN 978-3-11-047518-0, e-ISBN (PDF) 978-3-11-047623-1,
e-ISBN (EPUB) 978-3-11-047520-3, Set-ISBN 978-3-11-047624-8

Volume 33
Alexander M. Formalskii
Stabilisation and Motion Control of Unstable Objects, 2015
ISBN 978-3-11-037582-4, e-ISBN (PDF) 978-3-11-037589-3,
e-ISBN (EPUB) 978-3-11-039282-1, Set-ISBN 978-3-11-037590-9

Bei Fragen zur Produktsicherheit wenden Sie sich bitte an:
If you have any questions regarding product safety,
please contact:

Walter de Gruyter GmbH
Genthiner Straße 13
10785 Berlin
productsafety@degruyterbrill.com